Adaptive Filtering and Change Detection

Adaptive Filtering and Change Detection

Fredrik Gustafsson
Linkoping University, Linkoping, Sweden

JOHN WILEY & SONS, LTD
Chichester • Weinheim • New York • Brisbane • Singapore • Toronto

Baffins Lane, Chichester,
West Sussex, PO 19 1UD, England

National 01243 779777
International (+44) 1243 779777

e-mail (for orders and customer service enquiries): cs-books@wiley.co.uk

Visit our Home Page on http://www.wiley.co.uk
or
http://www.wiley.com

Reprinted September 2001

Other Wiley Editorial Offices

John Wiley & Sons, Inc., 605 Third Avenue,
New York, NY 10158-0012, USA

Wiley-VCH Verlag GmbH
Pappelallee 3, D-69469 Weinheim, Germany

Jacaranda Wiley Ltd, 33 Park Road, Milton,
Queensland 4064, Australia

John Wiley & Sons (Canada) Ltd, 22 Worcester Road
Rexdale, Ontario, M9W 1L1, Canada

John Wiley & Sons (Asia) Pte Ltd, 2 Clementi Loop #02-01,
Jin Xing Distripark, Singapore 129809

British Library Cataloguing in Publication Data

A catalogue record for this book is available from the British Library

ISBN 0 471 49287 6

Produced from PostScript files supplied by the authors
Printed and bound in Great Britain by Antony Rowe, Chippenham, Wilts
This book is printed on acid-free paper responsibly manufactured from sustainable forestry, in which at least two trees are planted for each one used for paper production.

Contents

Preface

This book is rather broad in that it covers many disciplines regarding both mathematical tools (algebra, calculus, statistics) and application areas (airborne, automotive, communication and standard signal processing and automatic control applications). The book covers all the theory an applied engineer or researcher can ask for: from algorithms with complete derivations, their properties to implementation aspects. Special emphasis has been placed on examples, applications with real data and case studies for illustrating the ideas and what can be achieved. There are more than 130 examples, of which at least ten are case studies that are reused at several occasions in the book. The practitioner who wants to get a quick solution to his problem may try the 'student approach' to learning, by studying standard examples and using pattern recognition to match them to the problem at hand.

There is a strong connection to MATLAB™ There is an accompanying *toolbox*, where each algorithm in the book is implemented as one function, each example is one demo, and where algorithm design, tuning, testing and learning are all preferably done in the graphical user interface. A demo version of the toolbox is available to download from the URL `http://www.sigmoid.se`. The demo toolbox makes it possible to reproduce all examples in the book in a simple way, for instance by typing `book('ex1.7')`, so all 250 figures or so are completely reproducible. Further, it might be instructive to tune the design parameters and compare different methods! The toolbox works under MATLAB™ , but to some extent also under the freeware clone *Octave*. From the home page, exercises can be downloaded, about half of which concern computer simulations, where the toolbox is useful. Further information can be found on the URLs `http://www.wiley.co.uk/commstech/gustafsson.html` and `http://www.comsys.isy.liu.se/books/adfilt`.

It might be interesting to note that the toolbox and its structure came before the first plans of writing a book. The development of the toolbox started during a sabbatical visit at Newcastle University 1993. The outline and structure of the book have borrowed many features from the toolbox.

This book was originally developed during several courses with major revisions in between them: mini-courses at the Nordic Matlab Conference 1997 (50 participants), a course at SAAB during summer 1998 (25 participants), ABB Corporate Research September 1998 (10 participants), and a graduate course for the graduate school Ecsel at Linköping University 1998 and 1999

(25 participants). Parts of the material have been translated into Swedish for the model-based part of a book on Digital Signal Processing, where about 70 undergraduate students participate each year at Linköping University.

My interest in this area comes from two directions: the theoretical side, beginning with my thesis and studies/lecturing in control theory, signal processing and mathematical statistics; and the practical side, from the applications I have been in contact with. Many of the examples in this book come from academic and professional consulting. A typical example of the former starts with an email request on a particular problem, where my reply is "Give me representative data and a background description, and I'll provide you with a good filter". Many of the examples herein are the result of such informal contacts. Professionally, I have consulted for the automotive and aerospace industries, and for the Swedish defense industry. There are many industries that have raised my interest in this area and fruitfully contributed to a set of benchmark examples. In particular, I would like to mention Volvo Car, SAAB Aircraft, SAAB Dynamics, ABB Corporate Research, Ericsson Radio, Ericsson Components and Ericsson-SAAB Avionics. In addition, a number of companies and helpful contacts are acknowledged at the first appearance of each real-data example. The many industrial contacts, acquired during the supervision of some 50 master's theses, at least half of them in target tracking and navigation, have also been a great source of inspiration.

My most challenging task at the time of finishing this book is to participate in bringing various adaptive filters and change detectors into vehicular systems. For NIRA Dynamics `http://www.niradynamics.se`), I have published a number of patents on adaptive filters, Kalman filters and change detection, which are currently in the phase of implementation and evaluation.

Valuable comments and proof reading are gratefully acknowledged to many of the course attendants, my colleagues and co-authors. There are at least 30 people who have contributed to the errata sheets during the years. I have received a substantial number of constructive comments that I believe improved the content. In general, the group of automatic control has a quite general competence area that has helped me a lot, for instance with creating a Latex style file that satisfies me. In particular, in alphabetical order, I wish to mention Dr. Niclas Bergman, Dr. Fredrik Gunnarsson, Dr. Johan Hellgren, Rickard Karlsson MSc, Dr. Magnus Larsson, Per-Johan Nordlund MSc, Lic. Jan Palmqvist, Niclas Persson MSc, Dr. Predrag Pucar, Dr. Anders Stenman, Mikael Tapio MSc, Lic. Fredrik Tjärnström and Måns Östring MSc. My co-authors of related articles are also acknowledged, from some of these I have rewritten some material.

Finally, a project of this kind would not be possible without the support of an understanding family. I'm indebted to Lena and to my daughters, Rebecca and Erica. Thanks for letting me take your time!

Part I: Introduction

1

Extended summary

1.1. About the book

1.1.1. Outlook

The areas of adaptive filtering and change (fault) detection are quite active fields, both in research and applications. Some central keywords of the book are listed in Table 1.1, and the figures, illustrated in Figure 1.1, give an idea of the relative activity in the different areas. For comparison, the two related and well established areas of adaptive control and system identification are included in the table. Such a search gives a quick idea of the size of the areas, but there are of course many shortcomings, and the comparison may be unfair at several instances. Still, it is interesting to see that the theory has reached many successful applications, which is directly reflected in the

Table 1.1. Keywords and number of hits (March 2000) in different databases. For ScienceDirect the maximum number of hits is limited to 2000. On some of the rows, the logical 'or' is used for related keywords like 'adaptive signal processing or adaptive estimation or adaptive filter'.

Keyword	IEL	ScienceDirect	IBM patent
Adaptive filter/estimation/SP	4661	952	871
Kalman filter	1921	1642	317
Adaptive equalizer (Eq)	479	74	291
Target tracking	890	124	402
Fault diagnosis (FDI)	2413	417	74
Adaptive control (AC)	4563	2000	666
(System) Identification (SI)	8894	2000	317
Total number of items	588683	856692	2582588

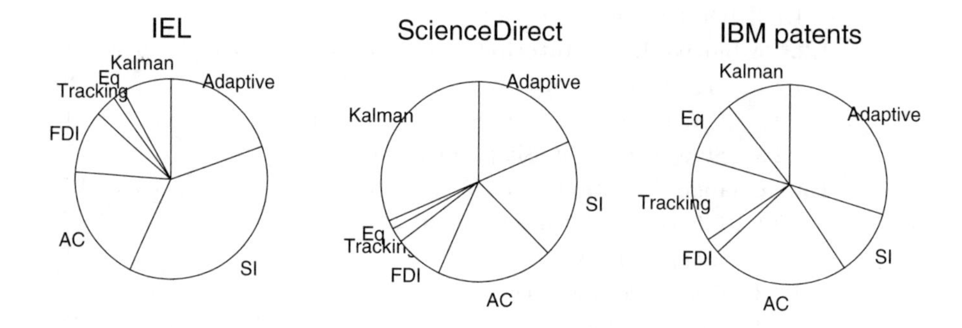

Figure 1.1. Relative frequency of keywords in different databases.

number of patents. Browsing the titles also indicates that many journal and conference publications concern applications. Figure 1.2 reveals the, perhaps well known, fact that the communication industry is more keen to hold patents (here: equalization). Algorithms aimed at real-time implementation are also, of course, more often subject to patents, compared to, for instance, system identification, which is a part of the design process.

Table 1.2 lists a few books in these areas. It is not meant to be comprehensive, only to show a few important monographs in the respective areas.

1.1.2. Aim

The aim of the book is to provide theory, algorithms and applications of adaptive filters with or without support from change detection algorithms. Applications in these areas can be divided into the the following categories:

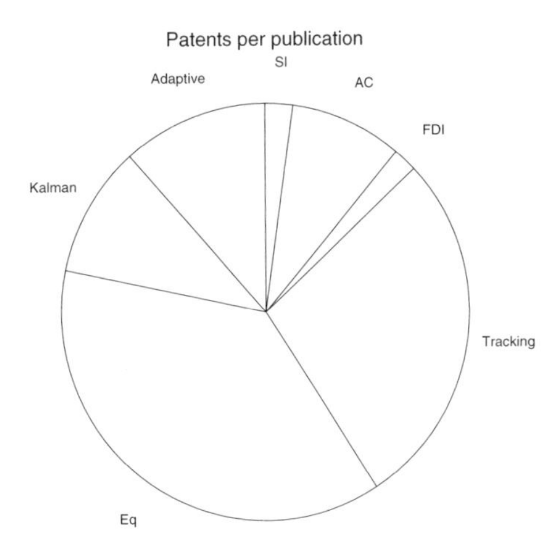

Figure 1.2. Relative ratio of number of patents found in the IBM database compared to publications in IEL for different keywords.

Table 1.2. Related books.

Keyword	Books
Adaptive filters	Haykin (1996), Mulgrew and Cowan (1988), Widrow and Stearns (1985), Cowan and Grant (1985)
Kalman filter	Kailath et al. (1998), Minkler and Minkler (1990), Anderson and Moore (1979), Brown and Hwang (1997), Chui and Chen (1987),
Adaptive equalizer	Proakis (1995), Haykin (1994), Gardner (1993), Mulgrew and Cowan (1988)
Target tracking	Bar-Shalom and Fortmann (1988), Bar-Shalom and Li (1993), Blackman (1986)
Fault diagnosis	Basseville and Nikiforov (1993), Gertler (1998), Chen and Patton (1999), Mangoubi (1998)
Adaptive control	Åström and Wittenmark (1989), Goodwin and Sin (1984)
System identification	Ljung (1999), Söderström and Stoica (1989), Johansson (1993)

- *Surveillance* and *parameter tracking.* Classical surveillance problems consist in filtering noisy measurements of physical variables as flows, temperatures, pressures etc, which will be called *signal estimation.* Model-based approaches, where (time-varying) parameters in a model of a *non-stationary signal* need to be estimated, is a problem of parameter tracking. *Adaptive control* belongs to this area. Another example is *blind equalization* in digital communication.

- *State estimation.* The Kalman filter provides the best linear state estimate, and change detection support can be used to speed up the response after disturbances and abrupt state changes. Feedback control using *state feedback*, such as Linear Quadratic Gaussian LQG control, belongs to this area. *Navigation* and *target tracking* are two particular application examples.

- *Fault detection.* Faults can occur in almost all systems. Change detection here has the role of locating the fault occurrence in time and to give a quick alarm. After the alarm, isolation is often needed to locate the faulty component. The combined task of *detection* and *isolation* is commonly referred to as *diagnosis.* Fault detection can be recast to one of parameter or state estimation. Faults in actuators and sensors are most easily detected in a state space context, while system dynamic changes often require parametric models.

These problems are usually treated separately in literature in the areas of signal processing, mathematical statistics, automatic control, communication systems and quality control. However, the tools for solving these problems have much in common, and the same type of algorithms can be used (C.R. Johnson, 1995). The close links between these areas are clearly under-estimated in literature.

The main difference of the problem areas above lies in the evaluation criteria. In surveillance the parameter estimate should be as close as possible to the true value, while in fault detection it is essential to get an alarm from the change detector as soon as possible after the fault, and at the same time generating few false alarms. In fault detection, *isolation* of the fault is also a main task. The combination of fault detection and isolation is often abbreviated to *FDI*, and the combined task can be referred to as *diagnosis.* More terminology used in this area is found in Appendix B.

The design usually consists of the following steps:

1. Modeling the signal or system.

2. Implementing an algorithm.

3. Tuning the algorithm with respect to certain evaluation criteria, either using real or simulated data.

The main focus is on algorithms and their properties, implementation, tuning and evaluation. Modeling is covered only briefly, but the numerous examples should give an idea of the possibilities of model-based signal processing.

1.1.3. Background knowledge

The derivations and analysis can be divided into the following areas, and some prior knowledge, or at least orientation, of these is required:

- Statistical theory: maximum likelihood, conditional distributions etc.
- Calculus: integrations, differentiations, equation solving etc.
- Matrix algebra: projections, subspaces, matrix factorizations etc.
- Signal modeling: transfer functions and state space models.
- Classical filter theory: the use of a low-pass filter for signal conditioning, poles and zeros etc. Transforms and frequency domain interpretations occur, but are relatively rare.

To use the methods, it is essential to understand the model and the statistical approach. These are explained in each chapter in a section called *'Basics'*. These sections should provide enough information for understanding and tuning the algorithms. A deeper understanding requires the reader to go through the calculus and matrix algebra in the derivations. The practitioner who is mainly interested in what kind of problems can be addressed is advised to start with the examples and applications sections.

1.1.4. Outline and reading advice

There are certain shortcuts to approaching the book, and advice on how to read the book is appropriate. Chapter 1 is a summary and overview of the book, while Chapter 2 overviews possible applications and reviews the basic mathematical signal models. These first two chapters should serve as an overview of the field, suitable for those who want to know *what* can be done rather than *how* it is done. Chapters 3, 5 and 8 – the first chapter in each part – are the core chapters of the book, where standard approaches to adaptive filtering are detailed. These can be used independently of the rest of the material. The other chapters start with a section called 'Basics', which can also be considered as essential knowledge. Part V is a somewhat abstract presentation of filter theory in general, without using explicit signal models. It is advisable to check the content at an early stage, but the reader should in no way spend too much time trying to digest all of the details. Instead, browse through and return to the details later. However, the ideas should be familiar before starting with the other parts. The material can be used as follows:

- Chapters 1 and 2 are suitable for people from within industry who want an orientation in what adaptive filtering is, and what change detection can add to performance. An important goal is to understand what kind of practical problems can be solved.
- Chapters 5, 8 and 13 are suitable for an undergraduate course in adaptive filtering.

Table 1.3. Organization of the book chapters.

Approach	Estimation of		
	Signal	Parameter	State
Adaptive filtering and whiteness based change detection	Chapter 3	Chapter 5	Chapter 8
Maximum likelihood based change detection	Chapter 3	Chapter 6	Chapter 9
Multiple-model based change detection	Chapter 4	Chapter 7	Chapter 10
Algebraic (parity space) change detection			Chapter 11

- Chapters 1, 2, 3, 5, 8, 12, 13 and the 'Basics' sections in the other chapters can be included in a graduate course on adaptive filtering with orientation of change detection, while a more thorough course for students specializing in the area would include the whole book.

This matrix organization is illustrated in Table 1.3. Part II on signal estimation has many interesting signal processing applications, but it also serves as a primer on the change detection chapters in Parts III and IV. The approach in Chapter 11 is algebraic rather than statistical, and can be studied separately. Appendix A overviews the signal models used in the book, and presents the main notation, while Appendix B summarizes notation used in the literature on fault detection. The only way in which the book should not be approached is probably a reading from cover to cover. The theory in the last part is important to grasp at an early stage, and so are the basics in change detection. Some of the parts on change detection will appear rather repetitive, since the basic ideas are quite similar for signal, parameter and state estimation. More specifically, Part II can be seen as a special case (or an illustrative first order example) of Part III.

1.2. Adaptive linear filtering

Three conceptually different (although algorithmically similar) cases exist:

- Signal estimation.
- Parameter estimation in an unknown model.
- State estimation in a known model.

The following sections will explain the basic ideas of these problems, and introduce one central example to each of them that will be used throughout the chapter.

1.2.1. Signal estimation

The basic signal estimation problem is to estimate the signal part θ_t in the noisy measurement y_t in the model

$$y_t = \theta_t + e_t. \tag{1.1}$$

An example of an adaptive algorithm is

$$\begin{aligned} \varepsilon_t &= y_t - \hat{\theta}_t, \\ \hat{\theta}_t &= (1 - \lambda_t)\hat{\theta}_{t-1} + \lambda_t \varepsilon_t. \end{aligned} \tag{1.2}$$

Here λ_t will be referred to as the *forgetting factor*. It is a design parameter that affects the tracking speed of the algorithm. As will become clear from the examples to follow, it is a trade-off between tracking speed and noise attenuation. The archetypical example is to use $\lambda_t = \lambda$, when this also has the interpretation of the pole in a first order low-pass filter. More generally, any (low-pass) filter can be used. If it is known that the signal level has undergone an abrupt change, as might be indicated by a change detection algorithm, then there is a possibility to momentarily forget all old information by setting $\lambda_t = 0$ once. This is an example of decision feedback in an adaptive filter, which will play an important role in change detection. An illustrative surveillance problem is given below.

Example 1.1 Fuel consumption

The following application illustrates the use of change detection for improving signal quality. The data consist of measurements of instantaneous fuel consumption available from the electronic injection system in a Volvo 850 GLT used as a test car. The raw data are pulse lengths of a binary signal, called t_q, which is the control signal from the electronic injection system to the cylinders. When $t_q = 1$, fuel is injected with roughly constant flow, so the length of the t_q pulses is a measure of fuel consumption. The measured signal contains a lot of measurement noise and needs some kind of filtering before being displayed to the driver on the dashboard. Intuitively, the actual fuel consumption cannot change arbitrarily fast, and the measured signal must be smoothed by a filter. There are two requirements on the filter:

- Good attenuation of noise is necessary to be able to tune the accelerator during cruising.
- Good tracking ability. Tests show that fuel consumption very often changes abruptly, especially in city traffic.

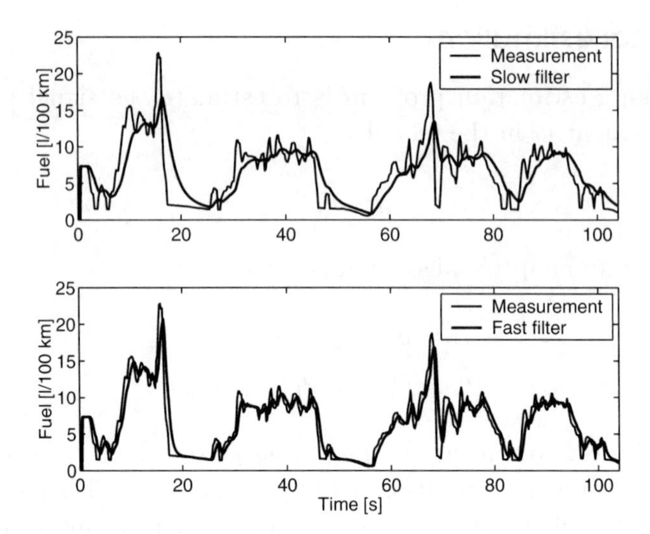

Figure 1.3. Measurements of fuel consumption and two candidate filters. Data collected by the author in a collaboration with Volvo.

These requirements are contradictory for standard linear filters. The thin lines in Figure 1.3 show measurements of fuel consumption for a test in city traffic. The solid lines show the result of (1.2) for two particular values of the forgetting factor λ. The fast filter follows the abrupt changes well, but attenuates the noise unsatisfactorily, and it is the other way around for the slow filter. The best compromise is probably somewhere in between these filters.

The fundamental trade-off between speed and accuracy is inherent in all linear filters. Change detectors provide a tool to design non-linear filters with better performance for the type of abruptly changing signal in Figure 1.3.

Figure 1.4 shows the raw data, together with a filter implemented by Volvo (not exactly the same filter, but the principal functionality is the same). Volvo uses a quite fast low-pass filter to get good tracking ability and then quantizes the result to a multiple of 0.3 to attenuate some of the noise. To avoid a rapidly changing value in the monitor, they update the monitored estimate only once a second. However, the quantization introduces a problem when trying to minimize fuel consumption manually, and the response time to changes of one second makes the feedback information to the driver less useful.

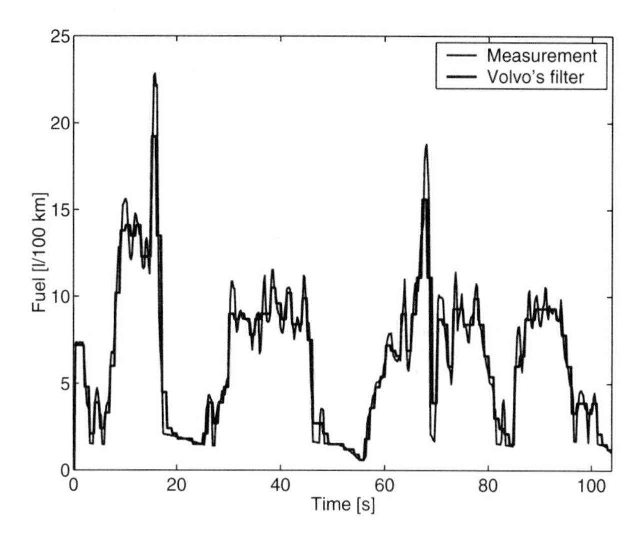

Figure 1.4. Measured fuel consumption and a filtered signal similar to Volvo's implemented filter.

1.2.2. Parameter estimation using adaptive filtering

A quite general parametric model of a linear system is

$$y_t = G(q;\theta)u_t + H(q;\theta)e_t. \tag{1.3}$$

Here $G(q;\theta)$ and $H(q;\theta)$ are two filters expressed in the delay operator q defined by $qu_t = u_{t+1}$. The parameters θ in the model are assumed to be time-varying, and are to be estimated recursively. Here and in the sequel u_t denotes measured inputs to the system, if available, and e_t is an unknown input supposed to be modeled as a stochastic noise disturbance.

The generic form of an adaptive algorithm is

$$\begin{aligned}\hat{\theta}_{t+1} &= \hat{\theta}_t + K_t\varepsilon_t,\\ \varepsilon_t &= y_t - \hat{y}_t.\end{aligned} \tag{1.4}$$

The output from the estimated model is compared to the system, which defines the residual ε_t. The adaptive filter acts as a system inverse, as depicted in Figure 1.5. One common filter that does this operation for linear in parameter models is the *recursive least squares* (RLS) filter. Other well known filters are *Least Mean Square* (LMS) (unnormalized or normalized), the *Kalman filter* and *least squares over sliding window*. This book focuses on parametric models that are linear in the parameters (not necessarily linear in the measurements). The reason for this is that the statistical approaches become optimal in this case. How to obtain sub-optimal algorithms for the general linear filter model will be discussed.

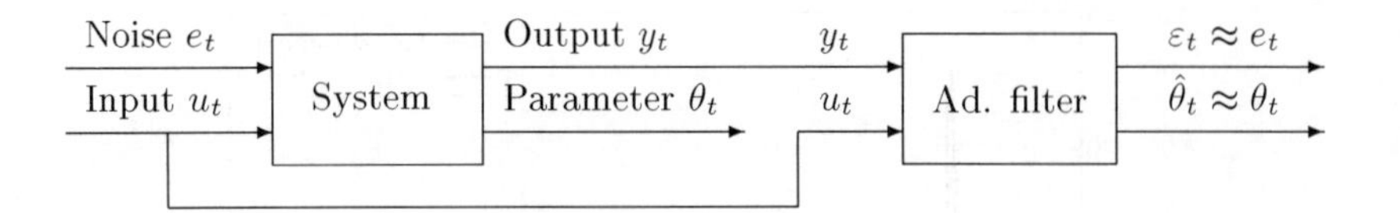

Figure 1.5. The interplay between the parameter estimator and the system.

Example 1.2 Friction estimation

This example introduces a case study that will be studied in Section 5.10.3. It has been noted (Dieckmann, 1992) that the dependence between the so-called wheel slip and traction force is related to friction. The slip is defined as the relative difference of a driven wheel's circumferential velocity, $\omega_w r_w$, and its absolute velocity, v_w:

$$s = \frac{\omega r_w - v_w}{v_w}, \tag{1.5}$$

where r_w is the wheel radius. We also define the *normalized traction force*, μ, (sometimes referred to as the friction coefficient) as the ratio of traction force (F_f) at the wheel and normal force (N) on one driven wheel,

$$\mu = \frac{F_f}{N}. \tag{1.6}$$

Here μ is computed from measurements, in this case a fuel injection signal, via an engine model. In the sequel we will consider s and μ as measurements.

Define the slip slope k as

$$k = \left.\frac{d\mu}{ds}\right|_{\mu=0}. \tag{1.7}$$

The hypothesis is that k is related to friction, and the problem is to adaptively estimate it.

The goal is to derive a linear regression model, i.e. a model linear in the parameters. The slip slope k we want to compute is defined in (1.7), which for small μ reads (including an offset term δ)

$$\mu = k(s - \delta), \tag{1.8}$$

where also δ is unknown. Although this is a model linear in the parameters, there are two good reasons for rewriting it as

$$s = \mu\frac{1}{k} + \delta. \tag{1.9}$$

That is, we consider s to be a function of μ rather than the other way around. The reasons are that s_m contains more noise than μ_m, and that the parameter δ is varying much slower as compared to $k\delta$. Both these arguments facilitate a successful filter design. Note that all quantities in the basic equation (1.9) are *dimensionless*.

We will apply a filter where $1/k$ and δ are estimated simultaneously. The design goals are

- to get accurate values on k while keeping the possibility to track slow variations in both k and δ, and at the same time
- to detect abrupt changes in k rapidly.

This case study will be approached by a Kalman filter supplemented by a change detection algorithm, detailed in Section 5.10.3.

A notationally simplified model that will be used is given by

$$y_t = \theta_t^{(1)} u_t + \theta_t^{(2)} + e_t, \tag{1.10}$$

where $\theta_t = (\theta_t^{(1)}, \theta_t^{(2)})^T$ is the parameter vector consisting of inverse slope and the offset, u_t is the input to the model (the engine torque) and y_t is the measured output (the wheel slip).

An example of measurements of y_t, u_t from a test drive is shown in Figure 1.6(a). The car enters a low friction area at sample 200 where the slope parameter $\theta_t^{(1)}$ increases. Figure 1.6(b) shows a scatter plot of the measurements together with a least squares fit of straight lines, according to the model 1.10, to the measurements before and after a friction change.

Two adaptive filters were applied, one fast and one slow. Figure 1.7 illustrates the basic limitations of linear filters: the slow filter gives a stable estimate, but is too slow in tracking the friction change, while the fast filter gives an output that is too noisy for monitoring purposes. Even the fast filter has problem in tracking the friction change in this example. The reason is poor excitation, as can be seen in Figure 1.6(b). There is only a small variation in u_t after the change (circles), compared to the first 200 samples. A thorough discussion on this matter is found in Section 5.10.3.

1.2.3. State estimation using Kalman filtering

For state estimation, we essentially run a model in parallel with the system and compute the residual as the difference. The model is specified in *state*

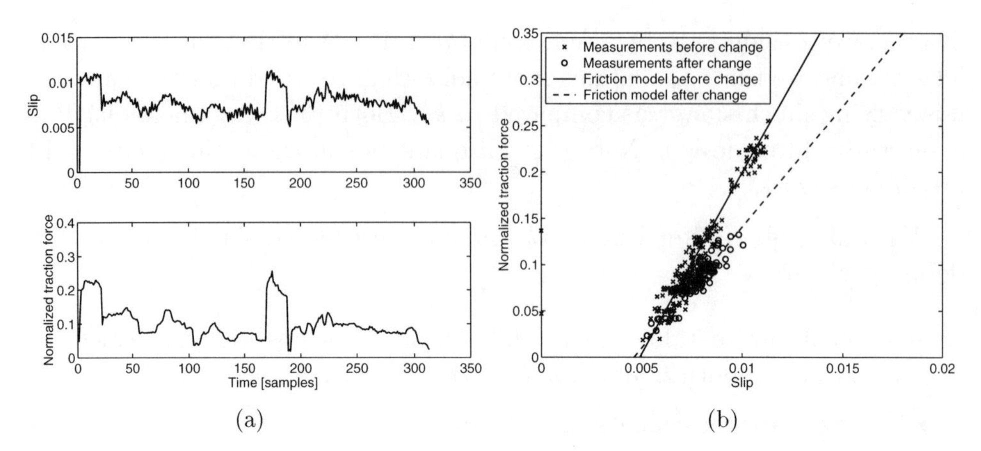

Figure 1.6. Measurements of y_t and u_t from a test drive going from dry asphalt to ice (a). A scatter plot (b) reveals the straight line friction model. The data were collected by the author in collaboration with Volvo.

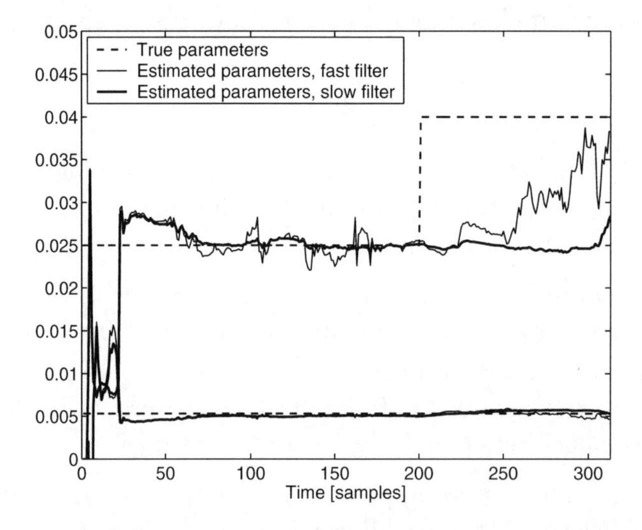

Figure 1.7. Estimated slope and offset as a function of sample number.

space form:

$$\begin{aligned} x_{t+1} &= Ax_t + B_u u_t + B_v v_t, \\ y_t &= Cx_t + Du_t + e_t. \end{aligned} \tag{1.11}$$

Here y_t is a measured signal, A, B, C, D are known matrices and x_t is an unknown vector (the state). There are three inputs to the system: the observable (and controllable) u_t, the non-observable process noise v_t and the measurement noise e_t.

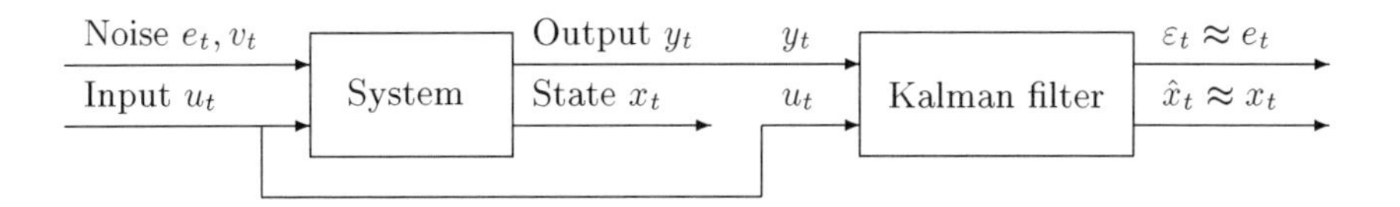

Figure 1.8. The interplay between the state estimator and the system.

In a statistical setting, state estimation is achieved via a *Kalman filter*, while in a deterministic setting, where the noises are neglected, the same device is called an *observer*. A generic form of a Kalman filter and an observer is

$$\begin{aligned} \hat{x}_{t+1} &= A\hat{x}_t + K_t \varepsilon_t \\ \varepsilon_t &= y_t - \hat{y}_t = y_t - C\hat{x}_t, \end{aligned} \tag{1.12}$$

where K_t is the filter gain specifying the algorithm. The Kalman filter provides the equations for mapping the model onto the gain as a function

$$K_t = K_t(A, B_u, B_v, C, D, Q, R). \tag{1.13}$$

The noise covariance matrices Q, R are generally considered to have some degrees of freedom that act as the design parameters. Figure 1.8 illustrates how the Kalman filter aims at 'inverting' the system.

Example 1.3 Target tracking

The typical application of target tracking is to estimate the position of aircraft. A civil application is air traffic control (ATC) where the traffic surveillance system at each airport wants to get the position and predicted positions of all aircraft at a certain distance from the airport. There are plenty of military applications where, for several reasons, the position and predicted position of hostile aircraft are crucial information. There are also a few other applications where the target is not an aircraft.

As an illustration of Kalman filtering, consider the problem of target tracking where a radar measures range R and bearing θ to an aircraft. The measurements are shown in Figure 1.9 as circles. These are noisy and the purpose of the Kalman filter is firstly to attenuate the noise and secondly to predict future positions of the aircraft. This application is described in Section 8.12.2.

One possible model for these tracking applications is the following state

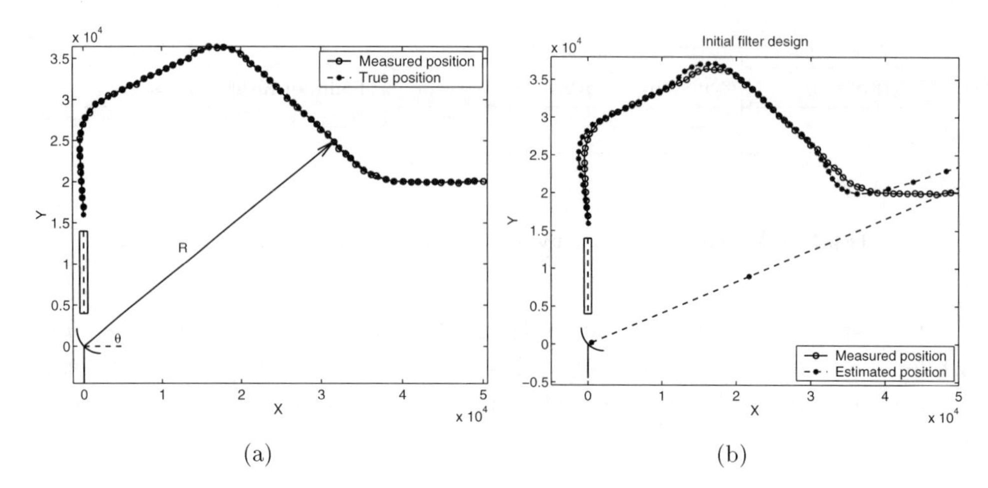

Figure 1.9. Radar measurements (a) and estimates from a Kalman filter (b) for an aircraft in-flight manoeuvre.

space model

$$
\begin{aligned}
x_{t+T} &= \begin{pmatrix} 1 & 0 & T & 0 \\ 0 & 1 & 0 & T \\ 0 & 0 & 1 & 0 \\ 0 & 0 & 0 & 1 \end{pmatrix} x_t + \begin{pmatrix} T^2/2 & 0 \\ 0 & T^2/2 \\ T & 0 \\ 0 & T \end{pmatrix} w_t \qquad (1.14) \\
y_t &= h(x_t) + e_t.
\end{aligned}
$$

The state vector used here is $x = (x_1, x_2, v_1, v_2)^T$, where x_i is the position in 2D, v_i the corresponding velocity. The state equation is one example of a *motion model* describing the dynamics of the object to be tracked. More examples are given in Chapter 8.

The measurements are transformed to this coordinate system and a Kalman filter is applied. The resulting estimates are marked with stars. We can note that the tracking is poor, and as will be demonstrated, the filter can be better tuned.

To summarize what has been said, conventional adaptive filters have the following well-known shortcoming:

> **Fundamental limitation of linear adaptive filters**
> The adaptation gain in a linear adaptive filter is a compromise between noise attenuation and tracking ability.

1.3. Change detection

Algorithmically, all proposed change detectors can be put into one of the following three categories:

- Methods using one filter, where a whiteness test is applied to the residuals from a linear filter.
- Methods using two filters, one slow and one fast one, in parallel.
- Methods using multiple filters in parallel, each one matched to certain assumption on the abrupt changes.

In the following subsections, these will be briefly described. Let us note that the computational complexity of the algorithm is proportional to how many filters are used. Before reviewing these methods, we first need to define what is meant by a residual in this context, and we also need a tool for deciding whether a result is significant or not – a stopping rule.

1.3.1. Filters as residual generators

A good understanding of the Kalman and adaptive filters requires a thorough reading of Chapters 5 and 8. However, as a shortcut to understanding statistical change detection, we only need to know the following property, also illustrated in Figure 1.10.

Residual generation
Under certain model assumptions, the Kalman and adaptive filters take the measured signals and transform them to a sequence of residuals that resemble white noise before the change occurs.

From a change detection point of view, it does not matter which filter we use and the modeling phase can be seen as a standard task. The filters also computes other statistics that are used by some change detectors, but more on this later.

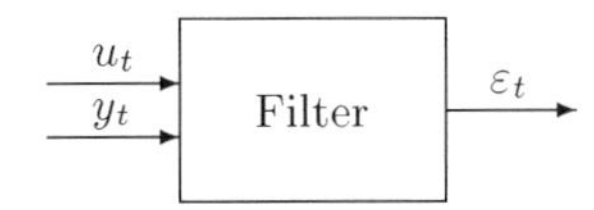

Figure 1.10. A whitening filter takes the observed input u_t and output y_t and transforms them to a sequence of residuals ε_t.

In a perfect world, the residuals would be zero before a change and non-zero afterwards. Since measurement noise and process disturbances are fundamental problems in the statistical approach to change detection, the actual value of the residuals cannot be predicted. Instead, we have to rely on their average behavior.

If there is no change in the system, and the model is correct, then the residuals are so-called white noise, that is a sequence of independent stochastic variables with zero mean and known variance.

After the change either the mean or variance or both changes, that is, the residuals become 'large' in some sense. The main problem in statistical change detection is to decide what 'large' is.

Chapter 11 reviews how state space models can be used for filtering (or *residual generation* as it will be referred to in this context). The idea is to find a set of residuals that is sensitive to the faults, such that a particular fault will excite different combinations of the residuals. The main approach taken in that chapter is based on parity spaces. The first step is to stack all variables into vectors. The linear signal model can then be expressed as

$$Y_t = H_u U_t + H_d D_t + H_f F_t,$$

where Y_t is a vector of outputs, U_t is a vector of inputs, D_t the disturbances and F_t the faults. The residual is then defined as a projection

$$\varepsilon_t = W^T (Y_t - H_u U_t).$$

With proper design of W, the residual will react to certain faults in specific patters, making *fault isolation* possible. A simple example is when the measurement is two-dimensional, and the state disturbance and the fault are both scalar functions of time. Then, under certain conditions, it is possible to linearly transform the measurement to $\varepsilon_t = (d_t, f_t)^T$. A projection that keeps only the second component can now be used as the residual to detect faults, and it is said that the disturbance is decoupled.

It should also be noted that the residual is not the only indicator of a change (that is, it is not a sufficient statistic) in all cases. So even though residual based change detection as outlined below is applicable in many cases, there might be improved algorithms. The simplified presentation in this chapter hides the fact that the multi-model approaches below actually use other statistics, but the residual still plays a very important role.

1.3.2. Stopping rules

Many change detection algorithms, among these algorithms in the classes of one-model and two-model approaches below, can be recast into the problem of deciding on the following two hypotheses:

$$H_0 : \mathrm{E}(s_t) = 0,$$
$$H_1 : \mathrm{E}(s_t) > 0.$$

A stopping rule is essentially achieved by low-pass filtering s_t and comparing this value to a threshold. Below, two such low-pass filters are given:

- The CUmulative SUM (CUSUM) test of Page (1954):

$$g_t = \max(g_{t-1} + s_t - \nu, 0), \quad \text{alarm if } g_t > h.$$

  The *drift parameter* ν influences the low-pass effect, and the *threshold* h (and also ν) influences the performance of the detector.

- The Geometric Moving Average (GMA) test in Roberts (1959)

$$g_t = \lambda g_{t-1} + (1 - \lambda)s_t, \quad \text{alarm if } g_t > h.$$

  Here, the forgetting factor λ is used to tune the low-pass effect, and the threshold h is used to tune the performance of the detector. Using no forgetting at all ($\lambda = 0$), corresponds to thresholding directly, which is one option.

1.3.3. One-model approach

Statistical whiteness tests can be used to test if the residuals are white noise as they should be if there is no change.

Figure 1.11 shows the basic structure, where the filter residuals are transformed to a *distance measure*, that measures the deviation from the no-change hypothesis. The *stopping rule* decides whether the deviation is significant or not. The most natural distance measures are listed below:

- Change in the mean. The residual itself is used in the stopping rule and $s_t = \varepsilon_t$.

- Change in variance. The squared residual subtracted by a known residual variance λ is used and $s_t = \varepsilon_t^2 - \lambda$.

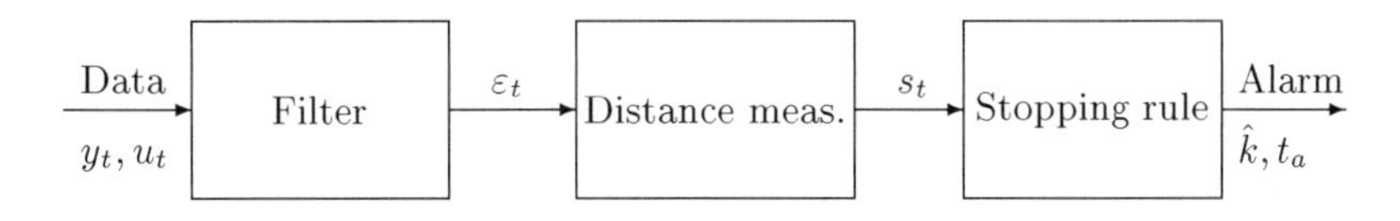

Figure 1.11. Change detection based on a whiteness test from filter residuals.

- Change in correlation. The correlation between the residual and past outputs and/or inputs are used and $s_t = \varepsilon_t y_{t-k}$ or $s_t = \varepsilon_t u_{t-k}$ for some k.

- Change in sign correlation. For instance, one can use the fact that white residuals should change sign every second sample in the average and use $s_t = \text{sign}(\varepsilon_t \varepsilon_{t-1})$. A variant of this sign test is given in Section 4.5.3.

Example 1.4 Fuel consumption

To improve on the filter in Example 1.1, the CUSUM test is applied to the residuals of a slow filter, like that in Figure 1.3. For the design parameters $h = 5$ and $\nu = 0.5$, the response in Figure 1.12 is obtained. The vertical lines illustrate the alarm times of the CUSUM algorithm. The lower plot shows how the test statistic exceeds the threshold level h at each alarm. The adaptive filter in this example computes the mean of the signal from the latest alarm to the current time. With a bad tuning of the CUSUM algorithm, we get either the total mean of the signal if there are no alarms at all, or we get the signal back as the estimate if the CUSUM test gives an alarm at each time instant. These are the two extreme points in the design. Note that nothing worse can happen, so the stability of the filter is not an issue here. To avoid the first situation where the estimate will converge to the overall signal mean, a better

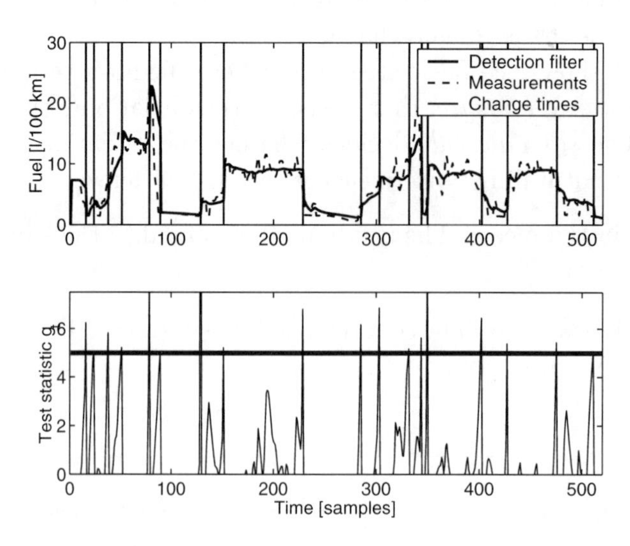

Figure 1.12. Response of an adaptive filter restarted each time a CUSUM test ($h = 5$, $\nu = 0.5$), fed with the filter residuals, gives an alarm. The lower plot shows the test statistic of the CUSUM test.

design is to use a slow adaptive filter of the type illustrated in Figure 1.3. To avoid the second degenerate case, an alarm can trigger a fast filter instead of a complete filter restart. That is, the algorithm alternates between slow and fast forgetting instead of complete or no forgetting.

In contrast to the example above, the next example shows a case where the user gets important information from the change detector itself.

Example 1.5 Friction estimation

Figure 1.13 shows how a whiteness test, used for restarting the filter, can improve the filtering result in Example 1.2 quite considerably.

Here the CUSUM test statistics from the residuals is used, and the test statistics g_t are shown in the lower plot in Figure 1.13. Note that the test statistics start to grow at time 200, but that the conservative threshold level of 3 is not reached until time 210.

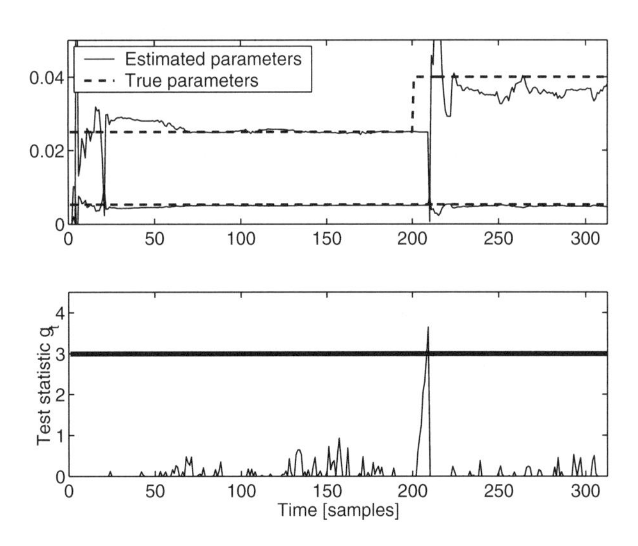

Figure 1.13. Estimated friction parameters (upper plot) and test statistics from the CUSUM test (lower plot). Note the improved tracking ability compared to Figure 1.7.

In the following example, change detection is a tool for improved tracking, and the changes themselves do not contain much information for the user.

Example 1.6 Target tracking

This example is a continuation of Example 1.3, and the application of the CUSUM test is analogous to Example 1.5. Figure 1.14(a) shows the estimated

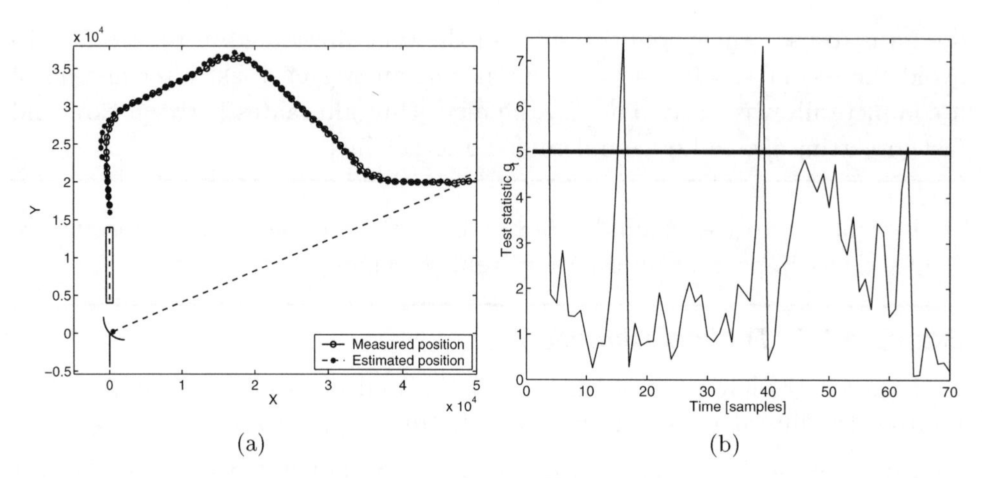

Figure 1.14. Radar measurements and estimates from a Kalman filter with feedback from a CUSUM detector (a), giving alarms at samples 19 and 39. The right plot (b) shows the CUSUM test statistics and the alarm level.

position compared to the true one, while Figure 1.14(b) shows the test statistics from the CUSUM test. We get an alarm in the beginning telling us that the position prediction errors are much larger than expected. After filter convergence, there are alarms at samples 16, 39 and 64, respectively. These occur in the manoeuvres and the filter gain is increased momentarily, implying that the estimates come back to the right track almost directly after the alarms.

1.3.4. Two-model approach

A model based on recent data only is compared to a model based on data from a much larger data window. By recent data we often mean data from a sliding window. The basic procedure is illustrated in (1.15) and Figure 1.15.

$$\text{Data}: \quad \overbrace{y_1, y_2, \ldots, y_{t-L}, \underbrace{y_{t-L+1}, \ldots, y_t}_{\text{Model } M_2}}^{\text{Model } M_1} \tag{1.15}$$

The model (M_2) is based on data from a sliding window of size L and is compared to a model (M_1) based on all data or a substantially larger sliding window. If the model based on the larger data window gives larger residuals,

$$\|\varepsilon_t^1\| > \|\varepsilon_t^2\|,$$

then a change is detected. The problem here is to choose a norm that corresponds to a relevant statistical measure. Some norms that have been proposed are:

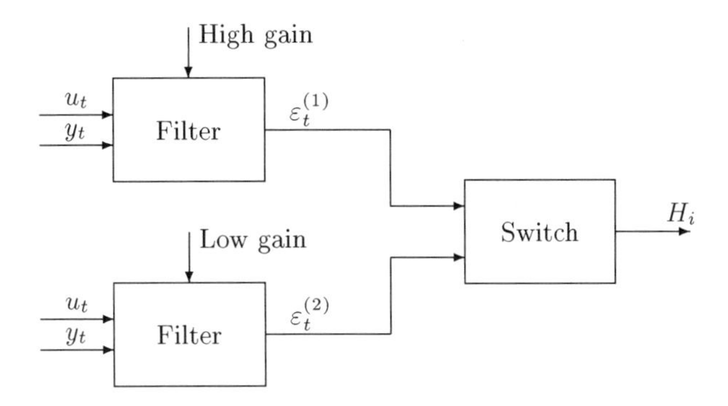

Figure 1.15. Two parallel filters, one slow to get good noise attenuation and one fast to get fast tracking. The switch decides hypothesis H_0 (no change) or H_1 (change).

- The Generalized Likelihood Ratio (GLR), see Appel and Brandt (1983).
- The divergence test, see Basseville and Benveniste (1983a).
- Change in spectral distance. There are numerous methods to measure the distance between two spectra. One approach, not discussed further here, would be to compare the spectral distance of the two models.

These criteria provide an s_t to be put into a stopping rule in Section 1.3.2, for instance, the CUSUM test.

The choice of window size L is critical here. On the one hand, a large value is needed to get an accurate model in the sliding window and, on the other hand, a small value is needed to get quick detection.

Example 1.7 Friction estimation

Consider the slow and fast filters in Example 1.2. As one (naïve) way to switch between these filters, we may compare their residuals. Figure 1.16 shows the ratio

$$\frac{|\varepsilon_t^{(2)}|}{\alpha + |\varepsilon_t^{(1)}|}$$

as a function of time. Here α is a small number to prevent division by zero. After time 200, the fast filter gives smaller residuals in the average. However, this simple strategy has certain shortcomings. For instance, the switch will change far too often between the two filter outputs.

1.3.5. Multi-model approach

An important property of the Kalman filter and RLS filter is the following. If the change time, or set of change times, is known, then the filter can be tailored

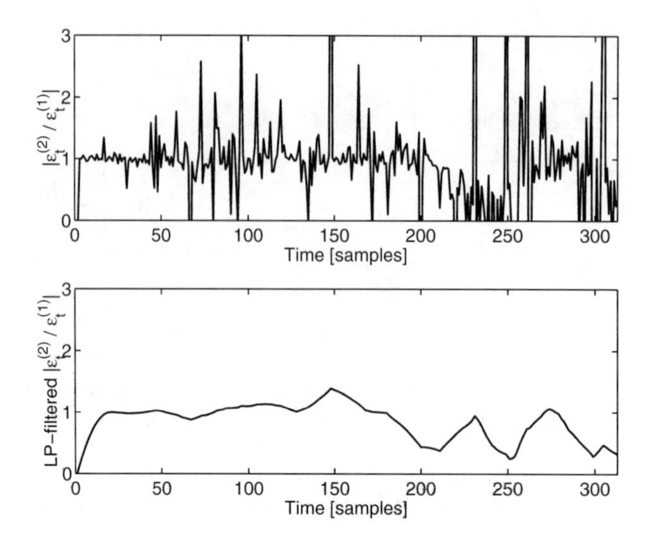

Figure 1.16. The residual ratio $|\varepsilon_t^{(2)}|/(\alpha + |\varepsilon_t^{(1)}|)$ from the fast and slow filters in Example 1.2. In the lower plot, the ratio is low-pass filtered before the magnitude is plotted.

to this knowledge so that it gives white residuals even after the change(s). Such filters are usually called *matched filters*, because they are matched to specific assumptions on the true system.

The idea in the multi-model approach is to enumerate all conceivable hypotheses about change times and compare the residuals from their matched filters. The filter with the 'smallest' residuals wins, and gives an estimate of the change times that usually is quite accurate. The setup is depicted in Figure 1.17. The formulation is in a sense off-line, since a batch of data is considered, but many proposed algorithms turn out to process data recursively, and they are consequently on-line.

Again, we must decide what is meant by 'small' residuals. By just taking the norm of the residuals, it turns out that we can make the residuals smaller and smaller by increasing the number of hypothesized change times. That is, a penalty on the number of changes must be builtin.

There can be a change or no change at each time instant, so there are 2^t possible matched filters at time t. Thus, it is often infeasible to apply all possible matched filters to the data. Much effort has been spent in developing intelligent search schemes that only keep a constant number of filters at each time.

To summarize, the fundamental trade-off inherent in all change detection algorithms is the following:

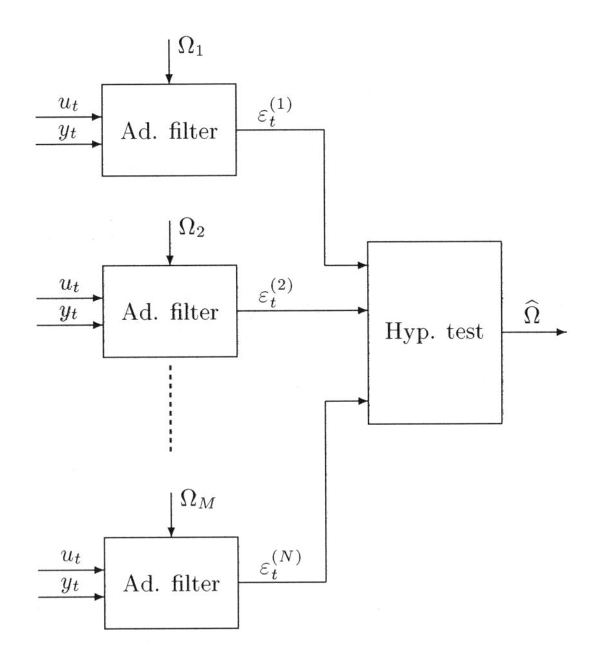

Figure 1.17. A bank of matched filters, each one based on a particular assumption on the set of change times $\Omega = \{k_i\}_{i=1}^n$, that are compared in a hypothesis test.

> **Fundamental limitation of change detection**
> The design is a compromise between detecting true changes and avoiding false alarms.

On the other hand, adaptive filters complemented with a change detection algorithms, as depicted in Figure 1.18 for the whiteness test principle, offer the following possibility:

> **Fundamental limitation of change detection for adaptive filtering**
> The design can be seen as choosing one slow nominal filter and one fast filter used after detections. The design is a compromise between how often the fast filter should be used. A bad design gives either the slow filter (no alarms from the detector) or the fast filter (many false alarms), and that is basically the worst that can happen.

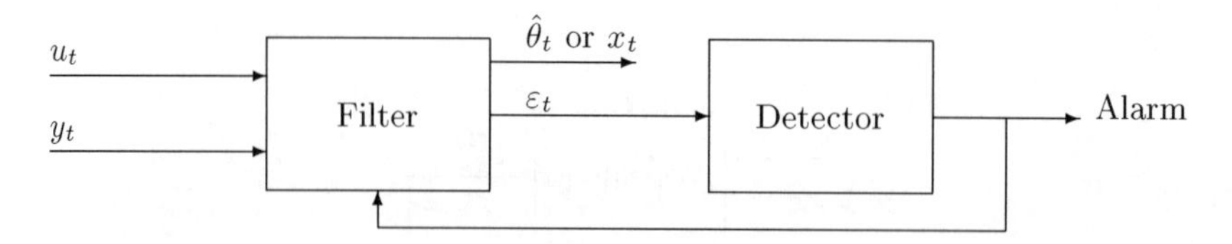

Figure 1.18. Basic feedback structure for adaptive filtering with a change detector. The linear filter switches between a slow and a fast mode depending on the alarm signal.

1.4. Evaluation and formal design

1.4.1. General considerations

Evaluation of the design can be done either on simulated data or on real data, where the change times or true parameters are known. There are completely different evaluation criteria for surveillance and fault detection:

- For signal, parameter or state estimation, the main performance measures are tracking ability and variance error in the estimated quantities. For linear adaptive filters there is an inherent trade-off, and the tuning consists in getting a fair compromise between these two measures. Change detectors are non-linear functions of data and it is in principle possible to get arbitrarily fast tracking and small variance error between the estimated change times, but the change time estimation introduces another kind of variance error. Measures for evaluating change detectors for estimation purposes include Root Mean Square (RMS) parameter estimation error in simulation studies and information measures of the obtained model which work both for simulated and real data.

- For fault detection, it is usually important to get the alarms as soon as possible – the delay for detection – while the number of false alarms should be small. The compromise inherent in all change detectors consists in simultaneously minimizing the false alarm rate and delay for detection. Other measures of interest are change time estimation accuracy, detectability (which faults can be detected) and probability of detection.

The final choice of algorithm should be based on a careful design, taking these considerations into account. These matters are discussed in Chapter 13.

The first example is used to illustrate how the design parameters influence the performance measures.

Example 1.8 Fuel consumption

The interplay of the two design parameters in the CUSUM test is complicated, and there is no obvious simple way to approach the design in Example

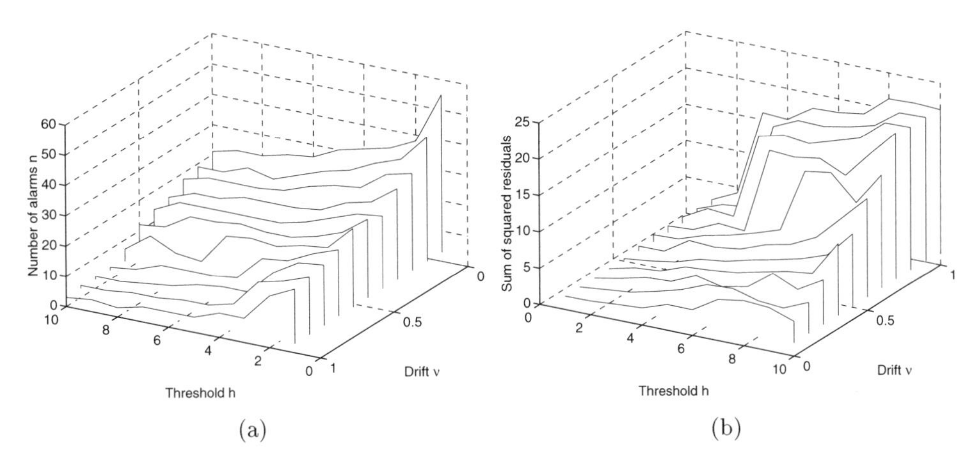

Figure 1.19. Examination of the influence of the CUSUM test's design parameters h and ν on the number of alarms (a) and the sum of squared residuals (b).

1.4 than to use experience and visual examination. If the true fuel consumption was known, formal optimization based design can be used. One examination that can be done is to compare the number of alarms for different design parameters, as done in Figure 1.19(a). We can see from the signal that 10-15 alarms seem to be plausible. We can then take any of the combinations giving this number as candidate designs. Again visual inspection is the best way to proceed. A formal attempt is to plot the sum of squared residuals for each combination, as shown in Figure 1.19(b). The residual is in many cases, like this one, minimized for any design giving false alarms all the time, like $h = 0$, $\nu = 0$, and a good design is given by the 'knee', which is a compromise between few alarms and small residuals.

The second example shows a case where joint evaluation of change detection and parameter tracking performance is relevant.

Example 1.9 Friction estimation

Consider the change detector in Example 1.5. To evaluate the over-all performance, we would like to have many signals under the same premises. These can be simulated by noting that the residuals are well modeled by Gaussian white noise, see the upper plot in Figure 1.20(a) for one example. Using the 'true' values of the friction parameters, the measured input μ_t and simulated measurement noise, we can simulate, say, 50 realizations of s_t. The mean performance of the filter is shown in Figure 1.20(b). This is the relevant plot for validation for the purpose of friction surveillance. However, if an alarm device for skid is the primary goal, then the delay for detection illustrated in

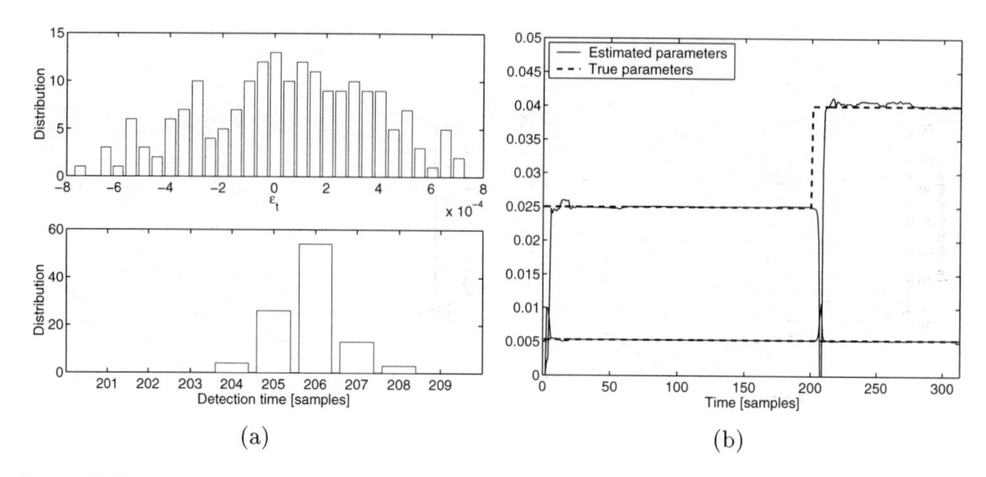

Figure 1.20. Illustration of residual distribution from real data and distribution of alarm times (a). Averaged parameter tracking (b) and alarm times are evaluated from 100 simulations with an abrupt friction change at sample 200.

the lower plot of Figure 1.20(a) is the relevant evaluation criterion. Here one can make statements like: the mean delay for alarm after hitting an ice spot is less than five samples (1 second)!

Finally, the last example illustrates how parameter tracking can be optimized.

Example 1.10 Target tracking

Let us consider Example 1.3 again. It was remarked that the tracking in Figure 1.9 is poor and the filter should be better tuned. The adaptation gain can be optimized with respect to the RMS error and an optimal adaptation gain is obtained, as illustrated in Figure 1.21(a). The optimization criterion has typically, as here, no local minima but is non-convex, making minimization more difficult. Note that the optimized design in Figure 1.21(b) shows much better tracking compared to Figure 1.9, at the price of somewhat worse noise attenuation.

1.4.2. Performance measures

A general *on-line* statistical change detector can be seen as a device that takes a sequence of observed variables and at each time makes a binary decision if the system has undergone a change. The following measures are critical:

- *Mean Time between False Alarms* ($MTFA$). How often do we get alarms when the system has not changed? The reciprocal quantity is called the *false alarm rate* (FAR).

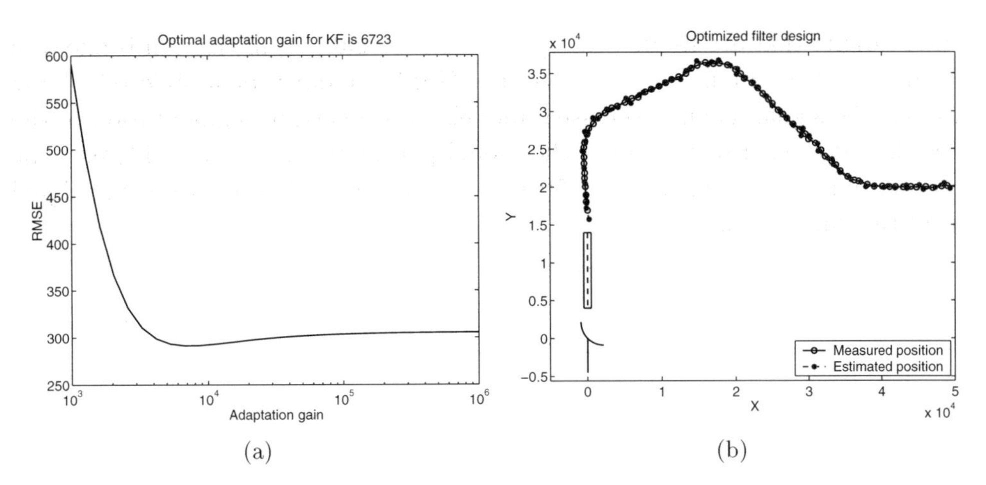

Figure 1.21. Total RMS error from the Kalman filter as a function of its gain (a). Note that there is a clear minimum. The optimized design is shown in (b).

- *Mean Time to Detection* (*MTD*). How long time do we have to wait after a change until we get the alarm?
- *Average Run Length function*, $ARL(\theta)$, which generalizes MTFA and MTD. The ARL function is defined as the mean time between alarms from the change detector as a function of the magnitude of the change. Hence, the MTFA and DFD are special cases of the ARL function. The ARL function answers the question of how long it takes before we get an alarm after a change of size θ.

In practical situations, either MTFA or MTD is fixed, and we optimize the choice of method and design parameters to minimize the other one. For instance, in an airborne navigation system, the MTFA might be specified to one false alarm per 10^5 flight hours, and we want to get the alarms as quickly as possible under these premises.

In *off-line* applications, we have a batch of data and want to find the time instants for system changes as accurately as possible. This is usually called *segmentation*. Logical performance measures are:

- Estimation accuracy. How accurately can we locate the change times?
- The Minimum Description Length (MDL). How much information is needed to store a given signal?

The latter measure is relevant in data compression and communication areas, where disk space or bandwidth is limited. MDL measures the number of binary digits that are needed to represent the signal. Segmentation is one tool for making this small. For instance, the GSM standard for mobile telephony models the signal as autoregressive models over a certain segmentation of the signal, and then transmits only the model parameters and the residuals. The change times are fixed to every 50 ms. The receiver then recovers the signal from this information.

2

Applications

This chapter provides background information and problem descriptions of the applications treated in this book. Most of the applications include real data and many of them are used as case studies examined throughout the book with different algorithms. This chapter serves both as a reference chapter and as a motivation for the area of adaptive filtering and change detection. The applications are divided here according to the model structures that are used. See the model summary in Appendix A for further details. The larger case studies on target tracking, navigation, aircraft control fault detection, equalization and speech coding, which deserve more background information, are not discussed in this chapter.

2.1. Change in the mean model

The fuel consumption application in Examples 1.1, 1.4 and 1.8 is one example of change in the mean model. Mathematically, the model is defined in equation (A.1) in Appendix A). Here a couple of other examples are given.

2.1.1. Airbag control

Conventional airbags explode when the front of the car is decelerated by a certain amount. In the first generation of airbags, the same pressure was used in all cases, independently of what the driver/passenger was doing, or their weight. In particular, the passenger might be leaning forwards, or may not even be present. The worst cases are when a baby seat is in use, when a child is standing in front of the seat, and when very short persons are driving and sitting close to the steering wheel. One idea to improve the system is to monitor the weight on the seat in order to detect the presence of a passenger

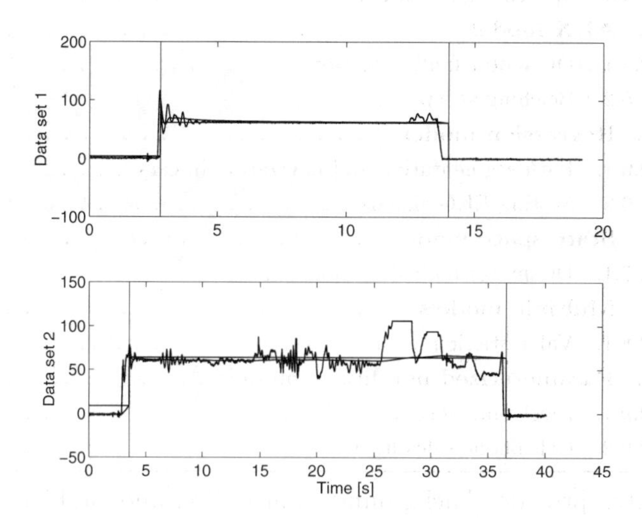

Figure 2.1. Two data sets showing a weight measurement on a car seat when a person enters and leaves the car. Also shown are one on-line and one off-line estimates of the weight as a function of time. Data provided by Autoliv, Linköping, Sweden.

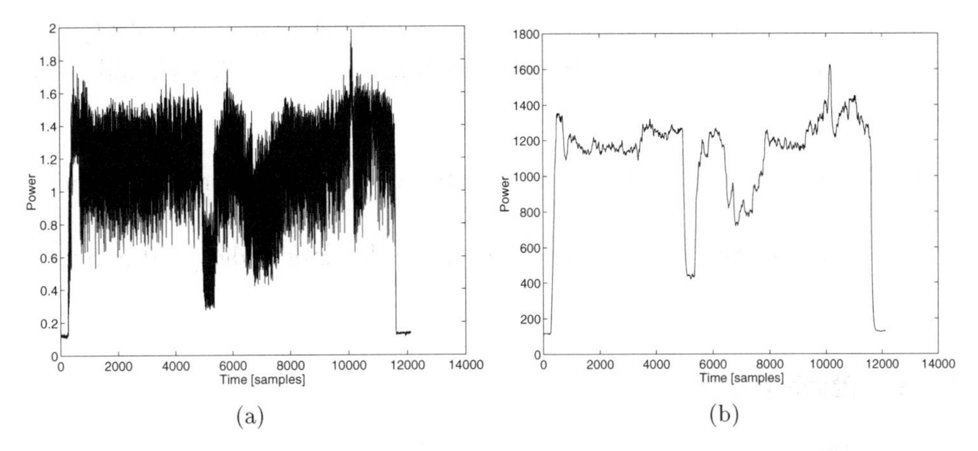

Figure 2.2. Power signal from a paper refinery and the output of a filter designed by the company. Data provided by Thore Lindgren at Sund's Defibrator AB, Sundsvall, Sweden.

and, in that case, his position, and to then use two or more different explosion pressures, depending on the result. The data in Figure 2.1 show weight measurements when a passenger is entering and shortly afterwards leaving the seat. Two different data sets are shown. Typically, there are certain oscillations after manoeuvres, where the seat can be seen as a damper-spring system. As a change detection problem, this is quite simple, but reliable functionality is still rather important. In Figure 2.1, change times from an algorithm in Chapter 3 are marked.

2.1.2. Paper refinery

Figure 2.2(a) shows process data from a paper refinery (M48 Refiner, Techboard, Wales; the original data have been rescaled). The refinery engine grinds tree fibers for paper production. The interesting signal is a raw engine power signal in kW, which is extremely noisy. It is used to compute the reference value in a feedback control system for quality control and also to detect engine overload. The requirements on the power signal filter are:

- The noise must be considerably attenuated to be useful in the feedback loop.
- It is very important to quickly detect abrupt power decreases to be able to remove the grinding discs quickly and avoid physical disc faults.

That is, both tracking and detection are important, but for two different reasons. An adaptive filter provides some useful information, as seen from the low-pass filtered squared residual in Figure 2.2(b).

- There are two segments where the power clearly decreases quickly. Furthermore, there is a starting and stopping transient that should be detected as change times.

- The noise level is fairly constant (0.05) during the observed interval.

These are the starting points when the change detector is designed in Section 3.6.2.

2.1.3. Photon emissions

Tracking the brightness changes of galactical and extragalactical objects is an important subject in astronomy. The data examined here are obtained from X-ray and γ-ray observatories. The signal depicted in Figure 2.3 consists of even integers representing the time of arrival of the photon, in units of microseconds, where the fundamental sampling interval of the instrument is 2 microseconds. More details of the application can be found in Scargle (1997). This is a typical queue process where a Poisson process is plausible. A Poisson process can be easily converted to a *change in the mean* model by computing the time difference between the arrival times. By definition, these differences will be independently exponentially distributed (disregarding quantization errors).

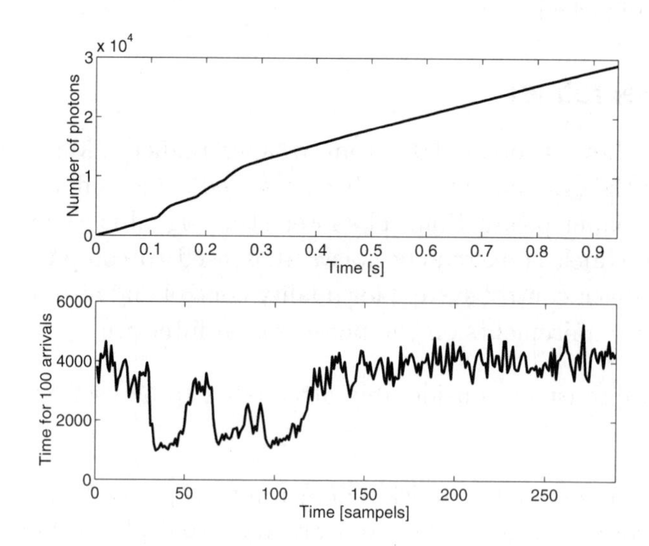

Figure 2.3. Photon emissions are counted in short time bins. The first plot shows the number of detected photons as a function of time. To recast the problem to a change in the mean, the time for 100 arrivals is computed, which is shown in the lower plot. Data provided by Dr. Jeffrey D. Scargle at NASA.

That is, the model is

$$y_t = \theta_t + e_t, \qquad p(y_t) = \frac{1}{\theta_t} e^{-\frac{y_t}{\theta_t}}, \qquad \mathrm{E}(y_t) = \theta_t.$$

Thus, e_t is white exponentially distributed noise.

The large number of samples make interactive design slow. One alternative is to study the number of arrivals in larger bins of, say, 100 fundamental sampling intervals. The sum of 100 Poisson variables is approximated well by a Gaussian distribution, and the standard Gaussian signal estimation model can be used.

2.1.4. Econometrics

Certain economical data are of the change in the mean type, like the sales figures for a particular product as seen in Figure 2.4. The original data have been rescaled. By locating the change points in these data, important conclusions on how external parameters influence the sale can be drawn.

2.2. Change in the variance model

The change in variance model (A.2) assumes that the measurements can be transformed to a sequence of white noise with time varying variance.

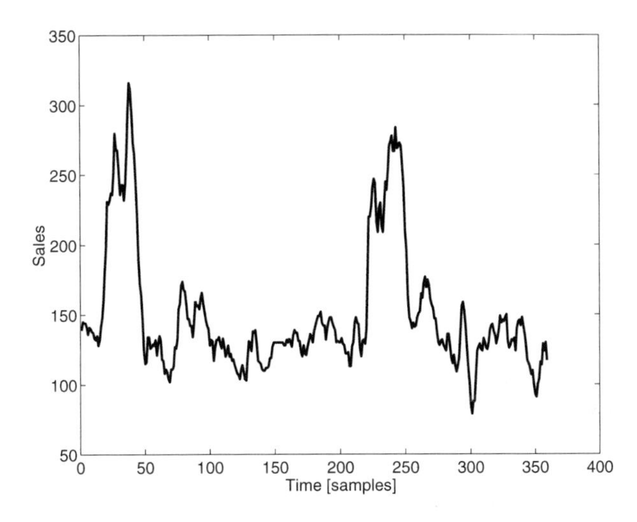

Figure 2.4. Sales figures for a particular product. (Data provided by Prof. Duncan, UMIST, UK).

2.2.1. Barometric altitude sensor in aircraft

A barometric air pressure sensor is used in airborne navigation systems for stabilizing the altitude estimate in the inertial navigation system. The barometric sensor is not very accurate, and gives measurements of height with both a bias and large variance error. The sensor is particularly sensitive to the so called transonic passage, that is, when Mach 1 (speed of sound) is passed. It is a good idea to detect for which velocities the measurements are useful, and perhaps also to try to find a table for mapping velocity to noise variance. Figure 2.5 shows the errors from a calibrated (no bias) barometric sensor, compared to the 'true' values from a GPS system. The lower plot shows low-pass filtered squared errors. The data have been rescaled.

It is desirable here to have a procedure to automatically find the regions where the data have an increased error, and to tabulate a noise variance as a function of velocity. With such a table at hand, the navigation system can weigh the information accordingly.

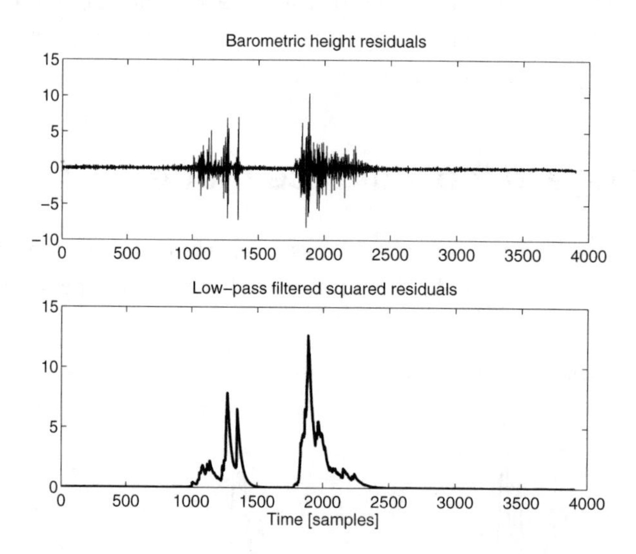

Figure 2.5. Barometric altitude measurement error for a test flight when passing the speed of sound (sample around 1600). The lower plot shows low-pass filtered errors as a very rough estimate of noise variance. (Data provided by Dr. Jan Palmqvist, SAAB Aircraft.)

2.2.2. Rat EEG

The EEG signal in Figure 2.6 is measured on a rat. The goal is to classify the signal into segments of so called "spindles" or background noise. Currently, researchers are using a narrow band filter, and then apply a threshold for the output power of the filter. That method gives

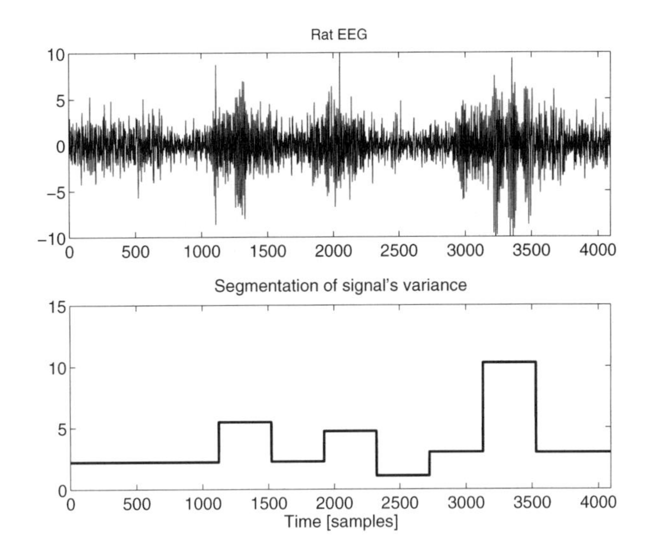

Figure 2.6. EEG for a rat and segmented signal variance. Three distinct areas of brain activity can be distinguished. (Data provided by Dr. Pasi Karjalainen, Dept. of Applied Physics, University of Kuopio, Finland.)

```
[1096 1543 1887 2265 2980 3455 3832 3934].
```

The lower plot of Figure 2.6 shows an alternative approach which segments the noise variance into piecewise constant intervals. The estimated change times are:

```
[1122 1522 1922 2323 2723 3129 3530]
```

The 'spindles' can be estimated to three intervals in the signal from this information.

2.3. FIR model

The *Finite Impulse Response* (*FIR*) model (see equation (A.4) in in Appendix A) is standard in real-time signal processing applications as in communication systems, but it is useful in other applications as well, as the following control oriented example illustrates.

2.3.1. Ash recycling

Wood has become an important fuel in district heating plants, and in other applications. For instance, Swedish district heating plants produce about 500 000 tons of ash each year, and soon there will be a \$30 penalty fee for depositing the ash as waste, which is an economical incentive for recycling. The ash from burnt wood cannot be recycled back to nature directly, mainly because of its volatility.

We examine here a recycling procedure described in Svantesson et al. (2000). By mixing ash with water, a granular material is obtained. In the water mixing process, it is of the utmost importance to get the right mixture. When too much water is added, the mixture becomes useless. The idea is to monitor the mixture's viscosity indirectly, by measuring the power consumed by the electric motor in the mixer. When the dynamics between input water and consumed power changes, it is important to stop adding water immediately.

A simple semi-physical model is that the viscosity of the mixture is proportional to the amount of water, where the initial amount is unknown. That is, the model with water flow as input is

$$\begin{aligned} y_t &= \theta^1 + \theta^2 \frac{1}{1-q^{-1}} u_t + e_t \\ &= \theta^1 + \theta^2 \bar{u}_t + e_t, \end{aligned}$$

where $\bar{u}_t$ is the integrated water flow. This is a simple model of the FIR type (see equation (A.4)) with an extra offset, or equivalently a linear regression (equation (A.3)). When the proportional coefficient θ^2 changes, the granulation material is ready.

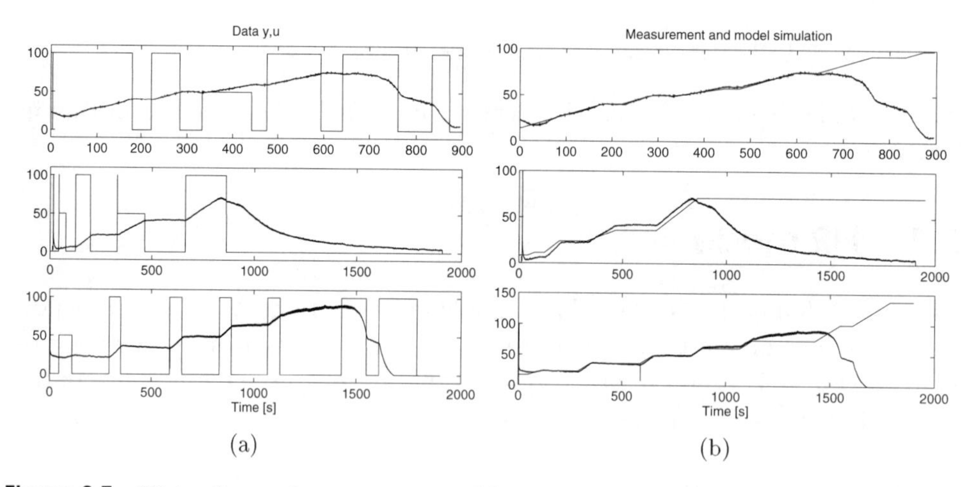

Figure 2.7. Water flow and power consumed by a mixer for making granules of ash from burnt wood (a). At a certain time instant, the dynamics changes and then the mixture is ready. A model based approach enables the simulation of the output (b), and the change point in the dynamics is clearly visible. (Data provided by Thomas Svantesson, Kalmar University College, Sweden.)

A more precise model that fits the data better would be to include a parameter for modeling that the mixture dries up after a while when no water is added.

2.4. AR model

The AR model defined below is very useful for modeling time series, of which some examples are provided in this section.

2.4.1. Rat EEG

The same data as in 2.2.2 can be analyzed assuming an AR(2) model. Figure 2.8 shows the estimated parameters from a segmentation algorithm (there seems to be no significant parameter change) and segmented noise variance. Compared to Figure 2.6, the variance is roughly one half of that here, showing that the model is relevant and that the result should be more accurate. The change times were estimated here as:

```
[1085 1586 1945 2363 2949 3632 3735]
```

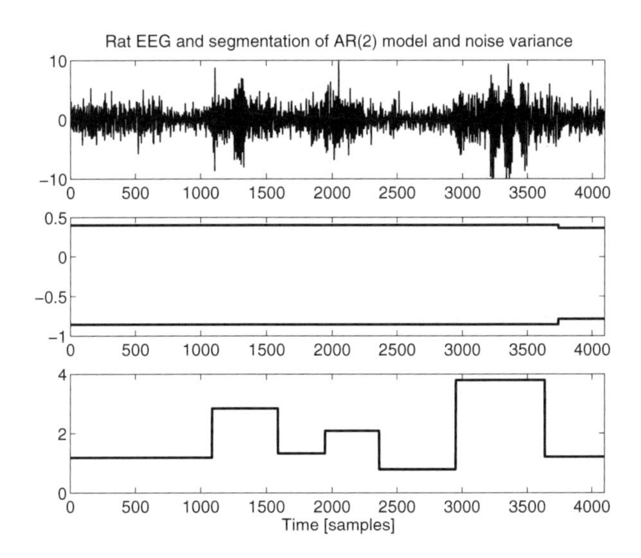

Figure 2.8. EEG for a rat and segmented noise variance from an AR(2) model.

2.4.2. Human EEG

The data shown in Figure 2.9 are measured from the human occipital area. Before time t_b the lights are turned on in a test room where a test person is looking at something interesting. The neurons are processing information in the visual cortex, and only noise is seen in the measurements. When the lights are turned off, the visual cortex is at rest. The neuron clusters start 10 Hz periodical 'rest rhythm'. The delay between t_b and the actual time when the rhythm starts varies strongly. It is believed that the delay correlates with, for example, Alzheimer disease, and methods for estimating the delay would be useful in *medical diagnosis*.

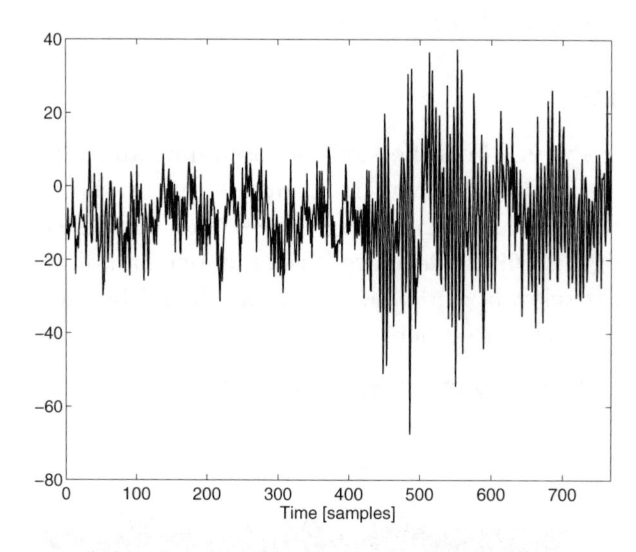

Figure 2.9. EEG for a human in a room, where the light is turned off at time 387. After a delay which varies for different test objects, the EEG changes character. (Data provided by Dr. Pasi Karjalainen, Dept. of Applied Physics, University of Kuopio, Finland.)

2.4.3. Earthquake analysis

Seismological data are collected and analyzed continuously all over the world. One application of analysis aims to detect and locate earth quakes. Figure 2.10 shows three of, in this case, 16 available signals, where the earthquake starts around sample number 600.

Visual inspection shows that both the energy and frequency content undergo an abrupt change at the onset time, and smaller but still significant changes can be detected during the quake.

As another example, Figure 2.11 shows the movements during the 1989 earth quake in San Francisco. This data set is available in MATLAB^{TM} as `quake`. Visually, the onset time is clearly visible as a change in energy and frequency content, which is again a suitable problem for an AR model.

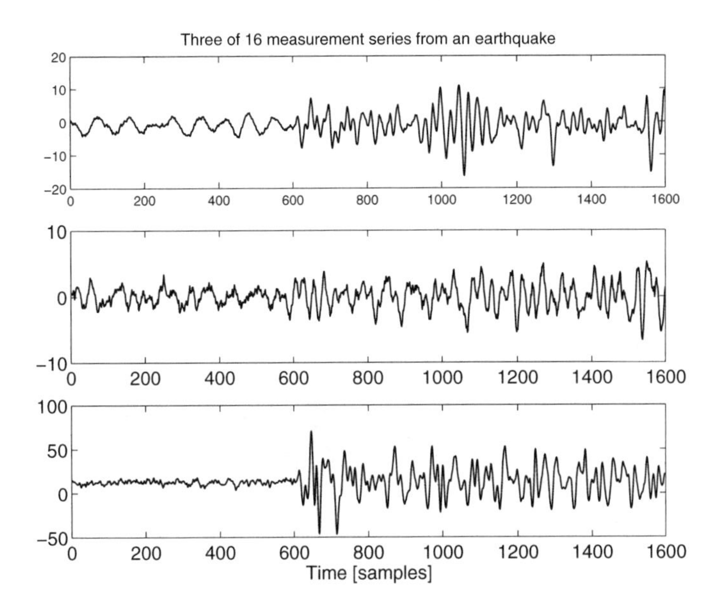

Figure 2.10. Example of logged data from an earthquake.

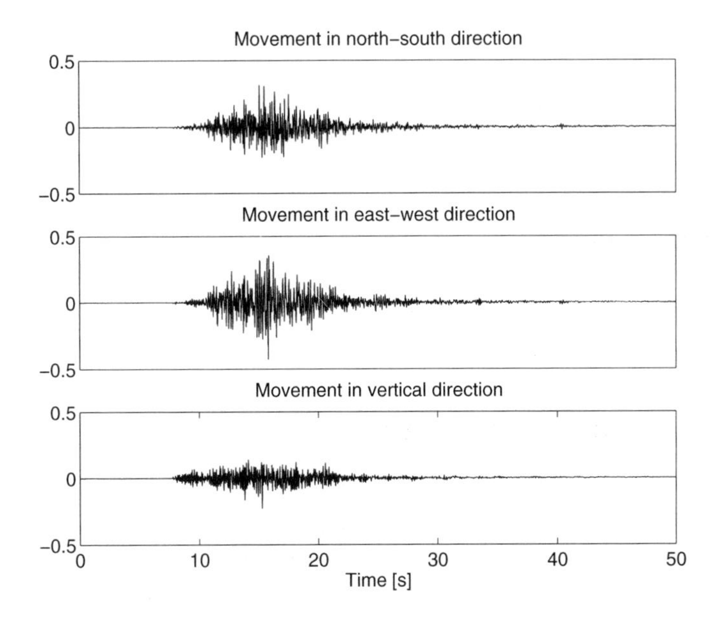

Figure 2.11. Movements for the 1989 San Francisco earthquake.

2.4.4. Speech segmentation

The speech signal is one of the most classical applications of the AR model. One reason is that it is possible to motivate it from physics, see Example 5.2. The speech signal shown in Figure 2.12, that will be analyzed later one, was recorded inside a car by the French National Agency for Telecommunications, as described in Andre-Obrecht (1988). The goal of segmentation might be speech recognition, where each segment corresponds to one phoneme, or speech coding (compare with Section 5.11).

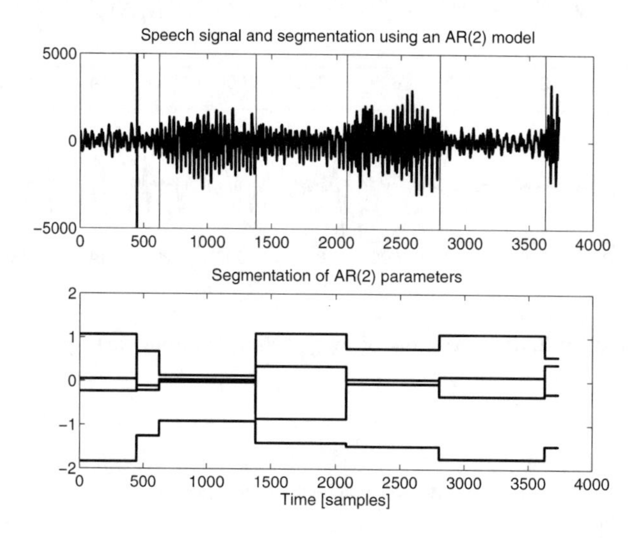

Figure 2.12. A speech signal and a possible segmentation. (Data provided by Prof. Michele Basseville, IRISA, France, and Prof. Regine Andre-Obrecht, IRIT, France.)

2.5. ARX model

The ARX model (see equation (A.12) in Appendix A) an extension of the AR model for dynamic systems driven by an input u_t.

2.5.1. DC motor fault detection

An application studied extensively in Part IV and briefly in Part III is based on simulated and measured data from a DC motor. A typical application is to use the motor as a servo which requires an appropriate controller designed to a model of the motor. If the dynamics of the motor change with time, we have an *adaptive control* problem. In that case, the controller needs to be redesigned, at regular time instants or when needed, based on the updated model. Here we are facing a fundamental isolation problem:

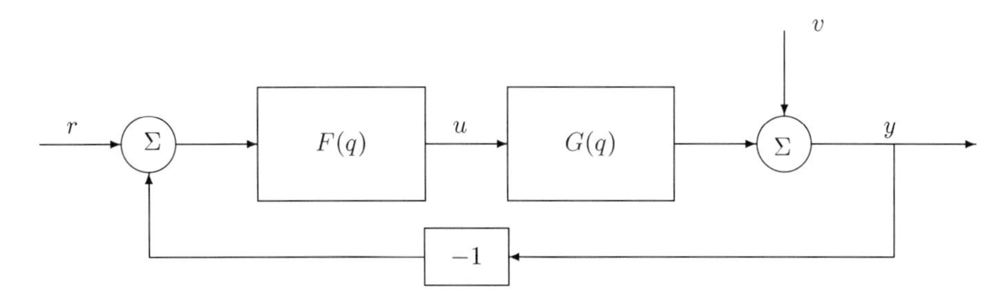

Figure 2.13. A closed loop control system. Here $G(q)$ is the DC motor dynamics, $F(q)$ is the controller, r the reference angle, u the controlled input voltage to the engine, y the actual angle and v is a disturbance.

Fundamental adaptive control problem

Disturbances and system changes must be isolated. An alarm caused by a system change requires that the controller should be re-designed, while an alarm caused by a disturbance (false alarm) implies that the controller should be frozen.

We here describe how data from a lab motor were collected as presented in Gustafsson and Graebe (1998). The data are collected in a closed loop, as shown in the block diagram in Figure 2.13. Below, the transfer functions $G(q)$ (using the discrete time shift operator q, see Appendix A) and the controller $F(q)$ are defined. A common form of a transfer function describing a DC motor in continuous time (using Laplace operator s) is

$$G(s) = \frac{b}{s(s+a)}.$$

The parameters were identified by a step response experiment to

$$b = 140, \quad a = 3.5.$$

The discrete time transfer function, assuming piecewise constant input, is for sampling time $T_s = 0.1$:

$$G(q) = \frac{0.625q + 0.5562}{q^2 - 1.705q + 0.7047} = \frac{0.62501(q+0.89)}{(q-1)(q-0.7047)}.$$

The PID (Proportional, Integrating and Differentiating) structured regulator is designed in a pole-placement fashion and is, in discrete time with $T_s = 0.1$,

$$F(q) = \frac{0.272q^2 - 0.4469q + 0.1852}{q^2 - 1.383q + 0.3829}.$$

Table 2.1. Data sets for the DC lab motor

Data set 1	No fault or disturbance.
Data set 2	Several disturbances.
Data set 3	Several system changes.
Data set 4	Several disturbances and one (very late) system change.

The reference signal is generated as a square wave pre-filtered by $1.67/(s+1.67)$ (to get rid of an overshoot due to the zeros of the closed loop system) with a small sinusoidal perturbation signal added ($0.02\sin(4.5 \cdot t)$).

The closed loop system from r to y is, with $T_s = 0.1$,

$$y_t = \frac{0.1459q^3 - 0.1137q^2 - 0.08699q + 0.0666}{q^4 - 2.827q^3 + 3.041q^2 - 1.472q + 0.2698} r_t. \tag{2.1}$$

An alternative model is given in Section 2.7.1.

Data consist of `y,r` from the process under four different experiments, as summarized in Table 2.1.

Disturbances were applied by physically holding the outgoing motor axle. *System changes* were applied in software by shifting the pole in the DC motor from 3.5 to 2 (by including a block $(s+3.5)/(s+2)$ in the regulator).

The transfer function model in (2.1) is well suited for the case of detecting model changes, while the state space model to be defined in (2.2) is better for detecting disturbances.

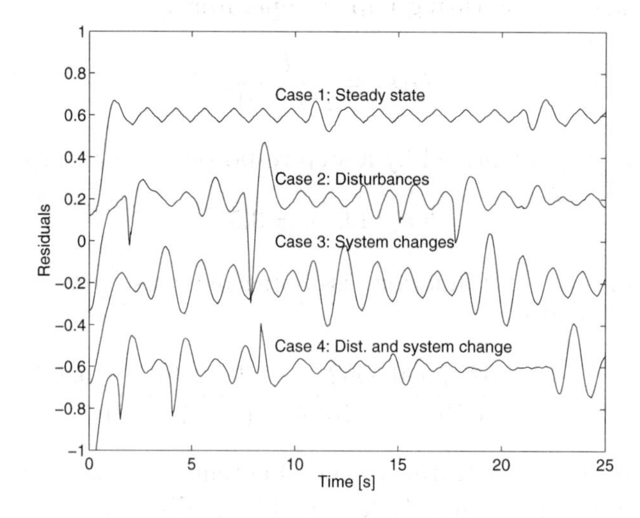

Figure 2.14. Model residuals defined as measurements subtracted by a simulated output. Torque disturbances and dynamical model changes are clearly visible, but they are hard to distinguish.

The goal with change detection is to compute residuals better suited for change detection than simply taking the error signal from the real system and the model. As can be seen from Figure 2.14, it is hard to distinguish disturbances from model changes (the *isolation* problem), generally.

This application is further studied in Sections 5.10.2, 8.12.1 and 11.5.2.

2.5.2. Belching sheep

The input u_t is the lung volume of a sheep and the output y_t the air flow through the throat, see Figure 2.15 (the data have been rescaled). A possible model is

$$A(q)y_t = B(q)u_t + e_t, \quad e_t \in \mathrm{N}(0, \sigma_t^2),$$

where the noise variance σ_t^2 is large under belches.

The goal is to get a model for how the input relates to the output, that is $B(q)$, $A(q)$, and how different medicines affect this relation. A problem with a straightforward system identification approach is that the sheep belches regularly. Therefore, belching segments must be detected before modeling. The approach here is that the residuals from an ARX model are segmented according to the variance level. This application is investigated in Section 6.5.2.

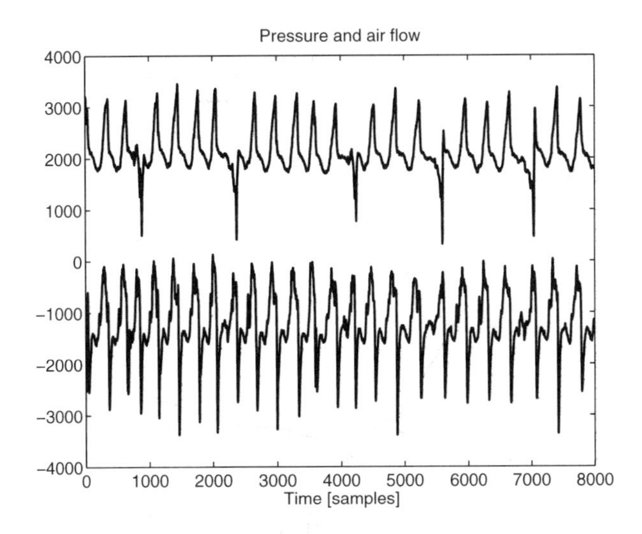

Figure 2.15. The air pressure in the stomach and air inflow through the throat of a sheep. The belches are visible as negative flow dips. (Data provided by Draco and Prof. Bo Bernardsson, Lund, Sweden.)

2.6. Regression model

The general form of the linear regression model (see equation (A.3) in Appendix A) includes FIR, AR and ARX models as special cases, but also appears in contexts other than modeling dynamical time series. We have already seen the friction estimation application in Examples 1.2, 1.5 and 1.9.

2.6.1. Path segmentation and navigation in cars

This case study will be examined in Section 7.7.3. The data were collected from test drives with a Volvo 850 GLT using sensor signals from the ABS system.

There are several commercial products providing guidance systems for cars. These require a position estimate of the car, and proposed solutions are based on expensive GPS (Global Positioning System) or a somewhat less expensive gyro. The idea is to compare the estimated position with a digital map. Digital street maps are available for many countries. The map and position estimator, possibly in combination with traffic information transmitted over the FM band or from road side beacons, are then used for guidance.

We demonstrate here how an adaptive filter, in combination with a change detector, can be used to develop an almost free position estimator, where no additional hardware is required. It has a worse accuracy than its alternatives, but on the other hand, it seems to be able to find relative movements on a map, as will be demonstrated.

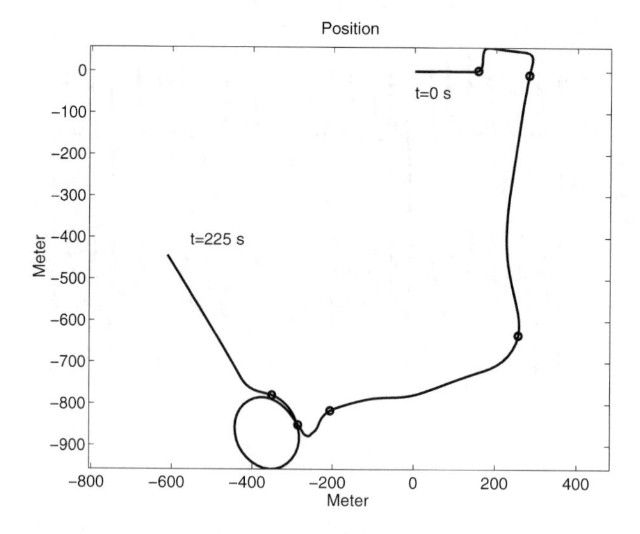

Figure 2.16. Driven path and possible result of manoeuvre detection. (Data collected by the author in collaboration with Volvo.)

Figure 2.16 shows an estimated path for a car, starting with three sharp turns and including one large roundabout. The velocities of the free-rolling wheels are measured using sensors available in the Anti-lock Braking System (ABS). By comparing the wheel velocities ω_r and ω_l on the right and left side, respectively, the velocity v and curve radius R can be computed from

$$\begin{aligned} v &= \frac{\omega_l + \omega_r}{2} r \\ R^{-1} &= \frac{1 - \frac{\omega_l}{\omega_r}(1+\varepsilon)}{L}, \end{aligned}$$

where L is the wheel base, r is the nominal wheel radius and ε is the relative difference in wheel radius on the left and right sides. The wheel radius difference ε gives an offset in heading angle, and is thus quite important for how the path looks (though it is not important for segmentation). It is estimated on a long-term basis, and is in this example $2.5 \cdot 10^{-3}$. The algorithm is implemented on a PC and runs on a Volvo 850 GLT.

The heading angle ψ_t and global position (X_t, Y_t) as functions of time can be computed from

$$\begin{aligned} \psi_{t+1} &= \psi_t + v_t T_s R_t^{-1} \\ X_{t+1} &= X_t + v_t T_s \cos(\psi_t) \\ Y_{t+1} &= Y_t + v_t T_s \sin(\psi_t) \end{aligned}$$

The sampling interval was chosen to $T_s = 1$ s. The approach requires that the initial position and heading angle X_0, Y_0, ψ_0 are known.

The path shown in Figure 2.16 fits a street map quite well, but not perfectly. The reason for using segmentation is to use corners, bends and roundabouts for updating the position from the digital map. Any input to the algorithm dependent on the velocity will cause a lot of irrelevant alarms, which is obvious from the velocity plot in Figure 2.17. The ripple on the velocity signal is caused by gear changes. Thus, segmentation using velocity dependent measurements should be avoided. Only the heading angle ψ_t is needed for segmentation.

The model is that the heading angle is piecewise constant or piecewise linear, corresponding to straight paths and bends or roundabouts. The regression model used is

$$\begin{aligned} \theta_{t+1} &= (1 - \delta_t)\theta_t + \delta_t v_t \\ \psi_t &= \theta_t^1 + \theta_t^2 t + e_t \\ \mathrm{E}\, e_t^2 &= \lambda_t. \end{aligned}$$

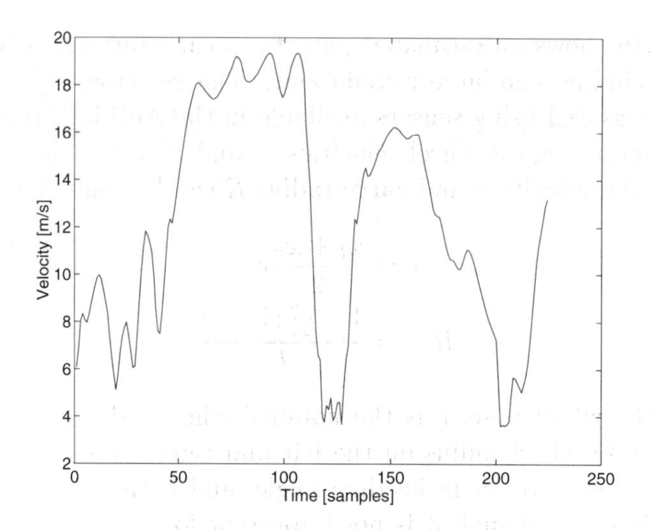

Figure 2.17. Velocity in the test drive.

2.6.2. Storing EKG signals

Databases for various medical applications are becoming more and more frequent. One of the biggest is the FBI fingerprint database. For storage efficiency, data should be compressed, without losing information. The fingerprint database is compressed by wavelet techniques. The EKG signal examined here will be compressed by polynomial models with piecewise constant parameters. For example, a linear model is

$$y_t = \theta_t^1 + \theta_t^2 t + e_t = (1,\ t)\theta_t + e_t.$$

Figure 2.18 shows a part of an EKG signal and a possible segmentation. For evaluation, the following statistics are interesting:

Model	Linear model
Error (%)	0.85
Compression rate (%)	10

The linear model gives a decent error rate and a low compression rate. The compression rate is measured here as the number of parameters (here 2) times the number of segments, compared to the number of data. It says how many real numbers have to be saved, compared to the original data. Details on the implementation are given in Section 7.7.1. There is a design parameter in the algorithm to trade off between the error and compression rates.

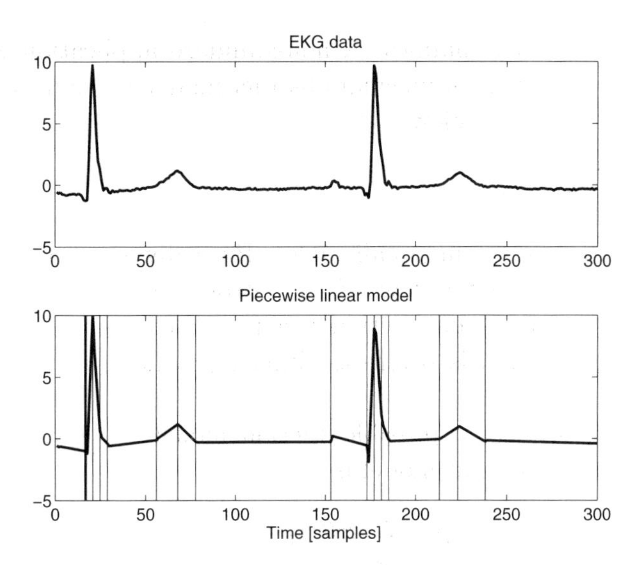

Figure 2.18. An EKG signal (upper plot) and a piecewise constant linear model (lower plot).

2.7. State space model

2.7.1. DC motor fault detection

Consider the DC motor in Section 2.5.1. For a particular choice of state vector, the transfer function (2.1) can be written as a state space model:

$$x_{t+1} = \begin{pmatrix} 2.8269 & -1.5205 & 0.7361 & -0.2698 \\ 2.0000 & 0 & 0 & 0 \\ 0 & 1.0000 & 0 & 0 \\ 0 & 0 & 0.5000 & 0 \end{pmatrix} x_t + \begin{pmatrix} 0.5 \\ 0 \\ 0 \\ 0 \end{pmatrix} u_t$$

$$y_t = \begin{pmatrix} 0.2918 & -0.1137 & -0.0870 & 0.1332 \end{pmatrix} x_t. \tag{2.2}$$

The state space model is preferable to the transfer function approach for detecting actuator and sensor faults and disturbances, which are all modeled well as additive changes in a state space model. This corresponds to case 2 in Figure 2.14. This model will be used in Sections 8.12.1 and 11.5.2.

2.8. Multiple models

A powerful generalization of the linear state space model, is the multiple model, where a discrete mode parameter is introduced for switching between a finite number of modes (or operating points). This is commonly used for approximating non-linear dynamics. A non-standard application which demonstrates the flexibility of the somewhat abstract model, given in equation (A.25) in Appendix A), is given below.

2.8.1. Valve stiction

Static friction, *stiction*, occurs in all valves. Basically, the valve position sticks when the valve movement is low. For control and supervision purposes, it is important to detect when stiction occurs. A control action that can be undertaken when stiction is severe is dithering, which forces the valve to go back and forth rapidly.

A block diagram over a possible stiction model is shown in Figure 2.19. Mathematically, the stiction model is

$$
\begin{aligned}
y_t &= G(q;\theta)x_t \\
x_t &= \begin{cases} u_t & \text{if } \delta_t = 0 \\ x_{t-1} & \text{if } \delta_t = 1 \end{cases}.
\end{aligned}
$$

Here δ_t is a discrete binary state, where $\delta_t = 0$ corresponds to the valve following the control input, and $\delta_t = 1$ is the stiction mode. Any prior can be assigned to the discrete state. For instance, a Markov model with certain transition probabilities is plausible. The parameters θ in the dynamical model for the valve dynamics are unknown, and should be estimated simultaneously.

Figure 2.20 shows logged data from a steam valve, together with the identified discrete state and a simulation of the stiction model, using an algorithm described in Chapter 10. We can clearly see in the lower plot that the valve position is in the stiction mode most of the time. Another approach based on monitoring oscillations of the closed loop system can be found in Thornhill and Hägglund (1997).

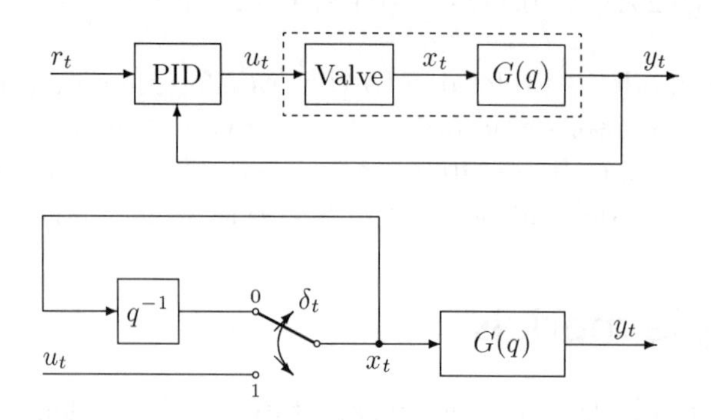

Figure 2.19. A control loop (a) and the the assumed stiction model (b).

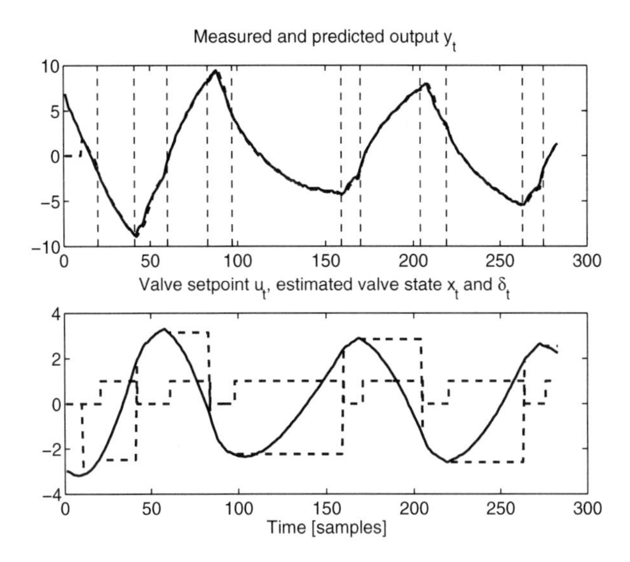

Figure 2.20. Valve input and output, and estimated valve position. (Data provided by Dr. Krister Forsman, ABB Automation.)

2.9. Parameterized non-linear models

Linear models, as linear regressions and state space models, can in many applications give a satisfactory result. However, in some cases tailored non-linear models with unknown parameters might give a considerable improvement in performance, or may even be the only alternative. We will give some examples of the latter case here.

One application of models of the type (A.17) is for innovation processes. Innovation models are common in various disciplines, for instance:

- Biology: growth processes in certain species or diseases or bacteria.
- Economy: increase and saturation of sales figures.

The two innovation processes used as illustration here are both based on a continuous time state space model of the kind:

$$\dot{x}_t = f(x_t, u_t, v_t; \theta), \tag{2.3}$$

$$y_t = g(x_t, u_t, e; \theta). \tag{2.4}$$

The problem can be seen as curve fitting to a given differential function.

2.9.1. Electronic nose

We start with a biological example of the innovation process presented in Holmberg et al. (1998). The data are taken from an artificial nose, which is using 15 sensors to classify bacteria. For feature extraction, a differential equation model is used to map the time series to a few parameters. The 'nose' works as follows. The bacteria sample is placed in a substrate. To start with, the bacteria consume substrate and they increase in number (the growth phase). After a while, when the substrate is consumed, the bacteria start to die. Let $f(t)$ denote the measurements. One possible model for these is

$$\hat{f}(t;\theta) = \theta^1 \frac{1}{1+e^{-\theta^2(\theta^3-t)}}\left(1-\frac{1}{1+e^{-\theta^4(\theta^5-t)}}\right),$$

with $y_t = x_t$ in equation (2.3). The physical interpretation of the parameters are as follows:

θ^1	Scaling (amplitude)
θ^2	Growth factor (steepness in descent)
θ^3	Growth initial time (defined by the 50% level)
θ^4	Death factor (steepness in ascent)
θ^5	Death initial time (defined by the 50% level)

The basic signal processing idea of minimizing a least squares fit means

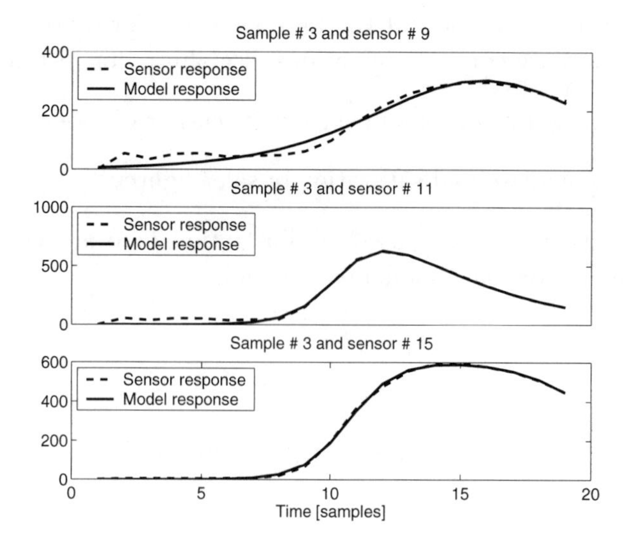

Figure 2.21. Example of least squares fit of parametric differential equation to data.

that:

$$V(\theta) = \sum_{t=1}^{N} (f(t) - \hat{f}(t;\theta))^2$$

is minimized with respect to θ. The measure of fit

$$\frac{\min V(\theta)}{\sum_k f(t)^2}$$

can be used to judge the performance. For the measurement responses in Figure 2.21, the fit values are 0.0120, 0.0062 and 0.0004, respectively.

The classification idea is as follows: estimate the parametric model to each of the 15 sensor signals for each nose sample, each one consisting of 19 time values. Then, use the parameters as features in classification. As described in Holmberg et al. (1998), a classification rate of 76% was obtained using leave-one-out validation, from five different bacteria. Details of the parameter identification are given in Example 5.10.

2.9.2. Cell phone sales figures

Economical innovation models are defined by an innovation that is communicated over certain channels in a social system (Rogers, 1983). The example in Figure 2.22 shows sales figures for telephones in the standards NMT 450 and NMT 900. The latter is a newer standard, so the figures have not reached the saturation level at the final time. A possible application is to use the

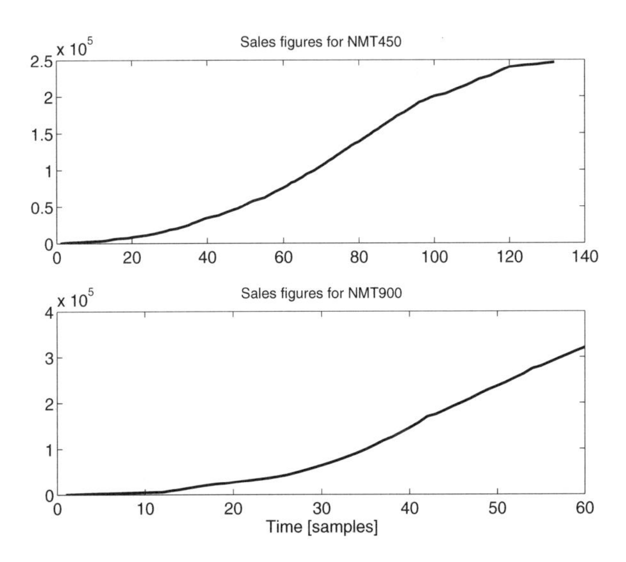

Figure 2.22. Sales figures for the telephone standards NMT 450 and NMT 900.

innovation model for the previous model to predict both saturation time and level from the current information.

Many models have been proposed, see Mahajan et al. (1990). Most of them rely on the exponential function for modeling the growth process and saturation phase. One particular model is the so-called *combined model* in Wahlbin (1982):

$$\dot{y} = a(N - y) + by(N - y).$$

In this case, N is the final number of NMT telephone owners, a is a growth parameter and b a saturation parameter. Details on the parameter identification and a fitted curve are presented in Example 5.11.

Part II: Signal estimation

3

On-line approaches

3.1. Introduction

The basic assumption in this part, signal estimation, is that the measurements y_t consist of a deterministic component θ_t – the signal – and additive white noise e_t,

$$y_t = \theta_t + e_t. \tag{3.1}$$

For change detection, this will be labeled as a *change in the mean model.* The task of determining θ_t from y_t will be referred to as *estimation*, and *change detection* or *alarming* is the task of finding abrupt, or rapid, changes in θ_t, which is assumed to start at time k, referred to as the *change time.* *Surveillance* comprises all these aspects, and a typical application is to monitor levels, flows and so on in industrial processes and alarm for abnormal values.

The basic assumptions about model (3.1) in *change detection* are:

- The deterministic component θ_t undergoes an abrupt change at time $t = k$. Once this change is detected, the procedure starts all over again to detect the next change. The alternative is to consider θ_t as piecewise constant and focus on a sequence of change times $k_1, k_2, \dots, k_n$, as shown in Chapter 4. This sequence is denoted k^n, where both k_i and n are free parameters. The *segmentation* problem is to find both the number and locations of the change times in k^n.

- In the statistical approaches, it will be assumed that the noise is white and Gaussian $e_t \in \mathrm{N}(0, R)$. However, the formulas can be generalized to other distributions, as will be pointed out in Section 3.A.

The change magnitude for a change at time k is defined as $\nu \triangleq \theta_{k+1} - \theta_k$. Change detection approaches can be divided into hypothesis tests and estimation/information approaches. Algorithms belonging to the class of hypothesis tests can be split into the parts shown in Figure 3.1.

For the change in the mean model, one or more of these blocks become trivial, but the picture is useful to keep in mind for the general model-based case. Estimation and information approaches do everything in one step, and do not suit the framework of Figure 3.1.

The alternative to the *non-parametric approach* in this chapter is to model the deterministic component of y_t as a parametric model, and this issue will

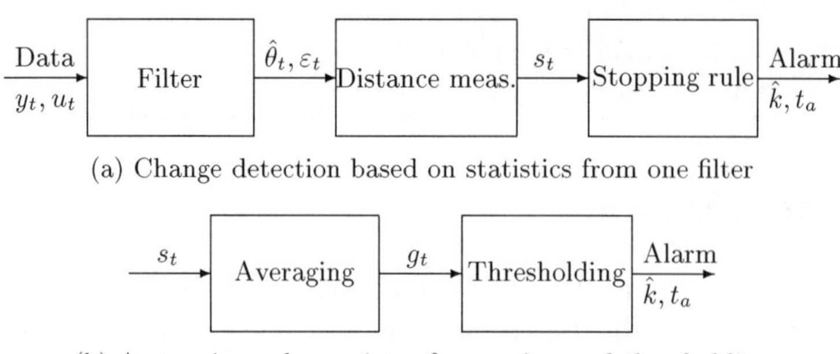

(a) Change detection based on statistics from one filter

(b) A stopping rule consists of averaging and thresholding

Figure 3.1. The steps in change detection based on hypothesis tests. The stopping rule can be seen as an averaging filter and a thresholding decision device.

be the dealt with in Part III. It must be noted that the signal model (3.1), and thus all methods in this part, are special cases of what will be covered in Part III.

This chapter presents a review of averaging strategies, stopping rules, and change detection ideas. Most of the ideas to follow in subsequent chapters are introduced here.

3.2. Filtering approaches

The standard approach in *signal processing* for separating the signal θ_t and the noise e_t is by (typically low-pass) *filtering*

$$\hat{\theta}_t = H(q)y_t. \tag{3.2}$$

The filter can be of Finite Impulse Response (FIR) or Infinite Impulse Response (IIR) type, and designed by any standard method (Butterworth, Chebyshev etc.). An alternative interpretation of filters is data *windowing*

$$\hat{\theta}_t = \sum_{k=0}^{\infty} w_k y_{t-k}, \tag{3.3}$$

where the weights should satisfy $\sum w_k = 1$. This is equal to the filtering approach if the weights are interpreted as the impulse response of the (low-pass) filter $H(q)$, i.e. $w_k = h_k$.

An important special case of these is the *exponential forgetting window*, or *Geometric Moving Average* (*GMA*)

$$w_k = (1-\lambda)\lambda^k, \quad 0 \leq \lambda < 1. \tag{3.4}$$

A natural, and for change detection fundamental, principle is to use a *sliding window*, defined by

$$w_k = \begin{cases} \frac{1}{L} & \text{if } 0 \leq k < L, \\ 0 & \text{if } k \geq L. \end{cases} \tag{3.5}$$

A more general approach, which can be labeled *Finite Moving Average* (*FMA*) is obtained by using arbitrary weights w_k in the sliding window with the constraint $\sum_{k=0}^{L-1} w_k = 1$, which is equivalent to a FIR filter.

3.3. Summary of least squares approaches

This section offers a summary of the adaptive filters presented in a more general form in Chapter 5.

A common framework for the most common estimation approaches is to let the signal estimate be the *minimizing argument* (arg min) of a certain loss function

$$\hat{\theta}_t = \arg\min_{\theta} V_t(\theta). \tag{3.6}$$

In the next four subsections, a number of loss functions are given, and the corresponding estimators are derived. For simplicity, the noise variance is assumed to be constant $\mathrm{E}\, e_t^2 = R$. We are interested in the signal estimate, and also its theoretical variance $P_t \triangleq \mathrm{E}(\theta_t - \hat{\theta}_t)^2$, and its estimate $\hat{P}_t$. For adaptive methods, the parameter variance is defined under the assumption that the parameter θ_t is time invariant.

3.3.1. Recursive least squares

To start with, the basic idea in *least squares* is to minimize the sum of squared errors:

$$V_t(\theta) \triangleq \sum_{i=1}^{t} (y_i - \theta)^2 \tag{3.7}$$

$$\hat{\theta}_t = \frac{1}{t} \sum_{i=1}^{t} y_i \tag{3.8}$$

$$P_t = \frac{1}{t^2} \sum_{i=1}^{t} R = \frac{R}{t} \tag{3.9}$$

$$\hat{P}_t = \frac{1}{t^2} V_t(\hat{\theta}) \tag{3.10}$$

$$\hat{R}_t = \frac{1}{t} V_t(\hat{\theta}). \tag{3.11}$$

This is an off-line approach which assumes that the parameter is time invariant. If θ_t is time-varying, adaptivity can be obtained by forgetting old measurements using the following loss function:

$$V_t(\theta) \triangleq \sum_{i=1}^{t} \lambda^{t-i} (y_i - \theta)^2$$

$$\hat{\theta}_t = \frac{1-\lambda}{1-\lambda^t} \sum_{i=1}^{t} \lambda^{t-i} y_i$$

$$P_t = \frac{1-\lambda}{1-\lambda^t} \frac{1+\lambda^t}{1+\lambda} R$$

$$\hat{P}_t = \frac{1-\lambda}{1-\lambda^t}\frac{1+\lambda^t}{1+\lambda}V_t(\hat{\theta})$$
$$\hat{R}_t = \frac{1-\lambda}{1-\lambda^t}V_t(\hat{\theta}).$$

Here λ is referred to as the *forgetting factor*. This formula yields the *recursive least squares* (*RLS*) estimate. Note that the estimate $\hat{\theta}_t$ is unbiased only if the true parameter is time invariant. A recursive version of the RLS estimate is

$$\begin{aligned}\hat{\theta}_t &= \lambda\hat{\theta}_{t-1} + (1-\lambda)y_t \\ &= \hat{\theta}_{t-1} + (1-\lambda)\varepsilon_t,\end{aligned} \tag{3.12}$$

where $\varepsilon_t = (y_t - \hat{\theta}_{t-1})$ is the prediction error. This latter formulation of RLS will be used frequently in the sequel, and a general derivation is presented in Chapter 5.

3.3.2. The least squares over sliding window

Computing the least squares loss function over a sliding window of size L gives:

$$V_t(\theta) \triangleq \sum_{i=t-L+1}^{t}(y_i-\theta)^2$$
$$\hat{\theta}_t = \frac{1}{L}\sum_{i=t-L+1}^{t} y_i$$
$$= \hat{\theta}_{t-1} + \frac{y_t - y_{t-L}}{L}$$
$$P_t = \frac{1}{L^2}\sum_{i=t-L+1}^{t} R$$
$$\hat{P}_t = \frac{1}{L^2}V_t(\hat{\theta})$$
$$\hat{R}_t = \frac{1}{L}V_t(\hat{\theta}).$$

This approach will be labeled the *Windowed Least Squares* (WLS) method. Note that a memory of size L is needed in this approach to store old measurements.

3.3.3. Least mean square

In the *Least Mean Square* (*LMS*) approach, the objective is to minimize

$$V_t(\theta) \triangleq E(y_t-\theta)^2 \tag{3.13}$$

by a *stochastic gradient algorithm* defined by

$$\hat{\theta}_t \triangleq \hat{\theta}_{t-1} - \mu \frac{1}{2} \left. \frac{dV_t(\theta)}{d\theta} \right|_{\theta = \hat{\theta}_{t-1}}. \tag{3.14}$$

Here, μ is the *step size* of the algorithm. The expectation in (3.13) cannot be evaluated, so the standard approach is to just ignore it. Differentiation then gives the LMS algorithm:

$$\hat{\theta}_t = \hat{\theta}_{t-1} + \mu \varepsilon_t. \tag{3.15}$$

That is, for signal estimation, LMS and RLS coincide with $\mu = 1 - \lambda$. This is not true in the general case in Chapter 5.

3.3.4. The Kalman filter

One further alternative is to explicitly model the parameter time-variations as a so called *random walk*:

$$\theta_{t+1} = \theta_t + v_t. \tag{3.16}$$

Let the variance of the noise v_t be Q. Then the *Kalman filter*, as will be derived in Chapter 13, applies:

$$\hat{\theta}_t = \hat{\theta}_{t-1} + \frac{P_{t-1}}{P_{t-1} + R} \varepsilon_t, \tag{3.17}$$

$$P_t = P_{t-1} - \frac{P_{t-1}^2}{P_{t-1} + R} + Q. \tag{3.18}$$

If the assumption (3.16) holds, the Kalman filter is the optimal estimator in the following meanings:

- It is the *minimum variance estimator* if v_t and e_t (and also the initial knowledge of θ_0) are independent and Gaussian. That is, there is no other estimator that gives a smaller variance error $\text{Var}(\hat{\theta}_t - \theta_t)$.
- Among all *linear estimators*, it gives the minimum variance error independently of the distribution of the noises.
- Since minimum variance is related to the least squares criterion, we also have a least squares optimality under the model assumption (3.16).
- It is the *conditional expectation* of θ_t, given the observed values of y_t.

This subject will be thoroughly treated in Chapter 8.

Example 3.1 Signal estimation using linear filters

Figure 3.2 shows an example of a signal, and the estimates from RLS, LMS and KF, respectively. The design parameters are $\lambda = 0.9$, $\mu = 0.1$ and $Q = 0.02$, respectively. RLS and LMS are identical and the Kalman filter is very similar for these settings.

3.4. Stopping rules and the CUSUM test

A *stopping rule* is used in surveillance for giving an alarm when θ_t has exceeded a certain threshold. Often, a stopping rule is used as a part of a change detection algorithm. It can be characterized as follows:

- The definition of a stopping rule here is that, in contrast to change detection, no statistical assumptions on its input are given.
- The change from $\theta_t = 0$ to a positive value may be abrupt, linear or incipient, whereas in change detection the theoretical assumption in the derivation is that the change is abrupt.
- There is prior information on how large the threshold is.

An auxiliary test statistic g_t is introduced, which is used for alarm decisions using a threshold h. The purpose of the stopping rule is to give an alarm when

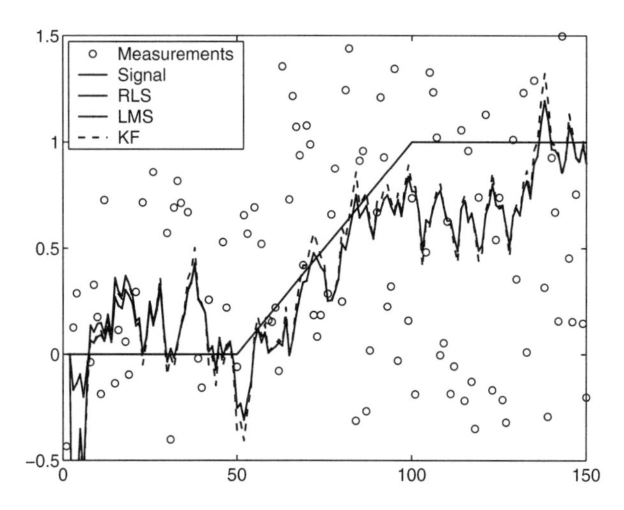

Figure 3.2. A signal observed with noise. The almost identical lines show the signal estimate from RLS, LMS and KF.

g_t exceeds a certain value,

$$\text{Alarm if } g_t > h. \tag{3.19}$$

Stopping rules will be used frequently when discussing change detection based on filter residual whiteness tests and model validation.

3.4.1. Distance measures

The input to a stopping rule is, as illustrated in Figure 3.3, a *distance measure* s_t. Several possibilities exist:

- A simple approach is to take the residuals

$$s_t = \varepsilon_t = y_t - \hat{\theta}_{t-1}, \tag{3.20}$$

  where $\hat{\theta}_{t-1}$ (based on measurements up to time $t-1$) is any estimate from Sections 3.2 or 3.3. This is suitable for the change in the mean problem, which should be robust to variance changes. A good alternative is to normalize to unit variance. The variance of the residuals will be shown to equal $R + P_t$, so use instead

$$s_t = \frac{\varepsilon_t}{\sqrt{R + P_t}} = \frac{y_t - \hat{\theta}_{t-1}}{\sqrt{R + P_t}}. \tag{3.21}$$

  This scaling facilitates the design somewhat, in that approximately the same design parameters can be used for different applications.

- An alternative is to square the residuals

$$s_t = \varepsilon_t^2. \tag{3.22}$$

  This is useful for detecting both variance and parameter changes. Again, normalization, now to unit expectation, facilitates design

$$s_t = \frac{\varepsilon_t^2}{P_t^2}. \tag{3.23}$$

Other options are based on likelihood ratios to be defined. For general filter formulations, certain correlation based methods apply. See Sections 5.6 and 8.10.

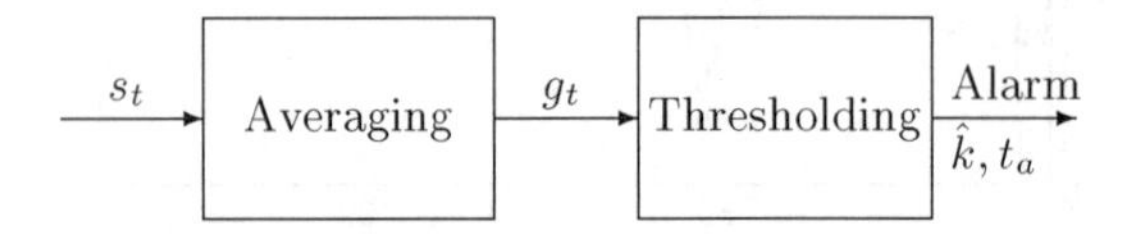

Figure 3.3. Structure of a stopping rule.

3.4.2. One-sided tests

As shown in Figure 3.3, the stopping rule first averages the inputs s_t to get a test statistic g_t, which is then thresholded. An alarm is given when $g_t = \hat{\theta}_t > h$ and the test statistic is reset, $g_t = 0$. Indeed, any of the averaging strategies discussed in Section 3.2 can be used. In particular, the sliding window and GMA filters can be used.

As an example of a wellknown combination, we can take the squared residual as distance measure from the no change hypothesis, $s_t = \varepsilon_t^2$, and average over a sliding window. Then we get a χ^2 test, where the distribution of g_t is $\chi^2(L)$ if there is no change. There is more on this issue in subsequent chapters. Particular named algorithms are obtained for the exponential forgetting window and finite moving average filter. In this way, stopping rules based on FMA or GMA are obtained.

The methods so far have been linear in data, or for the χ^2 test quadratic in data. We now turn our attention to a fundamental and historically very important class of non-linear stopping rules. First, the Sequential Probability Ratio Test (SPRT) is given.

Algorithm 3.1 SPRT

$$g_t = g_{t-1} + s_t - \nu \tag{3.24}$$

$$g_t = 0, \text{ and } \hat{k} = t \text{ if } g_t < a < 0 \tag{3.25}$$

$$g_t = 0, \text{ and } t_a = t \text{ and alarm if } g_t > h > 0. \tag{3.26}$$

Design parameters: Drift ν, threshold h and reset level a.
Output: Alarm time(s) t_a.

In words, the test statistic g_t sums up its input s_t, with the idea to give an alarm when the sum exceeds a threshold h. With a white noise input, the test statistic will drift away similar to a random walk. There are two mechanisms to prevent this natural fluctuation. To prevent positive drifts, eventually yielding a false alarm, a small drift term ν is subtracted at each time instant. To prevent a negative drift, which would increase the time to detection after a change, the test statistic is reset to 0 each time in becomes less than a negative constant a.

The level crossing parameter a should be chosen to be small in magnitude, and it has been thoroughly explained why $a = 0$ is a good choice. This important special case yields the cumulative sum (*CUSUM*) algorithm.

Algorithm 3.2 CUSUM

$$g_t = g_{t-1} + s_t - \nu \tag{3.27}$$
$$g_t = 0, \text{ and } \hat{k} = t \text{ if } g_t < 0 \tag{3.28}$$
$$g_t = 0, \text{ and } t_a = t \text{ and alarm if } g_t > h > 0. \tag{3.29}$$

Design parameters: Drift ν and threshold h.
Output: Alarm time(s) t_a and estimated change time $\hat{k}$.

Both algorithms were originally derived in the context of quality control (Page, 1954). A more recent reference with a few variations analyzed is Malladi and Speyer (1999).

In both SPRT and CUSUM, the alarm time t_a is the primary output, and the drift should be chosen as half of the critical level that must not be exceeded by the physical variable θ_t. A non-standard, but very simple, suggestion for how to estimate the change time is included in the parameter $\hat{k}$, but remember that the change is not necessarily abrupt when using stopping rules in general. The estimate of the change time is logical for the following reason (although the change location problem does not seem to be dealt with in literature in this context). When $\theta_t = 0$ the test statistic will be set to zero at almost every time instant (depending on the noise level and if $a < -\nu$ is used). After a change to $\theta_t > \nu$, g_t will start to grow and will not be reset until the alarm comes, in which case $\hat{k}$ is close to the correct change time. As a rule of thumb, the drift should be chosen as one half of the expected change magnitude.

Robustness and decreased false alarm rate may be achieved by requiring several $g_t > h$. This is in quality control called a *run test*.

Example 3.2 Surveillance using the CUSUM test

Suppose we want to make surveillance of a signal to detect if its level reaches or exceeds 1. The CUSUM test with $\nu = 0.5$ and $h = 5$ gives an output illustrated in Figure 3.4. Shortly after the level of the signal exceeds 0.5, the test statistic starts to grow until it reaches the threshold, where we get an alarm. After this, we continuously get alarms for level crossing. A run test where five stops in the CUSUM test generates an alarm would give an alarm at sample 150.

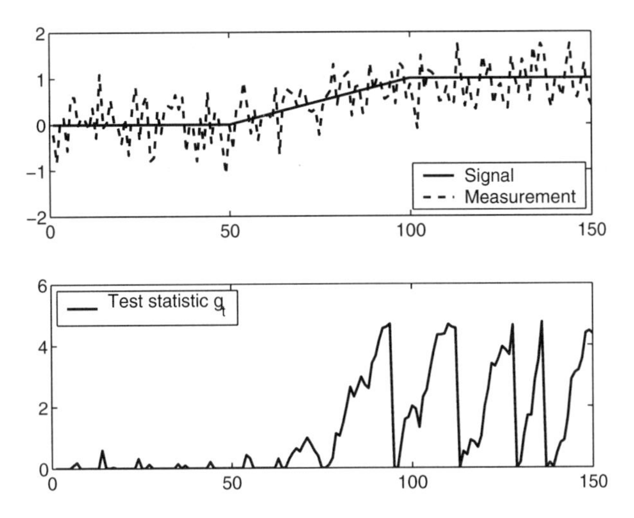

Figure 3.4. A signal observed with noise. The lower plot shows the test statistic g_t from (3.27)–(3.29) in the CUSUM test.

3.4.3. Two-sided tests

The tests in the previous section are based on the assumption that θ_t is positive. A two-sided test is obtained as follows:

- For the averaging and estimation approaches where g_t is a linear function of data, simply test if $g_t > h_1$ *or* $g_t < -h_2$.

- For the non-linear (in data) methods CUSUM and SPRT, apply two tests in parallel. The second one can be seen as having $-y_t$ as the input and h_2 as the threshold. We get an alarm when one of the single tests signals an alarm.

In a fault detection context, we here get a very basic diagnosis based on the sign of the change.

3.4.4. The CUSUM adaptive filter

To illustrate one of the main themes in this book, we here combine adaptive filters with the CUSUM test as a change detector, according to the general picture shown in Figure 1.18. The first idea is to consider the signal as piecewise constant and update the least squares estimate in between the alarm times. After the alarm, the LS algorithm is restarted.

Algorithm 3.3 The CUSUM LS filter

$$
\begin{aligned}
\hat{\theta}_t &= \frac{1}{t-t_0} \sum_{k=t_0+1}^{t} y_k \\
\varepsilon_t &= y_t - \hat{\theta}_{t-1} \\
s_t^{(1)} &= \varepsilon_t \\
s_t^{(2)} &= -\varepsilon_t \\
g_t^{(1)} &= \max(g_{t-1}^{(1)} + s_t^{(1)} - \nu, 0) \\
g_t^{(2)} &= \max(g_{t-1}^{(2)} + s_t^{(2)} - \nu, 0) \\
\text{Alarm if } & g_t^{(1)} > h \text{ or } g_t^{(2)} > h
\end{aligned}
$$

After an alarm, reset $g_t^{(1)} = 0$, $g_t^{(2)} = 0$ and $t_0 = t$.
Design parameters: ν, h.
Output: $\hat{\theta}_t$.

Since the output from the algorithm is a piecewise constant parameter estimate, the algorithm can be seen as a *smoothing* filter. Given the detected change times, the computed parameter estimates are the best possible off-line estimates.

Example 3.3 Surveillance using the CUSUM test

Consider the same signal as in Example 3.2. Algorithm 3.3 gives the signal estimate and the test statistics in Figure 3.5. We only get one alarm, at time 79, and the parameter estimate quickly adapts afterwards.

A variant is obtained by including a forgetting factor in the least squares estimation. Technically, this corresponds to an assumption of a signal that normally changes slowly (caught by the forgetting factor) but sometimes undergoes abrupt changes.

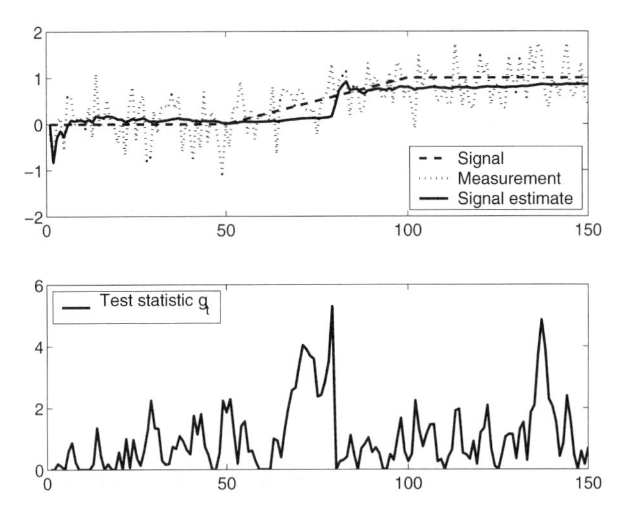

Figure 3.5. A signal and its measurements, with an estimate from the CUSUM filter in Algorithm 3.3. The solid line shows the test statistic $g(t)$ in the CUSUM test.

Algorithm 3.4 The CUSUM RLS filter

$$
\begin{aligned}
\hat{\theta}_t &= \lambda \hat{\theta}_{t-1} + (1-\lambda) y_t \\
\varepsilon_t &= y_t - \hat{\theta}_{t-1} \\
s_t^{(1)} &= \varepsilon_t \\
s_t^{(2)} &= -\varepsilon_t \\
g_t^{(1)} &= \max(g_{t-1}^{(1)} + s_t^{(1)} - \nu, 0) \\
g_t^{(2)} &= \max(g_{t-1}^{(2)} + s_t^{(2)} - \nu, 0) \\
\text{Alarm if } & g_t^{(1)} > h \text{ or } g_t^{(2)} > h
\end{aligned}
$$

After an alarm, reset $g_t^{(1)} = 0$, $g_t^{(2)} = 0$ and $\hat{\theta}_t = y_t$.
Design parameters: ν, h.
Output: $\hat{\theta}_t$.

The point with the reset $\hat{\theta}_t = y_t$ is that the algorithm forgets all old information instantaneously, while, at the same time, avoids bias and a transient.

Finally, some general advice for tuning these algorithms, which are defined by combinations of the CUSUM test and adaptive filters, are given.

Tuning of CUSUM filtering algorithms

Start with a very large threshold h. Choose ν to one half of the expected change, or adjust ν such that $g_t = 0$ more than 50% of the time. Then set the threshold so the required number of false alarms (this can be done automatically) or delay for detection is obtained.

- If faster detection is sought, try to decrease ν.
- If fewer false alarms are wanted, try to increase ν.
- If there is a subset of the change times that does not make sense, try to increase ν.

3.5. Likelihood based change detection

This section provides a compact presentation of the methods derived in Chapters 6 and 10, for the special case of signal estimation.

3.5.1. Likelihood theory

Likelihood is a measure of likeliness of what we have observed, given the assumptions we have made. In this way, we can compare, on the basis of observed data, different assumptions on the change time. For the model

$$y_t = \theta + e_t, \quad e_t \in \mathrm{N}(0, R), \tag{3.30}$$

the likelihood is denoted $p(y^t|\theta, R)$ or $l_t(\theta, R)$. This should be read as "the likelihood for data y^t given the parameters θ, R". Independence and Gaussianity give

$$l_t(\theta, R) \triangleq p(y^t|\theta, R) = \prod_{i=1}^{t} p(y_i|\theta, R) = (2\pi R)^{-t/2} e^{-\frac{1}{2R}\sum_{i=1}^{t}(y_i-\theta)^2} \tag{3.31}$$

The parameters are here *nuisance*, which means they are irrelevant for change detection. There are two ways to eliminate them if they are unknown: estimation or marginalization.

3.5.2. ML estimation of nuisance parameters

The *Maximum Likelihood* (*ML*) estimate of θ (or any parameter) is formally defined as

$$\hat{\theta}^{ML} \triangleq \arg\max_{\theta} l_t(\theta, R). \tag{3.32}$$

Rewrite the exponent of (3.31) as a quadratic form in θ,

$$\sum_{i=1}^{t}(y_i - \theta)^2 = t(\overline{y^2} - 2\theta\overline{y} + \theta^2) = t((\theta - \overline{y})^2 + \overline{y^2} - \overline{y}^2) \tag{3.33}$$

where

$$\overline{y} = \frac{1}{t}\sum_{i=1}^{t} y_i$$
$$\overline{y^2} = \frac{1}{t}\sum_{i=1}^{t} y_i^2$$

denote sample averages of y and y^2, respectively. From this, we see that

$$\hat{\theta}^{ML} = \overline{y} = \frac{1}{t}\sum_{i=1}^{t} y_i \tag{3.34}$$

which is, of course, a wellknown result. Similarly, the joint ML estimate of θ and R is given by

$$\begin{aligned}(\hat{\theta}, \hat{R})^{ML} &= \arg\max_{\theta,R} l_t(\theta, R)\\ &= (\hat{\theta}, \arg\max_{R} l_t(\hat{\theta}, R))\\ &= (\hat{\theta}, \arg\max_{R}(2\pi)^{-t/2}R^{-t/2}e^{-\frac{\overline{y^2}-\overline{y}^2}{2R}t}).\end{aligned}$$

By taking the logarithm, we get

$$-2\log l_t(\hat{\theta}, R) = t\log(2\pi) + t\log(R) + \frac{t}{R}(\overline{y^2} - \overline{y}^2).$$

Setting the derivative with respect to R equal to zero gives

$$\hat{R}^{ML} = \overline{y^2} - \overline{y}^2 = \frac{1}{t}\sum_{i=1}^{t}(y_i - \hat{\theta}^{ML})^2. \tag{3.35}$$

Compare to the formula $\text{Var}(X) = E(X^2) - (E(X))^2$. Note, however, that the ML estimate of R is not unbiased, which can be easily checked. An unbiased estimator is obtained by using the normalization factor $1/(t-1)$ instead of $1/t$ in (3.35).

Example 3.4 Likelihood estimation

Let the true values in (3.30) be $\theta = 2$ and $R = 1$. Figure 3.6 shows how the likelihood (3.31) becomes more and more concentrated around the true values as more observations become available. Note that the distribution is narrower along the θ axle, which indicates that successful estimation of the mean is easier than the variance. Marginalization in this example can be elegantly performed numerically by projections. Marginalization with respect to R and θ, respectively, are shown in Figure 3.7. Another twist in this example is to

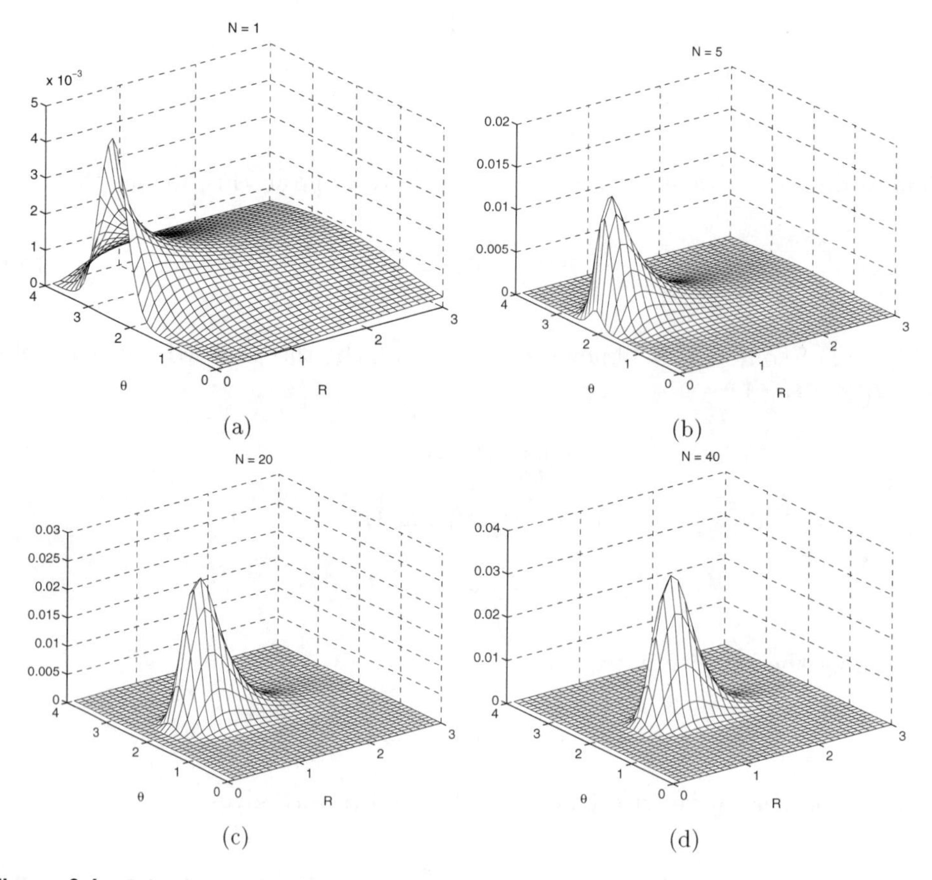

Figure 3.6. Likelihood of the mean θ and variance R from 1, 5, 20 and 40 observations, respectively.

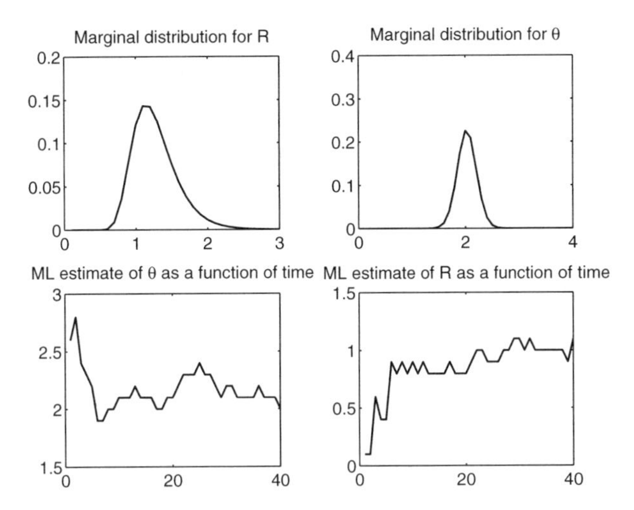

Figure 3.7. Upper row of plots shows the marginalized likelihood using 40 observations. Lower row of plots shows how the estimates converge in time.

study the time response of the maximum likelihood estimator, which is also illustrated in Figure 3.7.

The numerical evaluation is performed using the *point mass filter*, see Bergman (1999), where a grid of the parameter space is used.

3.5.3. ML estimation of a single change time

We will now extend ML to change time estimation. Let $p(y^t|k)$ denote the likelihood[1] for measurements $y^t = y_1, y_2, \ldots, y_t$, given the change time k. The change time is then estimated by the *maximum likelihood* principle:

$$\hat{k} = \arg\max_k p(y^t|k). \tag{3.36}$$

This is basically an off-line formulation. The convention is that $\hat{k} = t$ should be interpreted as no change. If the test is repeated for each new measurement, an on-line version is obtained, where there could be efficient ways to compute the likelihood recursively.

If we assume that θ before and after the change are independent, the likelihood $p(y^t|k)$ can be divided into two parts by using the rule $P(A, B|C) = P(A|C)P(B|C)$, which holds if A and B are independent. That is,

$$p(y^t|k) = p(y^k|k)p(y^t_{k+1}|k) = p(y^k)p(y^t_{k+1}). \tag{3.37}$$

[1] This notation is equivalent to that commonly used for conditional distribution.

Here $y_{k+1}^t = y_{k+1}, \dots, y_t$. The conditioning on a change at time k does not influence the likelihoods in the right-hand side and is omitted. That is, all that is needed is to compute the likelihood for data in all possible splits of data into two parts. The number of such splits is t, so the complexity of the algorithm increases with time. A common remedy to this problem is to use the sliding window approach and only consider $k \in [t-L, t]$.

Equation (3.37) shows that change detection based on likelihoods brakes down to computing the likelihoods for batches of data, and then combine these using (3.37). Section 3.A contains explicit formulas for how to compute these likelihoods, depending on what is known about the model. Below is a summary of the results for the different cases needed to be distinguished in applications. Here *MML* refers to *Maximum Marginalized Likelihood* and *MGL* refers to *Maximum Generalized Likelihood.* See Section 3.A or Chapters 7 and 10 for details of how these are defined.

1. The parameter θ is unknown and the noise variance R is known:

$$-2\log l_t^{MGL}(R) \approx t\log(2\pi R) + \frac{t\hat{R}}{R}, \tag{3.38}$$

$$-2\log l_t^{MML}(R) \approx (t-1)\log(2\pi R) + \log(t) + \frac{t\hat{R}}{R}. \tag{3.39}$$

   Note that $\hat{R}$ is a compact and convenient way of writing (3.35), and should not be confused with an estimate of what is assumed to be known.

2. The parameter θ is unknown and the noise variance R is unknown, but known to be constant. The derivation of this practically quite interesting case is postponed to Chapter 7.

3. The parameter θ is unknown and the noise variance R is unknown, and might alter after the change time:

$$\begin{aligned} -2\log l_t^{MGL} &\approx t\log(2\pi) + t + t\log(\hat{R}), && (3.40)\\ -2\log l_t^{MML} &\approx t\log(2\pi) + (t-5) + \log(t) - (t-3)\log(t-5) + \\ &\qquad (t-3)\log(t\hat{R}). && (3.41) \end{aligned}$$

4. The parameter θ is known (typically to be zero) and the noise variance R is unknown and abruptly changing:

$$-2\log l_t^{MGL}(\theta) \approx t\log(2\pi) + t + t\log(\hat{R}), \tag{3.42}$$

$$-2\log l_t^{MML}(\theta) \approx t\log(2\pi) + (t-4) - (t-2)\log(t-4) + (t-2)\log(t\hat{R}). \tag{3.43}$$

Note that the last case is for detection of *variance changes*. The likelihoods above can now be combined before and after a possible change, and we get the following algorithm.

Algorithm 3.5 Likelihood based signal change detection

Define the likelihood $l_{m:n} \triangleq p(y_m^n)$, where $l_t = l_{1:t}$. The log likelihood for a change at time k is given by

$$-\log l_t(k) = -\log l_{1:k} - \log l_{k+1:t},$$

where each log likelihood is computed by one of the six alternatives (3.38)–(3.43).

The algorithm is applied to the simplest possible example below.

Example 3.5 Likelihood estimation

Consider the signal

$$y_t = \begin{cases} 0 + e(t), & \text{for } 0 < t \leq 250 \\ 1 + e(t), & \text{for } 250 < t \leq 500. \end{cases}$$

The different likelihood functions as a function of change time are illustrated in Figure 3.8, for the cases:

- unknown θ and R,
- R known and θ unknown,
- θ known both before and after change, while R is unknown.

Note that MGL has problem when the noise variance is unknown.

The example clearly illustrates that marginalization is to be preferred to maximization of nuisance parameters in this example.

3.5.4. Likelihood ratio ideas

In the context of hypothesis testing, the likelihood ratios rather than the likelihoods are used. The *Likelihood Ratio* (*LR*) test is a multiple hypotheses test, where the different jump hypotheses are compared to the no jump hypothesis pairwise. In the LR test, the jump magnitude is assumed to be known. The hypotheses under consideration are

$$\begin{aligned} H_0 &: \text{ no jump} \\ H_1(k,\nu) &: \text{ a jump of magnitude } \nu \text{ at time } k. \end{aligned}$$

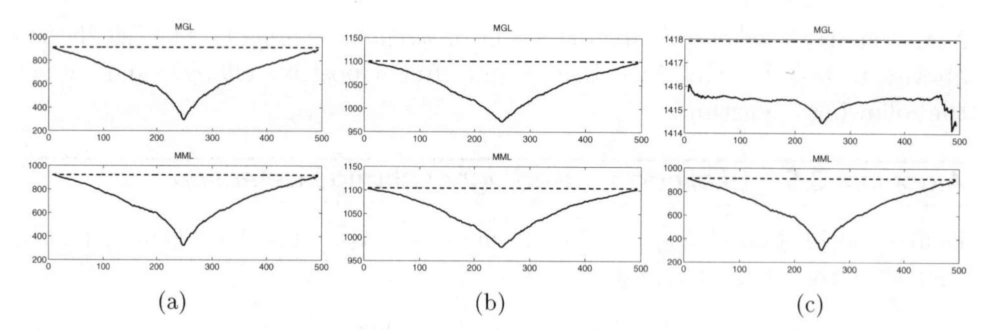

Figure 3.8. Maximum generalized and marginalized likelihoods, respectively (MGL above, MML below). Cases: unknown θ and R in (a), unknown θ in (b) and unknown R (θ known both before and after change) in (c).

The test is as follows. Introduce the log likelihood ratio for the hypotheses as the test statistic:

$$g_t(k,\nu) \triangleq 2\log\frac{p(y^t|H_1(k,\nu))}{p(y^t|H_0)} = 2\log\frac{p(y^t|k,\nu)}{p(y^t|k=t)}. \tag{3.44}$$

The factor 2 is just used for later notational convenience. The notation g_t has been chosen to highlight that this is a *distance measure* between two hypotheses. We use the convention that $H_1(t,\nu) = H_0$, so that $k=t$ means no jump. Then the LR estimate can be expressed as

$$\hat{k}^{LR} = \arg\max_k g_t(k,\nu), \tag{3.45}$$

when ν is known.

There are again two possibilities of how to eliminate the unknown nuisance parameter ν. Maximization gives the GLR test, proposed for change detection by Willsky and Jones (1976), and marginalization results in the MLR test (Gustafsson, 1996).

Starting with the likelihood ratio in (3.45), the GLR test is a double maximization over k and ν,

$$\hat{\nu}(k) = \arg\max_\nu 2\log\frac{p(y^t|k,\nu)}{p(y^t|k=t)},$$
$$\hat{k} = \arg\max_k 2\log\frac{p(y^t|k,\hat{\nu}(k))}{p(y^t|t)}.$$

By definition, $\hat{\nu}(k)$ is the maximum likelihood estimate of ν, given a jump at time k. The jump candidate $\hat{k}$ in the GLR test is accepted if

$$g_t(\hat{k},\hat{\nu}(\hat{k})) > h. \tag{3.46}$$

The idea in MLR is to assume that ν is a random variable with a certain prior distribution $p(\nu)$. Then

$$\hat{k} = \arg\max_k 2\log \int \frac{p(y^t|k,\nu)}{p(y^t|t)} p(\nu)d\nu.$$

As a comparison and summary of of GLR and MLR, it follows from the derivations in Section 3.A.2 that the two tests are

$$g_t^{GLR}(k) = \frac{\hat{\nu}^2(k)}{R/(t-k)} \gtrless h, \tag{3.47}$$

$$g_t^{MLR}(k) = \frac{\hat{\nu}^2(k)}{R/(t-k)} - \log\left(\frac{2\pi R}{t-k}\right) \gtrless 0. \tag{3.48}$$

Here the prior in MLR is assumed flat, $p(\nu) = 1$. The theoretical distribution of $g_t^{GLR}(k)$ is $\chi^2(1)$. The threshold can be taken from standard distribution tables. It can be remarked that the distribution is rather robust to the assumed Gaussian distribution. Since (1) $\hat{\nu}$ is an average, (2) averages converge according to the central limit theorem to a Gaussian distribution, and (3) the square of the Gaussian variable is $\chi^2(1)$ distributed, we can conclude that the the test statistic approaches $\chi^2(1)$ distribution asymptotically.

GLR and MLR are conceptually different, since they represent two different philosophies.

- GLR is a hypothesis test where $g_t(k, \hat{\nu}(k))$ is always positive. Thus a threshold is needed, and the size of the threshold determines how large a change is needed to get an alarm.

- MLR is an estimation approach, so there is no threshold. The threshold 0 in (3.48) is explicitely written out just to highlight the similarities in implementation (note, however, that the constant can be interpreted as a non-tunable threshold). No change is estimated if $g_t(k) < 0$ for all $k < t$, since $g_t(t) = 0$ and $k = t$ means that no change has occured yet. Here, the prior on ν can be used to specify what a sufficiently large change is.

More on these issues can be found in Chapters 7 and 10.

3.5.5. Model validation based on sliding windows

The basic idea in this *model validation* approach is to compare two models where one is obtained from a sliding window. Any averaging function can be used, but we will only discuss a rectangular sliding window here. This model is compared to a nominal model obtained from either a longer window or off-line analysis (system identification or a physical model). Equation (3.49) illustrate two possibilities of how to split data. The option in (3.49.a) is common in applications, but the one in (3.49.b) is somewhat simpler to analyze, and will be used here:

$$\begin{array}{ll} \text{a.} & \text{Data}: \overbrace{y_1, y_2, \ldots, y_{t-L}, \underbrace{y_{t-L+1}, \ldots, y_t}_{\text{Model } M_2}}^{\text{Model } M_1} \\ \text{b.} & \text{Data} \quad \underbrace{y_1, y_2, \ldots, y_{t-L}}_{\text{Model M}_1}, \underbrace{y_{t-L+1}, \ldots, y_t}_{\text{Model M}_2} \end{array} \qquad (3.49)$$

The slow filter, that estimates M_1, uses data from a very large sliding window, or even uses all past data (since the last alarm). Then two estimates, $\hat{\theta}_1$ and $\hat{\theta}_2$ with variances P_1 and P_2, are obtained. If there is no abrupt change in the data, these estimates will be consistent. Otherwise, a hypothesis test will reject H_0 and a change is detected.

There are several ways to construct such a hypothesis test. The simplest one is to study the difference

$$\hat{\theta}_1 - \hat{\theta}_2 \in \mathrm{N}(0, P_1 + P_2), \quad \text{under } H_0 \qquad (3.50)$$

or, to make the test single sided,

$$\frac{(\hat{\theta}_1 - \hat{\theta}_2)^2}{P_1 + P_2} \in \chi^2(1), \quad \text{under } H_0 \qquad (3.51)$$

from which a standard hypothesis test can be formulated.

From (3.49.b), we immediately see that

$$\hat{\theta}_1 = \frac{1}{t-L} \sum_{i=1}^{t-L} y_i, \; P_1 = \frac{R}{t-L}$$

$$\hat{\theta}_2 = \frac{1}{L} \sum_{i=t-L+1}^{t} y_i, \; P_2 = \frac{R}{L}$$

so that

$$P_1 + P_2 = \frac{R}{L} \frac{t}{t-L} \approx \frac{R}{L}$$

if t is large.

Example 3.6 Hypothesis tests: Gaussian and χ^2

Suppose we want to design a test with a probability of false alarm of 5%, that is,

$$P\left(\frac{|\hat{\theta}_1 - \hat{\theta}_2|}{\sqrt{P_1 + P_2}} > h\right) = 0.05.$$

A table over the Gaussian distribution shows that $P(X < 1.96) = 0.975$, so that $P(|X| < 1.96) = 0.95$, where $X \in \mathrm{N}(0,1)$. That is, the test becomes

$$\frac{|\hat{\theta}_1 - \hat{\theta}_2|}{\sqrt{P_1 + P_2}} \gtrless 1.96.$$

Similarly, using a squared test statistic and a table over the $\chi^2(1)$ distribution ($P(X^2 < 3.86) = 0.95$) gives

$$\frac{(\hat{\theta}_1 - \hat{\theta}_2)^2}{P_1 + P_2} \gtrless 3.86 = 1.96^2,$$

which of course is the same test as above.

The GLR test restricted to only considering the change time $t - L$ gives

$$g_t(t-L) = \max_{\nu} 2 \log \frac{p(y^t|t-L,\nu)}{p(y^t|t)}.$$

This version of GLR where no search for the change time is performed is commonly referred to as Brandt's GLR (Appel and Brandt, 1983). The sliding window approach is according to (3.49.a).

Algorithm 3.6 Brandt's GLR

Filter statistics:

$$\hat{\theta}_t^{(1)} = \frac{1}{t-t_0} \sum_{k=t_0+1}^{t} y_k \qquad \hat{\theta}_t^{(2)} = \frac{1}{t-L} \sum_{k=L+1}^{t} y_k$$

$$\hat{R}_t^{(1)} = \frac{1}{t-t_0} \sum_{k=t_0+1}^{t} (y_k - \hat{\theta}_t^{(1)})^2 \qquad \hat{R}_t^{(2)} = \frac{1}{t-L} \sum_{k=L+1}^{t} (y_k - \hat{\theta}_t^{(2)})^2$$

$$\varepsilon_t^{(1)} = y_t - \hat{\theta}_{t-1}^{(1)} \qquad \varepsilon_t^{(2)} = y_t - \hat{\theta}_{t-1}^{(2)}$$

Test statistics:

$$s_t = \log\left(\frac{\hat{R}_t^{(1)}}{\hat{R}_t^{(2)}}\right) + \frac{\left(\varepsilon_t^{(1)}\right)^2}{\hat{R}_t^{(1)}} - \frac{\left(\varepsilon_t^{(2)}\right)^2}{\hat{R}_t^{(2)}}$$

$$g_t = g_{t-1} + s_t$$

Alarm if $g_t > h$

After an alarm, reset $g_t = 0$ and $t_0 = t$.
Design parameters: h, L (also R_i if they are known).
Output: $\hat{\theta}_t^{(1)}$.

The follow general design rules for sliding windows apply here:

Design of sliding windows

The window size is coupled to the desired delay for detection. A good starting value is the specified or wanted mean delay for detection. Set the threshold to get the specified false alarm rate. Diagnosis:

- Visually check the variance error in the parameter estimate in the sliding window. If the variance is high, this may lead to many false alarms and the window size should be increased.
- If the estimated change times look like random numbers, too little information is available and the window size should be increased.
- If the change times make sense, the mean delay for detection might be improved by decreasing the window.

As a side comment, it is straightforward to make the updates in Algorithm 3.6 truly recursive. For instance, the noise variance estimate can be computed by $\hat{R}_t = \overline{y^2} - \overline{y}^2$ where both the signal and squared signal means are updated recursively.

More approaches to sliding window change detection will be presented in Chapter 6.

3.6. Applications

Two illustrative on-line applications are given here. Similar applications on the signal estimation problem, which are based on off-line analysis, are given in Section 4.6. In recent literature, there are applications of change detection in a heat exchanger (Weyer et al., 2000), Combined Heat and Power (CHP) unit (Thomson et al., 2000) and growth rate in a slaughter-pig production unit (Nejsum Madsen and Ruby, 2000).

3.6.1. Fuel monitoring

Consider the fuel consumption filter problem discussed in in Examples 1.1, 1.4 and 1.8. Here Algorithms 3.2, 3.3 and 3.6 are applied. These algorithms are only capable to follow abrupt changes. For incipient changes, the algorithm will give an alarm only after the total change is large or after a long time. In both algorithms, it is advisable to include data forgetting, as done in Algorithm 3.4, in the parameter estimation to allow for a slow drift in the mean of the signal.

Table 3.1 shows how the above change detection algorithms perform on this signal. Figure 3.9 shows the result using the CUSUM LS Algorithm 3.3 with threshold $h = 5$, drift $\nu = 0.5$, but in contrast to the algorithm, we here use the squared residuals as distance measure s_t. Compared to the existing filter, the tracking ability has improved and, even more importantly, the accuracy gets better and better in segments with constant fuel consumption.

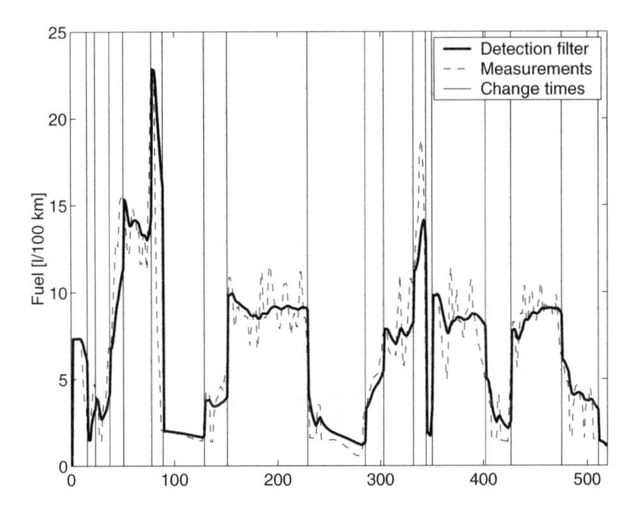

Figure 3.9. Filtering with the CUSUM LS algorithm.

Table 3.1. Simulation result for fuel consumption

Method	Design parameters	$\hat{n}$	MDL	kFlops
RLS	$R = 0.9$, quant $= 0.3$	–	–	19
CUSUM	$h = 3,\ \nu = 2$	14	8.39	20
Brandt's GLR	$h = 20,\ \nu = 3,\ L = 3$	13	8.40	60
ML	$\sigma^2 = 3$	14	8.02	256

3.6.2. Paper refinery

Consider now the paper refinery problem from Section 2.1.2, where the power signal to the grinder needs to be filtered. As a preliminary analysis, RLS provides some useful information, as seen from the plots in Figure 3.11.

- There are two segments where the power clearly decreases quickly. Furthermore, there is a starting and stopping transient that should be detected as change times.
- The noise level is fairly constant (0.05) during the observed interval.

The CUSUM LS approach in Algorithm 3.3 is applied, with $h = 10$ and $\nu = 1$. Figure 3.12 shows g_t which behaves well and reaches the threshold quickly after level changes. The resulting power estimate (both recursive and smoothed) are compared to the filter implemented in Sund in the right plots of Figure 3.10.

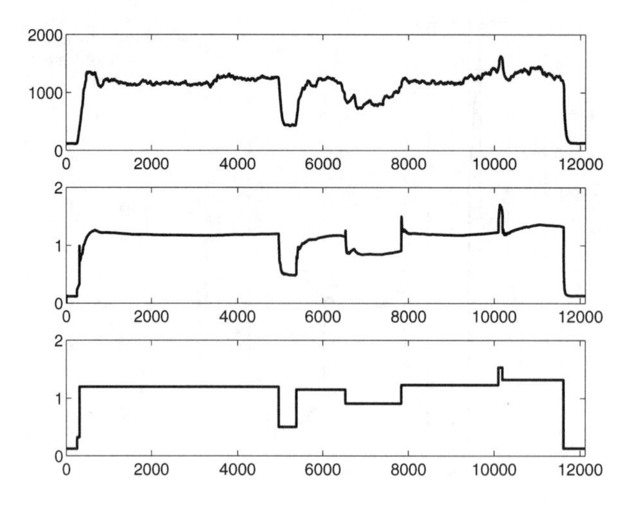

Figure 3.10. Estimated power signal using the filter currently used in Sund, the CUSUM LS filter and smoother, respectively.

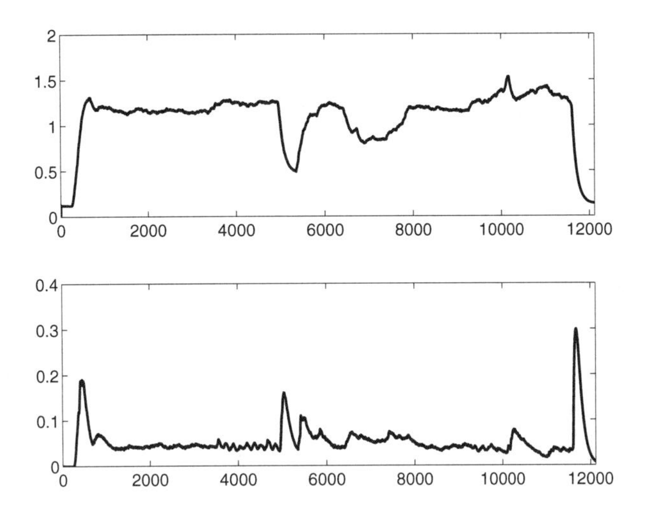

Figure 3.11. Filtered power signal (RLS with $\lambda = 0.99$) in Sund's defibrator. Lower plot is an estimate of noise variance.

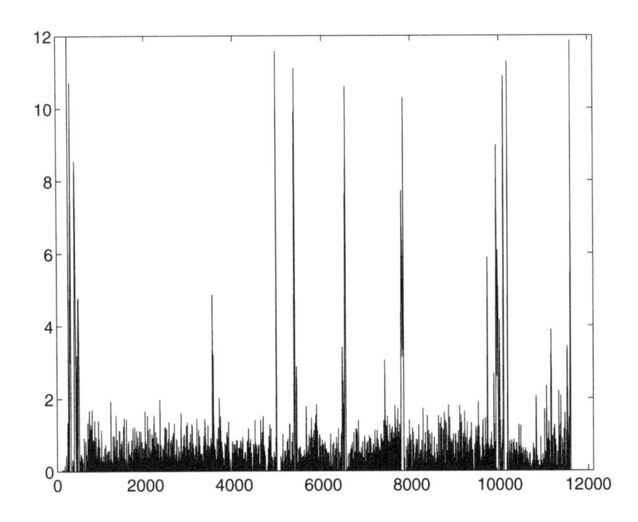

Figure 3.12. The CUSUM test statistic g_t for Sund's paper refinery data.

Both tracking and noise rejection are improved. A good alternative would be to make the algorithm adaptive and thus to include a small data forgetting to track slow variations. Here Algorithm 3.4 can be used.

3.A. Derivations

3.A.1. Marginalization of likelihoods

The idea in marginalization is to assign a prior to the unknown parameter and eliminate it.

Marginalization of θ

A flat prior $p(\theta) = 1$ is used here for simplicity. Other proper density functions, which integrate to one, can of course be used but only a few give explicit formulas. The resulting likelihood would only marginally differ from the expressions here. Straightforward integration yields

$$\begin{aligned} p(y^t|R) &= \int p(y^t|R,\theta)p(\theta)d\theta \\ &= \int (2\pi R)^{-t/2} e^{-\frac{1}{2R}\sum_{i=1}^{t}(y_i-\theta)^2} d\theta \\ &= \int (2\pi R)^{-t/2} e^{-\frac{1}{2R}t((\theta-\overline{y})^2+\overline{y^2}-\overline{y}^2)} d\theta, \end{aligned}$$

where the quadratic form (3.33) is used. Identification with the Gaussian distribution $x \in \mathrm{N}(\overline{y}, R/t)$, which integrates to one, gives

$$\begin{aligned} p(y^t|R) &= (2\pi R)^{-t/2}(2\pi R/t)^{1/2} e^{-\frac{1}{2R/t}(\overline{y^2}-\overline{y}^2)} \\ &= (2\pi R)^{-(t-1)/2} t^{-1/2} e^{-\frac{\hat{R}^{ML}}{2R/t}}. \end{aligned} \tag{3.52}$$

That is,

$$-2\log l_t(R) = -2\log p(y^t|R) = t\frac{\hat{R}^{ML}}{R} - \log\frac{R}{t} + t\log(2\pi R) - \log(2\pi). \tag{3.53}$$

Marginalization of R

In the same way, conditioning the data distribution on the parameter only gives

$$\begin{aligned} p(y^t|\theta) &= \int p(y^t|R,\theta)p(R)dR \\ &= \int (2\pi R)^{-t/2} e^{-\frac{1}{2R}\sum_{i=1}^{t}(y_i-\theta)^2} dR, \end{aligned}$$

where $p(R) = 1$ has been assumed. Identification with the inverse *Wishart distribution* (or *gamma distribution* in this scalar case),

$$p(x) = \frac{V^{m/2}}{2^{m/2}\Gamma(m/2)} \frac{e^{-\frac{V}{2x}}}{x^{(m+2)/2}}, \quad x > 0 \tag{3.54}$$

gives $m = t - 2$ and $V = \sum(y_i - \theta)$. The distribution integrates to one, so

$$p(y^t|\theta) = (2\pi)^{-\frac{t}{2}} 2^{\frac{t-2}{2}} \Gamma\left(\frac{t-2}{2}\right) \left(\sum_{i=1}^{t}(y_i - \theta)^2\right)^{-\frac{t-2}{2}}. \tag{3.55}$$

Hence,

$$\begin{aligned}-2\log l_t(\theta) = -2\log p(y^t|\theta) &= t\log(2\pi) - (t-2)\log(2) - 2\log\Gamma\left(\frac{t-2}{2}\right) \\ &+ (t-2)\log\left(\sum_{i=1}^{t}(y_i-\theta)^2\right).\end{aligned} \tag{3.56}$$

This result can be somewhat simplified by using *Stirling's formula*

$$\Gamma(n+1) \approx \sqrt{2\pi} n^{n+1/2} e^{-n}. \tag{3.57}$$

The *gamma function* is related to the *factorial* by $n! = \Gamma(n+1)$. It follows that the following expression can be used for reasonably large t ($t > 30$ roughly):

$$\begin{aligned}-2\log\Gamma\left(\frac{t-2}{2}\right) &\approx -\log(2\pi) + (t-4) - (t-5)\log\frac{t-4}{2} \\ &= (t-4) - (t-2)\log\frac{t-4}{2} + O(\log(t)).\end{aligned}$$

In the last equality, a term $3\log((t-4)/2)$ was subtracted and $O(\log(t))$ added. With this small trick, we get a simplified form of the likelihood as follows:

$$l_t(\theta) \approx t\log(2\pi) + (t-4) - (t-2)\log(t-4) + (t-2)\log\left(\sum_{i=1}^{t}(y_i-\theta)^2\right). \tag{3.58}$$

Marginalization of both θ and R

When eliminating both θ and R by marginalization, there is the option to start from either (3.55) or (3.52). It turns out that (3.52) gives simpler calculations:

$$\begin{aligned}p(y^t) &= \int p(y^t|R)p(R)dR \\ &= \int (2\pi R)^{-(t-1)/2} t^{-1/2} e^{-\frac{\hat{R}}{2R/t}} dR.\end{aligned}$$

Again, identification with the inverse Wishart distribution (3.54) gives $m = t-3$ and $V = t\hat{R}$, and

$$p(y^t) = (2\pi R)^{-(t-1)/2} t^{-1/2} (t\hat{R})^{-\frac{t-3}{2}} 2^{\frac{t-3}{2}} \Gamma\left(-\frac{t-3}{2}\right).$$

Using Stirling's approximation (3.57) finally gives

$$\begin{aligned} l_t &\triangleq -2\log p(y^t) \\ &\approx t\log(2\pi) + (t-5) + \log(t) - (t-3)\log(t-5) + (t-3)\log(t\hat{R}). \end{aligned} \tag{3.59}$$

3.A.2. Likelihood ratio approaches

The use of likelihood ratios is quite widespread in the change detection literature, and its main motivation is the Neyman Pearson lemma (Lehmann, 1991). For change detection, it says that the likelihood ratio is the optimal test statistic to test H_0 against $H_1(k,\nu)$ for a given change ν and a *given* change time k. It does, however, not say anything about the composite test when all change times k are considered.

We will, without loss of generality, consider the case of known $\theta_t = 0$ before the change and unknown $\theta_t = \nu$ after the change,

$$\theta_i = \begin{cases} 0 & \text{if } i < k, \\ \nu & \text{if } i > k. \end{cases}$$

The likelihood ratio

We denote the likelihood ratio with g_t to highlight that it is a *distance measure* between two hypothesis, which is going to be thresholded. The definition is

$$g_t(k,\nu) \triangleq 2\log\frac{p(y^t|H_1(k,\nu))}{p(y^t|H_0)}. \tag{3.60}$$

As can be seen from (3.31), all constants and all terms for $i < k$ will cancel, leaving

$$\begin{aligned} g_t(k,\nu) &= -\frac{1}{R}\sum_{i=k+1}^{t}(y_i-\nu)^2 + \frac{1}{R}\sum_{i=k+1}^{t} y_i^2 \\ &= -\frac{1}{R}\sum_{i=k+1}^{t}(\nu^2 - 2\nu y_i) \\ &= \frac{2\nu}{R}\sum_{i=k+1}^{t}\left(y_i - \frac{\nu}{2}\right). \end{aligned} \tag{3.61}$$

The generalized likelihood ratio

The *Generalized Likelihood Ratio* (*GLR*) is obtained by maximizing over the change magnitude:

$$g_t^{GLR}(k) = \max_{\nu} 2 \log \frac{p(y^t|H_1(k,\nu))}{p(y^t|H_0)} \tag{3.62}$$

From (3.61), the maximizing argument is seen to be

$$\hat{\nu}(k) = \frac{1}{t-k} \sum_{i=k+1}^{t} y_i. \tag{3.63}$$

which of course is the maximum likelihood estimate of the change, conditioned on the change time k. The GLR in (3.61) can then be written

$$g_t^{GLR}(k) \triangleq g_t^{LR}(k, \hat{\nu}(k)) = \frac{2\hat{\nu}(k)}{R}(t-k)(\hat{\nu}(k) - \frac{\hat{\nu}(k)}{2}) = \frac{\hat{\nu}^2(k)}{R/(t-k)}. \tag{3.64}$$

The latter form, the squared estimate normalized by the estimation variance, will be extended to more general cases in Chapter 7.

4

Off-line approaches

4.1. Basics

This chapter surveys off-line formulations of single and multiple change point estimation. Although the problem formulation yields algorithms that process data batch-wise, many important algorithms have natural on-line implementations and recursive approximations. This chapter is basically a projection of the more general results in Chapter 7 to the case of signal estimation. There are, however, some dedicated algorithms for estimating one change point off-line that apply to the current case of a scalar signal model. In the literature of mathematical statistics, this area is known as *change point estimation*.

In *segmentation*, the goal is to find a sequence $k^n = (k_1, k_2, \ldots, k_n)$ of time indices, where both the number n and the locations k_i are unknown, such that the signal can be accurately described as piecewise constant, i.e.,

$$y_t = \theta(i) + e_t, \quad \text{when } k_{i-1} < t \leq k_i, \tag{4.1}$$

is a good description of the observed signal y_t. The noise variance will be denoted $\mathrm{E}(e_t^2) = R$. The standard assumption is that $e_t \in \mathrm{N}(0, R)$, but there are other possibilities. Equation (4.1) will be the signal model used throughout this chapter, but it should be noted that an important extension to the case where the parameter is slowly varying within each segment is possible with minor modifications. However, equation (4.1) illustrates the basic ideas.

One way to guarantee that the best possible solution is found is to consider all possible segmentations k^n, estimate the mean in each segment, and then choose the particular k^n that minimizes an optimality criteria,

$$\widehat{k^n} = \arg \min_{n \geq 1, 0 < k_1 < \cdots < k_n = N} V(k^n).$$

The procedure is illustrated below:

$$\begin{array}{lcccc} \text{Data} & \underbrace{y_1, y_2, \ldots, y_{k_1}} & \underbrace{y_{k_1+1}, \ldots, y_{k_2}} & \cdots & \underbrace{y_{k_{n-1}+1}, \ldots, y_{k_n}} \\ \text{Segmentation} & \text{Segment } 1 & \text{Segment } 2 & \cdots & \text{Segment } n \\ \text{LS estimates} & \hat{\theta}(1), \hat{R}(1) & \hat{\theta}(2), \hat{R}(2) & \cdots & \hat{\theta}(n), \hat{R}(n) \end{array} . \tag{4.2}$$

Note that the segmentation k^n has $n-1$ degrees of freedom. Two types of optimality criteria have been proposed:

- Statistical criteria: The maximum likelihood or maximum *a posteriori* estimate of k^n is studied.

- Information based criteria: The information of data in each segment is $V(i)$ (the sum of squared residuals), and the total information is the sum of these. Since the total information is minimized for the degenerated solution $k^n = 1, 2, 3, \ldots, N$, giving $V(i) = 0$, a *penalty term* is needed. Similar problems have been studied in the context of model structure selection, and from this literature Akaike's AIC and BIC criteria have been proposed for segmentation.

The real challenge in segmentation is to cope with the *curse of dimensionality*. The number of segmentations k^n is 2^N (there can be either a change or no change at each time instant). Here, several strategies have been proposed:

- Numerical searches based on dynamic programming or Markov Chain Monte Carlo (MCMC) techniques.

- Recursive local search schemes.

A section is devoted to each of these two approaches. First, a summary of possible loss functions, or segmentation criteria, $V(k^n)$ is given. Chapter 7 is a direct continuation of this chapter for the case of segmentation of parameter *vectors*.

4.2. Segmentation criteria

This section describes the available statistical and information based optimization criteria.

4.2.1. ML change time sequence estimation

Consider first an off-line problem, where the sequence of change times $k^n = k_1, k_2, \ldots, k_n$ is estimated from the data sequence y^t. Later, on-line algorithms will be derived from this approach. We will use the likelihood for data, given that the vector of change points is $p(y^t|k^n)$.

$$\begin{array}{lcccc} \text{Data} & \underbrace{y_1, y_2, \ldots, y_{k_1},} & \underbrace{y_{k_1+1}, \ldots, y_{k_2}} & \cdots & \underbrace{y_{k_{n-1}+1}, \ldots, y_{k_n} = y_N} \\ \text{Segment} & p(y_1^{k_1}) & p(y_{k_1+1}^{k_2}) & \cdots & p(y_{k_{n-1}+1}^{k_n}) \end{array}. \tag{4.3}$$

Repeatedly using independence of θ in different segments gives

$$p(y^t|k^n) = \begin{cases} p(y^t), & \text{if } n = 0, \\ p(y_1^{k_1}) \prod_{i=1}^{n-1} p(y_{k_i+1}^{k_{i+1}}) p(y_{k_n+1}^t), & \text{if } n > 0. \end{cases} \tag{4.4}$$

The maximum likelihood estimate is

$$\widehat{(n, k^n)} = \arg \max_{(n,k^n)} p(y^t|k^n), \tag{4.5}$$

using the notation and segmentation in (4.3). The notational convention is that no change is estimated if $n = 1$ and $k_1 = t$. The likelihood in each segment is defined exactly as in Section 3.5.3:

- The advantage of dealing with multiple change times is that it provides an elegant approach to the start-up problem when a change is detected. A consequence of this is that very short segments can be found.
- The disadvantage is that the number of likelihoods (4.4) increases exponentially with time as 2^t.

A slight variation of this approach is provided in a Bayesian setting. In the Bayesian world, everything that is unknown is treated as a random variable. Here the change time has to be interpreted as a random variable, and one idea is to assign a probability q for a change at each time instant, and assuming independence. That is,

$$P(\text{change at time } i) = q, \quad 0 < q < 1.$$

The *a posteriori probability* for k is defined by $p(k|y^t)$. *Bayes' rule* $P(A|B) = \frac{P(A)}{P(B)}P(B|A)$ thus gives

$$p(k^n|y^t) = \frac{p(k^n)}{p(y^t)}p(y^t|k^n). \tag{4.6}$$

The denominator $p(y^t)$ is a scaling factor independent of the change times, which is uninteresting for our purposes. Its role is to make the integral of the probability density function equal to one. The last term is recognized as the likelihood (4.4), so the difference to the ML estimate of k^n stems from the last factor (the prior for k^n). The maximizing argument is called the *maximum a posteriori* (*MAP*) estimate, which is not influenced by the scaling factor $p(y^t)$,

$$\widehat{(n, k^n)} = \arg \max_{(n,k^n)} p(y^t|k^n)p(k^n) = \arg \max_{(n,k^n)} p(y^t|k^n)q^n(1-q)^{t-n}. \tag{4.7}$$

One advantage of the MAP estimator might be that we get a tuning knob to control the number of estimated change points. Note that with $q = 0.5$, the MAP and ML estimators coincide.

4.2.2. Information based segmentation

A natural estimation approach to *segmentation* would be to form a loss function similar to (3.7). An off-line formulation using N observations is

$$V_N(\theta^{n+1}, k^n) = \sum_{i=0}^{n} V_i(\theta_i) \tag{4.8}$$

$$V_i(\theta_i) = \sum_{t=k_i+1}^{k_{i+1}} (y_t - \theta_i)^2, \tag{4.9}$$

where the dummy variables $k_0 = 0$ and $k_{n+1} = N$ are used to define the first and last segments. Straightforward minimization of $V_i(\theta_i)$ gives

$$\widehat{(n, k^n)} = \arg \min_{(n,k^n)} V_N(k^n) \tag{4.10}$$

$$= \arg \min_{(n,k^n)} \sum_{i=0}^{n} (k_{i+1} - k_i)\hat{R}_i. \tag{4.11}$$

Unfortunately, it is quite easy to realize that this approach will fail. The more change points, the smaller loss function. The easiest way to see this is to consider the extreme case where the number of change points equals the number of data, $n = N$. Then $\hat{R}$ will equal zero in each segment (data point) because there is no error. In fact, the loss function is monotonously decreasing in n for all segmentations. This has motivated the use of a *penalty term* for the number of change points. This is in accordance with the *parsimonious principle*, which says that the best data description is a compromise between performance (small loss function) and complexity (few parameters). This is also sometimes referred to as *Ockham's razor.* The typical application of this principle is the choice of model structures in system identification. Penalty terms occuring in model order selection problems can also be used in this application, for instance:

- Akaike's AIC (Akaike, 1969) with penalty term $2n(d+1)$.

- The asymptotically equivalent criteria: Akaike's BIC (Akaike, 1977), Rissanen's *Minimum Description Length* (MDL) approach (Rissanen, 1989) and Schwartz criterion (Schwartz, 1978). The penalty term is $n(d+1)\log N$.

See also Gustafsson and Hjalmarsson (1995) for a list of likelihood-based penalty terms. Here, d refers to the number of parameters in the model, which is 1 in this part since θ is scalar. Other penalty terms are reviewed in Section 5.3.2.

AIC is proposed in Kitagawa and Akaike (1978) for auto-regressive models (see Chapter 5) with a changing noise variance. For change in the mean models, it would read

$$\widehat{k^n} = \arg\min_{k^n,n} \sum_{i=1}^{n} N(i) \log \frac{V(i)}{N(i)} + 6n. \tag{4.12}$$

BIC is suggested in Yao (1988) for a changing mean model and unknown constant noise variance, leading to

$$\widehat{k^n} = \arg\min_{k^n,n} N \log \frac{\sum_{i=1}^{n} V(i)}{N} + 3n \log N. \tag{4.13}$$

Both (4.12) and (4.13) are globally maximized for $n = N$ and $k_i = i$. That is, the wanted segmentation is a local (and not the global) minimum. This problem is solved in Yao (1988) by assuming that an upper bound on n is known, but it is not commented upon in Kitagawa and Akaike (1978).

The MDL theory provides a nice interpretation of the segmentation problem: Choose the segments such that the fewest possible data bits are used to describe the signal up to a certain accuracy, given that both the parameter vectors and the prediction errors are stored with finite accuracy.

Both AIC and BIC are based on an assumption on a large number of data, and its use in segmentation where each segment could be quite short is questioned in Kitagawa and Akaike (1978). Simulations in Djuric (1994) indicate that AIC and BIC tend to over-segment data in a simple example where marginalized ML works fine.

4.3. On-line local search for optimum

Computing the exact likelihood or information-based estimate is computationally intractable because of the exponential complexity. Here a recursive on-line algorithm is presented and illustrated with a simulation example, also used in the following section.

4.3.1. Local tree search

The exponential complexity of the segmentation problem can be illustrated with a tree as in Figure 4.1. Each branch marked with a '1' corresponds to an abrupt change, a jump, and '0' means no change in the signal level. In a local search, we make a time to time decision of which branches to examine. The other alternative using global search strategies, examined in the next section, decides which branches to examine on an off-line basis.

The algorithm below explores a finite memory property, which has much in common with the famous Viterbi algorithm in equalization; see Algorithm 5.5 and the references Viterbi (1967) and Forney (1973).

Algorithm 4.1 Recursive signal segmentation

1. Choose an optimality criterion. The options are likelihoods, *a posteriori* probabilities or information-based criteria.
2. Compute recursively the optimality criterion using a bank of least squares estimators of the signal mean value, each filter is matched to a particular segmentation.
3. Use the following rules for maintaining the hypotheses and keeping the number of considered sequences (M) fixed:
    a) Let only the most probable sequence split.
    b) Cut off the least probable sequence, so only M ones are left.
    c) Assume a minimum segment length: let the most probable sequence split *only if it is not too young.*

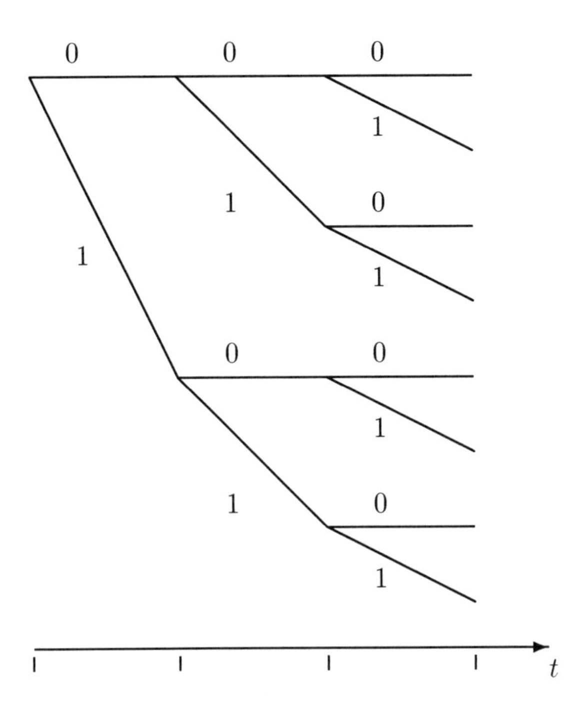

Figure 4.1. The tree of jump sequences. A path labeled 0 corresponds to no jump, while 1 corresponds to a jump.

d) Assure that sequences are not cut off immediately after they are born: cut off the least probable sequences *among those that are older than a certain minimum life-length*, until only M are left.

The last two restrictions are optional, but might be useful in some cases.

The most important design parameter is the number of filters. It will be shown in a more general framework in Appendix 7.A that the *exact* ML estimate is computed using $M = N$. That is, the algorithm has quadratic rather than exponential complexity in the data size, which would be the consequence of any straightforward approach. It should be noted, as argued in Chapter 7, that all design parameters in the algorithm can be assigned good default values *a priori*, given only the signal model.

4.3.2. A simulation example

Consider the change in the mean signal in Figure 4.2. There are three abrupt changes of magnitudes 2,3,4, respectively. The noise is white and Gaussian with variance 1. The local search algorithm is applied to minimize the max-

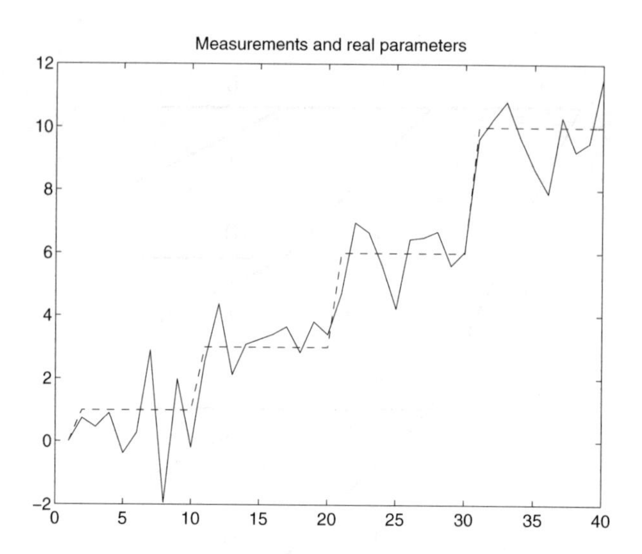

Figure 4.2. A change in the mean signal with three abrupt changes of increasing magnitude.

imum likelihood criterion with known noise variance in (4.10). The design parameters are M parallel filters with life-length $M-4$ and minimum segment length 0. The plots in Figure 4.3 show the cases $M=5$ and $M=8$, respectively. The plot mimics Figure 4.1 but is 'upside down'. Each line represents one hypothesis and shows how the number of change points evolves for that hypothesis. Branches that are cut off have an open end, and the most likely branch at each time instant is marked with a circle. This branch splits into a new change hypothesis. The upper plot shows that, in the beginning, there is one filter that performs best and the other filters are used to evaluate change points at each time instant. After having lived for three samples without becoming the most likely, they are cut off and a new one is started. At time 22 one filter reacts and at time 23 the correct change time is found. After the last change, it takes three samples until the correct hypothesis becomes the most likely.

Unfortunately, the first change is not found with $M=5$. It takes more than 3 samples to prove a small change in mean. Therefore, M is increased to 8 in the lower plot. Apparently, six samples are needed to prove a change at time 10.

Finally, we compute the *exact* ML estimate using $M=40=N$ (using a result from Appendix 7.A). Figure 4.4 shows how the hypotheses examine all branches that need to be considered.

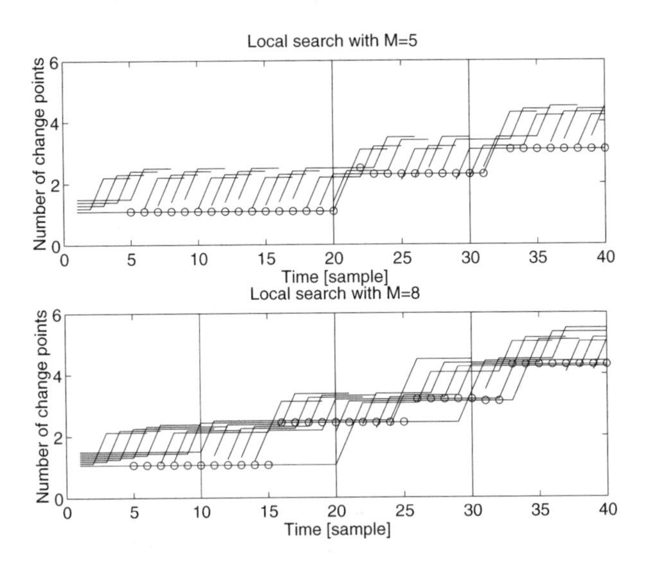

Figure 4.3. Evolution of change hypotheses for a local search with $M = 5$ and $M = 8$, respectively. A small offset is added to the number of change points for each hypothesis. Each line corresponds to one hypothesis and the one marked with 'o' is the most likely at this time. By increasing the number of parallel filters, the search becomes more efficient, and the first and smallest change at sample 10 can be found.

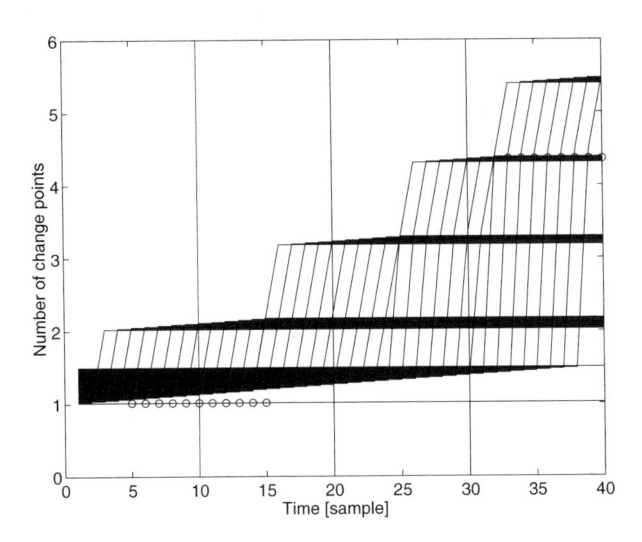

Figure 4.4. Evolution of change hypotheses for a local search with $M = 40$. A small offset is added to the number of change points for each hypothesis.

4.4. Off-line global search for optimum

There are essentially two possibilities to find the optimum of a loss function

$$\max_{k^n} V(k^n)$$

from N measurements:

- Gradient based methods where small changes in the current estimate k^n are evaluated.
- MCMC based methods, where random change points are computed.

4.4.1. Local minima

We discuss here why any global and numerical algorithm will suffer from local minima.

Example 4.1 Local minimum for one change point

The signal in Figure 4.5 has a change at the end of the data sequence. Assume that we model this noisefree signal as a change in the mean model with Gaussian noise. Assume we start with $n = 1$ and want to find the best possible change point. If the initial estimate of k is small, the likelihood in Figure 4.5 shows that we will converge to $k = 0$, which means no change!

This example shows that we must do an exhaustive search for the change point. However, this might not improve the likelihood if there are two change points as the following example demonstrates.

Example 4.2 Local minimum for two change points

The signal in Figure 4.6 is constant and zero, except for a small segment in the middle of the data record. The global minimum of the likelihood is attained for $k^n = (100, 110)$. However, there is no single change point which will improve the null hypotheses, since $-\log p(k) > -\log p(k = 0)$!

Such an example might motivate an approach where a complete search of one and two change points is performed. This will work in most cases, but as the last example shows, it is not guaranteed to work. Furthermore, the computational complexity will be cubic in time; there are $\binom{N}{2} = N(N-1)/2$ pairs (k_1, k_2) to evaluate (plus N single change points), and each sequence requires a filter to be run, which in itself is linear in time. The local search in Algorithm 4.1 will find the optimum with $N^2/2$ filter recursions (there are in average $N/2$ filters, each requiring N recursions).

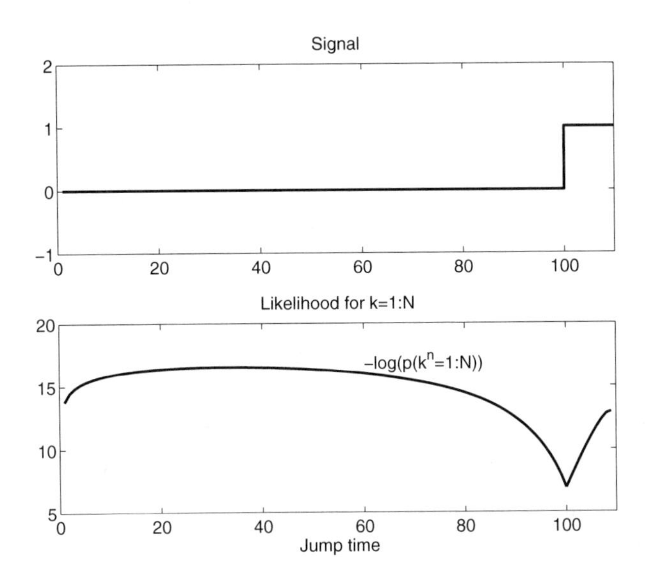

Figure 4.5. Upper plot: signal. Lower plot: negative log likelihood $p(k)$ with global minimum at $k = 100$ but local minimum at $k = 0$.

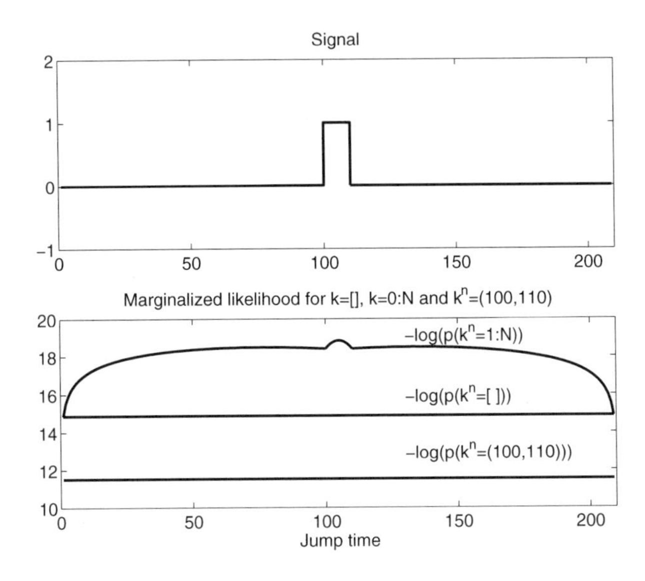

Figure 4.6. Upper plot: signal. Lower plot: negative log likelihood $p(k)$ with global minimum at $k^n = (100, 110)$ but local minimum at $k = 0$. No improvement for $k^n = m$, $m = 1, 2, \dots, N$

Example 4.3 Global search using one and two change points

Consider a change in the mean signal with two abrupt changes as shown in Figure 4.7. The likelihood as a function of two change times shows that the larger change at time 20 is more significant (the likelihood is narrower in one direction). Note that the likelihood is a symmetric function in k_1 and k_2.

In the example above, a two-dimensional exhaustive search finds the true maximum of the likelihood function. However, this does not imply that a complete search over all combinations of one or two changes would find the true change times generally.

Example 4.4 Counterexample of convergence of global search

Consider a signal with only two non-zero elements $+A$ and $-A$, respectively, surrounded by M zeros at each side. See Figure 4.8 for an example. Assume the following exponential distribution on the noise:

$$f(e) = Ce^{-\sqrt{|e|}}.$$

Then the negative log likelihood for no change point is

$$-\log p(y|k^n = 0) = \bar{C} + 2\sqrt{A}.$$

As usual, $k^n = 0$ is the notation for no change. The best way to place two change points is to pick up one of the spikes, and leave the other one as 'noise'.

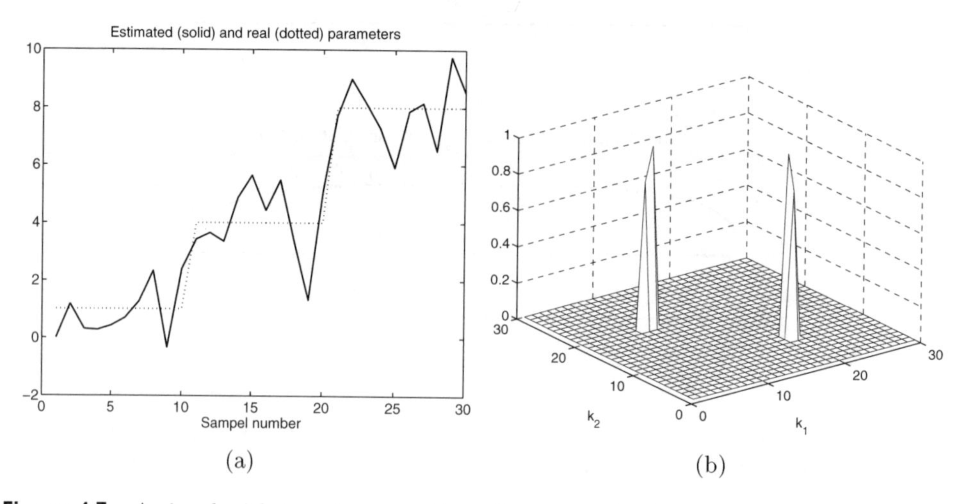

Figure 4.7. A signal with two changes with different amplitudes (a), and the likelihood as a function of k_1 and k_2 (b).

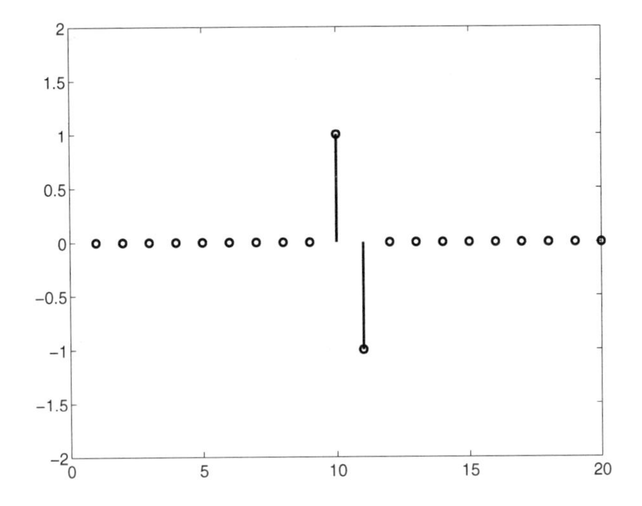

Figure 4.8. The signal in the counter example with $M = 10$.

The estimated mean in the segment that is left with a spike is then A/M. The residuals in two of the segments are identically zero, so their likelihood vanish (except for a constant). The likelihood for the whole data set is thus essentially the likelihood for the segment with a spike. For instance, we have

$$-\log p(y|k^n = (M, M+1)) = 3\bar{C} + \sqrt{A - \frac{A}{M}} + (M-1)\sqrt{\frac{A}{M}} \approx (1+\sqrt{M})\sqrt{A}.$$

That is, its negative log likelihood is larger than the likelihood given no change at all. Thus, we will never find the global optimum by trying all combinations of one and two change points. In this example, we have to make a complete search for three change points.

4.4.2. An MCMC approach

Markov Chain Monte Carlo (MCMC) approaches are surveyed in Chapter 12 and applied to particular models here and in Chapters 7 and 10. The MCMC algorithm proposed in Fitzgerald et al. (1994) for signal estimation is a combination of Gibbs sampling and the Metropolis algorithm. The algorithm below is based solely on the knowledge of the likelihood function for data given a certain segmentation.

It is the last step in the algorithm, where a random rejection sampling is applied, which defines the Metropolis algorithm: the candidate will be rejected with large probability if its value is unlikely.

Algorithm 4.2 MCMC signal segmentation

Decide the number of changes n, then:

1. Iterate Monte Carlo run i.
2. Iterate Gibbs sampler for component j in k^n, where a random number from

$$\bar{k}_j \sim p(k_j | k_1^n \text{ except } k_j)$$

is taken. Denote the new candidate sequence $\overline{k^n}$. The distribution may be taken as flat, or Gaussian centered around the previous estimate.

3. The candidate j is accepted with probability

$$\min\left(\frac{p(\overline{k^n})}{p(k^n)}, 1\right).$$

That is, the candidate sequence is always accepted if it increases the likelihood.

After the *burn-in time* (convergence), the distribution of change times can be computed by Monte Carlo techniques.

Note that there are no design parameters at all, except the number of changes and that one has to decide what the burn-in time is.

Example 4.5 MCMC search for two change points

Consider again the signal in Figure 4.7. The Algorithm 4.2 provides a histogram over estimated change times as shown in Figure 4.9. The left plot shows the considered jump sequence in each MC iteration. The burn-in time is about 60 iterations here. After this, the change at time 20 is very significant, while the smaller change at time 10 is sometimes estimated as 9.

4.5. Change point estimation

The detection problem is often recognized as *change point estimation* in the statistical literature. The assumption is that the mean of a white stochastic process changes at time k under $H_1(k)$:

$$\begin{aligned} y_t &= \theta_0 + e_t, \quad t \le k \\ y_t &= \theta_1 + e_t, \quad t > k. \end{aligned}$$

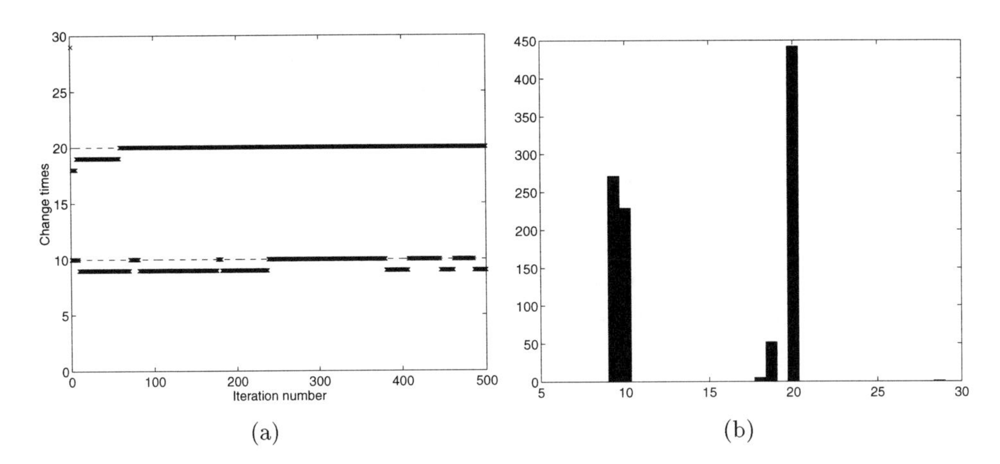

Figure 4.9. Result of the MCMC Algorithm 4.2. Left plot (a) shows the examined jump sequence in each iteration. The best encountered sequence is marked with dashed lines. The right plot (b) shows an histogram over all considered jump sequences. The burn-in time is not excluded in the histogram.

The following summary is based on the survey paper Sen and Srivastava (1975), where different procedures to test H_0 to H_1 are described. The methods are off-line and only one change point may exist in the data. The Bayesian and likelihood methods are closely related to the already described algorithms. However, the non-parametric approaches below are interesting and unique for this problem.

The following sub-problems are considered:

P1: $\theta_1 > \theta_0$, where θ_0 is unknown.
P2: $\theta_1 > \theta_0$, where $\theta_0 = 0$ is known.
P3: $\theta_1 \neq \theta_0$, where θ_0 is unknown.
P4: $\theta_1 \neq \theta_0$, where $\theta_0 = 0$ is known.

4.5.1. The Bayesian approach

A Bayesian approach, where the prior probability of all hypothesis are the same, gives:

$$\mathbf{P1} \quad g_1^B = \sum_{t=2}^{N} t(y_t - \bar{y})$$

$$\mathbf{P2} \quad g_2^B = \sum_{t=2}^{N} t y_t$$

$$\mathbf{P3} \quad g_3^B = \frac{1}{N^2} \sum_{k=1}^{N-1} \sum_{t=k+1}^{N} (y_t - \bar{y})^2$$

$$\mathbf{P4} \quad g_4^B = \frac{1}{N^2} \sum_{k=1}^{N-1} \sum_{t=k+1}^{N} y_t^2$$

where $\bar{y}$ is the sample mean of y_t. If $g > h$, where h is a pre-specified threshold, one possible estimate of the jump time (change point) is given by

$$\mathbf{P3} \quad \hat{k}_3^B = \arg \max_{1 \le k < N} \frac{1}{N-k} \sum_{t=k+1}^{N} (y_t - \bar{y})^2$$

for **P3** and similarly for **P4**.

4.5.2. The maximum likelihood approach

Using the ML method the test statistics are as follows:

$$\mathbf{P1} \quad g_1^{ML} = \max_k \frac{\bar{y}_{k+1,N} - \bar{y}_{1,k}}{\sqrt{k^{-1} + (N-k)^{-1}}}$$

$$\mathbf{P2} \quad g_2^{ML} = \max_k \sqrt{N-k}\bar{y}_{k+1,N}$$

$$\mathbf{P3} \quad g_3^{ML} = \max_k \frac{(\bar{y}_{k+1,N} - \bar{y}_{1,k})^2}{k^{-1} + (N-k)^{-1}}$$

$$\mathbf{P4} \quad g_4^{ML} = \max_k (N-k)\bar{y}_{k+1,N}^2$$

where $y_{m,n} = \frac{1}{n-m+1} \sum_{t=m}^{n} y_t$. If H_1 is decided, the jump time estimate is given by replacing $\max_k$ by $\arg\max_k$.

4.5.3. A non-parametric approach

The Bayesian and maximum likelihood approaches presumed a Gaussian distribution for the noise. Non-parametric tests for the first problem, assuming only whiteness, are based on the decision rule:

$$\mathbf{P1} \quad g_1^{NP} = \max_{1 \le k < N} \frac{s_k^i - \mathrm{E}s_k^i}{\mathrm{Var}s_k^i},$$

where the distance measure s_k^i is one of the following ones:

$$s_k^1 = \sum_{t=k+1}^{N} \operatorname{sign}(y_t - \operatorname{med}(y))$$

$$s_k^2 = \sum_{t=k+1}^{N} \sum_{m=1}^{N} I(y_t \le y_m).$$

Here, med denotes the median and sign is the sign function. The first method is a variant of a *sign test*, while the other one is an example of *ordered statistics*. These are a kind of whiteness test, based on the idea that under H_0, y_t is larger than its mean with probability 50%. Determining expectation and variance of s_k^i is a standard probability theory problem. For instance, the distribution of s_k^1 is hyper-geometric under H_0. Again, if H_1 is decided the jump time estimate is given by the maximizing argument.

Example 4.6 Change point estimation

To get a feeling of the different test statistics, we compare the statistics for the formulation **P1**. Two signals are simulated, one which is zero and one which undergoes an abrupt change in the middle of the data ($k = 50$) from zero to one. Both signals have white Gaussian measurement noise with variance 1 added. Figure 4.10 shows that all methods give a rather clear peak for the abruptly changing data (though the Bayesian statistic seems to have

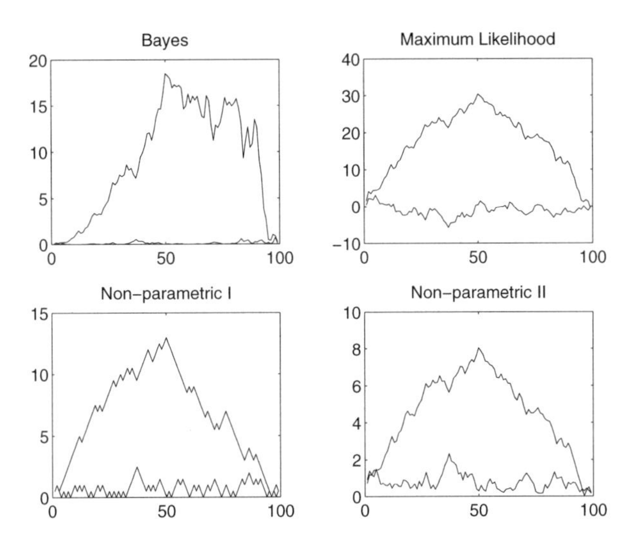

Figure 4.10. Comparison of four change point estimation algorithm for two signals: one with an abrupt change at time 50, and one without.

some probability of over-estimating the change time), and that there should be no problem in designing a good threshold.

The explicit formulas given in this section, using the maximum likelihood and Bayesian approaches, should in certain problem formulations come out as special cases of the more general formulas in Section 4.2, which can be verified by the reader.

4.6. Applications

4.6.1. Photon emissions

The signal model for photon emissions in Section 2.1.3 is

$$y_t = \theta_t + e_t, \quad p(y_t) = \frac{1}{\theta_t} e^{-\frac{y_t}{\theta_t}}.$$

Here $\mathrm{E}(e_t) = 0$ so e_t is white noise. There is no problem in modifying the likelihood based algorithms with respect to any distribution of the noise. The exponential distribution has the nice property of offering explicit and compact expressions for the likelihood after marginalization (MML) or maximization (MGL). The standard algorithm assumes Gaussian noise, but can still be used with a good result, though. This is an illustration of the robustness of the algorithms with respect to incorrectly modeled noise distribution. To save time, and to improve the model, the time for 100 arrivals is used as the measurement y_t in the model. A processed signal sample is, thus, a sum of

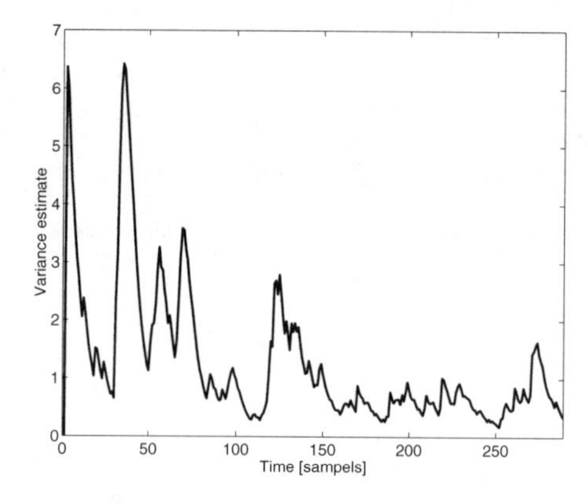

Figure 4.11. The noise variance in the photon data is estimated by applying a forgetting factor low-pass filter with $\lambda = 0.85$ to the measured arrival rates. The stationary values indicate a variance of slightly less than 1.

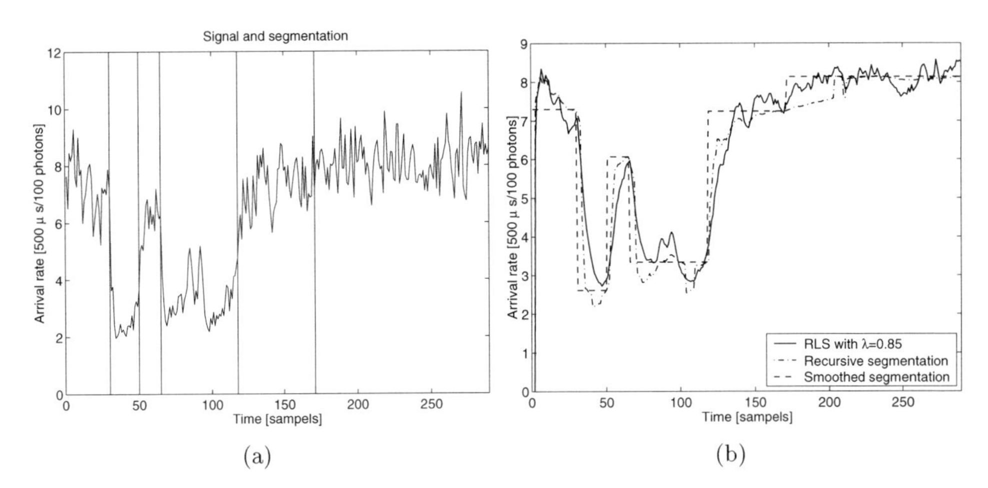

Figure 4.12. In (a) the time differences between 100 arrivals is shown together with a segmentation. The quantity is a scaled version of the mean time between arrivals (the intensity). In (b) the recursively estimated and segmented values of the intensity parameter are shown, with a low-pass filtered version of the arrival times.

100 exponentially distributed variables, which can be well approximated by a Gaussian variable according to the central limit theorem. To design the detector, a preliminary data analysis is performed. An adaptive filter with forgetting factor $\lambda = 0.85$ is applied to y_t. The squared residuals from this filter are filtered with the same filter to get an idea of the measurement noise R. The result is shown in Figure 4.11. Clearly, there are non-stationarities in the signal that should be discarded at this step, and the stationary values seems to be slightly smaller than 1. Now, the filter bank ML algorithm is applied with $M = 3$ parallel filter and an assumed measurement noise variance of $R = 1$. Figure 4.12 shows that a quite realistic result is obtained, and this algorithm might be used in real-time automatic surveillance.

4.6.2. Altitude sensor quality

The altitude estimate in an aircraft is computed mainly by deadreckoning using the inertial navigation system. To avoid drift in the altitude, the estimator is stabilized by air pressure sensors in pivot tubes as a barometric altitude sensor. This gets increased variance when Mach 1 is passed, as explained in Section 2.2.1. One problem is to detect the critical regions of variance increases from measured data. Figure 4.13 shows the errors from a calibrated (no bias) barometric sensor, and low-pass filtered squared errors. Here a forgetting factor algorithm has been used with $\lambda = 0.97$.

Figure 4.13 shows the same low-pass filtered variance estimate (in logarithmic scale) and the result from the ML variance segmentation algorithm. The

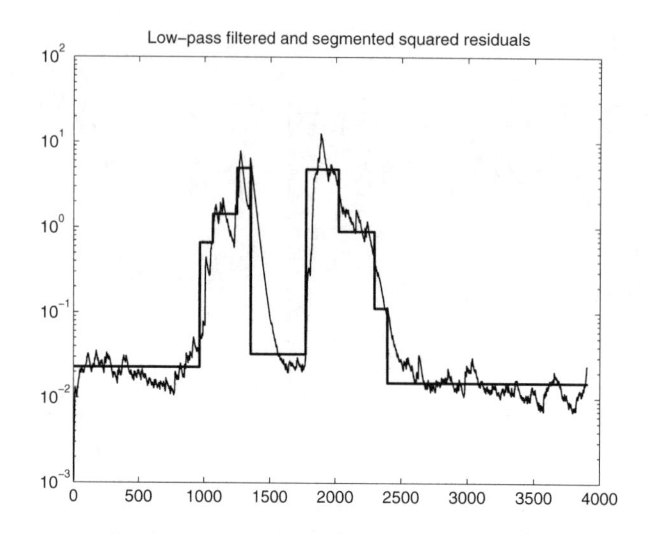

Figure 4.13. Low-pass filtered and segmented squared residuals for altitude data.

latter estimates the change times and noise variance as a function of sample number. That is, we know precisely where the measurements are useful.

4.6.3. Rat EEG

The rat EEG can be considered as a signal with piecewise constant noise variance, as discussed in Section 2.2.2. The method used by the researchers is based on a band pass filter and level thresholding on the output power. This

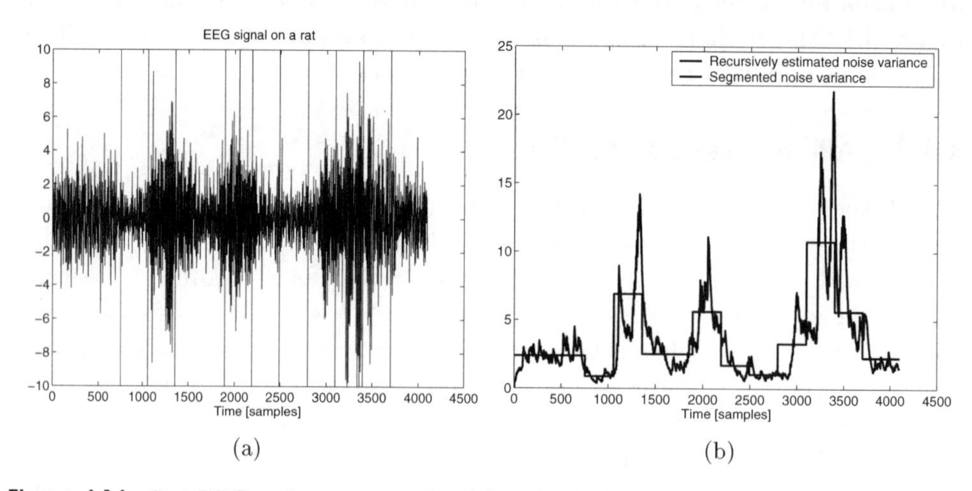

Figure 4.14. Rat EEG and a segmentation (a) and an adaptive and segmented estimate of the signal energy (b).

gives

```
[1096 1543 1887 2265 2980 3455 3832 3934].
```

This is a signal where the ML estimator for changes in noise variance can be applied. This gives

```
[754 1058 1358 1891 2192 2492 2796 3098 3398 3699].
```

It can be noted that the changes are hardly abrupt for this signal.

Part III: Parameter estimation

5

Adaptive filtering

5.1. Basics

The signal model in this chapter is, in its most general form,

$$y_t = G(q;\theta)u_t + H(q;\theta)e_t \tag{5.1}$$

$$A(q;\theta)y_t = \frac{B(q;\theta)}{D(q;\theta)}u_t + \frac{C(q;\theta)}{F(q;\theta)}e_t. \tag{5.2}$$

The noise is here assumed white with variance λ, and will sometimes be restricted to be Gaussian. The last expression is in a polynomial form, whereas G, H are filters. Time-variability is modeled by time-varying parameters θ_t. The adaptive filtering problem is to estimate these parameters by an adaptive filter,

$$\hat{\theta}_{t+1} = \hat{\theta}_t + K_t\varepsilon_t,$$

where ε_t is an application dependent error from the model.

We point out particular cases of (5.1) of special interest, but first, three archetypical applications will be presented:

- Consider first Figure 5.1(a). The main approach to *system identification* is to run a model in parallel with the true system, and the goal is to get $F(\theta) \approx G$. See Section 5.3.

- The radio channel in a *digital communication* system is well described by a filter $G(q;\theta)$. An important problem is to find an inverse filter, and this problem is depicted in Figure 5.1(b). This is also known as the *inverse system identification* problem. It is often necessary to include an overall delay. The goal is to get $F(\theta)G \approx q^{-D}$. In *equalization*, a feed-forward signal $q^{-D}u_t$, called training sequence, is available during a learning phase. In *blind equalization*, no training signal is available. The delay, as well as the order of the equalizer are design parameters. Both equalization and blind equalization are treated in Section 5.8.

- The *noise cancelation*, or *Acoustic Echo Cancelation* (*AEC*), problem in Figure 5.1(c) is to remove the noise component in $y = s+v$ by making use of an external sensor measuring the disturbance u in $v = Gu$. The goal is to get $F(\theta) \approx G$, so that $\hat{s} \approx s$. This problem is identical to system identification, which can be realized by redrawing the block diagram. However, there are some particular twists unique for noise cancelation. See Section 5.9.

Literature

There are many books covering the area of adaptive filtering. Among those most cited, we mention Alexander (1986), Bellanger (1988), Benveniste et al. (1987b), Cowan and Grant (1985), Goodwin and Sin (1984), Hayes (1996), Haykin (1996), C.R. Johnson (1988), Ljung and Söderström (1983), Mulgrew and Cowan (1988), Treichler et al. (1987), Widrow and Stearns (1985) and Young (1984).

Survey papers of general interest are Glentis et al. (1999), Sayed and Kailath (1994) and Shynk (1989). Concerning the applications, system identification is described in Johansson (1993), Ljung (1999) and Söderström and Stoica (1989), equalization in the books Gardner (1993), Haykin (1994), Proakis (1995) and survey paper Treichler et al. (1996), and finally acoustic echo cancelation in the survey papers Breining et al. (1999), Elliott and Nelson (1993) and Youhong and Morris (1999).

5.2. Signal models

In Chapter 2, we have seen a number of applications that can be recast to estimating the parameters in linear regression models. This section summarizes more systematically the different special cases of linear regressions and possible extensions.

5.2.1. Linear regression models

We here point out some common special cases of the general filter structure (5.1) that can be modeled as *linear regression* models, characterized by a regression vector φ_t and a parameter vector θ. The linear regression is defined

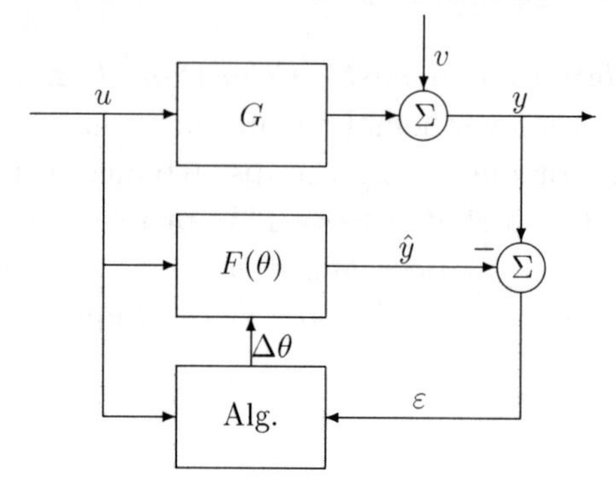

(a) System identification. The goal is to get a perfect system model $F(\theta) = G$.

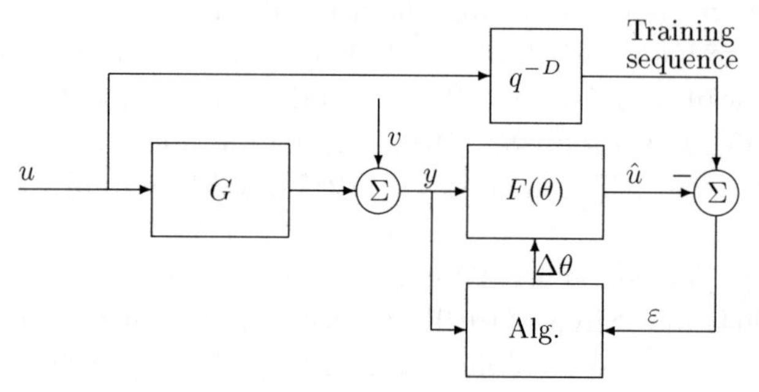

(b) Equalization. The goal is to get a perfect channel inverse $F(\theta) = G^{-1}$, in which case the transmitted information is perfectly recovered.

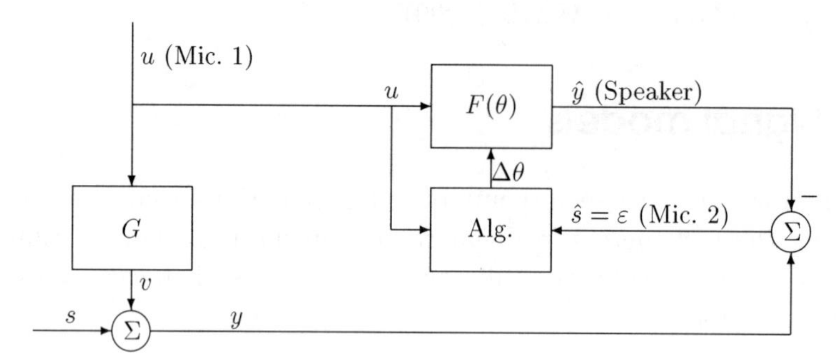

(c) Noise cancelation. The goal is to get a perfect model $F(\theta) = G$ of the acoustic path from disturbance to listener.

Figure 5.1. Adaptive filtering applications.

by

$$y_t = \varphi_t^T \theta + e_t. \tag{5.3}$$

The measurement y_t is assumed to be scalar, and a possible extension to the multi-variable case is given at the end of this section.

The most common model in communication applications is the *Finite Impulse Response* (*FIR*) model:

$$G(q;\theta) = B(q;\theta) \tag{5.4}$$
$$H(q;\theta) = 1 \tag{5.5}$$
$$\varphi_t = (u_{t-1}, \dots, u_{t-n})^T \tag{5.6}$$
$$\theta = (b_1, b_2, \dots, b_n)^T. \tag{5.7}$$

To explicitely include the model order, FIR(n) is a standard shorthand notation. It is natural to use this model for communication channels, where echoes give rise to the dynamics. It is also the dominating model structure in real-time signal processing applications, such as equalization and noise cancelling.

Example 5.1 Multi-path fading

In mobile communications, multi-path fading is caused by reflections, or echoes, in the environment, This *specular multi-path* is illustrated in Figure 5.2. Depending upon where the reflections occur, we get different phenomena:

- *Local scattering* occurs near the receiver. Here the difference in arrival time of the different rays is less than the symbol period, which means that no dynamic model can describe the phenomenon in discrete time. Instead, the envelope of the received signal is modeled as a stochastic variable with *Rayleigh distribution* or *Rice distribution.* The former distribution arises when the receiver is completely shielded from the transmitter, while the latter includes the effect of a stronger direct ray. The dynamical effects of this 'channel' are much faster than the symbol frequency and imply a distortion of the waveform. This phenomonen is called *frequency selective fading.*
- *Near-field scattering* occurs at intermediate distance between the transmitter and receiver. Here the difference in arrival time of the different rays is larger than the symbol period, so a discrete time echo model can be used to model the dynamic behaviour. First, in continuous time the scattering can be modeled as

$$y_t = h_t^1 u_{t-\tau_1} + h_t^2 u_{t-\tau_2} + \dots + h_t^n u_{t-\tau_n},$$

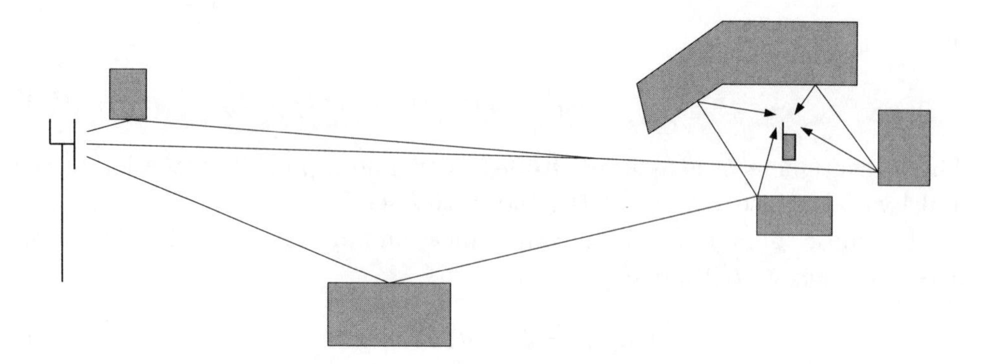

Figure 5.2. Multi-path fading is caused by reflections in environment.

where τ_i are real-valued time delays rather than multiples of a sample interval. This becomes a FIR model after sampling to discrete time.

- *Far-field scattering* occurs close to the transmitter. The received rays can be treated as just one ray.

Good surveys of multi-path fading are Ahlin and Zander (1998) and Sklar (1997), while identification of such a radio channel is described in Newson and Mulgrew (1994).

For modeling time series, an *Auto-Regressive* (*AR*) model is often used:

$$G(q;\theta) = 0 \tag{5.8}$$

$$H(q;\theta) = \frac{1}{A(q;\theta)} \tag{5.9}$$

$$\varphi_t = (-y_{t-1}, \dots, -y_{t-n})^T \tag{5.10}$$

$$\theta = (a_1, a_2, \dots, a_n)^T. \tag{5.11}$$

AR(n) is a shorthand notation. This is a flexible structure for many real-world signals like speech signals (Section 2.4.4), seismical data (Section 2.4.3) and biological data (Sections 2.4.1 and 2.4.2). One particular application is to use the model for spectral analysis as an alternative to transform based methods.

Example 5.2 Speech modeling

Speech is generated in three different ways. Voiced sound, like all vowels and 'm', is originating in the vocal chord. In signal processing terms, the vocal cord generates pulses which are modulated in the throat and mouth. Unvoiced sound, like 's' and 'v', is a modulated air stream, where the air pressure from the lungs can be modeled as white noise. Implosive sound, like 'k' and 'b', is generated by building up an air pressure which is suddenly released.

In all three cases, the human vocal system can be modeled as a series of cylinders and an excitation source (the 'noise' e_t) which is either a pulse train, white noise or a pulse. Each cylinder can be represented by a second order AR model, which leads to a physical motivation of why AR models are suitable for speech analysis and modeling. Time-variability in the parameters is explained by the fact that the speaker is continuously changing the geometry of the vocal tract.

In control and adaptive control, where there is a known control signal u available and where e represents measurement noise, the *Auto-Regressive model with eXogenous input* (*ARX*) is common:

$$G(q;\theta) = \frac{B(q;\theta)}{A(q;\theta)} \tag{5.12}$$

$$H(q;\theta) = \frac{1}{A(q;\theta)} \tag{5.13}$$

$$\varphi_t = (-y_{t-1}, \dots, -y_{t-n_a}, u_{t-1}, \dots, u_{t-n_b})^T \tag{5.14}$$

$$\theta = (a_1, a_2, \dots, a_{n_a}, b_1, b_2, \dots, b_{n_b})^T. \tag{5.15}$$

ARX(n_a, n_b, n_x) is a compact shorthand notation. This structure does not follow in a straightforward way from physical modeling, but is rather a rich structure whose main advantage is that there are simple estimation algorithms for it.

5.2.2. Pseudo-linear regression models

In system modeling, physical arguments often lead to the deterministic signal part of the measurements being expressed as a linear filter,

$$S_t = \frac{B(q;\theta)}{F(q;\theta)} u_t.$$

The main difference of commonly used model structures is how and where the noise enters the system. Possible model structures, that do not exactly fit the linear regression framework, are ARMA, OE and ARMAX models. These can be expressed as a *pseudo-linear regression*, where the regressor $\varphi_t(\theta)$ depends on the parameter.

The AR model has certain shortcomings for some other real world signals that are less resonant. Then the *Auto-Regressive Moving Average* (*ARMA*) model might be better suited,

$$H(q;\theta) = \frac{C(q;\theta)}{A(q;\theta)} \tag{5.16}$$

$$\varphi_t(\theta) = (-y_{t-1}, \dots, -y_{t-n_a}, e_{t-1}, \dots, e_{t-n_c})^T \tag{5.17}$$

$$\theta = (a_1, a_2, \dots, a_{n_a}, c_1, c_2, \dots, c_{n_c})^T. \tag{5.18}$$

The *Output Error* (*OE*) model, which is of the Infinite Impulse Response (IIR) type, is defined as additive noise to the signal part

$$G(q;\theta) = \frac{B(q;\theta)}{F(q;\theta)} \tag{5.19}$$

$$H(q;\theta) = 1 \tag{5.20}$$

$$\varphi_t(\theta) = (-y_{t-1} + e_{t-1}, \dots, -y_{t-n_f} + e_{t-n_f}, u_{t-1}, \dots, u_{t-n_b})^T \tag{5.21}$$

$$\theta = (f_1, f_2, \dots, f_{n_f}, b_1, b_2, \dots, b_{n_b})^T. \tag{5.22}$$

Note that the regressor contains the noise-free output, which can be written $y_t - e_t$. That is, the noise never enters the dynamics. The OE models follow naturally from physical modeling of systems, assuming only measurement noise as stochastic disturbance.

For modeling systems where the measurement noise is not white but still more correlated than that described by an ARX model, an *Auto-Regressive Moving Average model with eXogenous input* (*ARMAX*) model is often used:

$$G(q;\theta) = \frac{B(q;\theta)}{A(q;\theta)} \tag{5.23}$$

$$H(q;\theta) = \frac{C(q;\theta)}{A(q;\theta)} \tag{5.24}$$

$$\varphi_t = (-y_{t-1}, \dots, -y_{t-n_a}, u_{t-1}, \dots, u_{t-n_b}, e_{t-1}, \dots, e_{t-n_c})^T \tag{5.25}$$

$$\theta = (a_1, a_2, \dots, a_{n_a}, b_1, b_2, \dots, b_{n_b}, c_1, c_2, \dots, c_{n_c})^T. \tag{5.26}$$

This model has found a standard application in adaptive control.

The common theme in ARMA, ARMAX and OE models is that they can be written as a pseudo-linear regression

$$y_t = \varphi_t(\theta)\theta + e_t, \tag{5.27}$$

where the regressor depends on the true parameters. The parameter dependence comes from the fact that the regressor is a function of the noise. For an

ARMA model, the regressor in (5.17) contains e_t, which can be computed as

$$e_t = \frac{A(q;\theta)}{C(q;\theta)} y_t,$$

and similarly for ARMAX and OE models.

The natural approximation is to just plug in the latest possible estimate of the noise. That is, replace e_t with the residuals ε_t,

$$\hat{e}_t = \varepsilon_t = \frac{A(q;\hat{\theta})}{C(q;\hat{\theta})} y_t,$$

This is the approach in the *extended least squares* algorithm described in the next section. The adaptive algorithms and change detectors developed in the sequel are mainly discussed with respect to linear regression models. However, they can be applied to OE, ARMA and ARMAX as well, with the approximation that the noise e_t is replaced by the residuals.

Multi-Input Multi-Output (MIMO) models are usually considered to be built up as $n_y \times n_u$ independent models, where $n_y = \dim(y)$ and $n_u = \dim(u)$, one from each input to each output. MIMO adaptive filters can thus be considered as a two-dimensional array of *Single-Input Single-Output* (*SISO*) adaptive filters.

5.3. System identification

This section overviews and gives some examples of optimization algorithms used in system identification in general. As it turns out, these algorithms are fundamental for the understanding and derivation of adaptive algorithms as well.

5.3.1. Stochastic and deterministic least squares

The algorithms will be derived from a minimization problem. Let

$$\varepsilon_t(\theta) = y_t - \hat{y}_t = y_t - \varphi_t^T \theta. \tag{5.28}$$

Least squares optimization aims at minimizing a quadratic loss function $V(\theta)$,

$$\hat{\theta} = \arg\min_\theta V(\theta).$$

The (generally unsolvable) adaptive filtering problem can be stated as minimizing the loss function

$$V(\theta) = \varepsilon_t^2(\theta) \tag{5.29}$$

with respect to θ for each time instant. For system identification, we can distinguish two conceptually different formulations of the least squares criterion: the stochastic and deterministic least squares.

Stochastic least squares

The solution to the *stochastic least squares* is defined as the minimizing argument to

$$V(\theta) = \mathrm{E}[\varepsilon_t^2(\theta)]. \tag{5.30}$$

Substituting the residual (5.28) in (5.30), differentiating and equating to zero, gives the *minimum mean square error* solution

$$\frac{dV(\theta)}{d\theta} = -2E\varphi_t(y_t - \varphi_t^T\theta) = 0.$$

This equation defines the normal equations for the least squares problem. The solution to this problem will be denoted θ^* and is in case of invertible $\mathrm{E}[\varphi_t\varphi_t^T]$ given by

$$\theta^* = \mathrm{E}[\varphi_t\varphi_t^T]^{-1}\,\mathrm{E}[\varphi_t y_t]. \tag{5.31}$$

In practice, the expectation cannot be evaluated and the problem is how to estimate the expected values from real data.

Example 5.3 Stochastic least squares solution for FIR model

For a second order FIR model, (5.31) becomes

$$\theta^* = \begin{pmatrix} r_{uu}(0) & r_{uu}(1) \\ r_{uu}(1) & r_{uu}(0) \end{pmatrix}^{-1} \begin{pmatrix} r_{uy}(0) \\ r_{uy}(1) \end{pmatrix}.$$

In Section 13.3, this is identified as the solution to the *Wiener-Hopf equation* (13.10). The least squares solution is sometimes referred to as the Wiener filter.

Deterministic least squares

On the other hand, the solution to the *deterministic least squares* is defined as the minimizing argument to

$$V(\theta) = \sum_{k=1}^{t} \varepsilon_k^2(\theta). \tag{5.32}$$

The normal equations are found by differentiation,

$$\frac{dV(\theta)}{d\theta} = -2\sum_{k=1}^{t} \varphi_k(y_k - \varphi_k^T\theta) = 0,$$

and the minimizing argument is thus

$$\hat{\theta}_t = \left(\sum_{k=1}^{t} \varphi_k\varphi_k^T\right)^{-1} \sum_{k=1}^{t} \varphi_k y_k. \tag{5.33}$$

It is here assumed that the parameters are time-invariant, so the question is how to generalize the estimate to the time-varying case.

Example 5.4 Deterministic least squares solution for FIR model

For a second order FIR model, (5.33) becomes

$$\hat{\theta}_t = \begin{pmatrix} \hat{r}_{uu}(0) & \hat{r}_{uu}(1) \\ \hat{r}_{uu}(1) & \hat{r}_{uu}(0) \end{pmatrix}^{-1} \begin{pmatrix} \hat{r}_{uy}(0) \\ \hat{r}_{uy}(1) \end{pmatrix},$$

where the estimated covariances are defined as

$$\hat{r}_{uu}(\tau) = \frac{1}{t}\sum_{k=|\tau|+1}^{t} u_k u_{k-|\tau|}.$$

Note the similarity between stochastic and deterministic least squares. In the limit $t \to \infty$, we have convergence $\hat{\theta}_t \to \theta^*$ under mild conditions.

Example 5.5 AR estimation for rat EEG

Consider the rat EEG in Section 2.4.1, also shown in Figure 5.3. The least squares parameter estimate for an AR(2) model is

$$\hat{\theta}_{4096} = (-0.85, \quad 0.40)^T,$$

corresponding to two complex conjugated poles in $0.43 \pm i0.47$. The least squares loss function is $V(\hat{\theta}) = 1.91$, which can be interpreted as the energy in the model noise e_t. This figure should be compared to the energy in the signal itself, that is the loss function without model, $V(0) = 3.60$. This means that the model can explain roughly half of the energy in the signal.

We can evaluate the least squares estimate at any time. Figure 5.3 shows how the estimate converges. This plot must not be confused with the adaptive

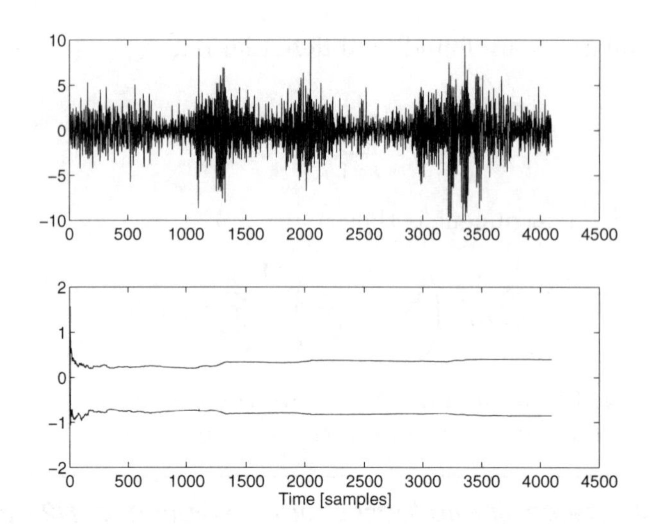

Figure 5.3. Rat EEG and estimated parameters of an AR(2) model for each time instant.

algorithms later on, since there is no forgetting of old information here. If we try a higher order model, say AR(4), the loss function only decreases to $V(\hat{\theta}) = 1.86$. This means that it hardly pays off to use higher order models in this example.

5.3.2. Model structure selection

The last comment in Example 5.5 generalizes to an important problem: which is the best model order for a given signal? One of the most important conclusions from signal modeling, also valid for change detection and segmentation, is that the more free parameters in the model, the better fit. In the example above, the loss function decreases when going from AR(2) to AR(4), but not significantly. That is the engineering problem: increase the model order until the loss function does not decrease *significantly*.

There are several formal attempts to try to get an objective measure of fit. All these can be interpreted as the least squares loss function plus a *penalty term*, that penalizes the number of parameters. This is in accordance with the *parsimonious principle* (or *Ockam's razor* after a greek philosoph). We have encountered this problem in Chapter 4, and a few penalty terms were listed in Section 4.2.2. These and some more approaches are summarized below, where d denotes the model order:

- Akaike's *Final Prediction Error* (*FPE*) (Akaike, 1971; Davisson, 1965):

$$\hat{d} = \arg\min_d V_N(d) \frac{1 + d/N}{1 - d/N}.$$

- Akaike's *Information Criterion A* (*AIC*) (Akaike, 1969):

$$\hat{d} = \arg\min_d \log(V_N(d)) + \frac{2d}{N}.$$

This is asymptotically the same as FPE, which is easily realized by taking the logarithm of FPE.

- The asymptotically equivalent criteria Akaike's *Information Criterion B* (*BIC*) (Akaike, 1977), Rissanen's *minimum description length* (*MDL*) approach (Rissanen, 1989), see Section 12.3.1, and *Schwartz criterion* (Schwartz, 1978).

$$\hat{d} = \arg\min_d \log(V_N(d)) + \frac{2d\log(N)}{N}.$$

- Mallow's C_p *criterion* (Mallows, 1973) is

$$\hat{d} = \arg\min_d \frac{V_N(d)}{R} + 2d - N,$$

which assumes known noise variance R.

- For time series with few data points, say 10-20, the aforementioned approaches do not work very well, since they are based on asymptotic arguments. In the field of econometrics, refined criteria have appeared. The corrected AIC (Hurvich and Tsai, 1989) is

$$\hat{d} = \arg\min_d \log(V_N(d)) + \frac{2d}{N} + \frac{2(d+1)(d+2)}{N-d-2}.$$

- The Φ *criterion* (Hannan and Quinn, 1979)

$$\hat{d} = \arg\min_d \log(V_N(d)) + \frac{2d\log\log N}{N}.$$

FPE and AIC tend to over-estimate the model order, while BIC and MDL are consistent. That is, if we simulate a model and then try to find its model order, BIC will find it when the number of data N tends to infinity with probability one. The Φ criterion is also consistent. A somewhat different approach, yielding a consistent estimator of d, is based on the *Predictive Least Squares* (*PLS*) (Rissanen, 1986). Here the unnormalized sum of squared residuals is used:

$$\hat{d} = \arg\min_d \sum_{m+1}^{N} (y_t - \varphi_t^T \hat{\theta}_{t-1}(d))^2,$$

where m is a design parameter to exclude the transient. Compare this to the standard loss function, where the final estimate is used. Using (5.96), the sum of squared residuals can be written as

$$V_N(d) = \sum_{1}^{N} \frac{(y_t - \varphi_t^T \hat{\theta}_N(d))^2}{R} = \sum_{1}^{N} \frac{(y_t - \varphi_t^T \hat{\theta}_{t-1}(d))^2}{R + \varphi_t^T P_{t-1} \varphi_t},$$

which is a smaller number than PLS suggests. This difference makes PLS parsimonious. Consistency and asymptotic equality with BIC are proven in Wei (1992).

5.3.3. Steepest descent minimization

The *steepest descent algorithm* is defined by

$$\hat{\theta}^i = \hat{\theta}^{i-1} - \mu \left. \frac{dV(\theta)}{d\theta} \right|_{\theta = \hat{\theta}^{i-1}}. \tag{5.34}$$

Hence, the estimate is modified in the direction of the negative gradient. In case the gradient is approximated using measurements, the algorithm is called a *stochastic gradient algorithm*.

Example 5.6 The steepest descent algorithm

Consider the loss function

$$V(x) = x_1^2 + x_1 x_2 + x_2^2.$$

The steepest descent algorithm in (5.34) becomes (replace θ by x)

$$x^{i+1} = x^i - \mu \begin{pmatrix} 2x_1 + x_2 \\ x_1 + 2x_2 \end{pmatrix}.$$

The left plot in Figure 5.4 shows the convergence (or *learning curves*) for different initializations with $\mu = 0.03$ and 100 iterations. A stochastic version is obtained by adding noise with variance 10 to the gradient, as illustrated in the right plot.

This example illustrates how the algorithm follows the gradient down to the minimum.

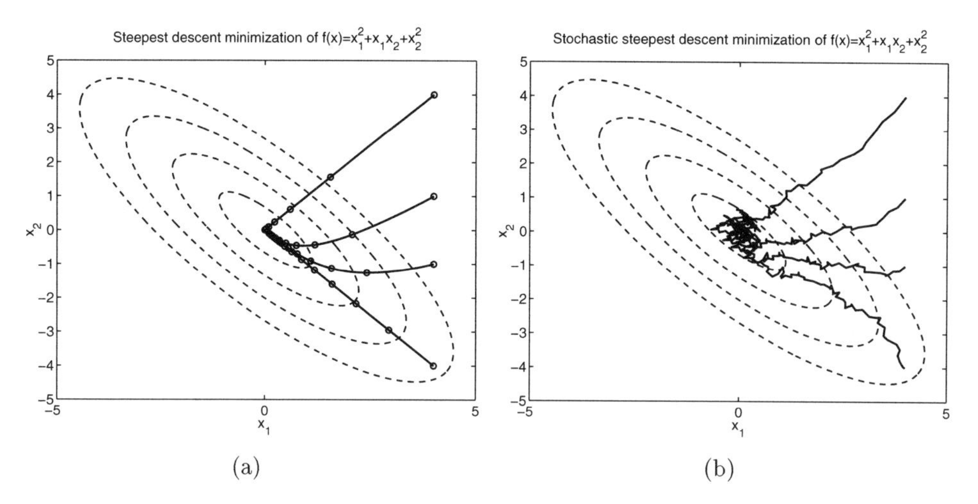

Figure 5.4. Deterministic (left) and stochastic (right) steepest descent algorithms.

5.3.4. Newton-Raphson minimization

The *Newton–Raphson algorithm*

$$\hat{\theta}^i = \hat{\theta}^{i-1} - \mu \left(\frac{d^2 V(\theta)}{d\theta^2} \right)^{-1} \frac{dV(\theta)}{d\theta} \bigg|_{\theta = \hat{\theta}^{i-1}} \tag{5.35}$$

usually has superior convergence properties, compared to the steepest descent algorithm. The price is a computational much more demanding algorithm, where the Hessian needs to be computed and also inverted. Sufficiently close to the minimum, all loss functions are approximately quadratic functions, and there the Newton–Raphson algorithm takes step straight to the minima as illustrated by the example below.

Example 5.7 The Newton-Raphson algorithm

Consider the application of the Newton–Raphson algorithm to Example 5.6 under the same premises. Figure 5.5 shows that the algorithm now finds the closest way to the minimum.

It is interesting to compare how the Hessian modifies the step-size. Newton–Raphson takes steps in more equidistant steps, while the gradient algorithm takes huge steps where the gradient is large.

Models linear in the parameters (linear regressions) give a quadratic least squares loss function, which implies that convergence can be obtained in one

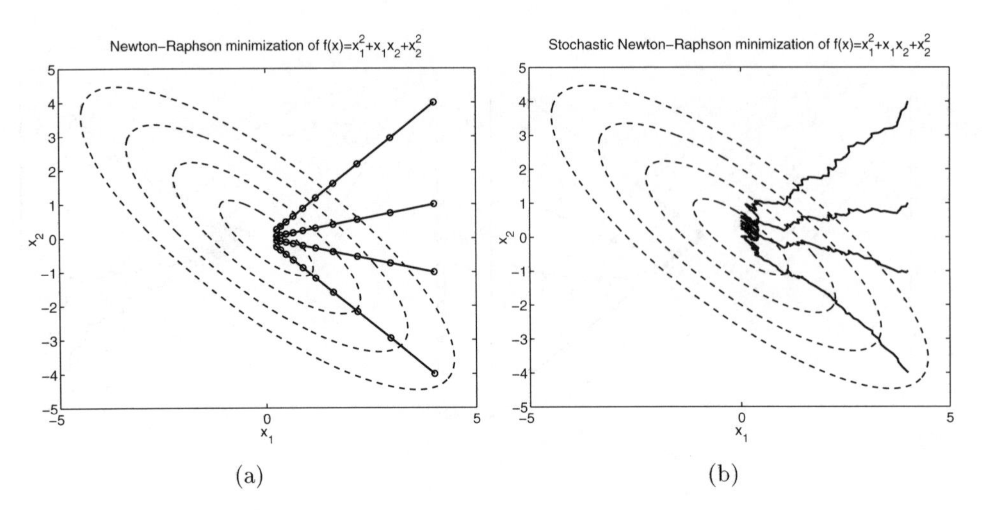

Figure 5.5. Deterministic (left) and stochastic (right) Newton–Raphson algorithms

iteration by Newton–Raphson using $\mu = 1$. On the other hand, model structures corresponding to pseudo-linear regressions can have loss functions with local minima, in which case initialization becomes an important matter.

Example 5.8 Newton-Raphson with local minima

The function

$$f(x) = x^5 - 6x^4 + 6x^3 + 20x^2 - 38x + 20$$

has a local and a global minimum, as the plot in Figure 5.6 shows. A few iterations of the Newton–Raphson algorithm (5.35) for initializations $x^0 = 0$ and $x^0 = 4$, respectively, are also illustrated in the plot by circles and stars, respectively.

5.3.5. Gauss-Newton minimization

Hitherto, the discussion holds for general optimization problems. Now, the algorithms will be applied to model estimation, or *system identification*. Notationally, we can merge stochastic and deterministic least squares by using

$$V(\theta) = \hat{\mathrm{E}}\varepsilon_t^2(\theta) = \hat{\mathrm{E}}(y_t - \varphi_t^T(\theta)\theta)^2, \tag{5.36}$$

where $\hat{\mathrm{E}}$ should be interpreted as a computable approximation to the expectation operator instead of (5.30) or an adaptive version of the averaging sum in (5.32), respectively.

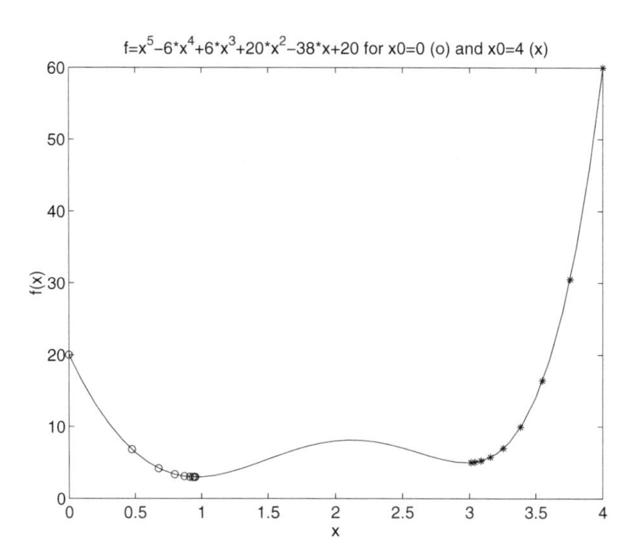

Figure 5.6. The Newton–Raphson algorithm applied to a function with several minima.

For generality, we will consider the pseudo-linear regression case. The gradient and Hessian are:

$$\begin{aligned}
\psi_t(\theta) &\triangleq -\frac{d\varepsilon_t(\theta)}{d\theta} \\
\frac{dV(\theta)}{d\theta} &= -2\hat{\mathrm{E}}\psi_t(\theta)\epsilon_t(\theta) \\
\frac{d^2V(\theta)}{d\theta^2} &= 2\hat{\mathrm{E}}\psi_t(\theta)\psi_t^T(\theta) - 2\hat{\mathrm{E}}\frac{d\psi_t(\theta)}{d\theta}\epsilon_t(\theta) \\
&\approx 2\hat{\mathrm{E}}\psi_t(\theta)\psi_t^T(\theta).
\end{aligned}$$

The last approximation gives the *Gauss–Newton algorithm*. The approximation is motivated as follows: first, the gradient should be uncorrelated with the residuals close to the minimum and not point in any particular direction. Thus, the expectation should be zero. Secondly, the residuals should, with any weighting function, average to something very small compared to the other term which is a quadratic form.

The gradient $\psi_t(\theta)$ depends upon the model. One approximation for pseudo-linear models is to use $\varphi_t(\hat{\theta})$, which gives the *extended least squares* algorithm. The approximation is to neglect one term in

$$\psi_t(\theta) = -\frac{d}{d\theta}(y_t - \varphi_t(\theta)\theta) = \varphi_t(\theta) + \underbrace{\frac{d\varphi_t(\theta)}{d\theta}}_{\approx 0}\theta$$

in the gradient. A related and in many situations better algorithm is the recursive maximum likelihood method given below without comments.

Algorithm 5.1 Gauss-Newton for ARMAX models

Consider the model

$$\begin{aligned} A(q;\theta)y_t &= B(q;\theta)u_t + C(q;\theta)e_t \\ y_t &= \varphi_t^T(\theta)\theta_t + e_t \\ \varphi_t &= (-y_{t-1}, \dots, -y_{t-n_a}, u_{t-1}, \dots, u_{t-n_b}, e_{t-1}, \dots, e_{t-n_c})^T \\ \theta &= (a_1, a_2, \dots, a_{n_a}, b_1, b_2, \dots, b_{n_b}, c_1, c_2, \dots, c_{n_c})^T, \end{aligned}$$

where the $C(q)$ polynomial is assumed to be monial with $c_0 = 1$. The Gauss–Newton algorithm is

$$\begin{aligned} \hat{\theta}^i &= \hat{\theta}^{i-1} + \mu \left(\frac{1}{N} \sum_{t=1}^{N} \psi_t \psi_t^T \right)^{-1} \left(\frac{1}{N} \sum_{t=1}^{N} \psi_t \varepsilon_t \right) \\ \psi_t(\theta) &= -\frac{d\varepsilon_t(\theta)}{d\theta} = \frac{dy_t(\theta)}{d\theta}. \end{aligned}$$

The extended least squares algorithm uses

$$\begin{aligned} \psi_t &= \varphi_t(\hat{\theta}_t) = (-y_{t-1}, \dots, -y_{t-n_a}, u_{t-1}, \dots, u_{t-n_b}, \hat{e}_{t-1}, \dots, \hat{e}_{t-n_c})^T \\ \hat{e}_t &= y_t - \varphi_t^T(\hat{\theta}_t)\hat{\theta}_t, \end{aligned}$$

while the *recursive maximum likelihood* method uses

$$\psi_t = \frac{1}{C(q;\hat{\theta}_t)} \varphi_t(\hat{\theta}_t).$$

Some practical implemenation steps are given below:

- A stopping criterion is needed to abort the iterations. Usually, this decision involves checking the relative change in the objective function V and the size of the gradient ψ.

- The step size μ is equal to unity in the original algorithm. Sometimes this is a too large a step. For objective functions whose values decrease in the direction of the gradient for a short while, and then start to increase, a shorter step size is needed. One approach is to always test if V decreases before updating the parameters. If not, the step size is halved, and the procedure repeated. Another approach is to always optimize the step size. This is a scalar optimization which can be done relatively efficient, and the gain can be a considerable reduction in iterations.

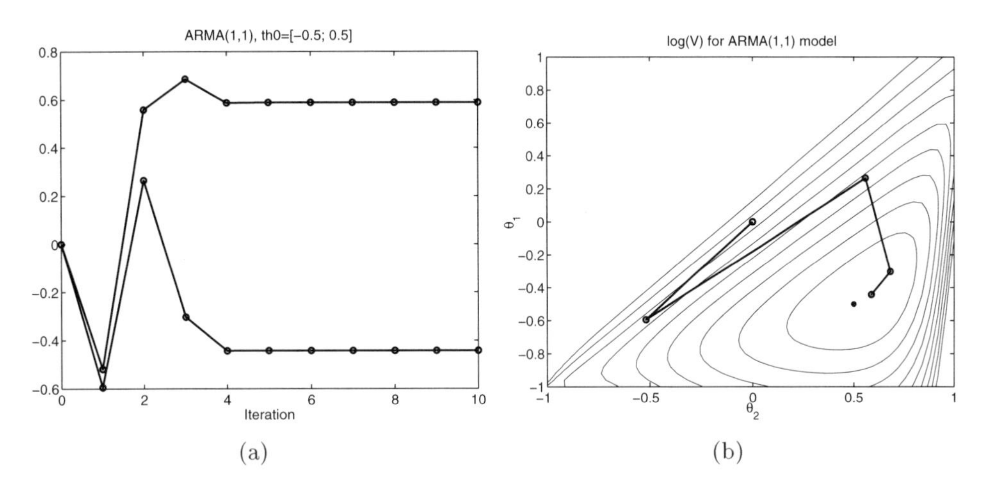

Figure 5.7. 10 iterations of Gauss–Newton (a). The logarithm of the least squares loss function with the GN iterations marked as a path (b).

A thorough treatment of such algorithms is given in Ljung and Söderström (1983) and Ljung (1999).

A revealing example with only one local minima but with a non-quadratic loss function is presented below.

Example 5.9 Gauss–Newton optimization of an ARMA(1,1) model

The ARMA(1,1) model below is simulated using Gaussian noise of length $N = 200$:

$$y(t) - 0.5y(t-1) = e(t) + 0.5e(t-1).$$

The Gauss–Newton iterations starting at the origin are illustrated both as an iteration plot and in the level curves of the loss function in Figure 5.7. The loss function is also illustrated as a mesh plot in Figure 5.8, which shows that there is one global minimum, and that the loss function has a quadratic behavior locally. Note that any ARMA model can be restricted to be stable and non-minimum phase, which implies that the intervals $[-1, 1]$ for the parameters cover all possible ARMA(1,1) models. The non-quadratic form far from the optimum explains why the first few iterations of Gauss–Newton are sensitive to noise. In this example, the algorithm never reaches the optimum, which is due to finite data length. The final estimation error decreases with simulation length.

To end this section, two quite general system identification examples are given, where the problem is to adjust the parameters of given ordinary differential equations to measured data.

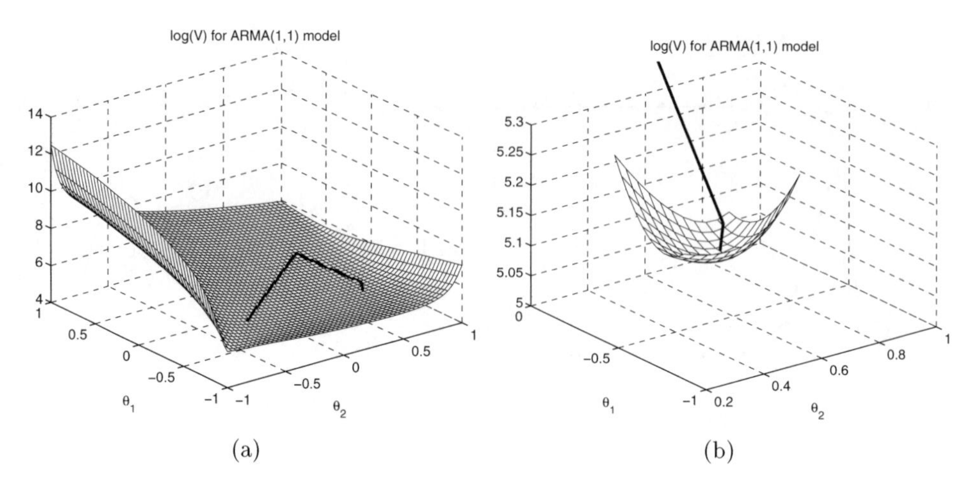

Figure 5.8. The logarithm of the least squares loss function.

Example 5.10 Gauss-Newton identification of electronic nose signals

Consider the electronic nose for classifying bacteria in Section 2.9.1. We illustrate here how standard routines for Gauss–Newton minimization can be applied to general signal processing problems. Recall the signal model for the sensor signals given in predictor form:

$$\hat{f}(t;\theta) = \theta(1)\frac{1}{1+e^{-\theta(2)(\theta(3)-t)}}\left(1-\frac{1}{1+e^{-\theta(4)(\theta(5)-t)}}\right).$$

A main feature in many standard packages, is that we do not have to compute the gradient ψ_t. The algorithm can compute it numerically. The result of an estimated model to three sensor signals was shown in Figure 2.21.

Example 5.11 Gauss-Newton identification of cell phone sales

Consider the sales figures for the cellular phones NMT in Section 2.9.2. The differential equation used as a signal model is

$$\dot{y} = \theta_1(\theta_3 - y) + \theta_2 y(\theta_3 - y).$$

This non-linear differential equation can be solved analytically and then discretized to suit the discrete time measurements. However, in many cases, there is either no analytical solution or it is very hard to find. Then one can

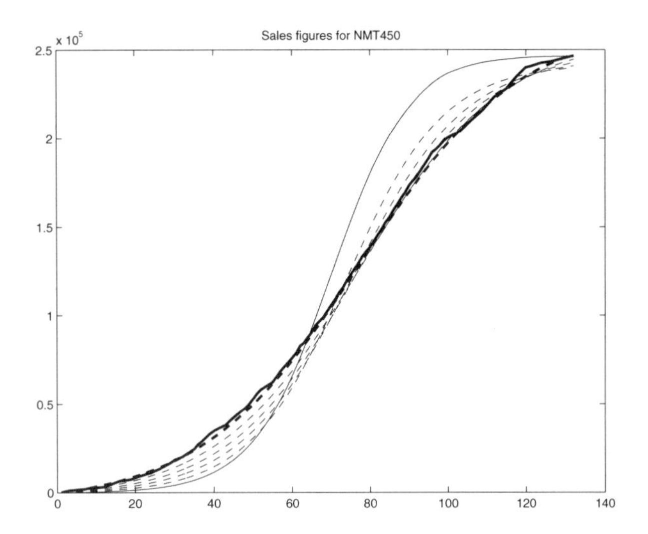

Figure 5.9. Convergence of signal modeling of the NMT 450 sales figures (thick solid line) using the Gauss–Newton algorithm. The initial estimate is the thin solid line, and then the dashed thin lines show how each iteration improves the result. The thick dashed line shows the final model.

use a numerical simulation tool to compute the mapping from parameters to predictions $\hat{y}_t(\theta)$, and then proceed as in Example 5.10.

Figure 5.9 shows how the initial estimate successively converges to a curve very close to the measurements. A possible problem is local minima. In this example, we have to start with a θ very close to the best values to get convergence. Here we used $\theta = (-0.0001,\ 0.1,\ \max(y))$, using the fact that the stationary solution must have $y_t \to \theta_3$ and then some trial and error for varying $\theta_1,\ \theta_2$. The final parameter estimate is $\hat{\theta} = (0.0011,\ 0.0444,\ 1.075)^T$.

5.4. Adaptive algorithms

System identification by using off-line optimization performs an *iterative* minimization of the type

$$\hat{\theta}^i = \hat{\theta}^{i-1} + \mu \sum_{t=1}^{N} K_t^i \varepsilon_t^i,$$

initiated at $\hat{\theta}^0$. Here K_t is either the gradient of the loss function, or the inverse Hessian times the gradient. As a starting point for motivating the adaptive algorithms, we can think of them as as an iterative minimization where one new data point is included in each iteration. That is, we let $N = i$ above. However, the algorithms will neither become recursive nor truly adaptive by this method (since they will probably converge to a kind of overall average). A better try is to use the previous estimate as the starting point in a new minimization,

$$\begin{aligned}
\hat{\theta}_t^i &= \hat{\theta}_t^{i-1} + \mu K_t^i \varepsilon_t^i, \quad i = 1, 2, \ldots n \\
\hat{\theta}_t^0 &= \hat{\theta}_{t-1}^n \\
\hat{\theta}_0^0 &= \hat{\theta}^0.
\end{aligned}$$

Taking the limited information in each measurement into account, it is logical to only make one iteratation per measurement. That is, a generic adaptive algorithm derived from an off-line method can be written

$$\begin{aligned}
\hat{\theta}_t &= \hat{\theta}_{t-1} + \mu K_t \varepsilon_t \\
\hat{\theta}_0 &= \hat{\theta}^0.
\end{aligned}$$

Here, only K_t needs to be specified.

5.4.1. LMS

The idea in the *Least Mean Square* (*LMS*) algorithm is to apply a steepest descent algorithm (5.34) to (5.29). Using (5.28), this gives the following algorithm.

Algorithm 5.2 LMS

For general linear regression models $y_t = \varphi_t^T \theta_t + e_t$, the LMS algorithm updates the parameter estimate by the recursion

$$\hat{\theta}_t = \hat{\theta}_{t-1} + \mu \varphi_t (y_t - \varphi_t^T \hat{\theta}_{t-1}). \tag{5.37}$$

The design parameter μ is a user chosen *step-size.* A good starting value of the step size is $\mu = 0.01/\operatorname{Std}(y)$.

The algorithm is applied to data from a simulated model in the following example.

Example 5.12 Adaptive filtering with LMS

Consider an AR(2) model

$$y_t = -a_1 y_{t-1} - a_2 y_{t-2} + e_t,$$

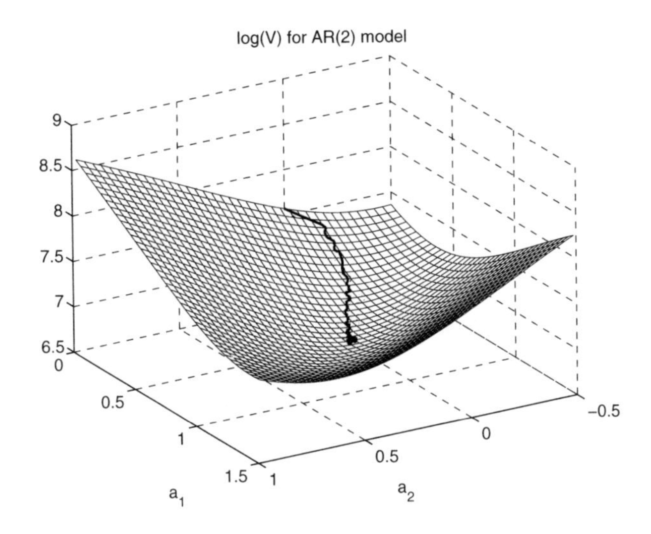

Figure 5.10. Convergence of $\log(V_t)$ to the global optimum using LMS for an AR(2) model.

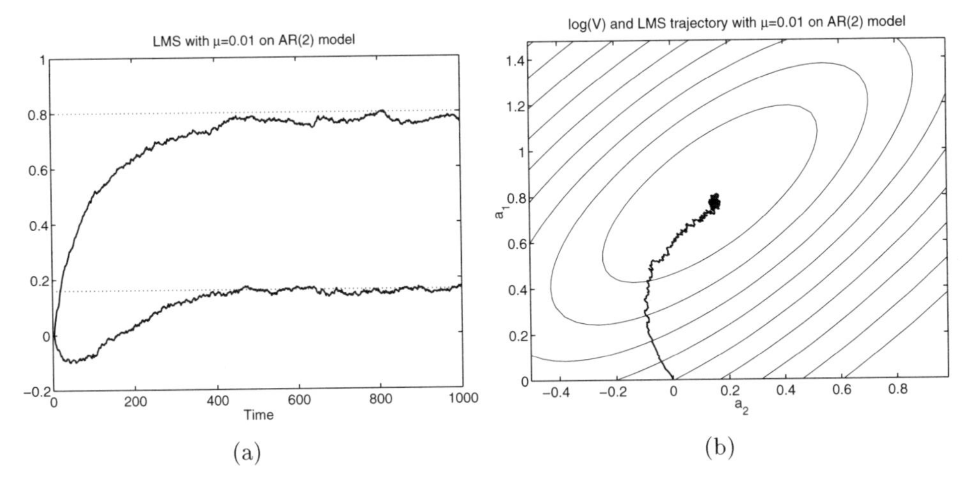

Figure 5.11. Convergence of LMS for an AR(2) model averaged over 25 Monte Carlo simulations and illustrated as a time plot (a) and in the loss function's level curves (b).

simulated with $a_1 = 0.8$ and $a_2 = 0.16$ (two poles in 0.4). Figure 5.10 shows the logarithm of the least squares loss function as a function of the parameters. The LMS algorithm with a step size $\mu = 0.01$ is applied to 1000 data items and the parameter estimates are averaged over 25 Monte Carlo simulations. Figure 5.11(a) shows the parameter convergence as a function of time, and Figure 5.11(b) convergence in the level curves of the loss function.

There are certain variants of LMS. The *Normalized LMS (NLMS)* is

$$\hat{\theta}_t = \hat{\theta}_{t-1} + \mu_t \varphi_t (y_t - \varphi_t^T \hat{\theta}_{t-1}), \tag{5.38}$$

where

$$\mu_t = \frac{\mu}{\varphi_t^T \varphi_t + \alpha}, \tag{5.39}$$

and α is a small number close to the machine precision. The main advantage of NLMS is that it gives simpler design rules and stabilizes the algorithm in case of energy increases in φ_t. The choice $\mu = 0.01$ should always give a stable algorithm independent of model structure and parameter scalings. An interpretation of NLMS is that it uses the *a posteriori* residual in LMS:

$$\hat{\theta}_t = \hat{\theta}_{t-1} + \mu \varphi_t (y_t - \varphi_t^T \hat{\theta}_t). \tag{5.40}$$

This formulation is implicit, since the new parameter estimate is found on both sides. Other proposed variants of LMS include:

- The *leaky LMS* algorithm regularizes the solution towards zero, in order to avoid instability in case of poor excitation:

  $$\hat{\theta}_t = (1-\gamma)\hat{\theta}_{t-1} + \mu \varphi_t (y_t - \varphi_t^T \hat{\theta}_t). \tag{5.41}$$

  Here $0 < \gamma \ll 1$ forces unexcited modes to approach zero.

- The *sign-error algorithm* where the residual is replaced by its sign, $\text{sign}(\varepsilon_t)$. The idea is to choose the step-size as a power of two $\mu = 2^{-k}$, so that the multiplications in $2^{-k}\varphi\, \text{sign}(\varepsilon_t)$ can be implemented as data shifts. In a DSP application, only additions are needed to implement. An interesting interpretation is that this is the stochastic gradient algorithm for the loss function $V(\theta) = E|\varepsilon_t|$. This is a criterion that is more robust to outliers.

- The *sign data algorithm* where φ_t is replaced by $\text{sign}(\varphi_t)$ (component-wise sign) is another way to avoid multiplications. However, the gradient is now changed and convergence properties are influenced.

- The *sign-sign algorithm*:

  $$\hat{\theta}_t = \hat{\theta}_{t-1} + \mu\, \text{sign}(\varphi_t)\, \text{sign}(y_t - \varphi_t^T \hat{\theta}_t). \tag{5.42}$$

  This algorithm is extremely simple to implement in hardware, which makes it interesting in practical situations where speed and hardware resources are critical parameters. For example, it is a part in the CCITT standard for 32 kbps modulation scheme ADPCM (adaptive pulse code modulation).

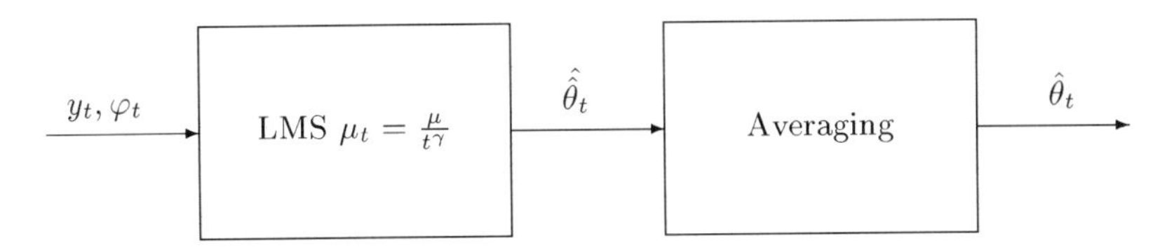

Figure 5.12. Averaging of a stochastic gradient algorithm implies asymptotically the same covariance matrix as for LS.

- Variable step size algorithms. Choices based on $\mu(t) = \mu/t$ are logical approximations of the LS solution. In the case of time-invariant parameters, the LMS estimate will then converge. This type of algorithm is sometimes referred to as a *stochastic gradient algorithm*.
- Filtered regressors are often used in noise cancelation applications.
- Many modifications of the basic algorithm have been suggested to get computational efficiency (Chen et al., 1999).

There are some interesting recent contributions to stochastic gradient algorithms (Kushner and Yin, 1997). One is based on averaging theory. First, choose the step size of LMS as

$$\mu_t = \frac{\mu}{t^\gamma}, \quad 0.5 < \gamma < 1.$$

The step size decays slower than for a stochastic gradient algorithm, where $\gamma = 1$. Denote the output of the LMS filter $\hat{\hat{\theta}}_t$. Secondly, the output vector is averaged,

$$\hat{\theta}_t = \frac{1}{t} \sum_{k=1}^{t} \hat{\hat{\theta}}_k.$$

The series of linear filters is illustrated in Figure 5.12. It has been shown (Kushner and Yang, 1995; Polyak and Juditsky, 1992) that this procedure is asymptotically efficient, in that the covariance matrix will approach that of the LS estimate as t goes to infinity. The advantage is that the complexity is $\mathcal{O}(dt)$ rather than $\mathcal{O}(d^2 t)$. An application of a similar idea of series connection of two linear filters is given in Wigren (1998). Tracking properties of such algorithms are examined in Ljung (1994).

One approach to *self-tuning* is to update the step-size of LMS. The result is two cross-coupled LMS algorithms relying on a kind of certainty equivalence; each algorithm relies on the fact that the other one is working. The gradient of the mean least square loss function with respect to μ is straightforward. Instead, the main problem is to compute a certain gradient which has to be done numerically. This algorithm is analyzed in Kushner and Yang (1995), and it is shown that the estimates of θ and μ converge weakly to a local minimum of the loss function.

5.4.2. RLS

The *Recursive Least Squares* (*RLS*) algorithm minimizes the criteria

$$V(\theta) = \sum_{k=1}^{t} \lambda^{t-k} (y_k - \hat{y}_k)^2 \tag{5.43}$$

as an approximation to (5.29). The derivation of the RLS algorithm below is straightforward, and similar to the one in Appendix 5.B.1.

Algorithm 5.3 RLS

For general linear regression models $y_t = \varphi_t^T \theta_t + e_t$, the RLS algorithm updates the parameter estimate by the recursion

$$\hat{\theta}_t = \hat{\theta}_{t-1} + K_t (y_t - \varphi_t^T \hat{\theta}_{t-1}) \tag{5.44}$$

$$K_t = \frac{P_{t-1}\varphi_t}{\lambda + \varphi_t^T P_{t-1} \varphi_t} \tag{5.45}$$

$$P_t = \frac{1}{\lambda}\left(P_{t-1} - \frac{P_{t-1}\varphi_t \varphi_t^T P_{t-1}}{\lambda + \varphi_t^T P_{t-1}\varphi_t}\right), \tag{5.46}$$

The design parameter λ (usually in $[0.9, 0.999]$) is called the *forgetting factor.* The matrix P_t is related to the covariance matrix, but $P_t \neq \operatorname{Cov} \hat{\theta}$.

The intuitive understanding of the size of the forgetting factor might be facilitated by the fact that the least squares estimate using a batch of N data can be shown to give approximately the same covariance matrix $\operatorname{Cov}\hat{\theta}$ as RLS if

$$N = \frac{1+\lambda}{1-\lambda} \approx = \frac{2}{1-\lambda}.$$

This can be proven by directly studying the loss function. Compare with the windowed least squares approach in Lemma 5.1.

Example 5.13 Adaptive filtering with RLS

Consider the same example as in Example 5.12. RLS with forgetting factor $\lambda = 0.99$ is applied to 1000 data and the parameter estimates are averaged over

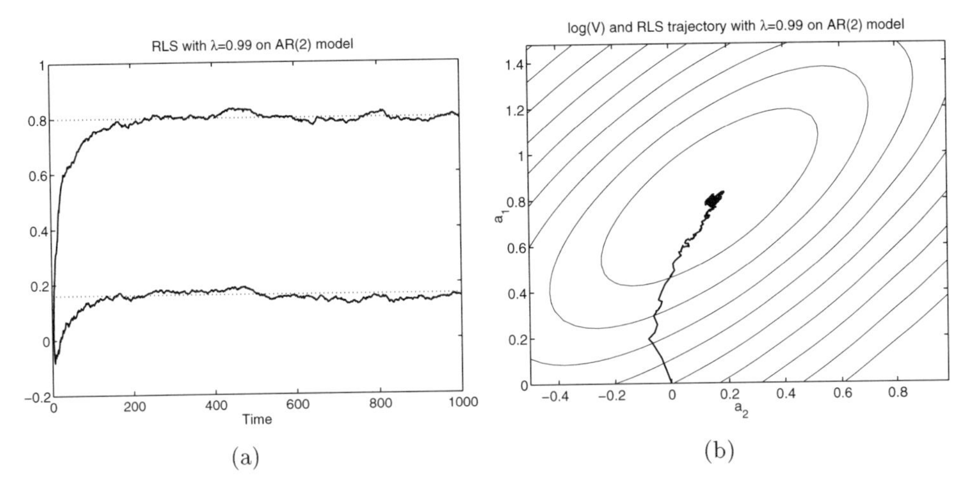

Figure 5.13. Convergence of RLS for an AR(2) model averaged over 25 Monte Carlo simulations and illustrated as a time plot (a) and in the loss function (b).

25 Monte Carlo simulations. Figure 5.13(a) shows the parameter convergence as a function of time, and Figure 5.13(b) convergence in the level curves of the loss function. To slow down the transient, the initial P_0 is chosen to $0.1I_2$. With a larger value of P_0, we will get convergence in the mean after essentially two samples. It can be noted that a very large value, say $P_0 = 100I_2$, essentially gives us NLMS (just simplify (5.45)) for a few recursions, until P_t becomes small.

Compared to LMS, RLS gives parameter convergence in the parameter plane as a straighter line rather than a steepest descent curve. The reason for not being completely straight is the incorrect initialization P_0.

As for LMS, there will be practical problems when the signals are not exciting. The covariance matrix will become almost singular and the parameter estimate may diverge. The solution is *regularization*, where the inverse covariance matrix is increased by a small scalar times the unit matrix:

$$R_t^{\varphi} = P_t^{-1} = \sum_{k=-\infty}^{t} \lambda^{t-k} \varphi_k \varphi_k^T + \delta I. \tag{5.47}$$

Note that this R_t^{φ} is not the same as the measurement covariance. Another problem is due to energy changes in the regressor. Speech signals modeled as AR models have this behavior. When the energy decreases in silent periods, it takes a long time for the matrix R_t^{φ} in (5.47) to adapt. One solution is to use the WLS estimator below.

In RLS, expectation in (5.30) is approximated by

$$\hat{\mathrm{E}}(x_t) = (1-\lambda)\sum_{k=-\infty}^{t} \lambda^{t-k} x_k \tag{5.48}$$

Another idea is to use a finite window instead of an exponential one,

$$\hat{\mathrm{E}}(x_t) = \frac{1}{L}\sum_{k=t-L+1}^{t} x_k \tag{5.49}$$

This leads to *Windowed Least Squares* (*WLS*), which is derived in Section 5.B, see Lemma 5.1. Basically, WLS applies two updates for each new sample, so the complexity increases a factor of two. A memory of the last L measurements is another drawback.

Example 5.14 Time-frequency analysis

Adaptive filters can be used to analyze the frequency content of a signal as a function of time, in contrast to spectral analysis which is a batch method. Consider the chirp signal which is often used a benchmark example:

$$y_t = \sin\left(\frac{2\pi t^2}{N}\right), \quad t = 0, 1, 2 \ldots N.$$

Defining momentaneous frequency as $\omega = d\arg(y_t)/dt$, the Fourier transform of a small neighborhood of t is $4\pi t/N$. Due to aliasing, a sampled version of the signal with sample interval 1 will have a folded frequency response, with a maximum frequency of π. Thus, the theoretical transforms assuming continuous time and discrete time measurements, respectively, are shown in Figure 5.14. A non-parametric method based on FFT *spectral analysis* of data over a sliding window is shown in Figure 5.15(a). As the window size L increases, the frequency resolution increases at the cost of decreased time resolution. This wellknown trade-off is related to *Heisenberg's uncertainty*: the product of time and frequency resolution is constant.

The parametric alternative is to use an AR model, which has the capability of obtaining better frequency resolution. An AR(2) model is adaptively estimated with WLS and $L = 20$, and Figure 5.15(b) shows the result. The frequency resolution is in theory infinite. The practical limitation comes from the variance error in the parameter estimate. There is a time versus frequency resolution trade-off for this parametric method as well. The larger time window L, the better parameter estimate and thus frequency estimate.

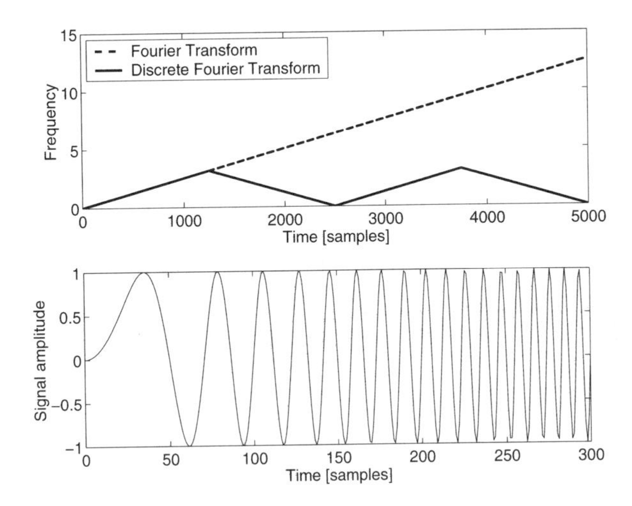

Figure 5.14. Time-frequency content (upper plot) of a chirp signal (lower plot).

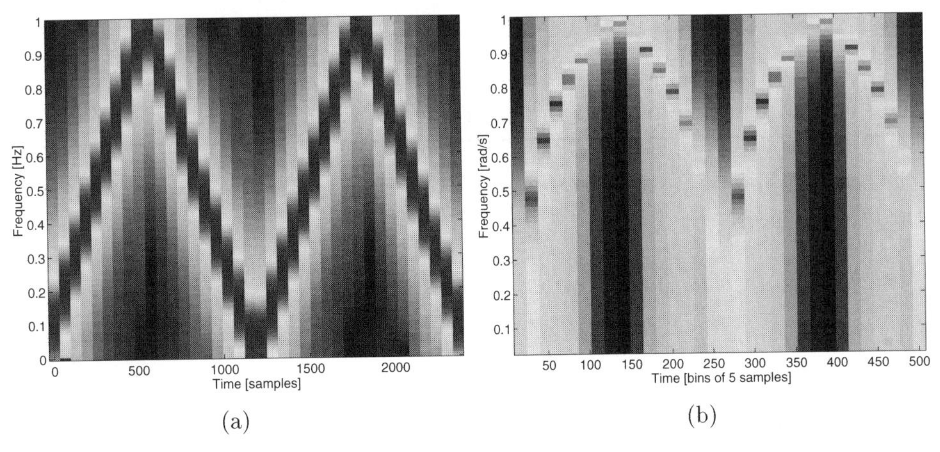

Figure 5.15. Time-frequency content of a chirp signal. Non-parametric (left) and parametric (right) methods, where the latter uses an AR(2) model and WLS with $L = 20$.

The larger AR model, the more frequency components can be estimated. That is, the model order is another critical design parameter. We also know that the more parameters, the larger uncertainty in the estimate, and this implies another trade-off.

5.4.3. Kalman filter

If the linear regression model is interpreted as the measurement equation in a state space model,

$$\theta_{t+1} = \theta_t + v_t, \quad \mathrm{Cov}(v_t) = Q_t \quad y_t = \varphi_t^T \theta_t + e_t, \quad \mathrm{Cov}(e_t) = R_t, \tag{5.50}$$

then the Kalman filter (see Chapters 13 and 8) applies.

Algorithm 5.4 Kalman filter for linear regressions

For general linear regression models $y_t = \varphi_t^T \theta_t + e_t$, the Kalman filter updates the parameter estimate by the recursion

$$\hat{\theta}_t = \hat{\theta}_{t-1} + K_t (y_t - \varphi_t^T \hat{\theta}_{t-1}) \tag{5.51}$$

$$K_t = \frac{P_{t-1}\varphi_t}{R_t + \varphi_t^T P_{t-1} \varphi_t} \tag{5.52}$$

$$P_t = P_{t-1} - \frac{P_{t-1}\varphi_t \varphi_t^T P_{t-1}}{R_t + \varphi_t^T P_{t-1} \varphi_t} + Q_t \tag{5.53}$$

The design parameters are Q_t and R_t. Without loss of generality, R_t can be taken as 1 in the case of scalar measurement.

There are different possibilities for how to interpret the physics behind the state model noise v_t:

- A *random walk* model, where v_t is white noise.
- *Abrupt changes*, where

$$v_t = \begin{cases} 0 & \text{with probability } 1-q \\ \nu & \text{with probability } q, \text{ where } \mathrm{Cov}(\nu) = \frac{1}{q} Q_t. \end{cases}$$

- *Hidden Markov models*, where θ switches between a finite number of values. Here one has to specify a transition matrix, with probabilities for going from one vector to another. An example is speech recognition, where each phoneme has its own *a priori* known parameter vector, and the transition matrix can be constructed by studying the language to see how frequent different transitions are.

These three assumptions are all, in a way, equivalent up to second order statistics, since $\mathrm{Cov}(v_t) = Q_t$ in all cases. The Kalman filter is the best possible conditional linear filter, but there might be better non-linear algorithms for the cases of abrupt changes and hidden Markov models. See Chapters 6 and 7 for details and Section 5.10 for examples.

In cases where physical knowledge is available about the time variations of the parameters, this can be included in a *multi-step algorithm*. For instance, it may be known that certain parameters have local or global trends, or abrupt changes lead to drifts etc. This can be handled by including this knowledge in the state model. The Kalman filter then gives a so called multi-step algorithm. See Section 8.2.4 and Examples 8.5 and 8.6.

5.4.4. Connections and optimal simulation

An interesting question is whether, for each adaptive filter, there does exist a signal for which the filter is optimal. The answer is yes for all linear filters, and this is most easily realized by interpreting the algorithms as special cases of the Kalman filter (Ljung and Gunnarsson, 1990).

The RLS algorithm can be written in a state space form with

$$\begin{aligned} Q_t &= \left(\frac{1}{\lambda} - 1\right)\left(P_{t-1} - \frac{P_{t-1}\varphi_t\varphi_t^T P_{t-1}}{\lambda + \varphi_t^T P_{t-1}\varphi_t}\right) \\ R_t &= \lambda, \end{aligned}$$

and NLMS corresponds to (let $\alpha = 1/\mu$ in (5.39))

$$\begin{aligned} Q_t &= \mu^2 \frac{\varphi_t\varphi_t^T}{1 + \mu\varphi_t^T\varphi_t} \\ R_t &= 1 \\ P_0 &= \mu I. \end{aligned}$$

The interpretation of these formulas are:

- Both RLS and NLMS can be seen as Kalman filters with particular assumptions on the random walk parameters. The results can be generalized so that all linear adaptive filters can be interpreted as Kalman filters. The relationship can be used to derive new algorithms lying in between RLS and KF (Haykin et al., 1995). This property is used in Gustafsson et al. (1997), where Q_t is designed to mimic the wavelets, and faster tracking of parameters affecting high frequencies is achieved.

- For each linear adaptive filter, the formulas define a state space model that can be simulated to generate signals for which it is impossible to outperform that particular filter.

The latter interpretation makes it possible to perform an *optimal simulation* for each given linear filter.

5.5. Performance analysis

The error sources for filters in general, and adaptive filters in particular, are:

- *Transient error* caused by incorrect initialization of the algorithms. For LMS and NLMS it depends on the initial parameter value, and for RLS and KF it also depends on the initial covariance matrix P_0. By making this very large, the transient can be reduced to a few samples.
- *Variance error*, caused by noise and disturbances. In simulation studies, this term can be reduced by Monte Carlo simulations.
- *Tracking errors* due to parameter changes.
- *Bias error* caused by a model that is not rich enough to describe the true signal. Generally, we will denote θ^* for the best possible parameter value within the considered model structure, and θ^0 the true parameters when available.

A standard design consists in the following steps:

Design of adaptive filters

1. Model structure selection from off-line experiments (for instance using BIC type criteria) or prior knowledge to reduce the bias error.
2. Include prior knowledge of the initial values θ_0 and P_0 or decide what a sufficiently large P_0 is from knowledge of typical parameter sizes to minimize the transient error.
3. Tune the filter to trade-off the compromise between tracking and variance errors.
4. Compare different algorithms with respect to performance, real-time requirements and implementational complexity.

We first define a formal performance measure. Let

$$V(\theta) = E[(y_t - \varphi_t^T \theta_t)^2] = V_{min} + V_{ex}, \tag{5.54}$$

where

$$V_{min} = V(\theta^0) = \mathrm{E}(e_t^2) = R, \tag{5.55}$$

assuming that the true system belongs to the model class. V_{ex} is the *excessive mean square error*. Define the *misadjustment*:

$$M(t) = \frac{V_{ex}}{V_{min}} \to \bar{M}, \quad \text{as } t \to \infty. \tag{5.56}$$

The assumptions used, for example, in Widrow and Stearns (1985), Bellanger (1988), Gunnarsson and Ljung (1989) and Gunnarsson (1991) are the following:

θ^0 exists	no bias error
$Z = E[\varphi_t \varphi_t^T]$	φ_t quasi-stationary process
$Q = E v_t v_t^T$ with $v_t = \theta_t^0 - \theta_{t-1}^0$	parameter changes quasi-stationary process
$e_t, \ v_t, \ \varphi_t$	independent white processes

That is, the true system can be exactly described as the modeled linear regression, the regressors are quasi-stationary and the parameter variations are a random walk. We will study the parameter estimation error:

$$\tilde{\theta}_t = \theta_t^0 - \hat{\theta}_t$$

5.5.1. LMS

The parameter error for LMS is

$$\tilde{\theta}_t = (I - \mu \varphi_t \varphi_t^T)(\tilde{\theta}_{t-1} + v_{t-1}) - \mu \varphi_t e_t.$$

Transient and stability for LMS

As a simple analysis of the transient, take the SVD of $Z = E[\varphi_t \varphi_t^T] = UDU^T$ and assume time-invariant true parameters $\theta_t^0 = \theta^0$. The matrix D is diagonal and contains the singular values σ_i of Z in descending order, and U satisfies $U^T U = I$. Let us also assume that $\hat{\theta}_0 = 0$:

$$\begin{aligned}
E\tilde{\theta}_t &= E(I - \mu \varphi_t \varphi_t^T) E\tilde{\theta}_{t-1} \\
&= (I - \mu Z) E\tilde{\theta}_{t-1} \\
&= (I - \mu Z)^t \theta^0 \\
&= U(I - \mu D)^t U^T \theta^0 \\
&= U \begin{pmatrix} (1-\mu\sigma_1)^t & 0 & 0 \\ 0 & \ddots & 0 \\ 0 & 0 & (1-\mu\sigma_n)^t \end{pmatrix} U^T \theta^0.
\end{aligned}$$

That is, LMS is stable only if

$$\mu < 2/\sigma_1.$$

More formally, the analysis shows that we get convergence in the mean if and only if $\mu < 2/\sigma_1$.

We note that the transient decays as $(1-\mu\sigma_n)^t$. If we choose $\mu = 1/\sigma_1$, so the first component converges directly, then the convergence rate is

$$(1-\frac{\sigma_n}{\sigma_1})^t = (1-\text{cond}(Z))^t.$$

That is, the possible convergence rate depends on the condition number of the matrix Z. If possible, the signals in the regressors should be pre-whitened to give $\text{cond}(Z) = 1$.

A practically computable bound on μ can be found by the following relations:

$$\sigma_1 \geq \frac{1}{n}\sum_{i=1}^{n}\sigma_i = \frac{1}{n}\,\text{tr}(Z) = \frac{1}{n}\,\text{E}(\varphi_t^T\varphi_t)$$

$$\Rightarrow \mu < \frac{2}{\sigma_1} \leq \frac{2n}{\text{E}(\varphi_t^T\varphi_t)}.$$

Here the expectation is simple to approximate in an off-line study, or to make it adaptive by exponential forgetting. Note the similarity to NLMS. It should be mentioned that μ in practice should be chosen to be about 100 times smaller than this value, to ensure stability.

Misadjustment for LMS

The stationary misadjustment for LMS can be shown to equal:

$$\bar{M} = \underbrace{\frac{\mu\,\text{tr}(Z)}{2}}_{\text{variance error}} + \underbrace{\frac{\text{tr}(Q)}{2\mu R}}_{\text{tracking error}}. \tag{5.57}$$

- The stationary misadjustment $\bar{M}$ splits into variance and tracking parts. The variance error is proportional to the adaptation gain μ while the tracking error is inversely proportional to the gain.
- The tracking error is proportional to the signal to noise ratio $\frac{\|Q\|}{R}$.
- The optimal step size is

$$\mu_{opt} = \sqrt{\frac{\text{tr}(Q)}{\text{tr}(Z)R}}.$$

5.5.2. RLS

The dynamics for the RLS parameter error is

$$\tilde{\theta}_t = (I - (R_t^{\varphi})^{-1}\varphi_t\varphi_t^T)(\tilde{\theta}_{t-1} + v_{t-1}) - \mu\varphi_t e_t.$$

Misadjustment and transient for RLS

As for LMS, the stationary misadjustment M splits into variance and tracking parts. The transient can be expressed in misadjustment as a function of time, $M(t)$ for constant θ_t^0. The following results can be shown to hold (see the references in the beginning of the section):

$$M(t) \approx \underbrace{\frac{n}{t+1}}_{\text{transient error}} + \overbrace{\underbrace{\frac{n(1-\lambda)}{2}}_{\text{variance error}} + \underbrace{\frac{\operatorname{tr}(ZQ)}{2(1-\lambda)R}}_{\text{tracking error}}}^{\bar{M}}. \tag{5.58}$$

- Transient and variance errors are proportional to the number of parameters n.
- As for LMS, the stationary misadjustment $\bar{M}$ splits into variance and tracking parts. Again, as for LMS, the variance error is proportional to the adaptation gain $1-\lambda$, while the tracking error is inversely proportional to the gain.
- As for LMS, the tracking error is proportional to the signal to noise ratio $\frac{\|Q\|}{R}$.
- By minimizing (5.58) w.r.t. λ, the optimal step size $1-\lambda$ is found to be

$$(1 - \lambda_{opt}) = \sqrt{\frac{\operatorname{tr}(ZQ)}{nR}}.$$

A refined analysis of the transient term in both RLS and LMS (with variants) is given in Eweda (1999).

5.5.3. Algorithm optimization

Note that the stationary expressions make it possible to optimize the design parameter to get the best possible trade-off between tracking and variance errors, as a function of the true time variability and covariances. For instance, we might ask which algorithm performs best for a certain Q and Z, in terms of excessive mean square error. Optimization of step size μ and forgetting factor λ in the expression for $\bar{M}$ in NLMS and RLS gives

$$\frac{\bar{M}^*_{NLMS}}{\bar{M}^*_{RLS}} = \sqrt{\frac{\mathrm{tr}Z\,\mathrm{tr}Q}{n\mathrm{tr}ZQ}}.$$

The trace operator can be rewritten as the sum of eigenvalues, $\mathrm{tr}(Q) = \sum \sigma_i(Q)$. If we put $Z = Q$, we get

$$Q = Z \Rightarrow \frac{\bar{M}^*_{NLMS}}{\bar{M}^*_{RLS}} = \sqrt{\frac{(\mathrm{tr}\, Z)^2}{n\, \mathrm{tr}\, Z^2}} = \sqrt{\frac{(\sum \sigma_i)^2}{n \sum \sigma_i^2}} \leq 1,$$

with equality only if $\sigma_i = \sigma_j$ for all i, j. As another example, take $Q = Z^{-1}$,

$$Q = Z^{-1} \Rightarrow \frac{\bar{M}^*_{NLMS}}{\bar{M}^*_{RLS}} = \sqrt{\frac{(\mathrm{tr}\, Z)(\mathrm{tr}\, Z^{-1})}{n\, \mathrm{tr}\, I_n}} = \sqrt{\frac{(\sum \sigma_i)(\sum \sigma_i^{-1})}{n^2}} \geq 1,$$

with equality only if $\sigma_i = \sigma_j$ for all i, j. That is, if $Z = Q$ then NLMS performs best and if $Z = Q^{-1}$, then RLS is better and we have by examples proven that no algorithm is generally better than the other one, see also Eleftheriou and Falconer (1986) and Benveniste et al. (1987b).

5.6. Whiteness based change detection

The basic idea is to feed the residuals from the adaptive filter to a change detector, and use its alarm as feedback information to the adaptive filter, see Figure 5.16. Here the detector is any scalar alarm device from Chapter 3 using a transformation $s_t = f(\varepsilon_t)$ and a stopping rule from Section 3.4. There are a few alternatives of how to compute a test statistic s_t, which is zero mean when there is no change, and non-zero mean after the change. First, note that if all noises are Gaussian, and if the true system is time-invariant and belongs to the modeled linear regression, then

$$\varepsilon_t \in \mathrm{N}(0, S_t), \quad S_t = R_t + \varphi_t^T P_t \varphi_t. \tag{5.59}$$

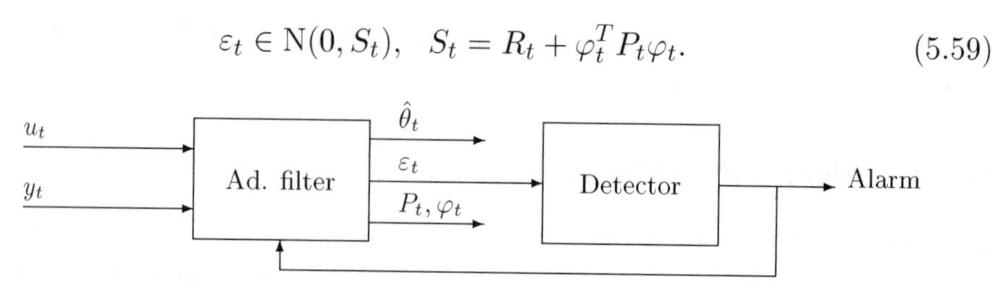

Figure 5.16. Change detection as a whiteness residual test, using e.g. the CUSUM test, for an arbitrary adaptive filter, where the alarm feedback controls the adaptation gain.

- The normalized residual

$$s_t = S_t^{-1/2} \varepsilon_t \in \mathrm{N}(0, I), \tag{5.60}$$

  is then a suitable candidate for change detection.

- The main alternative is to use the squared residual

$$s_t = \varepsilon_t^T S_t^{-1} \varepsilon_t \in \chi^2(n_y). \tag{5.61}$$

- Another idea is to check if there is a systematic drift in the parameter updates:

$$s_t = K_t \varepsilon_t. \tag{5.62}$$

  Here the test statistic is vector valued.

- Parallel update steps $\Delta\theta_t = K_t\varepsilon_t$ for the parameters in an adaptive algorithm is an indication of that a systematic drift has started. It is proposed in Hägglund (1983) to use

$$s_t = (\Delta\theta_t)^T \Delta\theta_{t-1} = \varepsilon_{t-1}^T K_{t-1}^T K_t \varepsilon_t \tag{5.63}$$

  as the residual. A certain filtering approach of the updates was also proposed.

The first approach is the original CUSUM test in Page (1954). The second one is usually labeled as just the χ^2 test, since the test statistic is χ^2 distributed under certain assumptions, while a variant of the third approach is called the *asymptotic local approach* in Benveniste et al. (1987a).

After an alarm, the basic action is to increase the gain in the filter momentarily. For LMS and KF, we can use a scalar factor α and set $\bar{\mu}_t = \alpha\mu_t$ and $\bar{Q}_t = \alpha Q_t$, respectively. For RLS, we can use a small forgetting factor, for instance $\lambda_t = 0$.

Applications of this idea are presented in Section 5.10; see also, for instance, Medvedev (1996).

5.7. A simulation example

The signal in this section will be a first order AR model,

$$y_t = -a_t y_{t-1} + e_t.$$

The noise variance is $Ee^2 = 1$ and $N = 200$ data are simulated. The AR parameter will be either constant, piecewise constant or slowly time-varying.

The parameter is estimated by LS, RLS, LMS and LS with a change detector, respectively. The latter will be refered to as Detection LS, and the detector is the two-sided CUSUM test with the residuals as input. For each method and design parameter, the loss function and code length are evaluated on all but the 20 first data samples, to avoid possible influence of transients and initialization. The RLS and LMS algorithms are standard, and the design parameters are the forgetting factor λ and step size μ, respectively.

5.7.1. Time-invariant AR model

Consider first the case of a constant AR parameter $a = -0.5$. Figure 5.17 shows *MDL*, as described in Section 12.3.1, and the loss function

$$V = \frac{1}{N}\sum_{t=1}^{N} \varepsilon_t^2$$

as a function of the design parameter and the parameter tracking, where the true parameter value is indicated by a dotted line. Table 5.1 summarizes the optimal design parameters and code lengths according to the MDL measure for this particular example.

Note that the optimal design parameter in RLS corresponds to the LS solution and that the step size of LMS is very small (compared to the ones to follow). All methods have approximately the same code length, which is logical.

5.7.2. Abruptly changing AR model

Consider the piecewise constant AR parameter

$$a_t = \begin{cases} -0.5 & \text{if } t \leq 100 \\ 0.5 & \text{if } t > 100 \end{cases}$$

Table 5.1. Optimal code length and design parameters for RLS, LMS and whiteness detection LS, respectively, for constant parameters in simulation model. For comparison, the LS result is shown.

Method	Optimal par.	MDL	V
LS	—	1.11	1.08
RLS	$\lambda = 1$	1.12	1.11
LMS	$\mu = 0.002$	1.13	1.13
Detection LS	$\nu = 1.1$	1.11	1.08

Table 5.2. Optimal code length and design parameters for RLS, LMS and whiteness detection LS, respectively, for abruptly changing parameters in simulation model. For comparison, the LS result is shown.

Method	Optimal par.	MDL	V
LS	—	1.40	1.37
RLS	$\lambda = 0.9$	1.07	1.06
LMS	$\mu = 0.07$	1.05	1.05
Detection LS	$\nu = 0.16$	1.25	0.93

Figure 5.18 shows MDL and the loss function V as a function of the design parameter and the parameter tracking, where the true parameter value is indicated by a dotted line. Table 5.2 summarizes the optimal design parameters and code lengths for this particular example. Clearly, an adaptive algorithm is here much better than a fixed estimate. The updates $\Delta\theta_t$ are of much smaller magnitude than the residuals. That is, for coding purposes it is more efficient to transmit the small parameter updates then the much larger residuals for a given numerical accuracy. This is exactly what MDL measures.

5.7.3. Time-varying AR model

The simulation setup is exactly as before, but the parameter vector is linearly changing from -0.5 to 0.5 over 100 samples. Figure 5.19 and Table 5.3 summarize the result. As before, the difference between the adaptive algorithms is insignificant and there is no clear winner. The choice of adaptation mechanism is arbitrary for this signal.

Table 5.3. Optimal code length and design parameters for RLS, LMS and whiteness detection LS, respectively, for slowly varying parameters in simulation model. For comparison, the LS result is shown.

Method	Optimal par.	MDL	V
LS	—	1.39	1.36
RLS	$\lambda = 0.92$	1.15	1.15
LMS	$\mu = 0.05$	1.16	1.16
Detection LS	$\nu = 0.46$	1.24	1.13

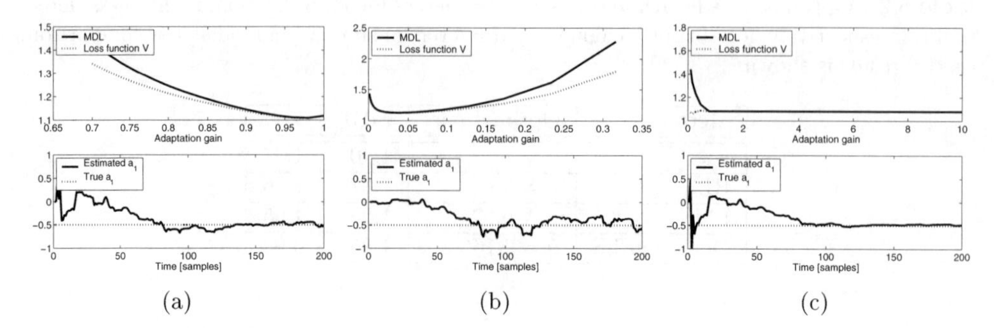

Figure 5.17. MDL and V as a function of design parameter and parameter tracking for RLS (a), LMS (b) and detection LS (c), respectively, for constant parameters in simulation model.

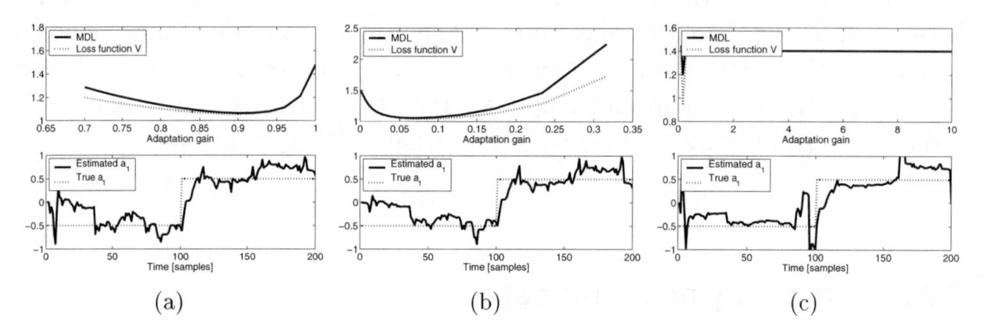

Figure 5.18. MDL and V as a function of design parameter and parameter tracking for RLS (a), LMS (b) and detection LS (c), respectively, for abruptly changing parameters in simulation model.

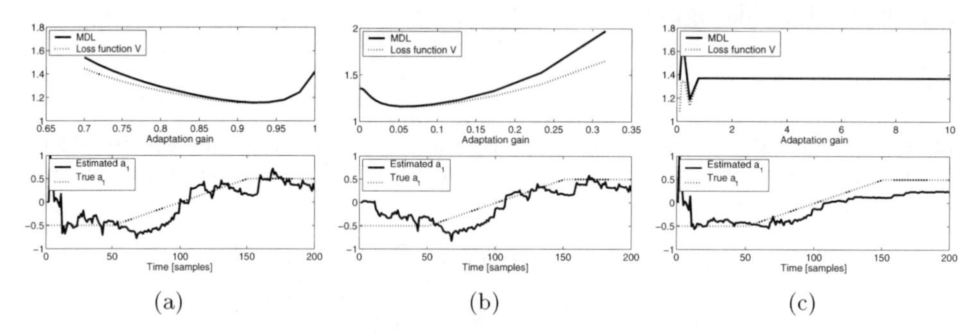

Figure 5.19. MDL and V as a function of design parameter and parameter tracking for RLS (a), LMS (b) and detection LS (c), respectively, for slowly varying parameters.

5.8. Adaptive filters in communication

Figure 5.20 illustrates the equalization problem. The transmitted signal u_t is distorted in a channel and the receiver measures its output y_t with additive noise e_t. The distortion implies that the output may be far away from the transmitted signal, and is thus not useful directly. The received output is passed to a filter (*equalizer*), with the objective to essentially invert the channel to recover u_t.

In general, it is hard to implement the inversion without accepting a delay D of the overall system. For example, since the channel might encompass time delays, a delayless inversion would involve predictions of u_t. Furthermore, the performance may be improved if accepting additional time delay. The delay D can be determined beforehand based on expected time delays and filter order of the channel. An alternate approach is to include the parameter in the channel inversion optimization.

The input signal belongs to a finite alphabet in digital communication. For the discussion and most of the examples, we will assume a binary input (in modulation theory called *BPSK*, *Binary Phase Shift Keying*),

$$u_t = \pm 1.$$

In this case, it is natural to estimate the transmitted signal with the sign of the received signal, $\hat{u}_{t-D} = \text{sign}(y_t)$.

If the time constant of the channel dynamics is longer than the length of one symbol period, the symbols will reach the receiver with overlap. The phenomenon is called *Inter-Symbol Interference* (*ISI*), and it implies that $\hat{u}_{t-D} = \text{sign}(y_t) \neq u_{t-D}$. If incorrect decisions occur too frequently, the need for *equalization* arises.

Example 5.15 Inter-symbol interference

As a standard example in this section, the following channel will be used:

$$B(q) = 1 - 2.2q^{-1} + 1.17q^{-2}$$
$$u_t = \pm 1.$$

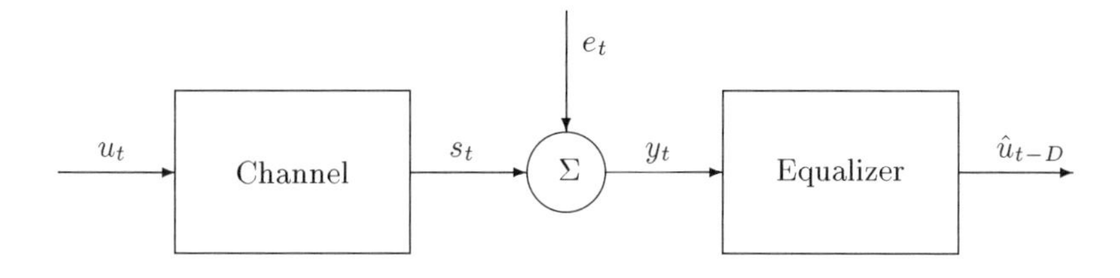

Figure 5.20. The equalization principle.

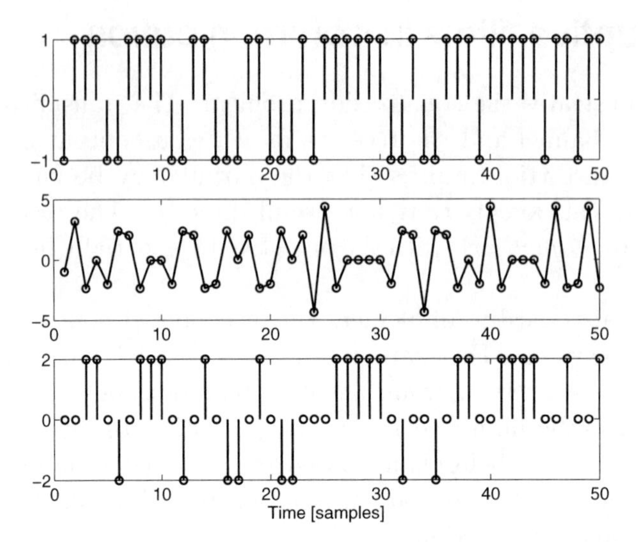

Figure 5.21. The transmitted signal (first plot) is distorted by the channel and the received signal (second plot) does not always have the correct sign, implying decision errors $\hat{u}_t - \text{sign}(y_t)$ (third plot).

$N = 50$ binary inputs are simulated according to the first plot in Figure 5.21. The output y_t after the channel dynamics is shown in the second plot. If a decision is taken directly $\hat{u}_t = \text{sign}(y_t)$, without equalization, a number of errors will occur according to the last plot which shows $u_t - \text{sign}(y_t)$. Similar results are obtained when plotting $u_{t-D} - \text{sign}(y_t)$ for $D = 1$ and $D = 2$.

Example 5.16 ISI for 16-QAM

The ISI is best illustrated when complex modulation is used. Consider, for instance, 16-QAM (Quadrature Amplitude Modulation), where the transmitted symbol can take on one of 16 different values, represented by four equidistant levels of both the real and imaginary part. The received signal distribution can then look like the left plot in Figure 5.22. After successful equalization, 16 well distinguished clusters appear, as illustrated in the right plot of Figure 5.22. One interpretation of the so called *open-eye condition*, is that it is satisfied when all clusters are well separated.

For evaluation, the *Bit Error Rate* (*BER*) is commonly used, and it is defined as

$$\text{BER} = \frac{\text{number of non-zero } (u_t - \hat{u}_t)}{N} \tag{5.64}$$

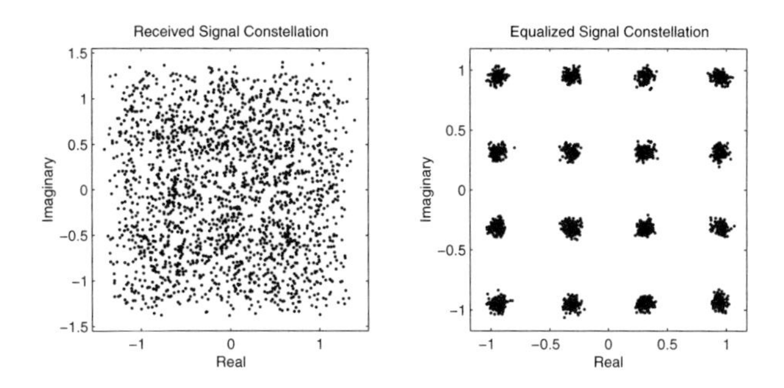

Figure 5.22. Signal constellation for 16-QAM for the recieved signal (left) and equalized signal (right).

where trivial phase shifts (sign change) of the estimate should be discarded. It is naturally extended to *Symbol Error Rate* (*SER*) for general input alphabets (not only binary signals).

Another important parameter for algorithm evaluation is the *Signal-to-Noise Ratio* (*SNR*):

$$\mathrm{SNR} = \frac{\mathrm{E}(s_t^2)}{\mathrm{E}(e_t^2)}. \tag{5.65}$$

A typical evaluation shows BER as a function of SNR. For high SNR one should have BER = 0 for an algorithm to be meaningful, and conversely BER will always approach 0.5 when SNR decreases.

Now, when the most fundamental quantities are defined, we will survey the most common approaches. The algorithms belong to one of two classes:

- Equalizers that assume a known model for the channel. The equalizer is either a linear filter or two filters, where one is in a feedback loop, or a filter bank with some discrete logics.
- Blind equalizers that simultaneous estimate the transmitted signal and the channel parameters, which may even be time-varying.

5.8.1. Linear equalization

Consider the *linear equalizer* structure in Figure 5.23. The linear filter tries to invert the channel dynamics and the decision device is a static mapping, working according to the nearest neighbor principle.

The underlying aassumption is that the transmission protocol is such that a *training sequence*, known to the reciever, is transmitted regularly. This sequence is used to estimate the inverse channel dynamics $C(q)$ according to the least squares principle. The dominating model structure for both channel

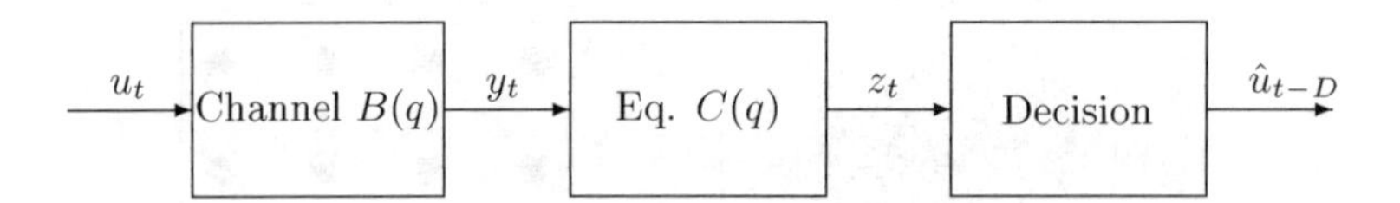

Figure 5.23. A linear equalizer consists of a linear filter $C(q)$ followed by a decision device ($\hat{u}_{t-D} = \text{sign}(z_t)$ for a binary signal). The equalizer is computed from knowledge of a training sequence of u_t.

and linear equalizer is FIR filters. A few attempts exist for using other more flexible structures (Grohan and Marcos, 1996). The FIR model for the channel is motivated by physical reasons; the signal is subject to multi-path fading or echoes, which implies delayed and scaled versions of the signal at the receiver. The FIR model for equalizer structures, where the equalizer consists of a linear filter in series with the channel, is motivated by practical reasons. The FIR model of the channel is, most likely, non-minimum phase, so the natural AR model for inverting the channel would be unstable.

An equalizer of order n, C_n, is to be estimated from L training symbols u_t aiming at a total time delay of D. Introduce the loss function

$$V_L(C_n, D) = \frac{1}{L}\sum_{t=1}^{L}(u_{t-D} - C_n(q)y_t)^2.$$

The least squares estimate of the equalizer is now

$$\hat{C}_n(D) = \arg\min_{C_n} V_L(C_n, D).$$

The designer of the communication system has three degrees of freedom. The first, and most important, choice for performance and spectrum efficiency is the length of the training sequence, L. This has to be fixed at an early design phase when the protocol, and for commercial systems the standard, is decided upon. Then, the order n and delay D have to be chosen. This can be done by comparing the loss function in the three-dimensional discrete space L, n, D. All this has to be done together with the design of channel coding (error correcting codes that accept a certain amount of bit errors). The example below shows a smaller study.

Example 5.17 Linear equalization: structure

As a continuation of Example 5.15, let the channel be

$$\begin{aligned} B(q) &= 1 - 2.2q^{-1} + 1.17q^{-2} \\ L &= 25 \\ u_t &= \pm 1, \end{aligned}$$

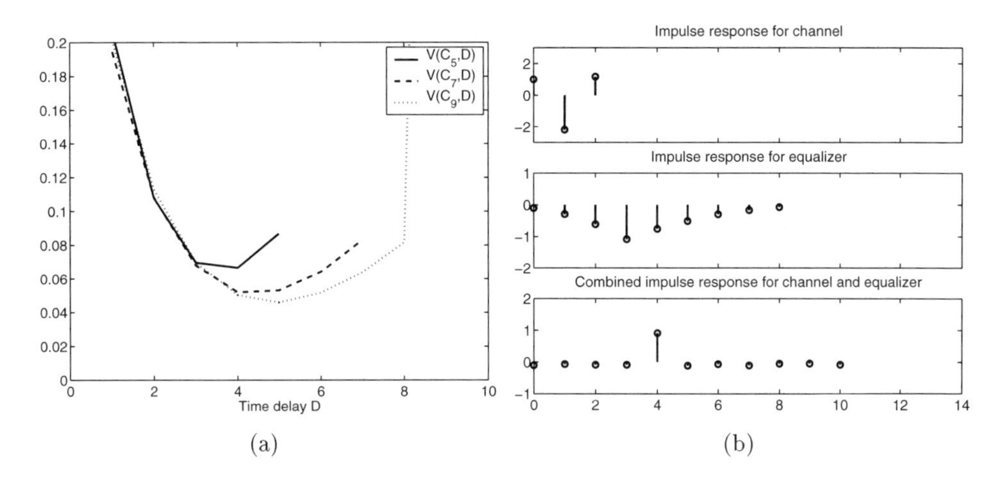

Figure 5.24. The loss function $V_{200}(\hat{C}_n, D)$ as a function of time delay $D = 1, 2, \ldots, n$ for $n = 5, 7, 9$, respectively (a). In (b), the impulse response of the channel, equalizer and combined channel equalizer, is shown, respectively.

where the transmitted signal is white. Compute the linear equalizer $C(q) = c_0 + c_1 q^{-1} + \cdots + c_n q^{-n}$ and its loss function $V_L(C_n, D)$ as a function of the delay D. Figure 5.24(a) shows the result for $n = 5, 7, 9$, respectively. For each n, there is a clear minimum for $D \approx n/2$.

The more taps n in the equalizer, the better result. The choise is a trade-off between complexity and performance. Since BER is a monotone function of the loss function, the larger n, the better performance. In this case, a reasonable compromize is $n = 7$ and $D = 4$.

Suppose now that the structure of the equalizer L, n, D is fixed. What is the performance under different circumstances?

Example 5.18 Linear equalization: SNR vs. BER

A standard plot when designing a communication system is to plot BER as a function of SNR.

For that reason, we add channel noise to the simulation setup in Example 5.17:

$$
\begin{aligned}
y_t &= B(q)u_t + e_t, \quad \mathrm{Var}(e_t) = \sigma^2 \\
B(q) &= 1 - 2.2q^{-1} + 1.17q^{-2} \\
N &= 200 \\
L &= 25 \\
u_t &= \pm 1
\end{aligned}
$$

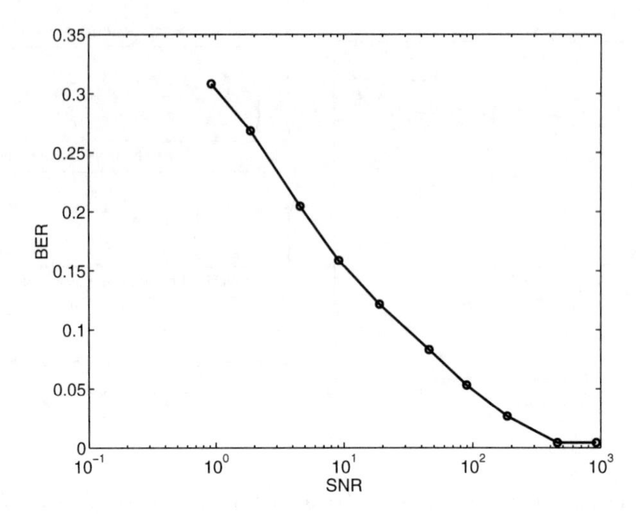

Figure 5.25. BER (Bit Error Rate) as a function of signal-to-noise ratio (SNR) for a linear equalizer.

SNR is now a function of the noise variance σ^2. Let the training sequence consists of the $L = 25$ first symbols in the total sequnce of length $N = 200$. Suppose we have chosen $n = 8$ and $D = 4$, and we use 100 Monte Carlo simulations to evaluate the performance. Figure 5.25 shows BER as a function of SNR under these premises.

If we can accept a 5% error rate, then we must design the transmission link so the SNR is larger than 100, which is quite a high value.

For future comparison, we note that this particular implementation uses $4 \cdot 10^4$ floating point operations, which include the design of the equalizer.

5.8.2. Decision feedback equalization

Figure 5.26 shows the structure of a *decision feedback equalizer*. The upper part is identical to a linear equalizer with a linear feed-forward filter, followed by a decision device. The difference lies in the feedback path from the non-linear *decisions*.

One fundamental problem with a linear equalizer of FIR type, is the many taps that are needed to approximate a zero close to the unit circle in the channel. For instance, to equalize $1 + 0.9q^{-1}$, the inverse filter $1 - 0.9q^{-1} + 0.9^2q^{-2} - 0.9^3q^{-3} \ldots$ is needed. With the extra degree of freedom we now have, these zeros can be put in the feedback path, where no inversion is needed. In theory, $D(q) = B(q)$ and $C(q) = 1$ would be a perfect equalizer. However, if the noise induces a *decision error*, then there might be a recovery problem

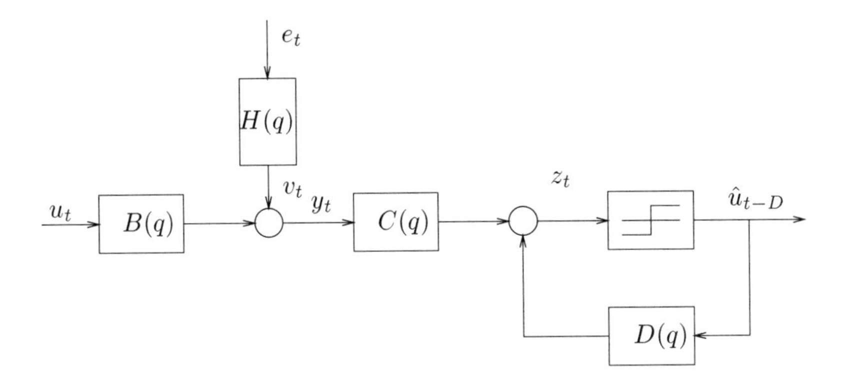

Figure 5.26. The decision feedback equalizer has one feedforward filter and a decision device (for binary input a sign function, or relay), exactly as the linear equalizer, and one feedback filter from past decisions.

for the DFE equalizer. There is the fundamental design trade-off: split the dynamics of the channel between $C(q)$ and $D(q)$ so few taps and robustness to decision errors are achieved.

In the design, we assume that the channel is known. In practice, it is estimated from a training sequence. To analyse and design a non-linear system is generally very difficult, and so is this problem. A simplifying assumption, that dominates the design described in literature, is the one of so called *Correct Past Decisions* (*CPD*). The assumption implies that we can take the input to the feedback filter from the true input and we get the block diagram in Figure 5.27. The assumption is sound when the SNR is high, so a DFE can only be assumed to work properly in such systems.

We have from Figure 5.27 that, if there are no decision errors, then

$$z_t = (C(q)B(q) + D(q))u_t + C(q)H(q)e_t.$$

For the CPD assumption to hold, the estimation errors must be small. That is, choose $C(q)$ and $D(q)$ to minimize

$$\begin{aligned}\tilde{u}_{t-D} &\triangleq u_{t-D} - z_t\\ &= (q^{-D} - C(q)B(q) - D(q))u_t - C(q)H(q)e_t.\end{aligned}$$

There are two principles described in the literature:

- The *zero forcing equalizer.* Neglect the noise in the design (as is quite common in equalization design) and choose $C(q)$ and $D(q)$ so that $q^{-D} - B(q)C(q) - D(q) = 0$.

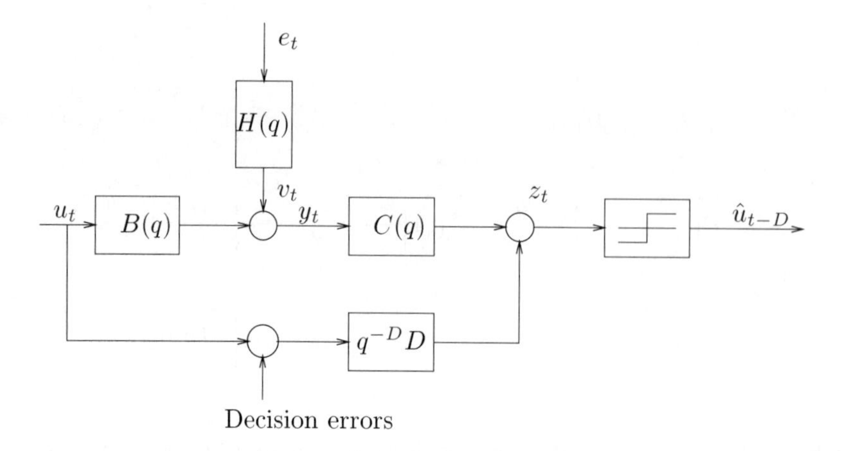

Figure 5.27. Equivalent DFE under the correct past decision (CPD) which implies that the input to the feedback filter can be taken from the real input.

- The *minimum variance equalizer*

$$
\begin{aligned}
(\hat{C}, \hat{D}) &= \arg \min_{C(q),D(q)} \mathrm{E}\, \tilde{u}_{t-D}^2 \\
&= \arg \min_{C(q),D(q)} \left| e^{-i\omega D} - C(e^{i\omega})B(e^{i\omega}) - D(e^{i\omega}) \right|^2 \Phi_u(e^{i\omega}) \\
&+ \left| C(e^{i\omega})H(e^{i\omega}) \right|^2 \Phi_e(e^{i\omega}).
\end{aligned}
$$

  Here we have used *Parseval's formula* and an independence assumption between u and e.

In both cases, a constraint of the type $c_0 = 1$ is needed to avoid the trivial minimum for $C(q) = 0$, in case the block diagram in Figure 5.27 does not hold.

The advantage of DFE is a possible considerable performance gain at the cost of an only slightly more complex algorithm, compared to a linear equalizer. Its applicability is limited to cases with high SNR.

As a final remark, the introduction of equalization in very fast modems, introduces a new kind of implementation problem. The basic reason is that the data rate comes close to the clock frequency in the computer, and the feedback path computations introduce a significant time delay in the feedback loop. This means that the DFE approach collapses, since no feedback delay can be accepted.

5.8.3. Equalization using the Viterbi algorithm

The structure of the *Viterbi equalizer* is depicted in Figure 5.28. The idea in Viterbi equalization is to enumerate all possible input sequences, and the

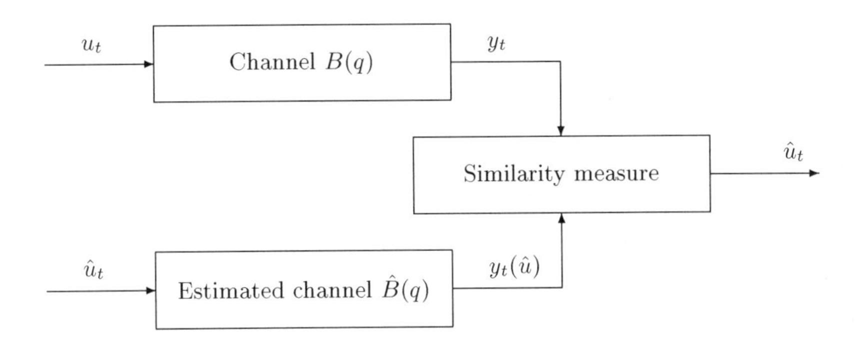

Figure 5.28. Equalization principle.

one that in a simulation using $\hat{B}(q)$ produces the output most similar to the measured output is taken as the estimate.

Technically, the similarity measure is the maximum likelihood criterion, assuming Gaussian noise. This is quite similar to taking the sum of squared residuals $\sum_t (y_t - y_t(\hat{u}))^2$ for each possible sequence $\hat{u}$. Luckily, not all sequences have to be considered. It turns out that only S^{n_b} sequences have to be examined, where S is the number of symbols in the finite alphabet, and n_b is the channel order.

There is an optimal search algorithm for a finite memory channel (where FIR is a special case), namely the *Viterbi algorithm*.

Algorithm 5.5 Viterbi

Consider a channel described by $y_t = f(u^t, e^t)$, where the input u_t belongs to a finite alphabet of size S and e_t is white noise. If the channel has a finite memory n_b, so $y_t = f(u_{t-n_b+1}^t, e^t)$, the search for the ML estimate of u^t can be restricted to S^{n_b} sequences. These sequences are an enumeration of all possible sequences in the interval $[t - n_b + 1, t]$. For the preceding inputs only the most likely one has to be considered.

Derivation: an induction proof of the algorithm is given here. From Bayes' rule we have

$$\begin{aligned} p(y^t|u^t) &= p(y_t|u^t, y^{t-1})p(y^{t-1}|u^t) \\ &= p(y_t|u_{t-n_b+1}^t, y^{t-1})p(y^{t-1}|u_{t-n_b+1}^{t-1}, u^{t-n_b}). \end{aligned}$$

In the second equality the finite memory property is used. Suppose that we have computed the most likely sequence at time $t-1$ for each sequence

$u_{t-n_b+1}^{t-1}$, so we have $p(y^{t-1}|u_{t-n_b+1}^{t-1}, u^{t-n_b})$ as a function of $u_{t-n_b+1}^{t-1}$. From the expression above, it follows that, independently of what value y_t has, the most likely sequence at time t can be found just by maximizing over $u_{t-n_b+1}^{t}$, since the maximization over u^{t-n_b} has already been done. $\square$

The proof above is non-standard. Standard references like Viterbi (1967), Forney (1973) and Haykin (1988) identify the problem with a *shortest route* problem in a *trellis* diagram and apply *forward dynamic programming*. The proof here is much more compact.

In other words, the Viterbi algorithm 'inverts' the channel model $\hat{B}(q)$ without applying $\hat{B}^{-1}(q)$, which is likely to be unstable. This works due to the finite alphabet of the input sequence.

1. Enumerate all possible input sequences within the sliding window of size n (the channel model length). Append these candidates to the estimated sequence $\hat{u}^{t-n}$.

2. Filter all candidate sequences with $\hat{B}(q)$.

3. Estimate the input sequence by the best possible candidate sequence. That is,

$$\hat{u} = \arg \min_{u_1, u_2, \ldots, u_N} \sum_{t=1}^{N} (\hat{B}(q)u_t - y_t)^2.$$

Example 5.19 BER vs. SNR evaluation

Consider the BER vs. SNR plot in Example 5.18, obtained using a training sequence $L = 25$. Exactly the same simulation setup is used here as for the Viterbi algorithm.

Figure 5.29(a) shows the plot of BER vs. SNR for $L = 25$. The performance is improved, and we can in principle tolerate 100 times more noise energy. The number of flops is here in the order $2 \cdot 10^5$, so five times more computations are needed compared to the linear equalizer.

Alternatively, we can use the performance gain to save bandwidth by decreasing the length of the training sequence. Figure 5.29(b) plots SNR vs. BER for different lengths of the training sequence. As can be seen, a rather short training sequence, say 7, gives good performance. We can thus 'buy' bandwidth by using a more powerful signal processor.

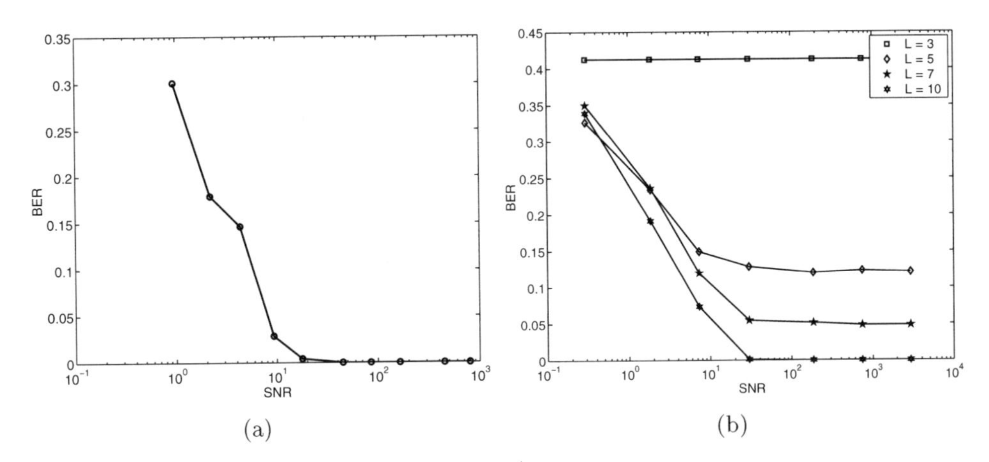

Figure 5.29. Bit error rate versus signal-to-noise ratio for $L = 25$ (a) and some other different lengths L of the training sequence (b).

5.8.4. Channel estimation in equalization

We describe here how to estimate the channel parameters from a training sequence in an efficient way, suitable for hardware implementation. As pointed out before, an equalizer works in two phases:

1. Parameter estimation.
2. Equalization.

The standard model for channels is the FIR model,

$$y_t = \sum_{k=0}^{n} b_t^k u_{t-k} = \varphi_t^T b$$

where

$$b = (b^0, b^1, b^2, \ldots, b^n)^T$$
$$\varphi_t = (u_t, u_{t-1}, \ldots, u_{t-n})^T.$$

Parameter estimation is a standard least squares problem, with the solution

$$\hat{b} = \left(\sum_{t=1}^{N} \varphi_t \varphi_t^T \right)^{-1} \sum_{t=1}^{N} \varphi_t y_t$$
$$= \sum_{t=1}^{N} \bar{\varphi}_t y_t.$$

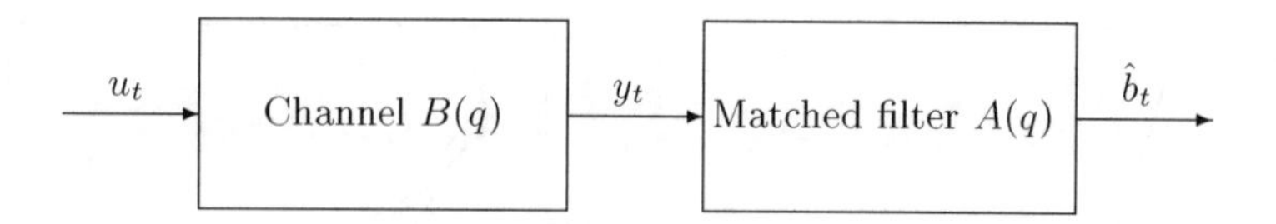

Figure 5.30. Matched filter for estimating the channel from a training sequence.

The second expression is useful since the training sequence is known beforehand so $\bar{\varphi}_t$ can be precomputed. This gives an efficient implementation usually referred to as a *matched filter*. Generally, a filter $A(q)$ which is matched to a signal u has the property $(u * a)_t \approx \delta_0$, so that

$$\hat{b}_t = (a * y)_t = (a * b * u)_t = (b * (a * u))_t \approx b_t.$$

In our case, we want a filter a such that

$$(a * u)_t = \sum_{k=1}^{N} a_{t-k} u_k \approx \delta_0,$$

and it can be taken as

$$a_{t-k} = \bar{\varphi}_t^{(k)}.$$

That is, each tap in the channel model is estimated by applying a linear filter to the training sequence, see Figure 5.30.

Example 5.20 Equalization and training in GSM

In mobile communication the transmitted signal is reflected in the terrain, and reaches the receiver with different delays and complex amplitudes. Mobility of the terminal makes the channel impulse response time-varying. We will here detail the solution implemented in the GSM system.

The standard burst consists of 148 bits, see Section 5.11. Of these, 26 bits located in the middle of the burst are used solely for the equalizer. The training sequence is of length 16 bits, and the other 5 bits on each side are chosen to get a quasi-stationary sequence. The training sequence is the same all the time, and its design is based on optimization, by searching through all 2^{16} sequences and taking the one that gives the best pulse approximation, $(a * u)_t \approx \delta_0$.

From the 16 training bits, we can estimate 16 impulse response coefficients by a matched filter. However, only 4 are used in the GSM system. This is considered sufficient to equalize multi-path fading for velocities up to 250 km/h. The normal burst will be described in detail in Section 5.11.

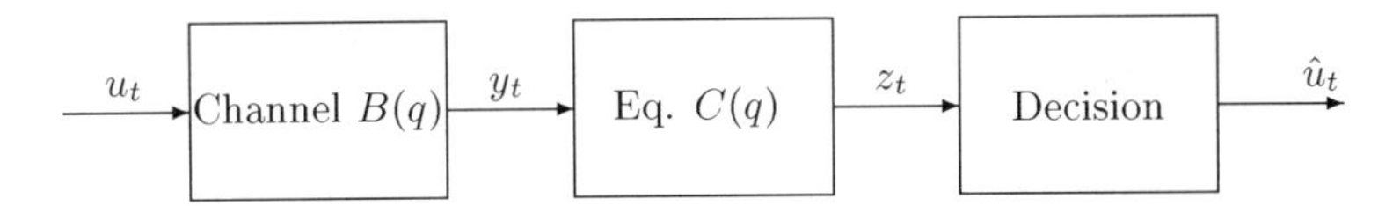

Figure 5.31. Structure of a blind equalizer, which adapts itself for knowledge of y_t and z_t, only..

5.8.5. Blind equalization

The best known application of blind deconvolution is to remove the distortion caused by the channel in digital communication systems. The problem also occurs in seismological and underwater acoustics applications. Figure 5.31 shows a block diagram for the adaptive blind equalizer approach. The channel is as usual modeled as a FIR filter

$$B(q) = b_t^1 q^{-1} + b_t^2 q^{-2} \cdots + b_t^{n_b} q^{-n_b}, \tag{5.66}$$

and the same model structure is used for the blind equalizer

$$C(q) = c_t^1 q^{-1} + c_t^2 q^{-2} \cdots + c_t^{n_c} q^{-n_c}. \tag{5.67}$$

The impulse response of the combined channel and equalizer, assuming FIR models for both, is

$$h_k = (b * c)_k,$$

where $*$ denotes convolution. Again, the best one can hope for is $h_k \approx m\delta_{k-D}$, where D is an unknown time-delay, and m with $|m| = 1$ is an unknown modulus. For instance, it is impossible to estimate the sign of the channel. The modulus and delay do not matter for the performance, and can be ignored in applications.

Assume binary signal u_t (BPSK). For this special case, the two most popular loss functions defining the adaptive algorithm are given by:

$$V = \frac{1}{2}\mathrm{E}[(1 - z^2)^2] \quad \textit{modulus restoral (Godard)} \tag{5.68}$$

$$V = \frac{1}{2}\mathrm{E}[(\mathrm{sign}(z) - z)^2] \quad \textit{decision directed (Sato)} \tag{5.69}$$

The modulus restoral algorithm also goes under the name *Constant Modulus Algorithm (CMA)*. Note the convention that decision directed means that the decision is used in a parameter update equation, whereas decision feedback means that the decisions are used in a linear feedback path. The steepest descent algorithm applied to these two loss functions gives the following two algorithms, differing only in the definition of residual:

Algorithm 5.6 Blind equalization

A stochastic gradient algorithm for minimizing one of (5.68) or (5.69) in the case of binary input is given by

$$\varphi_t = (y_{t-1}, y_{t-2}, \ldots, y_{t-n})^T \tag{5.70}$$

$$z_t = \varphi_t^T \hat{\theta}_{t-1} \tag{5.71}$$

$$\varepsilon_t^{Sato} = \text{sign}(z_t) - z_t \tag{5.72}$$

$$\varepsilon_t^{Godard} = z_t(1 - z_t^2) \tag{5.73}$$

$$\hat{\theta}_t = \hat{\theta}_{t-1} + \mu \varphi_t \varepsilon_t \tag{5.74}$$

$$\hat{u}_{t-D} = \text{sign}(z_t). \tag{5.75}$$

The extension to other input alphabets than ± 1 is trivial for the decision directed approach. Modulus restoral can be generalized to any phase shift coding.

Consider the case of input alphabet $u_t = \pm 1$. For successful demodulation and assuming no measurement noise, it is enough that the largest component of h_k is larger than the sum of the other components. Then the decoder $\hat{u}_{t-D} = \text{sign}(z_t)$ would give zero error. This condition can be expressed as $m_t > 0$, where

$$m_t = 2 - \frac{\sum(|h_k|)}{\max_k |h_k|}.$$

If the equalizer is a perfect inverse of the channel (which is impossible for FIR channel and equalizer), then $m_t = 1$. The standard definition of a so called *open-eye condition* corresponds to $m_t > 0$, when perfect reconstruction is possible, if there is no noise. The larger m_t, the larger noise can be tolerated.

Example 5.21 Blind equalization

Look at the plot in Figure 5.32 for an example of how a blind equalizer improves the open-eye measure with time. In this example, the channel $B(q) = 0.3q^{-1} + 1q^{-2} + 0.3q^{-3}$ is simulated using an input sequence taken from the alphabet $[-1, 1]$. The initial equalizer parameters are quite critical for the performance. They should at least satisfy the open-eye condition $m_t > 0$ (here $m_0 \approx 0.5$). The algorithms are initiated with $C(q) = 0 - 0.1q^{-1} + 1q^{-2} - 0.1q^{-3} + 0q^{-4}$. At the end, the equalizer is a good approximation of the channel impulse response, as seen from Figure 5.33.

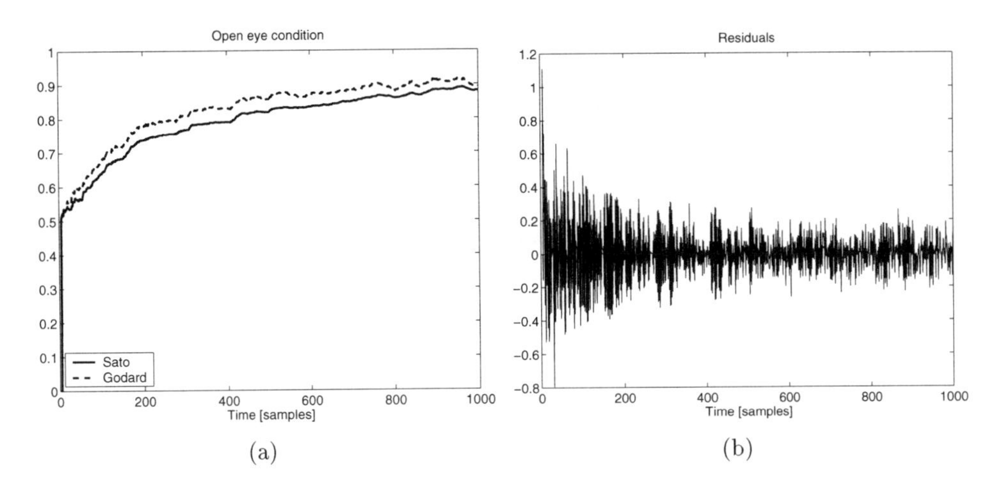

Figure 5.32. Open-eye measure (a) and residual sequences (b) for Sato's and Godard's algorithms, as a function of number of samples.

5.9. Noise cancelation

Noise cancelation, comprising *Acoustic Echo Cancelation* (*AEC*) as one application, has found many applications in various areas, and typical examples include:

- Loudspeaker telephone systems in cars, conference rooms or hands-free systems connected to a computer, where the feedback path from speaker to microphone should be suppressed.
- Echo path cancelation in telephone networks (Homer et al., 1995).

The standard formulation of noise cancelation as shown in Figure 5.1 falls into the class of general adaptive filtering problems. However, in practice there are certain complications worth special attention, discussed below.

5.9.1. Feed-forward dynamics

In many cases, the dynamics between speaker and microphone are not negligible. Then, the block diagram must be modified by the feed-forward dynamics $H(q)$ in Figure 5.34(a). This case also includes the situation when the transfer function of the speaker itself is significant. The signal estimate at the microphone is

$$\hat{s}_t = s_t + (G(q) - F(q;\theta)H(q))u_t. \tag{5.76}$$

For perfect noise cancelation, we would like the adaptive filter to mimic $G(q)/H(q)$. We make two immediate reflections:

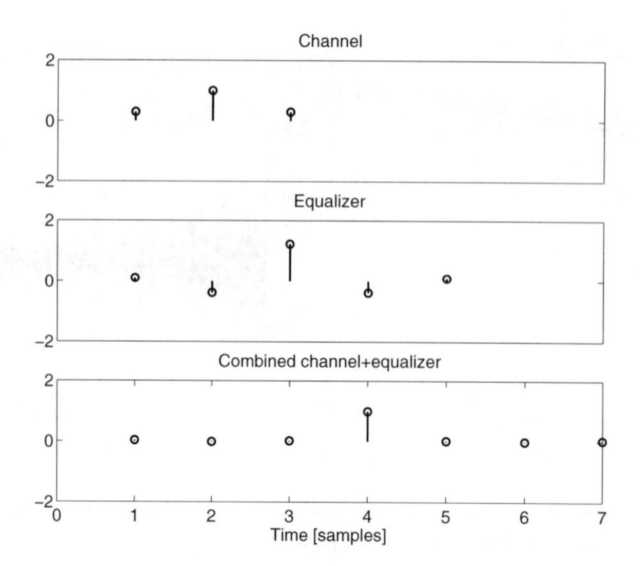

Figure 5.33. Impulse response of channel, equalizer and combined channel–equalizer. The overall response is close to an impulse function with a delay of $D = 4$ samples.

1. A necessary requirement for successful cancelation is that there must be a causal and stable realization of the adaptive filter $F(q;\theta)$ for approximating $G(q)/H(q)$. This is guaranteed if $H(q)$ is minimum phase. If $H(q)$ contains a time delay, then $G(q)$ must also contain a time delay.

2. Least squares minimization of the residuals ε_t in Figure 5.34(a) will generally not give the answer $F(q;\hat{\theta}) \approx G(q)/H(q)$. The reason is that the gradient of (5.76) is not the usual one, and may point in any direction.

However, with a little modification of the block diagram we can return to the standard problem. A very simple algebraic manipulation of (5.76) yields

$$\hat{s}_t = s_t + (G(q) - F(q;\theta)H(q))u_t = s_t + \left(\frac{G(q)}{H(q)} - F(q;\theta)\right)\underbrace{H(q)u_t}_{\bar{u}_t}. \quad (5.77)$$

This means that we should pre-filter the input to get back to the standard least squares framework of system identification. The improved alternative structure with pre-filter is illustrated in Figure 5.34(b) and (c)

For a FIR structure on $F(q;\theta)$, this means that we should pre-filter the regressors in a linear regression. This input pre-filtering should not be confused with the pre-filtering often done in system identification (Ljung, 1999), in which case the identified system does not change. The pre-filtering is the background for the so called *filtered–input LMS*, or *filtered–X LMS*, algorithm. The algorithm includes two phases:

1. Estimation of the feed-forward dynamics. This can be done off-line if H is time-invariant. Otherwise, it has to be estimated adaptively on-line and here it seems inevitably to use a perturbation signal, uncorrelated with u and s, to guarantee convergence. With a perturbation signal, we are facing a standard adaptive filtering problem with straightforward design for estimating H.

2. With H known, up to a given accuracy, the input u can be pre-filtered. In our notation, a more suitable name would be filtered-u LMS, or filtered regressor LMS.

Generally, time-varying dynamics like F and H in Figure 5.34(a) do not commute. If, however, the time variations of both filters are slow compared to the time constant of the filter, we can change their order and obtain the block diagram in Figure 5.34(b) and (c).

Example 5.22 Adaptive noise cancelation in the SAAB 2000

As a quite challenging application, we mention the adaptive noise cancelation system implemented in all SAAB 2000 aircraft around the world. The goal is to attenuate the engine noise in the cabin at the positions of all the passenger heads. By using 64 microphones and 32 loudspeakers, the overall noise attenuation is 10 dB on the average for all passengers. This system contributes to making SAAB 2000 the most quiet propeller aircraft on the market.

The disturbance u is in this system artificially generated by using the engine speed. Non-linearities introduce harmonics so the parametric signal model is a linear regression of the form

$$\begin{aligned} y_t = {} & \theta_t^{(1)} \sin(\omega t) + \theta_t^{(2)} \cos(\omega t) \\ & + \theta_t^{(3)} \sin(2\omega t) + \theta_t^{(4)} \cos(2\omega t) + \ldots \\ & + \theta_t^{(2n-1)} \sin(n\omega t) + \theta_t^{(2n)} \cos(n\omega t) + e_t, \end{aligned}$$

rather than the conventional dynamic FIR filter fed by a microphone input u_t. The multi-variable situation in this application implies that the measured residual vector is 64-dimensional and the speaker output is 32-dimensional.

The speaker dynamics, as a 64×32 dimensional filter, is identified off-line on the ground, when the engines are turned off.

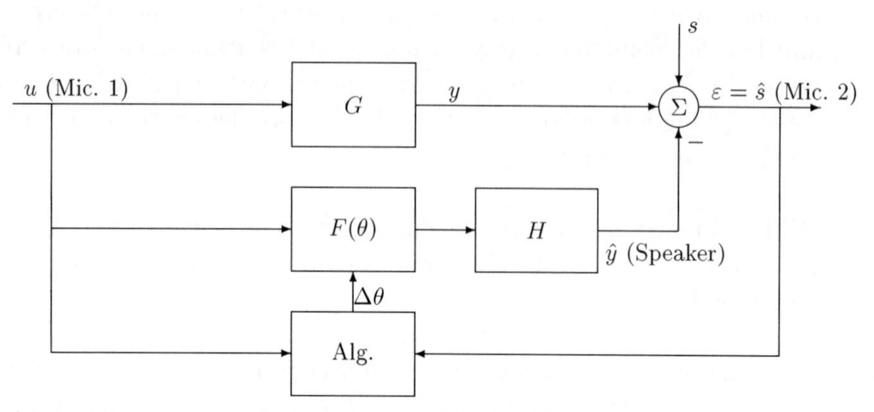

(a) Feed-forward dynamics in noise cancelling implies a convergence problem.

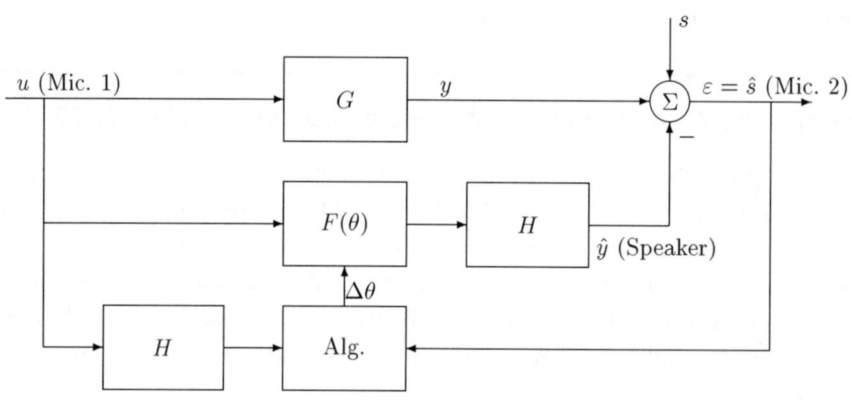

(b) Alternative structure with pre-filtering.

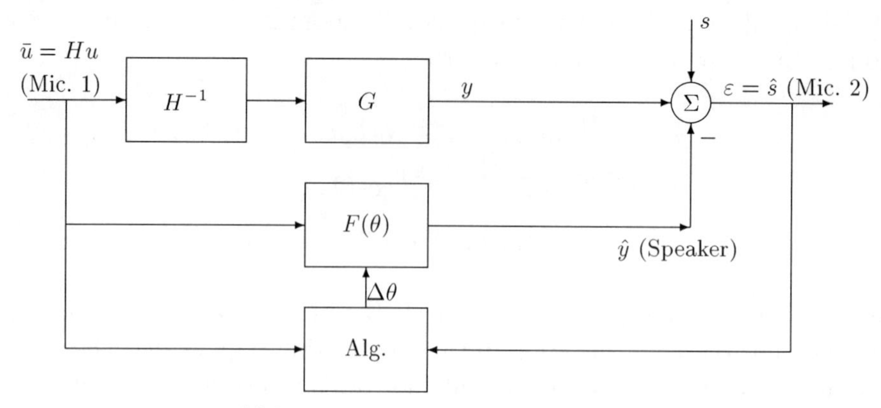

(c) Equivalent alternative structure.

Figure 5.34. Feed-forward dynamics in noise cancelation

5.9.2. Feedback dynamics

Acoustic feedback is a major problem in many sound applications. The most wellknown example is in presentations and other performances when the sound amplification is too high, where the feedback can be a painful experience. The hearing aid application below is another example.

The setup is as in Figure 5.35. The amplifier has a gain $G(q)$, while the acoustic feedback block is included in Figure 5.35 to model the dynamics $H(q)$ between speaker and microphone. In contrast to noise cancelation, there is only one microphone here. To counteract the acoustic feedback, an internal and artificial feedback path is created in the adaptive filter $F(q;\theta)$.

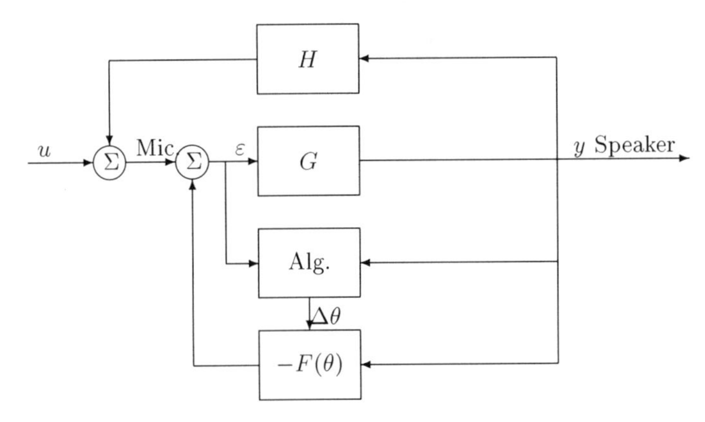

Figure 5.35. Feedback dynamics in noise cancelation: the speaker signal goes to the microphone.

The least squares goal of minimizing the residual ε_t from the filter input y_t (speaker signal) means that the filter minimizes

$$\varepsilon_t = u_t - F(q;\theta)y_t + H(q)y_t.$$

When F is of FIR type, this works fine. Otherwise there will be a bias in the parameters when the input signal u_t is colored, see Hellgren (1999), Siqueira and Alwan (1998) and Siqueira and Alwan (1999). This is a wellknown fact in closed loop system identification, see Ljung (1999) and Forssell (1999): colored noise implies bias in the identified system. Here we should interpret H as the system to be identified, G is the known feedback and u the system disturbance.

Example 5.23 Hearing aids adaptive noise cancelation

Hearing aids are used to amplify external sound by a possibly frequency shaped gain. Modern hearing aids implement a filter customized for the user. However, acoustic feedback caused by leakage from the speaker to the microphone is very annoying to hearing-aid users. In principle, if the feedback

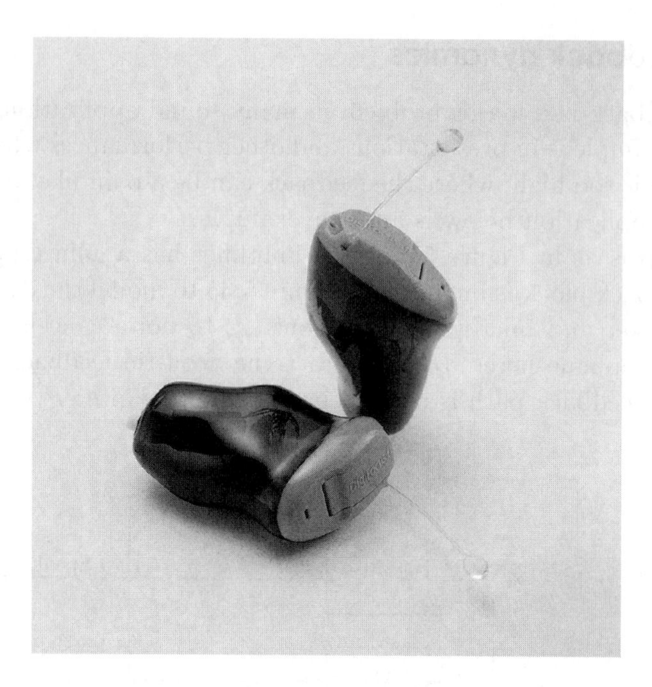

Figure 5.36. An 'in-the-ear' hearing aid with a built in adaptive filter. (Reproduced by permission of Oticon.)

transfer function was known, it can be compensated for in the hardware. One problem here is the time variability of the dynamics, caused by a change in interference characteristics. Possible causes are hugs or objects like a telephone coming close to the ear. The solution is to implement an adaptive filter in the hearing aid.

There is a number of different companies producing such devices. Figure 5.36 shows a commercial product from Oticon, which is the result of a joint research project (Hellgren, 1999). Here it should be noted that the feedback path differs quite a lot between the types of hearing aids. The 'behind-the-ear' model has the worst feedback path. The 'in-the-ear' models attenuate the direct path by their construction. For ventilation, there must always be a small air channel from the inner to the outer ear, and this causes a feedback path.

5.10. Applications

5.10.1. Human EEG

The data described in Section 2.4.2 are measurements from human occipital area. The experiment aims at testing the reaction delay for health diagnosis. Before a time $t_b = 387$ the lights are on in a test room and the test person is looking at something interesting. The neurons in the brain are processing information in visual cortex, and only noise is seen in measurements. When lights are turned off, the visual cortex has nothing to do. The neuron clusters start 10 Hz periodical 'rest rhythm'. The delay between t_b and the actual time when the rhythm starts varies strongly between healthy people and, for instance, those with Alzheimer's decease.

Figure 5.37 shows a quite precise location of the change point. The delay for detection can thus be used as one feature in health diagnosis. That is, medical diagnosis can be, at least partly, automized.

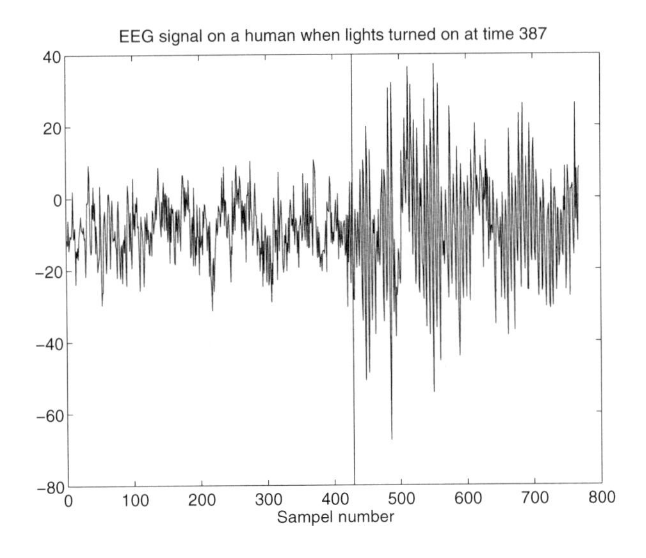

Figure 5.37. Human EEG, where the brain is excited at sample 387.

5.10.2. DC motor

Consider the DC motor lab experiment described in Section 2.5.1. Here we apply the RLS algorithm with a whiteness residual test with the goal to detect system changes while being insensitive to disturbances. The squared residuals are fed to the CUSUM test with $h = 7$ and $\nu = 0.1$. The physical model structure is from (2.1) an ARX(4,4,1) model (with appropriate noise assumptions), which is also used here. All test cases in Table 2.1 are considered.

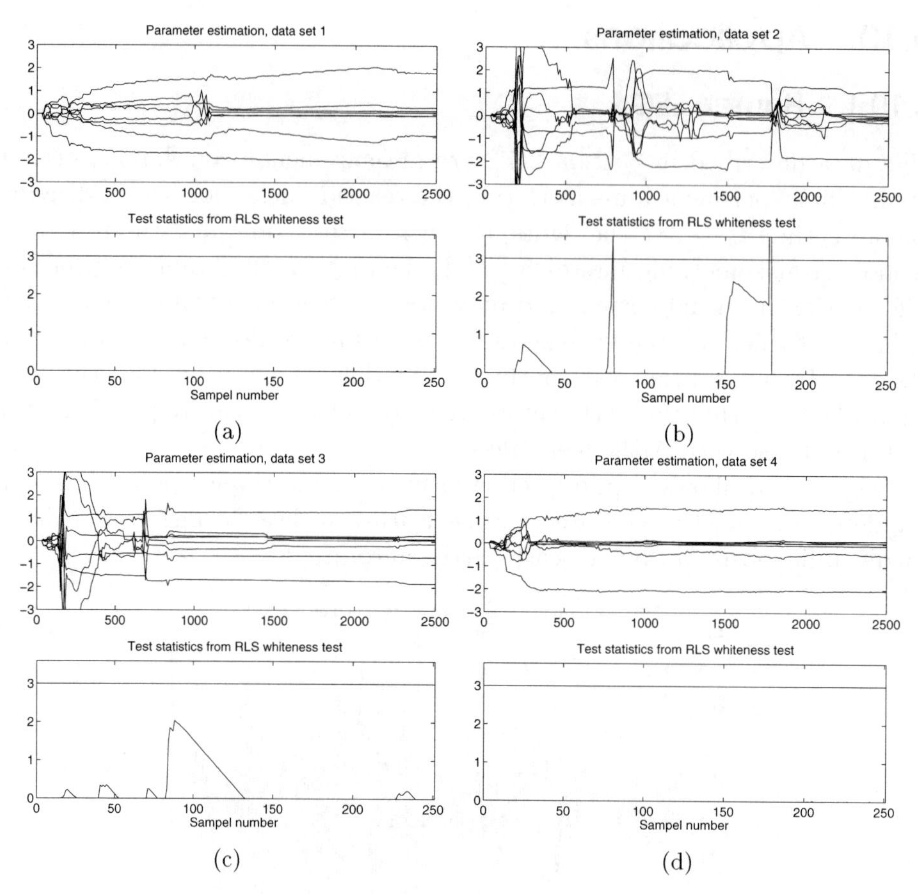

Figure 5.38. Recursive parameter estimation with change detection of a DC motor. Nominal system (a), with torque disturbances (b), with change in dynamics (c) and both disturbance and change (d).

From Figure 5.38 we can conclude the following:

- Not even in the nominal case, the parameters converge to the vicinity of the values of the theoretical expressions in (2.1). The explanation is that the closed loop system is over-modeled, so there are poles and zeros that cancel and thus introduce ambiguity. An ARX(2,2,1) model gives very similar test statistics but better parameter plots.

- It is virtually impossible to tune the CUSUM test to give an alarm for system changes while not giving alarms for the non-modeled disturbances.

5.10.3. Friction estimation

This section continues the case study discussed in Chapter 1 in the series of Examples 1.2, 1.5 and 1.9. The relations of interest are

$$\begin{aligned} k &= \frac{F_f}{sN} = \frac{\mu}{s} \\ s_m &= \frac{\omega_d}{\omega_n} - 1 = s + \delta_w + \delta_R + e_s \\ \mu_m &= \mu + e_\mu \end{aligned}$$

N is a computable normal force on the tire, δ_w is an unknown offset in slip that depends on a small difference in wheel radii – which will be referred to as slip offset – and e is an error that comes from measurement errors in wheel velocities. As before, s_t is the measured wheel slip, and μ_t is a measured normalized traction force. Collecting the equations, gives a linear regression model

$$\underbrace{s_m - \delta_R}_{y(t)} = \underbrace{\frac{1}{k}}_{\theta_1(t)} \underbrace{\mu_m}_{\varphi_1(t)} + \underbrace{1}_{\varphi_2(t)} \cdot \underbrace{\delta_w}_{\theta_2(t)} . \tag{5.78}$$

The theoretical relation between slip and traction force is shown in Figure 5.39. The classical tire theory does not explain the difference in initial slope, so empirical evidence is included in this schematic picture. During normal driving, the normalized traction force is in the order of 0.1. That is, measurements are normally quite close to the origin.

Figure 5.40 summarizes the estimation problem. Two estimation approaches are used, for different purposes.

- The **least squares** method is used for off-line analysis of data, for testing the data quality, debugging of pre-calculations, and most importantly for constructing the mapping between slip slope and friction. Here one single value of the slip slope is estimated for each batch of data, which is known to correspond to a uniform and known friction.

- The **Kalman filter** is used for on-line implementation. The main advantage of the Kalman filter over RLS and LMS is that it can be assigned different tracking speeds for the slip slope and slip offset.

Least squares estimation

The least squares estimate is as usually computed by

$$\hat{\theta}_N = \left(\sum_{t=1}^{N} \varphi_t \varphi_t^T \right)^{-1} \sum_{t=1}^{N} \varphi_t y_t .$$

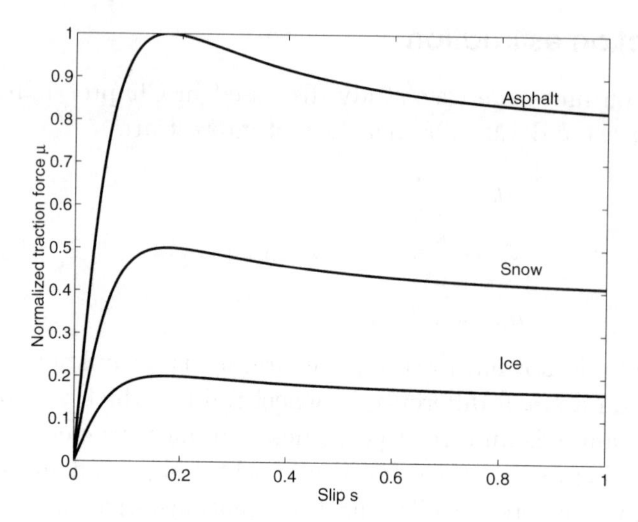

Figure 5.39. Theoretical and empirical relation between wheel slip and traction force.

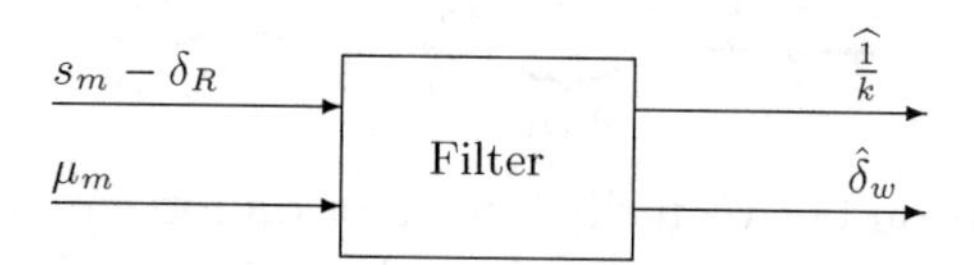

Figure 5.40. Filter for friction estimation.

There are some challenging practical problems worth mentioning:

1. Outliers. Some measurements just do not make sense to the model. There may be many reasons for this, and one should be careful which measurements are put into the filter.

2. The variation in normalized traction force determines the stochastic uncertainty in the estimates, as shown by the following result for the parameter covariance $P_N = \text{Cov}\hat{\theta}(N)$ from N data, assuming constant parameters:

$$\frac{1}{\det(P_N)} \sim \det\left(\sum_{t=1}^{N} \varphi_t \varphi_t^T\right) = N^2 \overline{\text{Var}}(\mu_t). \tag{5.79}$$

   That is, if the variation in normalized traction force is small during the time constant N of the filter, the parameter uncertainty will be large.

3. A wellknown problem in the least squares theory occurs in the case of errors in the regression vector. A general treatment on this matter, usually referred to as *errors in variables* or the *total least squares* problem,

is given in van Huffel and Vandewalle (1991). In our case, we have measurement and computation errors in the normalized traction force μ_t. Assume that

$$\mu_t^m = \mu_t + v_t^\mu, \quad \mathrm{Var}(v_t^\mu) = \lambda_\mu \tag{5.80}$$

is used in φ_t. Here, $\mathrm{Var}(v_t^\mu)$ means the variance of the error in the measurement μ_t^m. Some straightforward calculations show that the noise in μ_t^m leads to a positive bias in the slip slope,

$$\hat{k} \approx k \frac{\overline{\mathrm{Var}}(\mu) + \lambda_\mu}{\overline{\mathrm{Var}}(\mu)} > k. \tag{5.81}$$

Here, $\overline{\mathrm{Var}}(\mu)$ is the variation of the normalized traction force defined as

$$\overline{\mathrm{Var}}(\mu) = \frac{1}{N}\sum_{t=1}^{N} \mu_t^2 - \left(\frac{1}{N}\sum_{t=1}^{N} \mu_t\right)^2 . \tag{5.82}$$

This variation is identical to how one estimates the variance of a stochastic variable. Normally, the bias is small because $\overline{\mathrm{Var}}(\mu) \gg \lambda_\mu$. That is, the variation in normalized traction force is much larger than its measurement error.

An example of a least squares fit of a straight line to a set of data, when the friction undergoes an abrupt change at a known time instant, is shown in Figure 5.41.

The Kalman filter

We will here allow time variability in the parameters, k_t and δ_t. In this application, the Kalman filter is the most logical linear adaptive filter, because it is easily tuned to track parameters with different speeds. Equation (1.9) is extended to a state space model where the parameter values vary like a random walk:

$$\begin{aligned} \theta(t+1) &= \theta_t + v_t \\ y_t &= \varphi_t^T \theta_t + e_t, \end{aligned} \tag{5.83}$$

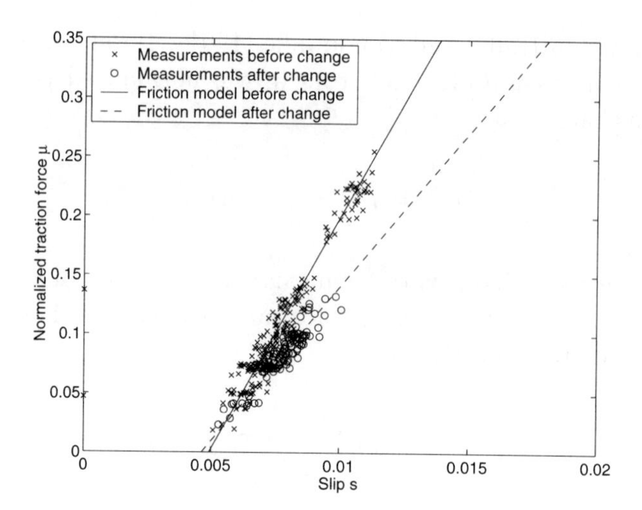

Figure 5.41. Least squares estimation of friction model parameters to a set of data where the friction changes at a known time instant, when the least squares estimation is restarted.

where

$$\begin{aligned}
Q_t &= \mathrm{E} v_t v_t^T \\
R_t &= \mathrm{E} e_t e_t^T \\
y_t &= s_t^m \\
\varphi_t &= (\mu_m, \ 1)^T \\
\theta_t &= \left(\frac{1}{k_t}, \ \delta_t\right)^T .
\end{aligned}$$

Here v_t and e_t are considered as independent white noise processes. The basic problems of lack of excitation, outliers and bias in the slip slope during cruising are the same as for least squares estimation, but explicit expressions are harder to find. A good rule of thumb, is to relate the time constant of the filter, to the time window N in the least squares method, which can lead to useful insights.

The filter originally presented in Gustafsson (1997b) was extended to a dynamic model in Yi et al. (1999). The idea here is to include a complete model of the driveline in the Kalman filter, instead of using the static mapping from engine torque to μ_t, as above.

Practical experience from test trials

This section summarizes results in Gustafsson (1997b) and Gustafsson (1998). The filter has been running for more than 10000 km, and almost 1000 documented tests have been stored. The overall result is that the linear regression model and Kalman filter are fully validated. The noise e_t in (5.83) has been analyzed by studying the residuals from the filter, and can be well approximated by white Gaussian noise with variance 10^{-7}. The offset δ is slowly time-varying and (almost) surface independent. Typical values of the slip slope k are also known, and most importantly, they are found to differ for different surfaces. The theoretical result that filter performance depends on excitation in μ_t is also validated, and it has also turned out that measurements for small slip and μ ($\mu_t < 0.05$) values are not reliable and should be discarded.

A simulation study

The general knowledge described in Section 5.10.3 makes it possible to perform realistic Monte Carlo simulations for studying the filter response, which is very hard to do for real data. What is important here is to use real measured μ_t, below denoted μ_t^m. The slip data is simulated from the slip model:

$$s_t = \frac{1}{k_t}\mu_t^m + \delta_t + e_t. \tag{5.84}$$

The noise e_t is simulated according to a distribution estimated from field tests, while real data are used in μ_t^m. In the simulation,

$$\begin{aligned} \delta_t &= 0.005, \quad k_t = 40, \quad t < 200 \\ \delta_t &= 0.005, \quad k_t = 30, \quad t \geq 200. \end{aligned}$$

This corresponds to a change from asphalt to snow for a particular tire. The slip offset is given a high, but still normal value. The noise has variance 10^{-7}, which implies that the signal is of the order 8 times larger than the noise. The used normalized traction force μ_t and one realization of s_t is shown in Figure 5.42.

As in the real-time implementation, the sampling interval is chosen to 0.2s. Note that the change time at 200 is carefully chosen to give the worst possible condition in terms of excitation, since the variation in normalized traction force is very small after this time, see Section 5.10.3.

First, the response of the Kalman filter presented in the previous section is examined. The design parameters are chosen as

$$\begin{aligned} Q_t &= \begin{pmatrix} 0.3 & 0 \\ 0 & 0 \end{pmatrix} \\ R_t &= 1. \end{aligned}$$

This implies that the slip slope is estimated adaptively, while the offset has no adaptation mechanism. In practice, the (2,2) element of Q is assigned

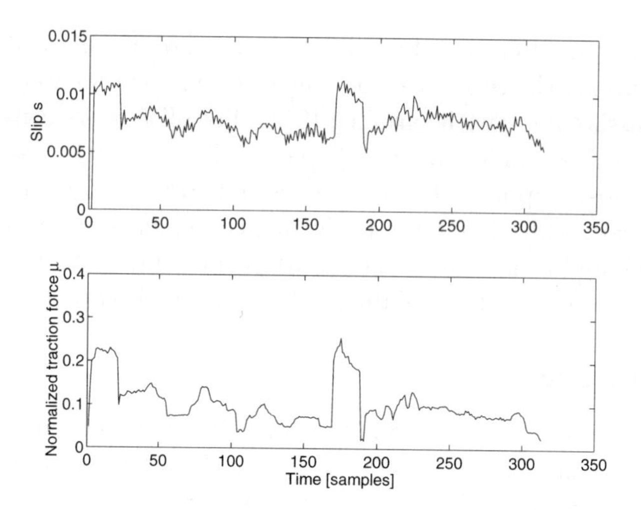

Figure 5.42. The measured normalized traction force μ_t^m (lower plot) and one realization of a simulated s_t (upper plot). Data collected by the author in collaboration with Volvo.

a small value. The result from the filter using the data in Figure 5.42 are shown in Figure 5.43. It is evident that the convergence time of 50 samples, corresponding to 10 seconds, is too long.

In the real-time implementation, the Kalman filter is supplemented by a CUSUM detector. For the evaluation purpose in this section, we use an RLS filter without adaptation ($\lambda = 1$). That is, we are not able to follow slow variations in the parameters, but can expect excellent performance for piece-wise constant parameters (see Algorithm 3.3). The LS filter (see Algorithm 5.7) residuals are fed into a two-sided CUSUM detector, and the LS filter is restarted after an alarm. The design parameters in the CUSUM test are a threshold $h = 3 \cdot 10^{-3}$ and a drift $\nu = 5 \cdot 10^{-4}$. The drift parameter effectively makes changes smaller than 5% impossible to detect. After a detected change, the RLS filter is restarted.

The performance measures are:

- Delay for detection (DFD).
- Missed detection rate (MDR).
- Mean time to detection (MTD).
- The sum of squared prediction errors from each filter,

$$V_N = \frac{1}{0.9N} \sum_{0.1N}^{N} (y_t - \varphi_t^T \hat{\theta}_t)^2.$$

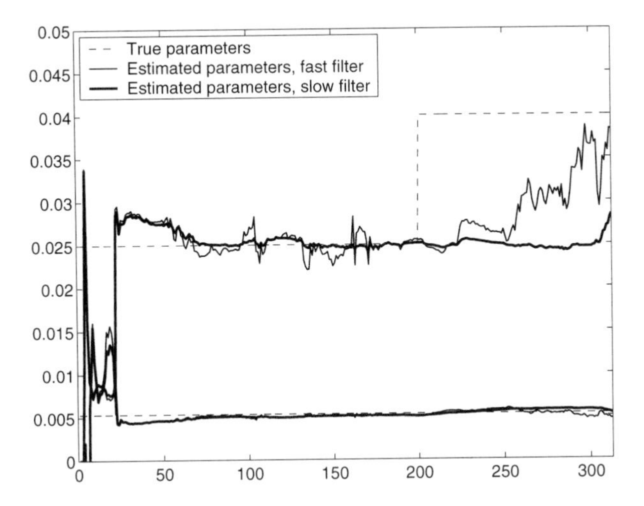

Figure 5.43. Illustration of parameter tracking after an abrupt friction change at sample 200 using the data in Figure 5.42 and a Kalman filter.

Here a delayed start index is used in the sum to rule out effects from the transient.

- As argued in Gustafsson (1997a), the *Minimum Description Length* (MDL) can be interpreted as a norm for each method.
- The root mean square parameter error

$$PE = \frac{1}{N}\sum_{0.1N}^{N} \|\theta_t^0 - \hat{\theta}_t\|^2.$$

- Algorithmic simplicity, short computation time and design complexity. This kind of algorithm is implemented in C or even Assembler, and the hardware is shared with other functionalities. There should preferably be no design parameters if the algorithm has to be re-designed, for instance for another car model.

The parameter norm PE is perhaps a naïve alternative, since the size of the parameter errors is not always a good measure of performance. Anyway, it is a measure of discrepancy in the parameter plots to follow.

Figures 5.44 and 5.45 show the parameter estimates from these filters averaged over 100 realizations and the histogram of alarm times, respectively. The main problem with the CUSUM test is the transient behavior after an alarm. In this implementation, a new identification is initiated after each alarm, but there might be other (non-recursive) alternatives that are better.

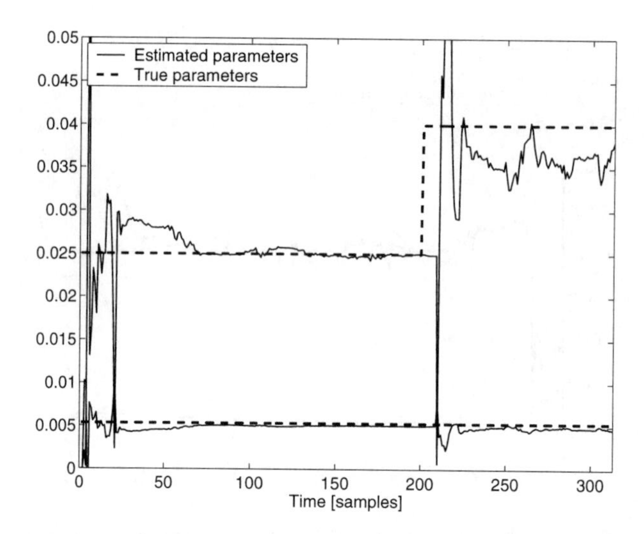

Figure 5.44. Illustration of parameter tracking after an abrupt friction change at sample 200 using the data in Figure 5.42 and CUSUM LS filter.

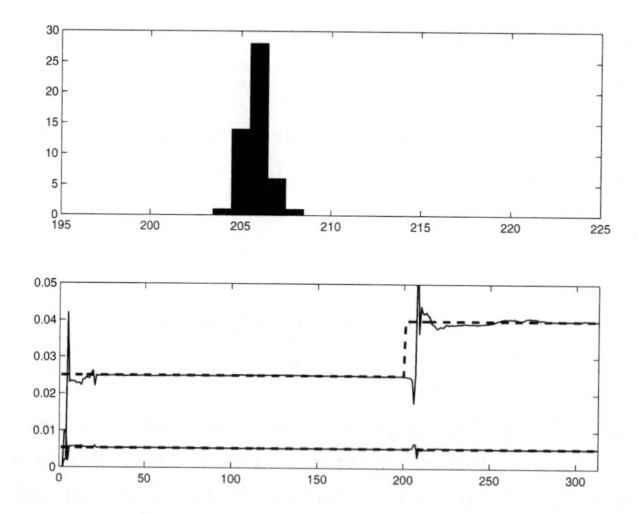

Figure 5.45. Histogram over delay for detection in number of samples ($T_s = 0.2$s) for CUSUM LS filter for 100 realizations. The same estimates as in Figure 5.44 averaged over 100 realizations are also shown.

Table 5.4. Comparison of some norms for the examined filters. The smaller the norm, the better the performance.

Method	MTD	MDR	FAR	V_N	MDL	PE	Time
RLS	–	–	–	$1.57 \cdot 10^{-7}$	$5.1 \cdot 10^{-7}$	0.0043	1
CUSUM	14.7	0	0	$1.41 \cdot 10^{-7}$	$6.3 \cdot 10^{-7}$	0.0058	1.11

The quantitative results are summarized in Table 5.4. The forgetting factor in RLS is optimized here to minimize the loss function V_N. The other parameter is chosen to give zero false alarm rate, and the best possible performance.

Real-time implementation

The design goals for the real-time application are as follows:

- It must be computationally fast.
- The mean time to detection should be in the order of a few seconds, while the mean time between false alarms should be quite large in case the driver is informed about the alarms. If the system is only used for friction surveillance, the false alarm rate is not critical.

The CUSUM test satisfies both conditions, and was chosen for real-time implementation and evaluation. The next sub-sections describe the chosen implementation.

The CUSUM test

The tracking ability of the Kalman filter is proportional to the size of Q. The Kalman filter is required to give quite accurate values of the slip slope and must by necessity have a small Q, see Anderson and Moore (1979). On the other hand, we want the filter to react quickly to sudden decreases in k due to worse friction conditions. This is solved by running the CUSUM detector in parallel with the Kalman filter. If it indicates that something has changed, then the diagonal elements of Q corresponding to the slip slope are momentarily increased to a large value. This allows a very quick parameter convergence (typically one or two samples).

In words, the CUSUM test looks at the prediction errors $\varepsilon_t = s_t^m - \varphi_t^T \hat{\theta}_t$ of the slip value. If the slip slope has actually decreased, we will get predictions that tend to underestimate the real slip. The CUSUM test gives an alarm when the recent prediction errors have been sufficiently positive for a while. Another test, where the sign of ε is changed, is used for detecting increases in slip slope.

Example of filter response

In this section, an example is given to illustrate the interaction between the Kalman filter and the change detector, using data collected at the test track CERAM outside Paris. A road with a sudden change from asphalt to gravel

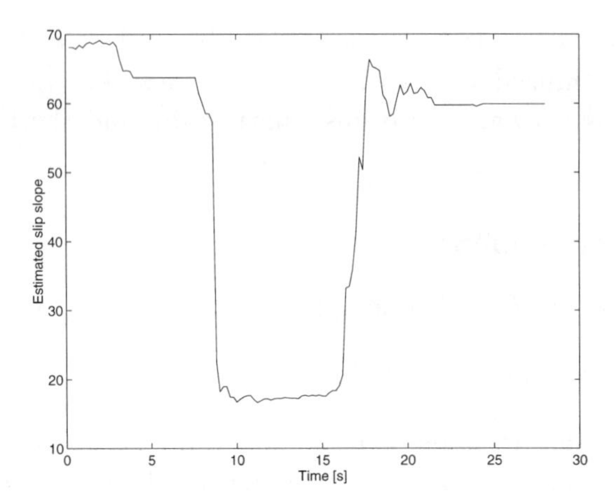

Figure 5.46. Illustration of tracking ability in the estimated slip slope as a function of time. After 8 s an asphalt road changes to a short gravel path that ends at 16 s, so the true slip slope is expected to have abrupt changes at these time instants.

and then back to asphalt is considered. Figure 5.46 shows the estimated slip slope as a function of time in one of the tests, where the gravel path starts after 8 seconds and ends after 16 seconds. Gravel roads have a much smaller slip slope than asphalt roads.

Note that the Kalman filter first starts to adapt to a smaller slip slope at $t = 8$ s, but after three samples (0.6 s) something happens. This is where the CUSUM detector signals for an alarm and we have convergence after one more sample. Similar behavior is noted at the end of the gravel path. The Kalman filter first increases k slightly, and after some samples it speeds up dramatically and then takes a couple of seconds to converge to a stable value.

Final remarks

A summary of the project is shown in Figure 5.47. This application study shows the similarity between the use of adaptive (linear) filters and change detectors for solving parameter estimation problems. A linear adaptive filter can be tailored to the prior information of parameter changes, but the performance was still not good enough. A non-linear adaptive filter as a combination of a linear adaptive filter and a change detector was able to meet the requirements of noise attenuation and fast tracking. As a by-product, the skid alarm functionality was achieved, which is a relative measure of abrupt friction changes. It turned out that tire changes and weariness of tires made it necessary to calibrate the algorithm. If an absolute measure of friction is needed, an automatic calibration routine is needed. However, as a skid alarm unit, it works without calibration.

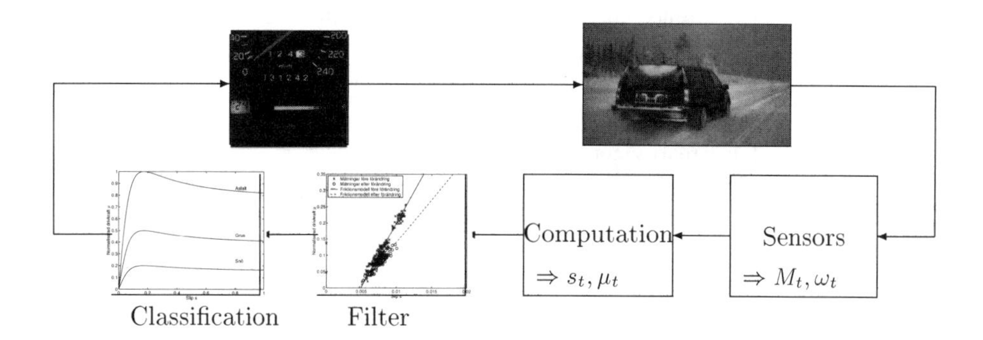

Figure 5.47. Overview of the friction estimation algorithm.

5.11. Speech coding in GSM

Figure 5.48 shows an overview of the signal processing in a second generation digital mobile telephone system. In this section, we will present the ideas in speech coding. Equalization is dealt with in Example 5.20.

Generally, signal modeling consists of a filter and an input model. For the GSM speech coder, the filter is an estimated AR(8) model (parameterized with a Lattice structure) and the input is a pulse train. There are many tricks to get an efficient quantization and robustness to bit errors, which will be briefly mentioned. However, the main purpose is to present the adaptive estimator for coding. Decoding is essentially the inverse operation.

The algorithm outlined below is rather compact, and the goal is to get a flavour of the complexity of the chosen approach.

Filter model

Denote the sampled (sampling frequency 8192 Hz) speech signal by s_t. The main steps in adaptive filter estimation are:

1. Pre-filter with
$$s_t^f = (1 - 0.9q^{-1})s_t.$$
This high pass filter highlights the subjectively important high frequencies around 3 kHz.

2. Segment the speech in fixed segments of 160 samples, corresponding to 20 ms speech. Each segment is treated independently from now on.

3. Apply the Hamming window
$$s_t^{fsw} = cw_t s_t^{fs}$$

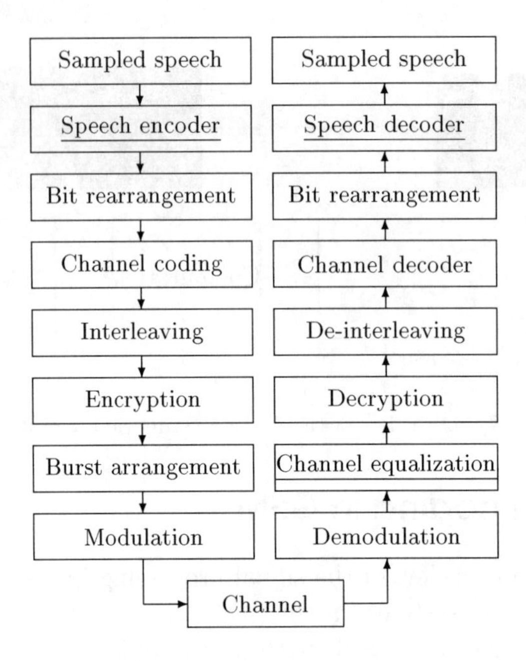

Figure 5.48. Overview of a digital mobile telephone system, e.g. GSM. The underlined blocks are described in detail in this book.

to avoid any influence from the signal tails. The constant $c = 1.59$ makes the energy invariant to the windowing, that is $c^2 \sum w_t^2 = 1$.

Estimation of AR(8) model

4. The AR(8) model is determined by first estimating the covariance function

$$\hat{R}_k = \sum_{t=1}^{160} s_t^{fsw} s_{t+k}^{fsw}$$

for $k = 0, 1, 2, \ldots, 8$.

5. The AR coefficients could be estimated from the Wiener-Hopf equations. For numerical purposes, another parameterization using reflection coefficients $\rho_1, \rho_2, \ldots, \rho_8$ is used. The algorithm is called a Schur recursion.

6. The reflection coefficients are not suitable to quantize, since the resolution needs to be better when their numerical values are close to ± 1. Instead the so called logarithmic area ratio is used

$$LAR(i) = \log_{10} \frac{1 + \rho_i}{1 - \rho_i}.$$

The number of bits spent on each parameter is summarized below:

LAR	1	2	3	4	5	6	7	8
Bits	6	6	5	5	4	4	3	3

The decoder interpolates the LAR parameters to get a softer transition between the segments.

Input model

The input is essentially modeled as a pulse train motivated by the vocal chord, which generates pulses that are modulated in the throat and mouth. The coder works as follows:

7. Decode the filter model. That is, invert the steps 4-6. This gives $\hat{s}_t^{fsw}$ and

$$\varepsilon_t = A(q; \hat{\theta}_{quant}) s_t^{fsw}.$$

This is essentially the prediction error, but it also contains quantization effects.

8. Split each segment into four subsegments of length 40. These are treated independently for each input model. Initialize a buffer with the 128 latest prediction errors.

9. Estimate the covariances

$$\hat{R}_k = \sum_{t=1}^{40} \varepsilon_t \varepsilon_{t-k}$$

for $k = 41, 42, \ldots, 128$.

10. Compute

$$D = \arg \max_{41 \leq k \leq 128} \hat{R}_k$$
$$G = \max_{41 \leq k \leq 128} \hat{R}_k,$$

where D corresponds to the pulse interval and G is the amplification of the pulse size compared to the previous subsegment. G is quantized with 2 bits and D with 7 bits ($D \in [1, 128]$).

11. Each pulse size is saved. To get efficient quantization, a *long term prediction* filter models the pulse size variations,

$$\tilde{\varepsilon}_t = \varepsilon_t - G\varepsilon_{t-D}.$$

Since $D > 40$ the last term is taken from the buffer. This can be seen as an AR(1) model with sampling interval D units and $a_1 = G$. Note that the overall filter operation is now a cascade of two filters on two different time scales.

12. Decimate the sequence $\tilde{\varepsilon}_t$ a factor of 3. That is LP filter with cut-off frequency $\omega_g = 4096/3$ Hz and pick out every third sample. This gives 13 samples.

13. Depending on at which sample (1,2 or 3) the decimation starts, there are three possible sequences. This offset is chosen to maximize the energy in the pulse sequence. The decoder constructs the pulse train by using this offset and inserting two zeros between each two pulses.

14. Normalize the pulses for a more efficient quantization:

$$v = \max p_i$$
$$\bar{p}_i = \frac{p_i}{v}.$$

Because of the structure, the input model is sometimes referred to as the long time prediction and the filter as the short time prediction.

The number of bits used for modeling each subsegment of the input is shown below:

Parameter	k	$\log v$	$\bar{p}_i$
Bits	2	6	3

Code length

All in all, each segment of 160 sample (20 ms) requires:

Parameters	# bits
8 $LAR(i)$	36
4 G	8
4 D	28
4 k	8
4 $\log v$	24
4× 13 p_i	156
Sum	260
Bit rate	13 kbit/s

Channel coding

The 260 bits to each segment are sorted after priority of importance.

50 most important	132 important	78 less important

To the most important, three parity bits are added to detect bit errors. To these 53 bits and the next 132 a cyclical convolution code is applied (denoted CC(2,1,5), which is a fifth order FIR filter). This gives $2\times$ 189= 378 bits. The less important bits are sent as they are.

Thus, in total 456 bits are sent to each segment. These are transmitted in four independent bursts, each consisting of 116 bits.

They are complemented by a training sequence of 26 bits used for equalization (see Example 5.20), and three zeros at each side. This results in a so called *normal burst*, constructed as

3 zeros	58	26 training bits	58	3 zeros

The code rate is thus

$$\frac{8192/160 \cdot 4 \cdot 148}{8192 \cdot 8} = 0.46,$$

so the coded speech signal requires 46% of the bandwidth of what an uncoded signal would require.

5.A. Square root implementation

To motivate the problem, consider the linear regression problem in matrix notation $Y = \Phi^T \theta + E$. One way to derive the LS estimate is to solve the over-determined system of equations

$$Y = \Phi^T \hat{\theta}. \tag{5.85}$$

Multiple both sides with Φ and solving for θ gives

$$\Phi Y = \Phi \Phi^T \hat{\theta} \Rightarrow \hat{\theta} = \left(\Phi \Phi^T\right)^{-1} \Phi Y.$$

It is the inversion of $\Phi\Phi^T$ which may be numerically ill-conditioned, and the problem occurs already when the matrix product is formed. One remedy is to apply the *QR-factorization* to the regression matrix, which yields

$$\Phi^T = Q \begin{pmatrix} R \\ 0 \end{pmatrix},$$

where Q is a square orthonormal matrix ($Q^TQ = QQ^T = I$) and R is an upper triangular matrix. In MATLAB™ , it is computed by `[Q,R]=qr(Phi')`.

Now, multiply (5.85) with $Q^{-1} = Q^T$:

$$\begin{pmatrix} Y_0 \\ Y_\varepsilon \end{pmatrix} \triangleq Q^T Y = Q^T \Phi^T \hat{\theta} = Q^T Q \begin{pmatrix} R \\ 0 \end{pmatrix} \hat{\theta} = \begin{pmatrix} R \\ 0 \end{pmatrix} \hat{\theta}.$$

The solution is now computed by solving the triangular system of equations given by

$$R\hat{\theta} = Y_0,$$

which has good numerical properties. Technically, the condition number of $\Phi\Phi^T$ is the square of the condition number of R, which follows from

$$\Phi\Phi^T = \begin{pmatrix} R^T & 0 \end{pmatrix} Q^T Q \begin{pmatrix} R \\ 0 \end{pmatrix} = R^T R.$$

It should be noted that MATLAB™ 's backslash operator performs a QR factorization in the call `thhat=Phi'\Y`. As a further advantage, the minimizing loss function is easily computed as

$$V(\hat{\theta}) = (Y - \Phi\hat{\theta})^T (Y - \Phi\hat{\theta}) = Y_\varepsilon^T Y_\varepsilon.$$

Square root implementations of recursive algorithms also exist. For RLS, *Bierman's UD factorization* can be used. See, for instance, Friedlander (1982), Ljung and Söderström (1983), Sayed and Kailath (1994), Shynk (1989), Slock (1993) and Strobach (1991) for other fast implementations.

5.B. Derivations

In this section, different aspects of the least squares solution will be highlighted. The importance of these results will become clear in later chapters, in particular for likelihood-based change detection. First a recursive version of weighted least squares is derived together with the LS over a sliding window (WLS). Then it is shown how certain off-line expressions that will naturally appear in change detection algorithms can be updated on-line. Some asymptotic approximations are derived, which are useful when implementing change detectors. Finally, marginalization corresponding to the computations in Chapter 3 for linear regression models are given. Here the parameter vector θ generalizes the mean in a changing mean model.

5.B.1. Derivation of LS algorithms

We now derive a recursive implementation of the weighted LS algorithms. In contrast to a common practice in the literature when RLS is cited, no forgetting factor will be used here.

Weighted least squares in recursive form

We begin by deriving a recursive solution to the LS estimator. The weighted multi-variable recursive least squares algorithm, without forgetting factor (see Ljung and Söderström (1983)), minimizes the loss function

$$V_t(\theta) = \sum_{k=1}^{t} (y_k - \varphi_k^T \theta)^T R_k^{-1} (y_k - \varphi_k^T \theta) \tag{5.86}$$

with respect to θ. Here R_k should be interpreted as the measurement covariance matrix, which in this appendix can be a matrix corresponding to a vector valued measurement. Differentiation of (5.86) gives

$$V_t'(\theta) = 2\sum_{k=1}^{t} (-\varphi_k R_k^{-1} y_k + \varphi_k R_k^{-1} \varphi_k^T \theta). \tag{5.87}$$

Equating (5.87) to zero gives the least squares estimate:

$$\begin{aligned} \hat{\theta}_t &= \underbrace{\left(\sum_{k=1}^{t} \varphi_k R_k^{-1} \varphi_k^T \right)^{-1}}_{R_t^{\varphi}} \underbrace{\sum_{k=1}^{t} \varphi_k R_k^{-1} y_k}_{f_t} \\ &= (R^{\varphi})_t^{-1} f_t. \end{aligned} \tag{5.88}$$

Note that $R_t^{\varphi} = \sum_{k=1}^{t} \varphi_k R_k^{-1} \varphi_k^T$ must not be confused with the covariance matrix of the measurement noise. Now

$$\begin{aligned} \hat{\theta}_t &= (R_t^{\varphi})^{-1} f_t \\ &= (R_t^{\varphi})^{-1} (f_{t-1} + \varphi_t R_t^{-1} y_t) \\ &= (R_t^{\varphi})^{-1} (R_{t-1}^{\varphi} \hat{\theta}_{t-1} + \varphi_t R_t^{-1} y_t) \\ &= (R_t^{\varphi})^{-1} (R_t^{\varphi} \hat{\theta}_{t-1} - \varphi_t R_t^{-1} \varphi_t^T \hat{\theta}_{t-1} + \varphi_t R_t^{-1} y_t) \\ &= \hat{\theta}_{t-1} + (R_t^{\varphi})^{-1} \varphi_t R_t^{-1} (y_t - \varphi_t^T \hat{\theta}_{t-1}). \end{aligned} \tag{5.89}$$

This is the recursive update for the LS estimate, together with the obvious update of R_t^{φ}:

$$R_t^{\varphi} = R_{t-1}^{\varphi} + \varphi_t R_t^{-1} \varphi_t^T. \tag{5.90}$$

It is convenient to introduce $P_t = (R_t^\varphi)^{-1}$. The reasons are to avoid the inversion of R_t^φ, and so that P_t can be interpreted as the covariance matrix of $\hat{\theta}$. The *matrix inversion lemma*

$$[A+BCD]^{-1} = A^{-1} - A^{-1}B[DA^{-1}B + C^{-1}]^{-1}DA^{-1} \tag{5.91}$$

applied to (5.90) gives

$$\begin{aligned} P_t &= [P_{t-1}^{-1} + \varphi_t R_t^{-1}\varphi_t^T]^{-1} \\ &= P_{t-1} - P_{t-1}\varphi_t[\varphi_t^T P_{t-1}\varphi_t + R_t]^{-1}\varphi_t^T P_{t-1}. \end{aligned} \tag{5.92}$$

With this update formula for P_t we can rewrite $(R_t^\varphi)^{-1}\varphi_t$ in (5.89) in the following manner:

$$\begin{aligned} (R_t^\varphi)^{-1}\varphi_t &= P_{t-1}\varphi_t - P_{t-1}\varphi_t[\varphi_t^T P_{t-1}\varphi_t + R_t]^{-1}\varphi_t^T P_{t-1}\varphi_t \\ &= P_{t-1}\varphi_t[\varphi_t^T P_{t-1}\varphi_t + R_t]^{-1}R_t. \end{aligned} \tag{5.93}$$

We summarize the recursive implementation of the LS algorithm:

Algorithm 5.7 Weighted LS in recursive form

Consider the linear regression (5.3) and the loss function (5.86). Assume that the LS estimate at time $t-1$ is $\hat{\theta}_{t-1}$ with covariance matrix P_{t-1}. Then a new measurement gives the update formulas:

$$\hat{\theta}_t = \hat{\theta}_{t-1} + P_{t-1}\varphi_t[\varphi_t^T P_{t-1}\varphi_t + R_t]^{-1}(y_t - \varphi_t^T\hat{\theta}_{t-1}) \tag{5.94}$$

$$P_t = P_{t-1} - P_{t-1}\varphi_t[\varphi_t^T P_{t-1}\varphi_t + R_t]^{-1}\varphi_t^T P_{t-1}. \tag{5.95}$$

Here the initial conditions are given by the prior, $\theta \in \mathrm{N}(\theta_0, P_0)$. The *a posteriori* distribution of the parameter vector is

$$\theta_t \in \mathrm{N}(\hat{\theta}_t, P_t).$$

The interpretation of P_t as a covariance matrix, and the distribution of θ_t follows from the Kalman filter. Note that the matrix that has to be inverted above is usually of lower dimension than R_t^φ (if $\dim y < \dim \theta$).

Windowed LS

In some algorithms, the parameter distribution given the measurements in a sliding window is needed. This can be derived in the following way. The trick is to re-use old measurements with negative variance.

Lemma 5.1 (Windowed LS (WLS))
The mean and covariance at time t for the parameter vector in (5.3), given the measurements $y_{t-k+1}^t = \{y_{t-k+1}, y_{t-k+2}, \dots, y_t\}$, are computed by applying the RLS scheme in Algorithm 5.7 to the linear regression

$$\begin{pmatrix} y_t \\ y_{t-k} \end{pmatrix} = \begin{pmatrix} \varphi_t^T \\ \varphi_{t-k}^T \end{pmatrix} \theta + \bar{e}_t,$$

where the (artificial) covariance matrix of $\bar{e}_t$ is

$$\begin{pmatrix} R_t & 0 \\ 0 & -R_{t-k} \end{pmatrix}.$$

Proof: This result follows directly by rewriting the loss function being minimized:

$$\begin{aligned}
V_t(\theta) &= \sum_{t=1}^{N} \left(\begin{pmatrix} y_t \\ y_{t-k} \end{pmatrix} - \begin{pmatrix} \varphi_t^T \\ \varphi_{t-k}^T \end{pmatrix} \theta \right)^T \begin{pmatrix} R_t & 0 \\ 0 & -R_{t-k} \end{pmatrix}^{-1} \\
&\quad \times \left(\begin{pmatrix} y_t \\ y_{t-k} \end{pmatrix} - \begin{pmatrix} \varphi_t^T \\ \varphi_{t-k}^T \end{pmatrix} \theta \right) \\
&= \sum_{t=1}^{N} (y_t - \varphi_t^T \theta)^T R_t^{-1} (y_t - \varphi_t^T \theta) \\
&\quad - \sum_{t=1}^{N} (y_{t-k} - \varphi_{t-k}^T \theta)^T R_{t-k}^{-1} (y_{t-k} - \varphi_{t-k}^T \theta) \\
&= \sum_{t=N-k+1}^{N} (y_t - \varphi_t^T \theta)^T R_t^{-1} (y_t - \varphi_t^T \theta).
\end{aligned}$$

Thus, the influence of the k first measurements cancels, and a minimization gives the distribution for θ conditioned on the measurements y_{t-k+1}^t. □

5.B.2. Comparing on-line and off-line expressions

In this section, some different ways of computing the parameters in the density function of the data are given. The following interesting equalities will turn out to be fundamental for change detection, and are summarized here for convenience.

The density function for the measurements can equivalently be computed from the residuals. With a little abuse of notation, $p(y^N) = p(\varepsilon^N)$. The sum of squared on-line residuals relates to the sum of squared off-line residuals as:

$$\sum_{t=1}^{N}(y_t - \varphi_t^T\hat{\theta}_{t-1})^T \left(\varphi_t^T P_{t-1}\varphi_t + R_t\right)^{-1} (y_t - \varphi_t^T\hat{\theta}_{t-1}) \tag{5.96}$$

$$= \sum_{t=1}^{N}(y_t - \varphi_t^T\hat{\theta}_N)^T R_t^{-1}(y_t - \varphi_t^T\hat{\theta}_N) + (\theta - \hat{\theta}_N)^T P_0^{-1}(\theta - \hat{\theta}_N).$$

The on-line residual covariances relate to the *a posteriori* parameter covariance as:

$$\sum_{t=1}^{N} \log\det\left(\varphi_t^T P_{t-1}\varphi_t + R_t\right)$$

$$= \sum_{t=1}^{N} \log\det R_t + \log\det P_0 - \log\det P_N. \tag{5.97}$$

Together, this means that the likelihood given all available measurements can be computed from off-line statistics as:

$$p(y^N) \sim p(y^N|\hat{\theta}_N)p_\theta(\hat{\theta}_N)\left(\det P_N\right)^{1/2}, \tag{5.98}$$

which holds for both a Gaussian prior $p_\theta(x)$ and a flat non-informative one $p_\theta(x) = 1$.

The matrix notation

$$Y = \Phi^T\theta + E \tag{5.99}$$

will be convenient for collecting N measurements in (5.3) into one equation. Here

$$\begin{aligned}
Y &= (y_1^T, y_2^T, \ldots, y_N^T)^T \\
\Phi &= (\varphi_1, \varphi_2, \ldots, \varphi_N) \\
E &= (e_1^T, e_2^T, \ldots, e_N^T)^T \\
\Lambda &= \operatorname{diag}(R_1, R_2, \ldots, R_N).
\end{aligned}$$

Thus, Λ is a block diagonal matrix. The off-line version of RLS in Algorithm 5.7 can then be written

$$\begin{aligned}\hat{\theta}_N &= (R_0^\varphi + R_N^\varphi)^{-1}(f_N + f_0)\\ P_N &= (R_0^\varphi + R_N^\varphi)^{-1},\end{aligned} \tag{5.100}$$

where

$$\begin{aligned}f_0 &= R_0^\varphi \theta_0\\ f_N &= \sum_{t=1}^{N} \varphi_t R_t^{-1} y_t\\ R_0^\varphi &= P_0^{-1}\\ R_N^\varphi &= \sum_{t=1}^{N} \varphi_t R_t^{-1} \varphi_t^T,\end{aligned}$$

or, in matrix notation,

$$\begin{aligned}f_N &= \Phi \Lambda^{-1} Y\\ R_N &= \Phi \Lambda^{-1} \Phi^T.\end{aligned}$$

Here θ_0 and R_0^φ are given by the prior. With a Gaussian prior we have that $\theta \in \mathrm{N}(\theta_0, (R_0^\varphi)^{-1})$ and a flat non-informative prior is achieved by letting $R_0^\varphi = 0$. Note that a flat prior cannot be used on-line (at least not directly).

First, an on-line expression for the density function of the data is given. This is essentially a sum of squared prediction errors. Then, an off-line expression is given, which is interpreted as the sum of squared errors one would have obtained if the final parameter estimate was used as the true one from the beginning.

Lemma 5.2 (On-line expression for the density function)
The density function of the sequence $y^N = \{y_k\}_{k=1}^N$ from (5.3) is computed recursively by

$$\begin{aligned}\log p(y^N) &= \log p(y^{N-1}) - \frac{p}{2}\log 2\pi - \frac{1}{2}\log\det\left(\varphi_N^T P_{N-1}\varphi_N + R_N\right)\\ &\quad - \frac{1}{2}(y_N - \varphi_N^T\hat{\theta}_{N-1})^T\left(\varphi_N^T P_{N-1}\varphi_N + R_N\right)^{-1}(y_N - \varphi_N^T\hat{\theta}_{N-1})\\ &= -\frac{Np}{2}\log 2\pi - \sum_{t=1}^{N}\frac{1}{2}\log\det\left(\varphi_t^T P_{t-1}\varphi_t + R_t\right)\\ &\quad - \frac{1}{2}\sum_{t=1}^{N}(y_t - \varphi_t^T\hat{\theta}_{t-1})^T\left(\varphi_t^T P_{t-1}\varphi_t + R_t\right)^{-1}(y_t - \varphi_t^T\hat{\theta}_{t-1})\\ &= \log p(\varepsilon^N).\end{aligned}$$

Here $\hat{\theta}_t$ and P_t are given by the RLS scheme in Algorithm 5.7 with initial conditions θ_0 and P_0.

Proof: Iteratively using Bayes' law gives

$$p(y^N) = \prod_{t=1}^{N} p(y_t|y^{t-1}).$$

Since $\theta_t|y^{t-1} \in \mathrm{N}(\hat{\theta}_{t-1}, P_{t-1})$, where $\hat{\theta}_{t-1}$ and P_{t-1} are the RLS estimate and covariance matrix respectively, it follows that

$$y_t|y^{t-1} \in \mathrm{N}(\varphi_t^T \hat{\theta}_{t-1}, \varphi_t^T P_{t-1}\varphi_t + R_t),$$

and the on-line expression follows from the definition of the Gaussian PDF. □

Lemma 5.3 (Off-line expression for the density function)
Consider the linear regression (5.3) with a prior $p_\theta(\theta)$ of the parameter vector θ, which can be either Gaussian with mean θ_0 and covariance $(R_0^\varphi)^{-1}$ or non-informative ($p_\theta(\theta) = 1$). Then the density function of the measurements can be calculated by

$$p(y^N) = p(y^N|\hat{\theta}_N)p_\theta(\hat{\theta}_N)\,(\det P_N)^{1/2}\,(2\pi)^{\frac{d}{2}} \tag{5.101}$$

if the prior is Gaussian or

$$p(y^N) = p(y^N|\hat{\theta}_N)\,(\det P_N)^{1/2} \tag{5.102}$$

if the prior is non-informative. Here $d = \dim\theta$.

Proof: Starting with the Gaussian prior we have

$$\begin{aligned} p(y^N) = \textstyle\int_{-\infty}^{\infty} p(y^N|\theta)p(\theta)d\theta = (2\pi)^{-\frac{Np+d}{2}}\,(\det\Lambda)^{-\frac{1}{2}}\,(\det R_0^\varphi)^{\frac{1}{2}} \\ \times \textstyle\int_{-\infty}^{\infty} \exp\left(-\frac{1}{2}\left((Y-\Phi^T\theta)^T\Lambda^{-1}(Y-\Phi^T\theta) + (\theta-\theta_0)^T R_0^\varphi(\theta-\theta_0)\right)\right) d\theta \end{aligned} \tag{5.103}$$

By completing the squares the exponent can be rewritten as (where $\hat{\theta} = \hat{\theta}_N$)

$$
\begin{aligned}
&(Y - \Phi^T\theta)^T\Lambda^{-1}(Y - \Phi^T\theta) + (\theta - \theta_0)^T R_0^\varphi(\theta - \theta_0) \\
=&Y^T\Lambda^{-1}Y - 2\theta^T\Phi\Lambda^{-1}Y + \theta^T\Phi\Lambda^{-1}\Phi^T\theta + \theta^T R_0^\varphi\theta - 2\theta^T R_0^\varphi\theta_0 + \theta_0^T R_0^\varphi\theta_0 \\
=&\theta^T(R_0^\varphi + R_N^\varphi)\theta - 2\theta^T(f_0 + f_N) + Y^T\Lambda^{-1}Y + \theta_0^T R_0^\varphi\theta_0 \\
=& \left(\theta - (R_0^\varphi + R_N^\varphi)^{-1}(f_0 + f_N)\right)^T (R_0^\varphi + R_N^\varphi)\left(\theta - (R_0^\varphi + R_N^\varphi)^{-1}(f_0 + f_N)\right) \\
&- (f_0 + f_N)^T(R_0^\varphi + R_N^\varphi)^{-1}(f_0 + f_N) + Y^T\Lambda^{-1}Y + \theta_0^T R_0^\varphi\theta_0 \\
=& \left(\theta - \hat{\theta}\right)^T (R_0^\varphi + R_N^\varphi)\left(\theta - \hat{\theta}\right) - \hat{\theta}^T(f_0 + f_N) + Y^T\Lambda^{-1}Y + \theta_0^T R_0^\varphi\theta_0 \\
=& \left(\theta - \hat{\theta}\right)^T (R_0^\varphi + R_N^\varphi)\left(\theta - \hat{\theta}\right) + (Y - \Phi^T\hat{\theta})^T\Lambda^{-1}(Y - \Phi^T\hat{\theta}) + \hat{\theta}^T f_N - \hat{\theta}^T R_N^\varphi\hat{\theta} \\
&- \hat{\theta}^T f_0 + \theta_0 R_0^\varphi\theta_0 \\
=& \left(\theta - \hat{\theta}\right)^T (R_0^\varphi + R_N^\varphi)\left(\theta - \hat{\theta}\right) + (Y - \Phi^T\hat{\theta})^T\Lambda^{-1}(Y - \Phi^T\hat{\theta}) + \hat{\theta}^T(f_0 + f_N) - 2\hat{\theta}^T f_0 \\
&- \hat{\theta}^T(R_0^\varphi + R_N^\varphi)\hat{\theta} + \hat{\theta}^T R_0^\varphi\hat{\theta} + \theta_0^T R_0^\varphi\theta_0 \\
=& \left(\theta - \hat{\theta}\right)^T (R_0^\varphi + R_N^\varphi)\left(\theta - \hat{\theta}\right) + (Y - \Phi^T\hat{\theta})^T\Lambda^{-1}(Y - \Phi^T\hat{\theta}) \\
&+ (\hat{\theta} - \theta_0)^T R_0^\varphi(\hat{\theta} - \theta_0).
\end{aligned}
$$

Here, the relations in (5.100) are frequently used. Hence, the integral (5.103) can be evaluated by using the Gaussian PDF for $\hat{\theta} \in \mathrm{N}(\theta, R_0^\varphi + R_N^\varphi)$,

$$
\begin{aligned}
p(y^N) =& (2\pi)^{-\frac{Np}{2}}(\det\Lambda)^{-\frac{1}{2}}\left(\frac{\det R_0^\varphi}{\det R_0^\varphi + R_N^\varphi}\right)^{\frac{1}{2}} \\
&\times \exp\left(-\frac{1}{2}\left((Y - \Phi^T\hat{\theta}_N)^T\Lambda^{-1}(Y - \Phi^T\hat{\theta}_N) + (\hat{\theta}_N - \theta_0)^T R_0^\varphi(\hat{\theta}_N - \theta_0)\right)\right) \\
&\times \int (2\pi)^{-\frac{d}{2}}(\det R_0^\varphi + R_N^\varphi)^{\frac{1}{2}} \exp\left(-\frac{1}{2}\left(\theta - \hat{\theta}_N\right)^T (R_0^\varphi + R_N^\varphi)\left(\theta - \hat{\theta}_N\right)\right) d\theta \\
=& p(y^N|\hat{\theta}_N)\left(\frac{\det R_0^\varphi}{\det R_0^\varphi + R_N^\varphi}\right)^{1/2} \exp\left(-\frac{1}{2}(\hat{\theta}_N - \theta_0)^T R_0^\varphi(\hat{\theta}_N - \theta_0)\right) \\
=& p(y^N|\hat{\theta}_N)p_\theta(\hat{\theta}_N)\left(\det P_N\right)^{1/2}(2\pi)^{\frac{d}{2}}.
\end{aligned}
$$

The case of non-informative prior, $p_\theta(\theta) = 1$ in (5.103), is treated by letting $R_0^\varphi = 0$ in the exponential expressions, which gives

$$
\begin{aligned}
p(y^N) =& p(y^N|\hat{\theta}_N)\int (2\pi)^{-\frac{d}{2}} \exp\left(-\frac{1}{2}\left(\theta - \hat{\theta}_N\right)^T (R_0^\varphi + R_N^\varphi)\left(\theta - \hat{\theta}_N\right)\right) d\theta \\
=& p(y^N|\hat{\theta}_N)\left(\det P_N\right)^{1/2}.
\end{aligned}
$$

□

It is interesting to note that the density function can be computed by just inserting the final parameter estimate into the prior densities $p(y^N|\theta)$ and

$p_\theta(\theta)$, apart from a data independent correcting factor. The prediction errors in the on-line expression is essentially replaced by the smoothing errors.

Lemma 5.4

Consider the covariance matrix P_t given by the RLS algorithm 5.7. The following relation holds:

$$\sum_{t=1}^{N} \log\det\left(\varphi_t^T P_{t-1}\varphi_t + R_t\right) = \sum_{t=1}^{N} \log\det R_t + \log\det P_0 - \log\det P_N.$$

Proof: By using the alternative update formula for the covariance matrix, $P_t^{-1} = P_{t-1}^{-1} + \varphi_t R_t^{-1}\varphi_t^T$ and the equality $\det(I+AB) = \det(I+BA)$ we have

$$\begin{aligned}
&\sum_{t=1}^{N} \log\det\left(\varphi_t^T P_{t-1}\varphi_t + R_t\right)\\
&= \sum_{t=1}^{N} \left(\log\det R_t + \log\det\left(I + R_t^{-1}\varphi_t^T P_{t-1}\varphi_t\right)\right)\\
&= \sum_{t=1}^{N} \left(\log\det R_t + \log\det\left(I + \varphi_t R_t^{-1}\varphi_t^T P_{t-1}\right)\right)\\
&= \sum_{t=1}^{N} \left(\log\det R_t - \log\det P_{t-1}^{-1} + \log\det\left(P_{t-1}^{-1} + \varphi_t R_t^{-1}\varphi_t^T\right)\right)\\
&= \sum_{t=1}^{N} \left(\log\det R_t - \log\det P_{t-1}^{-1} + \log\det P_t^{-1}\right)\\
&= \sum_{t=1}^{N} \log\det R_t + \log\det P_0 - \log\det P_N
\end{aligned}$$

and the off-line expression follows. In the last equality we used, most of the terms cancel. □

Actually, the third equality (5.98) we want to show follows from the first two (5.97) and (5.96), as can be seen by writing out the density functions in

Lemmas 5.2 and 5.3,

$$
\begin{aligned}
&2\log p(y^N) \\
&= -Np\log 2\pi - \sum_{t=1}^{N} \log\det\left(\varphi_t^T P_{t-1}\varphi_t + R_t\right) \\
&\quad - \sum_{t=1}^{N} (y_t - \varphi_t^T\hat{\theta}_{t-1})^T \left(\varphi_t^T P_{t-1}\varphi_t + R_t\right)^{-1} (y_t - \varphi_t^T\hat{\theta}_{t-1}) \\
&= -Np\log 2\pi - \sum_{t=1}^{N} (y_t - \varphi_t^T\hat{\theta}_N)^T R_t^{-1} (y_t - \varphi_t^T\hat{\theta}_N) \\
&\quad - \sum_{t=1}^{N} \log\det R_t - \log\det P_0 + \log\det P_N - (\theta - \hat{\theta}_N)^T P_0^{-1}(\theta - \hat{\theta}_N).
\end{aligned}
$$

Thus, we do not have to prove all of them. Since the first (5.97) and third (5.98) are proved the second one (5.96) follows. A direct proof of the second equality is found in Ljung and Söderström (1983), p 434.

5.B.3. Asymptotic expressions

The next lemma explains the behavior of the other model dependent factor $\log\det P_N$ (besides the smoothing errors) in Lemma 5.3.

Lemma 5.5 (Asymptotic parameter covariance)
Assume that the regressors satisfy the condition that

$$
Q = \overline{\mathrm{E}}\varphi_t R_t^{-1}\varphi_t^T = \lim_{N\to\infty} \frac{1}{N}\sum_{t=1}^{N} \mathrm{E}[\varphi_t R_t^{-1}\varphi_t^T]
$$

exists and is non-singular. Then

$$
\frac{\log\det P_N}{\log N} \to -d, \quad N \to \infty
$$

with probability one. Thus, for large N, $\log\det P_N$ approximately equals $-d\log N$.

Proof: We have from the definition of R_N^φ

$$\begin{aligned}
-\log\det P_N &= \log\det(R_0^\varphi + R_N^\varphi) \\
&= \log\det\left(R_0^\varphi + N\frac{1}{N}\sum_{t=1}^{N}\varphi_t R_t^{-1}\varphi_t^T\right) \\
&= \log\det N I_d + \log\det\left(\frac{1}{N}R_0^\varphi + \frac{1}{N}\sum_{t=1}^{N}\varphi_t R_t^{-1}\varphi_t^T\right).
\end{aligned}$$

Here the first term equals $d\log N$ and the second one tends to $\log\det Q$, and the result follows. □

5.B.4. Derivation of marginalization

Throughout this section, assume that

$$y_t = \varphi_t^T\theta + e_t,$$

where $\{e_t\}$ is Gaussian distributed with zero mean and variance λ and where θ is a stochastic variable of dimension d. The regression vector φ_t is known at time $t-1$.

This means that the analysis is conditioned on the model. The fact that all the PDF:s are conditioned on a particular model structure will be suppressed.

The simplest case is the joint likelihood for the parameter vector and the noise variance.

Theorem 5.6 (No marginalization)
For given λ and θ the PDF of y^N is

$$p(y^N|\theta,\lambda) = (2\pi\lambda)^{-N/2}\, e^{-\frac{1}{2\lambda}V_N(\theta)}, \tag{5.104}$$

where $V_N(\theta) = \sum_{t=1}^{N}(y_t - \varphi_t^T\theta)^2$.

Proof: The proof is trivial when $\{\varphi_t\}$ is a known sequence. In the general case when φ_t may contain past disturbances (through feedback), Bayes' rule gives

$$p(y^N) = p(y_N|y^{N-1})p(y^{N-1}) = \Pi_{t=1}^{N} p(y_t|y^{t-1})$$

and the theorem follows by noticing that

$$p(y_t|y^{t-1}) = \frac{1}{\sqrt{2\pi\lambda}}e^{-\frac{1}{2\lambda}(y_t - \varphi_t^T\theta)}.$$

since φ_t contains past information only. □

Next, continue with the likelihood involving only the noise variance. Since

$$p(y^N|\lambda) = \int p(y^N|\theta,\lambda)p(\theta|\lambda)d\theta,$$

the previous theorem can be used.

Theorem 5.7 (*Marginalization of θ*)
The PDF of y^N conditioned on the noise variance λ when

$$(\theta|\lambda) \in \mathrm{N}(\theta_0, \lambda(R_0^{\varphi})^{-1}) \tag{5.105}$$

is given by

$$p(y^N|\lambda) = \sqrt{\frac{(2\pi)^d}{\det(R_0^{\varphi}+R_N^{\varphi})}}p(y^N|\hat{\theta}_N,\lambda)p(\hat{\theta}_N|\lambda) \tag{5.106}$$

where $\hat{\theta}_N$ is given by

$$\hat{\theta}_N = \left(R_0^{\varphi}+R_N^{\varphi}\right)^{-1}\left(R_0^{\varphi}\theta_0 + f_N\right). \tag{5.107}$$

Here $R_N^{\varphi} = \sum_{t=1}^N \varphi_t\varphi_t^T$ and $f_N = \sum_{t=1}^N \varphi_t y_t$.

Proof: Consider the following relation

$$\begin{aligned}
&\sum_{t=1}^N (y_t - \varphi_t^T\theta)^2 + (\theta-\theta_0)^T R_0^{\varphi}(\theta-\theta_0) - \\
&\sum_{t=1}^N (y_t - \varphi_t^T\hat{\theta}_N)^2 - \left(\hat{\theta}_N-\theta_0\right)^T R_0^{\varphi}\left(\hat{\theta}_N-\theta_0\right) \\
&= -2f_N\left(\hat{\theta}_N-\theta_0\right)^T + \theta^T R_N^{\varphi}\theta - \hat{\theta}_N^T R_N^{\varphi}\hat{\theta}_N \\
&+ (\theta-\theta_0)^T R_0^{\varphi}(\theta-\theta_0) - \left(\hat{\theta}_N-\theta_0\right)^T R_0^{\varphi}\left(\hat{\theta}_N-\theta_0\right) \\
&= -2\left\{\hat{\theta}_N^T\left(R_0^{\varphi}+R_N^{\varphi}\right) - \theta_0^T R_0^{\varphi}\right\}\left(\theta-\hat{\theta}_N\right) \\
&+ \theta^T\left(R_0^{\varphi}+R_N^{\varphi}\right)\theta - \hat{\theta}_N^T\left(R_0^{\varphi}+R_N^{\varphi}\right)\hat{\theta}_N - 2\theta_0^T R_0^{\varphi}\left(\theta-\hat{\theta}_N\right) \\
&= -2\hat{\theta}_N^T\left(R_0^{\varphi}+R_N^{\varphi}\right)\theta + \theta^T\left(R_0^{\varphi}+R_N^{\varphi}\right)\theta + \hat{\theta}_N^T\left(R_0^{\varphi}+R_N^{\varphi}\right)\hat{\theta}_N \\
&= \left(\theta-\hat{\theta}_N\right)^T\left(R_0^{\varphi}+R_N^{\varphi}\right)\left(\theta-\hat{\theta}_N\right).
\end{aligned} \tag{5.108}$$

Using this together with (5.104) gives

$$
\begin{aligned}
p(y^N|\theta,\lambda) =& p(y^N|\hat{\theta}_N,\lambda)\frac{p(y^N|\theta,\lambda)}{p(y^N|\hat{\theta}_N,\lambda)} \\
=& p(y^N|\hat{\theta}_N,\lambda)e^{-\frac{1}{2\lambda}(\theta-\hat{\theta}_N)^T(R_0^\varphi+R_N^\varphi)(\theta-\hat{\theta}_N)} \times \\
& e^{-\frac{1}{2\lambda}(\hat{\theta}_N-\theta_0)^T R_0^\varphi(\hat{\theta}_N-\theta_0)} \times e^{\frac{1}{2\lambda}(\theta-\theta_0)^T R_0^\varphi(\theta-\theta_0)} \\
=& p(y^N|\hat{\theta}_N,\lambda)e^{-\frac{1}{2\lambda}(\theta-\hat{\theta}_N)^T(R_0^\varphi+R_N^\varphi)(\theta-\hat{\theta}_N)} \\
& e^{-\frac{1}{2\lambda}(\hat{\theta}_N-\theta_0)^T R_0^\varphi(\hat{\theta}_N-\theta_0)} \cdot \frac{\sqrt{\det(R_0^\varphi)}}{(2\pi)^{d/2}\lambda^{d/2}} \cdot \frac{1}{p(\theta|\lambda)},
\end{aligned}
\tag{5.109}
$$

and thus

$$
\begin{aligned}
p(y^N|\lambda) =& \int p(y^N|\theta,\lambda)p(\theta|\lambda)d\theta \\
=& \int \frac{\sqrt{\det(R_0^\varphi+R_N^\varphi)}}{(2\pi)^{d/2}\lambda^{d/2}} e^{-\frac{1}{2\lambda}(\theta-\hat{\theta}_N)^T(R_0^\varphi+R_N^\varphi)(\theta-\hat{\theta}_N)}d\theta \\
& \times \sqrt{\frac{\det(R_0^\varphi)}{\det(R_0^\varphi+R_N^\varphi)}} p(y^N|\hat{\theta}_N,\lambda)e^{-\frac{1}{2\lambda}(\hat{\theta}_N-\theta_0)^T R_0^\varphi(\hat{\theta}_N-\theta_0)} \\
=& \sqrt{\frac{\det(R_0^\varphi)}{\det(R_0^\varphi+R_N^\varphi)}} p(y^N|\hat{\theta}_N,\lambda)e^{-\frac{1}{2\lambda}(\hat{\theta}_N-\theta_0)^T R_0^\varphi(\hat{\theta}_N-\theta_0)} \\
=& \sqrt{\frac{(2\pi)^d}{\det(R_0^\varphi+R_N^\varphi)}} p(y^N|\hat{\theta}_N,\lambda)p(\hat{\theta}_N|\lambda),
\end{aligned}
\tag{5.110}
$$

which proves the theorem. □

The likelihood involving the parameter vector θ is given by

$$
p(y^N|\theta) = \int p(y^N|\theta,\lambda)\cdot p(\lambda|\theta)d\lambda.
$$

Computing this integral gives the next theorem.

Theorem 5.8 (Marginalization of λ)
The PDF of y^N conditioned on θ when λ is $W^{-1}(m,\sigma)$ is given by

$$
p(y^N|\theta) = \frac{1}{p(\theta)}\frac{\sigma^{m/2}}{\pi^{\frac{N+d}{2}}}\frac{\Gamma(\frac{N+m+d}{2})}{\Gamma(\frac{m}{2})}(\det(R_0^\varphi))^{1/2}\overline{V}_N^{-\frac{N+m+d}{2}}(\theta),
\tag{5.111}
$$

where

$$
\overline{V}_N(\theta) = V_N(\theta) + (\theta-\theta_0)^T R_0^\varphi(\theta-\theta_0) + \sigma
\tag{5.112}
$$

and where $p(\theta)$ is the PDF corresponding to a $t(\theta_0, \sigma(R_0^\varphi)^{-1}, m)$ distribution.

Proof: In order to prove (5.111), write

$$
\begin{aligned}
p(y^N|\theta) &= \int p(y^N|\theta,\lambda)\cdot p(\lambda|\theta)d\lambda \\
&= \int p(y^N|\theta,\lambda)\cdot \frac{p(\theta|\lambda)}{p(\theta)}p(\lambda)d\lambda \\
&= \frac{1}{p(\theta)}\int p(y^N|\theta,\lambda)p(\theta|\lambda)p(\lambda)d\lambda.
\end{aligned}
\tag{5.113}
$$

Then (5.111) follows by identifying the integral with a multiple of an integral of an $W^{-1}(\overline{V}_N(\theta), N+m+d)$ PDF. □

The distribution of the data conditioned only on the model structure is given by Theorem 5.9.

Theorem 5.9 (*Marginalization of θ and λ*)
The PDF of y^N when λ is $W^{-1}(m,\sigma)$ and θ is distributed as in (5.105) is given by

$$
p(y^N) = \frac{\sigma^{m/2}}{\pi^{N/2}}\frac{\Gamma(\frac{N+m}{2})}{\Gamma(\frac{m}{2})}\left(\frac{\det(R_0^\varphi)}{\det(R_0^\varphi+R_N^\varphi)}\right)^{1/2}\overline{V}_N^{-\frac{N+m}{2}}(\hat{\theta}_N),
\tag{5.114}
$$

where $\overline{V}_N(\theta)$ and $\hat{\theta}_N$ are defined by (5.112) and (5.107), respectively.

Remark 5.10
Remember that the theorem gives the posterior distribution of the data conditioned on a certain model structure, so $p(y^N|\mathcal{M})$ is more exact.

Proof: From (5.106) and the definition of the Wishart distribution (3.54) it follows

$$
p(y^N) = \int p(y^N|\lambda)p(\lambda)d\lambda \tag{5.115}
$$

$$
= \int \sqrt{\frac{(2\pi)^d}{\det(R_0^\varphi+R_N^\varphi)}}p(y^N|\hat{\theta}_N,\lambda)p(\hat{\theta}_N|\lambda)p(\lambda)d\lambda
$$

(5.116)

$$
\begin{aligned}
&= \int \sqrt{\frac{(2\pi)^d}{\det(R_0^\varphi + R_N^\varphi)}} \frac{1}{(2\pi\lambda)^{N/2}} e^{-\frac{1}{2\lambda}\sum_{t=1}^N (y_t - \varphi_t^T \hat{\theta}_N)^2} \\
&\times \frac{\sqrt{\det(R_0^\varphi)}}{(2\pi)^{d/2}} e^{-\frac{1}{2\lambda}(\hat{\theta}_N - \theta_0)^T R_0^\varphi (\hat{\theta}_N - \theta_0)} \\
&\times \frac{\sigma^{m/2} e^{-\frac{\sigma}{2\lambda}}}{2^{m/2}\Gamma(m/2)\lambda^{(m+2)/2}} d\lambda \\
&= \frac{\sigma^{m/2}}{\pi^{N/2}} \frac{\Gamma(\frac{N+m}{2})}{\Gamma(\frac{m}{2})} \left(\frac{\det(R_0^\varphi)}{\det(R_0^\varphi + R_N^\varphi)} \right)^{1/2} \overline{V}_N^{-\frac{N+m}{2}}(\hat{\theta}_N) \\
&\times \int \frac{|\overline{V}_N(\hat{\theta}_N)|^{\frac{N+m}{2}}}{\lambda^{\frac{N+m+2}{2}} \Gamma(\frac{N+m}{2}) 2^{\frac{N+m}{2}}} e^{-\frac{1}{2\lambda}\overline{V}_N(\hat{\theta}_N)} d\lambda \\
&= \frac{\sigma^{m/2}}{\pi^{N/2}} \frac{\Gamma(\frac{N+m}{2})}{\Gamma(\frac{m}{2})} \left(\frac{\det(R_0^\varphi)}{\det(R_0^\varphi + R_N^\varphi)} \right)^{1/2} \overline{V}_N^{-\frac{N+m}{2}}(\hat{\theta}_N),
\end{aligned}
$$

which proves the theorem. □

6

Change detection based on sliding windows

6.1. Basics

Model validation is the problem of deciding whether observed data are consistent with a nominal model. Change detection based on model validation aims at applying a consistency test in one of the following ways:

- The data are taken from a *sliding window*. This is the typical application of *model validation*.
- The data are taken from an increasing window. This is one way to motivate the *local approach*. The detector becomes more sensitive when the data size increases, by looking for smaller and smaller changes.

The nominal model will be represented by the parameter vector θ_0. This may be obtained in one of the following ways:

- θ_0 is recursively identified from past data, except for the ones in the sliding window. This will be our typical case.
- θ_0 corresponds to a nominal model, obtained from physical modeling or system identification.

The standard setup is illustrated in (6.1):

$$\text{Data}: \quad \underbrace{y_1, y_2, \ldots, y_{t-L}}_{\text{Model } \theta_0}, \underbrace{y_{t-L+1}, \ldots, y_t}_{\text{Model } \theta}. \tag{6.1}$$

A model (θ) based on data from a sliding window of size L is compared to a model (θ_0) based on all past data or a substantially larger sliding window. Let us denote the vector of L measurements in the sliding window by $Y = (y_{t-L+1}^T, \ldots, y_{t-1}^T, y_t^T)^T$. Note the convention that Y is a column vector of dimension Ln_y ($n_y = \dim(y)$). We will, as usual in this part, assume scalar measurements. In a linear regression model, Y can be written

$$Y = \Phi^T\theta + E,$$

where $E = (e_{t-L+1}^T, \ldots, e_{t-1}^T, e_t^T)^T$ is the vector of noise components and the regression matrix is $\Phi = (\varphi_{t-L+1}, \ldots, \varphi_{t-1}, \varphi_t)$. The noise variance is $\mathrm{E}(e_t^2) = \lambda$.

We want to test the following hypotheses:

H_0 : The parameter vectors are the same, $\theta = \theta_0$. That is, the model is validated.

H_1 : The parameter vector θ is significantly different from θ_0, and the null hypothesis can be rejected.

We will argue, from the examples below, that all plausible change detection tests can be expressed in one of the following ways:

1. The parameter estimation error is $\tilde{\theta} = MY - \theta_0$, where M is a matrix to be specified. Standard statistical tests can be applied from the Gaussian assumption on the noise

$$\tilde{\theta} = MY - \theta_0 \in \mathrm{N}(0, P), \quad \text{under } H_0.$$

Both M and P are provided by the method.

2. The test statistic is the norm of the simulation error, which is denoted a *loss function* V in resemblance with the weighted least squares loss function:
$$V = \|Y - Y_0\|_Q^2 \triangleq (Y - Y_0)^T Q (Y - Y_0).$$

The above generalizations hold for our standard models only when the noise variance is known. The case of unknown or changing variance is treated in later sections, and leads to the same kind of projection interpretations, but with non-linear transformations (logarithms).

In the examples below, there is a certain geometric interpretation in that Q turns out to be a projection matrix, *i.e.*, $QQ = Q$. The figure below (adopted from Section 13.1) is useful in the following calculations for illustrating the geometrical properties of the least squares solution ($Y = Y_0 + E$ below):

$$\begin{aligned}
Q &= \Phi^T \left(\Phi\Phi^T\right)^{-1}\Phi \\
Y_0 &= \Phi^T\theta_0 \\
QY_0 &= Y_0, \\
QY &= Y_0 + QE, \\
\hat{Y} &= QY = \Phi^T\hat{\theta}.
\end{aligned}$$

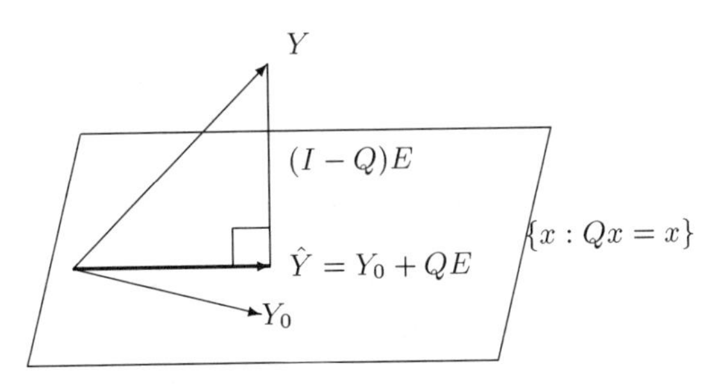

Example 6.1 Linear regression and parameter norm

Standard calculations give

$$\hat{\theta} - \theta_0 = \underbrace{\left(\Phi\Phi^T\right)^{-1}\Phi}_{M} Y - \theta_0 \in \mathrm{N}(0, \underbrace{\lambda\left(\Phi\Phi^T\right)^{-1}}_{P}), \quad \text{under } H_0.$$

The Gaussian distribution requires that the noise is Gaussian. Otherwise, the distribution is only asymptotically Gaussian. A logical test statistic is

$$\|\hat{\theta} - \theta_0\|_{P^{-1}}^2 \in \chi^2(d), \quad \text{under } H_0,$$

which is as indicated χ^2 distributed with $d = \dim(\theta)$ degrees of freedom. A standard table can be used to design a threshold, so the test becomes

$$\|\hat{\theta} - \theta_0\|^2_{P^{-1}} \gtrless h_\alpha.$$

The alternative formulation is derived by using a simulated signal $Y_0 = \Phi^T \theta_0$, which means we can write $\theta_0 = \left(\Phi\Phi^T\right)^{-1} \Phi Y_0$ and $P = \lambda \left(\Phi\Phi^T\right)^{-1}$, so

$$\begin{aligned}
V^{PN} &= \|\hat{\theta} - \theta_0\|^2_{P^{-1}} \\
&= (Y - Y_0)^T \Phi^T \left(\Phi\Phi^T\right)^{-1} \lambda^{-1} \left(\Phi\Phi^T\right) \left(\Phi\Phi^T\right)^{-1} \Phi(Y - Y_0) \\
&= \lambda^{-1}(Y - Y_0)^T \underbrace{\Phi^T \left(\Phi\Phi^T\right)^{-1} \Phi}_{Q}(Y - Y_0) \\
&= \lambda^{-1}(Y - Y_0)^T Q (Y - Y_0).
\end{aligned}$$

Note that $Q = \Phi^T \left(\Phi\Phi^T\right)^{-1} \Phi$ is a projection matrix $QQ = Q$. Here we have assumed persistent excitation so that $\Phi\Phi^T$ is invertible.

It should be noted that the alternative norm

$$\|\hat{\theta} - \theta_0\|^2,$$

without weighting with the inverse covariance matrix, is rather naïve. One example being an AR model, where the last coefficient is very small (it is the product of all poles, which are all less than one), so a change in it will not affect this norm very much, although the poles of the model can change dramatically.

Example 6.2 Linear regression and GLR

The *Likelihood Ratio* (*LR*) for testing the hypothesis of a new parameter vector and the nominal one is

$$LR = \frac{P(Y|\theta)}{P(Y|\theta_0)}.$$

Assuming Gaussian noise with constant variance we get the *log likelihood ratio* (*LLR*)

$$\begin{aligned}
2\lambda \cdot LLR &= 2\lambda \cdot \log \left(\frac{(2\pi\lambda)^{-L/2} e^{-\frac{\|Y - \Phi^T\theta\|^2}{2\lambda}}}{(2\pi\lambda)^{-L/2} e^{-\frac{\|Y - \Phi^T\theta_0\|^2}{2\lambda}}} \right) \\
&= \|Y - \Phi^T \theta_0\|^2 - \|Y - \Phi^T \theta\|^2.
\end{aligned}$$

Replacing the unknown parameter vector by its most likely value (the maximum likelihood estimate $\hat{\theta}$), we get the *Generalized Likelihood Ratio* (*GLR*)

$$\begin{aligned} V^{GLR} &= 2\lambda \cdot GLR \\ &= \|Y - \Phi^T\theta_0\|^2 - \|Y - \Phi^T\hat{\theta}\|^2 \\ &= \|Y - Y_0\|^2 - \|Y - \hat{Y}\|^2 \\ &= Y^TY - 2Y^TY_0 + Y_0^TY_0 - Y^TY + 2Y^T\hat{Y} - \hat{Y}^T\hat{Y} \\ &= -2Y^TY_0 + Y_0^TY_0 + 2Y^T\hat{Y} - \hat{Y}^T\hat{Y} \\ &= -2Y^TQY_0 + Y_0^TQY_0 + 2Y^TQY - Y^TQY \\ &= -2Y^TQY_0 + Y_0^TQY_0 + Y^TQY \\ &= (Y - Y_0)^TQ(Y - Y_0) \\ &= \|Y - Y_0\|_Q^2. \end{aligned}$$

This idea of combining GLR and a sliding window was proposed in Appel and Brandt (1983).

Example 6.3 Linear regression and model differences

A loss function, not very common in the literature, is the sum of model differences rather than prediction errors. The loss function based on *model differences* (*MD*) is

$$\begin{aligned} V^{MD} &= \|\Phi^T\theta_0 - \Phi^T\hat{\theta}\|_2^2 \\ &= \|QY_0 - QY\|_2^2 \\ &= \|Y_0 - Y\|_Q^2, \end{aligned}$$

which is again the same norm.

Example 6.4 Linear regression and divergence test

The *divergence test* was proposed in Basseville and Benveniste (1983b), and is reviewed in Section 6.2. Assuming constant noise variance, it gives

$$\begin{aligned} V^{DIV} &= \|Y - \Phi^T\theta_0\|_2^2 - (Y - \Phi^T\theta_0)^T(Y - \Phi^T\hat{\theta}) \\ &= \|Y - Y_0\|_2^2 - (Y - Y_0)^T(Y - QY) \\ &= Y^TY + Y_0^TY_0 - 2Y_0^TY - Y^TY + Y^TQY + Y_0^TY - Y_0^TQY \\ &= Y_0^TQY_0 - 2Y_0^TQY + Y^TQY + Y_0^TQY - Y_0^TQY \\ &= \|Y - Y_0\|_Q^2. \end{aligned}$$

Again, the same distance measure is obtained.

Example 6.5 Linear regression and the local approach

The test statistic in the *local approach* reviewed in Section 6.2.5 is

$$\eta_N = \frac{1}{\sqrt{N}} \sum_{k=1}^{N} \varphi_k \varepsilon_k \in \text{AsN}\left(\frac{1}{\sqrt{N}} P^{-1}\tilde{\theta}, \frac{1}{N} P^{-1}\right).$$

See Section 6.2.5 for details. Since a test statistic does not lose information during linear transformation, we can equivalently take

$$V^{LA} = \sqrt{N} P \eta \in \text{AsN}(\tilde{\theta}, P),$$

and we are essentially back to Example 6.1.

To summarize, the test statistic is (asymptotically) the same for all of the linear regression examples above, and can be written as the two-norm of the projection $Q(Y - Y_0)$,

$$\|Y - Y_0\|_Q^2 \in \chi^2(d)$$

or as the parameter estimation error

$$\hat{\theta} - \theta_0 \in \text{N}(0, P).$$

These methods coincide for Gaussian noise with known variance, otherwise they are generally different.

There are similar methods for state space models. A possible approach is shown in the example below.

Example 6.6 State space model with additive changes

State space models are discussed in the next part, but this example defines a parameter estimation problem in a state space framework:

$$x_{t+1} = A_t x_t + B_{u,t} u_t + B_{v,t} v_t + B_{\theta,t} \theta \tag{6.2}$$

$$y_t = C_t x_t + e_t + D_{u,t} u_t + D_{\theta,t} \theta. \tag{6.3}$$

The Kalman filter applied to an augmented state space model

$$\bar{x}_{t+1} = \begin{pmatrix} x_{t+1} \\ \theta_{t+1} \end{pmatrix} = \begin{pmatrix} A_t & B_{\theta,t} \\ 0 & I \end{pmatrix} \bar{x}_t + \begin{pmatrix} B_{u,t} \\ 0 \end{pmatrix} u_t + \begin{pmatrix} B_{v,t} \\ 0 \end{pmatrix} v_t \tag{6.4}$$

$$y_t = \begin{pmatrix} C_t & D_{\theta,t} \end{pmatrix} \bar{x}_t + e_t + D_{u,t} u_t \tag{6.5}$$

gives a parameter estimator

$$\hat{\theta}_{t+1|t} = \hat{\theta}_{t|t-1} + K_t^{\theta}(y_t - C_t\hat{x}_{t|t-1} - D_{\theta,t}\hat{\theta}_{t|t-1} - D_{u,t}u_t),$$

which can be expanded to a linear function of data, where the parameter estimate after L measurements can be written

$$\hat{\theta}_L = L^y Y + L^u U \in \mathrm{N}(0, P_L^{\theta\theta}), \quad \text{under } H_0.$$

Here we have split the Kalman filter quantities as

$$K_t = \begin{pmatrix} K_t^x \\ K_t^{\theta} \end{pmatrix}, \quad P_t = \begin{pmatrix} P_t^{xx} & P_t^{x\theta} \\ P_t^{\theta x} & P_t^{\theta\theta} \end{pmatrix}.$$

In a general and somewhat abstract way, the idea of a consistency test is to compute a residual vector as a linear transformation of a *batch of data*, for instance taken from a sliding window, $\varepsilon = A_i Y + b_i$. The transformation matrices depend on the approach. The norm of the residual can be taken as the *distance measure*

$$g = \|A_i Y + b_i\|$$

between the hypothesis H_1 and H_0 (no change/fault). The statistical approach in this chapter decides if the size of the distance measure is statistically significant, and this test is repeated at each time instant. This can be compared with the approach in Chapter 11, where algebraic projections are used to decide significance in a non-probabilistic framework.

6.2. Distance measures

We here review some proposed distance functions. In contrast to the examples in Section 6.1, the possibility of a changing noise variance is included.

6.2.1. Prediction error

A test statistic proposed in Segen and Sanderson (1980) is based on the prediction error

$$V^{PE} = \frac{\|Y - \Phi^T\theta_0\|^2}{\lambda_0} - N. \tag{6.6}$$

Here λ_0 is the nominal variance on the noise before the change. This statistic is small if no jump occurs and starts to grow after a jump.

6.2.2. Generalized likelihood ratio

In Basseville and Benveniste (1983c), two different test statistics for the case of two different models are given. A straightforward extension of the generalized likelihood ratio test in Example 6.2 leads to

$$V^{GLR} = N \log \frac{\lambda_0}{\lambda_1} + \frac{\|Y - \Phi^T \theta_0\|^2}{\lambda_0} - \frac{\|Y - \Phi^T \theta_1\|^2}{\lambda_1}. \tag{6.7}$$

The test statistic (6.7) was at the same time proposed in Appel and Brandt (1983), and will in the sequel be referred to as *Brandt's GLR method.*

6.2.3. Information based norms

To measure the distance between two models, any norm can be used, and we will here outline some general statistical information based approaches, see Kumamaru et al. (1989) for details and a number of alternatives. First, the *Kullback discrimination information* between two probability density functions p_1 and p_2 is defined as

$$I(1,2) = \int p_1(x) \log \frac{p_1(x)}{p_2(x)} dx \geq 0,$$

with equality only if $p_1(x) = p_2(x)$. In the special case of Gaussian distribution we are focusing on, we get

$$p_i(x) = \mathrm{N}(\hat{\theta}_i, P_i)$$
$$\Rightarrow I(1,2) = \frac{1}{2} \operatorname{tr}(P_2^{-1} P_1 - I) + \frac{1}{2} (\hat{\theta}_1 - \hat{\theta}_2)^T P_2^{-1} (\hat{\theta}_1 - \hat{\theta}_2) - \frac{1}{2} \log \left(\frac{\det P_1}{\det P_2} \right).$$

The Kullback information is not a norm and thus not suitable as a distance measure, simply because it is not symmetric $I(1,2) \neq I(2,1)$. However, this minor problem is easily resolved, and the *Kullback divergence* is defined as

$$V(1,2) = I(1,2) + I(2,1) \geq 0.$$

6.2.4. The divergence test

From the Kullback divergence, the divergence test can be derived and it is an extension of the ideas leading to (6.6). It equals

$$\begin{aligned} V^{DIV} &= N \left(\frac{\lambda_0}{\lambda_1} - 1 \right) + \left(1 + \frac{\lambda_0}{\lambda_1} \right) \frac{\|Y - \Phi^T \theta_0\|^2}{\lambda_0} \\ &\quad - 2 \frac{(Y - \Phi^T \theta_0)^T (Y - \Phi^T \theta_1)}{\lambda_1}. \end{aligned} \tag{6.8}$$

Table 6.1. Estimated change times for different methods.

Signal	Method	n_a	Estimated change times							
Noisy	Divergence	16	451	611	1450	1900	2125	2830		3626
Noisy	Brandt's GLR	16	451	611	1450	1900	2125	2830		3626
Noisy	Brandt's GLR	2		593	1450		2125	2830		3626
Filtered	Divergence	16	445	645	1550	1800	2151	2797		3626
Filtered	Brandt's GLR	16	445	645	1550	1800	2151	2797		3626
Filtered	Brandt's GLR	2	445	645	1550	1750	2151	2797	3400	3626

The corresponding algorithm will be called *the divergence test.* Both these statistics start to grow when a jump has occured, and again the task of the stopping rule is to decide whether the growth is significant. Some other proposed distance measures, in the context of speech processing, are listed in de Souza and Thomson (1982).

These two statistics are evaluated on a number of real speech data sets in Andre-Obrecht (1988) for the growing window approach. A similar investigation with the same data is found in Example 6.7 below.

Example 6.7 Speech segmentation

To illustrate an application where the divergence and GLR tests have been applied, a speech recognition system for use in cars is studied. The first task of this system, which is the target of our example, is to segment the signal.

The speech signal under consideration was recorded inside a car by the French National Agency for Telecommunications as described by Andre-Obrecht (1988). The sampling frequency is 12.8 kHz, and a part of the signal is shown in Figure 6.1, together with a high-pass filtered version with cut-off frequency 150 Hz, and the resolution is 16 bits.

Two segmentation methods were applied and tuned to these signals in Andre-Obrecht (1988). The methods are the divergence test and Brandt's GLR algorithm. The sliding window size is $L = 160$, the threshold $h = 40$ and the drift parameter $\nu = 0.2$. For the pre-filtered signal, a simple detector for finding voiced and unvoiced parts of the speech is used as a first step. In the case of unvoiced speech, the design parameters are changed to $h = 80$ and $\nu = 0.8$. A summary of the results is given in Table 6.1, and is also found in Basseville and Nikiforov (1993), for the same part of the signal as considered here. In the cited reference, see Figure 11.14 for the divergence test and Figures 11.18 and 11.20 for Brandt's GLR test. A comparison to a filter bank approach is given in Section 7.7.2.

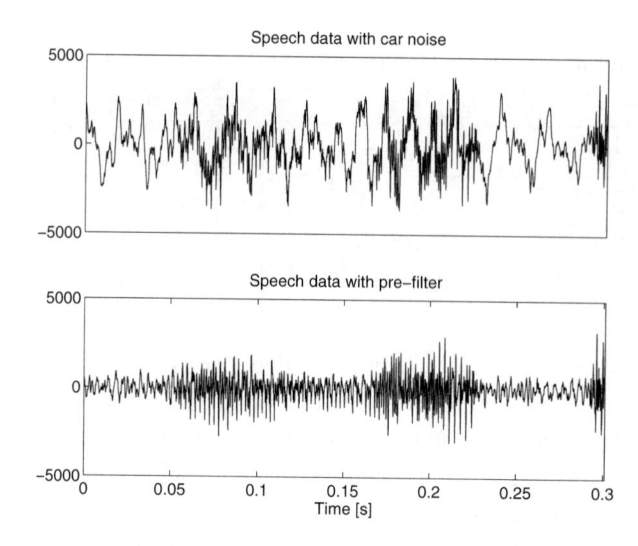

Figure 6.1. A speech signal recorded in a car (upper plot) and a high-pass filtered version (lower plot).

6.2.5. The asymptotic local approach

The *asymptotic local approach* was proposed in Benveniste et al. (1987a) as a means for monitoring any adaptive parameter estimation algorithm for abrupt parameter changes. The method is revisited and generalized to non-linear systems in Zhang et al. (1994).

The size of the data record L will be kept as an index in this section. The hypothesis test is

$$H_0 : \text{No change } \theta_L = \theta_0 \tag{6.9}$$

$$H_1 : \text{Change} \quad \theta_L = \theta_0 + \frac{1}{\sqrt{L}}\nu. \tag{6.10}$$

The assumed alternate hypothesis may look strange at first glance. Why should the change by any physical reason become less when time evolves? There is no reason. The correct interpretation is that the hypothesis makes use of the fact that the test can be made more sensitive when the number of data increases. In this way, an estimate of the change magnitude will have a covariance of constant size, rather than decreasing like $1/L$. Other approaches described in Section 6.1 implicitly have this property, since the covariance matrix P decays like one over L. The main advantages of this hypothesis test are the following:

- The asymptotic local approach, which is standard in statistics, can be applied. Thus, asymptotic analysis is facilitated. Note, however, from

Example 6.5 that algorithmically it is asymptotically the same as many other approaches when it comes to a standard model structure.

- The problem formulation can be generalized to, for example, non-linear models.

Let Z_k be the available data at time k. Assume we are given a function $K(Z_k, \theta_0)$. If it satisfies

$$\begin{aligned} \mathrm{E}(K(Z_k, \theta_0)) &= 0, \quad \theta = \theta_0 \\ \mathrm{E}(K(Z_k, \theta_0)) &\neq 0, \quad \theta \neq \theta_0 \end{aligned}$$

then it is called a *primary residual.*

Define what is called an improved residual, or *quasi-score*, as

$$\eta_L(\theta_0) \triangleq \frac{1}{\sqrt{L}} \sum_{k=1}^{L} K(Z_k, \theta_0).$$

Assume that it is differentiable and the following quantities exist:

$$\begin{aligned} M(\theta_0) &= \mathrm{E}\left(\frac{d}{d\theta} K(Z_k, \theta) \bigg|_{\theta=\theta_0} \right) \\ \Sigma(\theta_0) &= \lim_{L\to\infty} \Sigma_L(\theta_0) = \lim_{L\to\infty} \mathrm{E}(\eta_L(\theta_0)\eta_L^T(\theta_0)). \end{aligned}$$

One way to motivate the improved residual follows from a first order Taylor expansion

$$\begin{aligned} \eta_L(\theta) &\approx \eta_L(\theta_0) + \frac{1}{\sqrt{L}} L M(\theta_0) \frac{1}{\sqrt{L}} \nu \\ &= \eta_L(\theta_0) + M(\theta_0)\nu. \end{aligned}$$

By neglecting the rest term, it follows from the asymptotic distribution and a variant of the central limit theorem that

$$\eta_L(\theta_0) = \begin{cases} \mathrm{AsN}(0, \Sigma(\theta_0)) & \text{under } H_0 \\ \mathrm{AsN}(M(\theta_0)\nu, \Sigma(\theta_0)) & \text{under } H_1. \end{cases}$$

From the asymptotic distribution, standard tests can be applied as will be outlined below. A more formal proof is given in Benveniste et al. (1987a) using the ODE method.

Example 6.8 Asymptotic local approach for linear regression model

Consider as a special case the linear regression model, for which these definitions become quite intuitive. For a linear regression model with the following standard definitions:

$$\begin{aligned}
y_k &= \varphi_k^T(\theta - \theta_0) + e_k \\
\hat{\theta}_L &= \underbrace{\left(\sum_{k=1}^{L} \varphi_k \varphi_k^T\right)^{-1}}_{R_L} \underbrace{\left(\sum_{k=1}^{L} \varphi_k y_k\right)}_{f_L} \\
P_L &= \lambda R_L^{-1} \\
\hat{\theta}_L &\in \mathrm{AsN}(\theta - \theta_0, P_L).
\end{aligned}$$

the data, primary residual and improved residual (quasi-score) are defined as follows:

$$\begin{aligned}
Z_k &= (\varphi_k^T, y_k^T)^T \\
K(Z_k, \theta_0) &= \varphi_k(y_k - \varphi_k^T\theta_0) \\
\eta_L(\theta_0) &= \frac{1}{\sqrt{L}} \sum_{k=1}^{N} \varphi_k(y_k - \varphi_k^T\theta_0).
\end{aligned}$$

Using the least squares statistics above, we can rewrite the definition of $\eta_L(\theta_0)$ as

$$\begin{aligned}
\eta_L(\theta_0) &= \frac{1}{\sqrt{L}} \left(\sum_{k=1}^{L} \varphi_k y_k - \sum_{k=1}^{L} \varphi_k \varphi_k^T \theta_0\right) \\
&= \frac{1}{\sqrt{L}} \left(R_L\hat{\theta}_L - R_L\theta_0\right) \\
&= \frac{1}{\sqrt{L}} \lambda P_L^{-1} \left(\hat{\theta}_L - \theta_0\right).
\end{aligned}$$

Thus, it follows that the asymptotic distribution is

$$\eta_L(\theta_0) \in \mathrm{AsN}\left(\frac{\lambda}{\sqrt{L}} P_L^{-1}(\hat{\theta}_L - \theta_0), \frac{\lambda^2}{L} P_L^{-1}\right).$$

Note that the covariance matrix $\frac{1}{L}P_L^{-1}$ tends to a constant matrix $\Sigma(\theta_0)$ whenever the elements in the regressor are quasi-stationary.

The last remark is one of the key points in the asymptotic approach. The scaling of the change makes the covariance matrix independent of the sliding window size, and thus the algorithm has constant sensitivity.

The primary residual $K(Z_k, \theta_0)$ resembles the update step in an adaptive algorithm such as RLS or LMS. One interpretation is that under the no change hypothesis, then $K(Z_k, \theta_0) \approx \Delta\theta$ in the adaptive algorithm. The detection algorithm proposed in Hägglund (1983) is related to this approach, see also (5.63).

Assume now that $\eta \in \text{AsN}(M\nu, \Sigma)$, where we have dropped indices for simplicity. A standard Gaussian hypothesis test to test $H_0: \ \nu = 0$ can be used in the case that η is scalar (see Example 3.6). How can we obtain a hypothesis test in the case that η is not scalar? If M is a square matrix, a χ^2 test is readily obtained by noting that

$$
\begin{aligned}
M^{-1}\eta &\in \text{AsN}(\nu, M^{-1}\Sigma M^{-T}) \\
\underbrace{\left(M^{-1}\Sigma M^{-T}\right)^{-1/2} M^{-1}\eta}_{v} &\in \text{AsN}(\left(M^{-1}\Sigma M^{-T}\right)^{-1/2}\nu, I_{n_\nu}) \\
v^T v &\in \text{As}\chi^2(n_\nu),
\end{aligned}
$$

where the last distribution holds when $\nu = 0$. A hypothesis test threshold is taken from the χ^2 distribution. The difficulty occurs when M is a thin matrix, when a projection is needed. Introduce the test statistic

$$
v = \left(M^T\Sigma^{-1}M\right)^{-1/2} M^T\Sigma^{-1}\eta.
$$

We have

$$
\begin{aligned}
\text{E}(v) &= \left(M^T\Sigma^{-1}M\right)^{-1/2} M^T\Sigma^{-1}M\nu \\
&= \left(M^T\Sigma^{-1}M\right)^{1/2}\nu \\
\text{Cov}(v) &= \left(M^T\Sigma^{-1}M\right)^{-1/2} M^T\Sigma^{-1}M \left(M^T\Sigma^{-1}M\right)^{-1/2} = I_{n_\nu}.
\end{aligned}
$$

We have now verified that the test statistic satisfies $v^T v \in \chi^2(n_\nu)$ under H_0, so again a standard test can be applied.

6.2.6. General parallel filters

The idea of parallel filtering is more general than doing model validation on a batch of data. Figure 6.2 illustrates how two adaptive linear filters with different adaptation rates are run in parallel. For example, we can take one RLS filter with forgetting factor 0.999 as the slow filter and one LS estimator over a sliding window of size 20 as the fast filter. The task of the slow filter

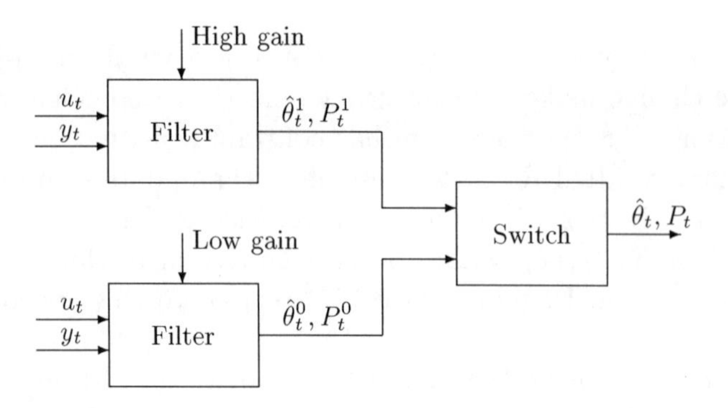

Figure 6.2. Two parallel filters. One slow to get good noise attenuation and one fast to get fast tracking.

is to produce low variance state estimates in normal mode, and the fast filter takes over after abrupt of fast changes. The *switch* makes the decision. Using the guidelines from Section 6.1, we can test the size of

$$(\hat{\theta}^1 - \hat{\theta}^0)^T (P^1)^{-1} (\hat{\theta}^1 - \hat{\theta}^0).$$

The idea is that P^1 is a valid measure of the uncertainty in θ^1, while the covariance P^0 of the slow filter is negligible.

6.3. Likelihood based detection and isolation

6.3.1. Diagnosis

Diagnosis is the combined task of detection and *isolation*. Isolation is the problem of deciding what has happened after the change detector has noticed that something has happened.

Let the parameter vector be divided into two parts,

$$Y = \Phi^T \theta = \begin{pmatrix} \Phi^a \\ \Phi^b \end{pmatrix}^T \begin{pmatrix} \theta^a \\ \theta^b \end{pmatrix}.$$

Here the parameter might have been reordered such that the important ones to diagnose come first. The difference between change detection and isolation can be stated as

$$\begin{array}{lll} \text{Detection} & H_0 : \theta = \theta_0 & H_1 : \theta \neq \theta_0 \\ \text{Isolation} & H_0^a : \theta^a = \theta_0^a & H_1^a : \theta^a \neq \theta_0^a \end{array}$$

Note that there might be several isolation tests. For instance, we may want to test H_0^b at the same time, or another partitioning of the parameter vector. There are two ways to treat the part θ^b of the parameter vector not included in the test:

- $\theta^b = \theta_0^b$, so the fault in θ^a does not influence the other elements in the parameter vector.
- θ^b is a nuisance, and its value before and after a fault in θ^a is unknown and irrelevant.

The notation below is that Y denotes the vector of measurements and $Y_0 = \Phi^T\theta_0$ is the result of a simulation of the nominal model. The first alternative gives the following GLR (where $\theta^a = \theta_0^a + \Delta\theta^a$ is used):

$$\begin{aligned}
V^1 &= -2\log\left(\frac{\max_{\theta^a} p\left(Y\middle|\begin{pmatrix}\theta^a\\ \theta_0^b\end{pmatrix}\right)}{p\left(Y\middle|\begin{pmatrix}\theta_0^a\\ \theta_0^b\end{pmatrix}\right)}\right)\\
&= -2\log\left(\frac{\max_{\Delta\theta^a} p\left(Y-Y_0\middle|\begin{pmatrix}\Delta\theta^a\\ 0\end{pmatrix}\right)}{p\left(Y-Y_0\middle|\begin{pmatrix}0\\ 0\end{pmatrix}\right)}\right)\\
&= (Y-Y_0)^T\underbrace{(\Phi^a)^T\left(\Phi^a(\Phi^a)^T\right)^{-1}\Phi^a}_{Q^a}(Y-Y_0)\\
&= (Y-Y_0)^T Q^a (Y-Y_0).
\end{aligned}$$

The second alternative is

$$\begin{aligned}
V^2 &= -2\log\left(\frac{\max_{\theta^a,\theta^b} p\left(Y\middle|\begin{pmatrix}\theta^a\\ \theta^b\end{pmatrix}\right)}{\max_{\theta^b} p\left(Y\middle|\begin{pmatrix}\theta_0^a\\ \theta^b\end{pmatrix}\right)}\right)\\
&= -2\log\left(\frac{\max_{\Delta\theta^a,\Delta\theta^b} p\left(Y-Y_0\middle|\begin{pmatrix}\Delta\theta^a\\ \Delta\theta^b\end{pmatrix}\right)}{\max_{\Delta\theta^b} p\left(Y-Y_0\middle|\begin{pmatrix}0\\ \Delta\theta^b\end{pmatrix}\right)}\right)\\
&= -(Y-Y_0)^T Q^b (Y-Y_0) + (Y-Y_0)^T Q (Y-Y_0)\\
&= (Y-Y_0)^T (Q-Q^b)(Y-Y_0).
\end{aligned}$$

We now make some comments on these results:

- For detection, we use the test statistic $V = (Y-Y_0)^T Q (Y-Y_0)$. This test is sensitive to all changes in θ.

- For isolation, we compute either V^1 or V^2 (depending upon the philosophy used) for different sub-vectors of θ — θ^a and θ^b being two possibilities — corresponding to different errors. The result of isolation is that the sub-vector with smallest loss V^j has changed.

- The spaces $\{x : Q^a x = x\}$ and $\{x : Q^b x = x\}$ can be interpreted as subspaces of the $\{x : Qx = x\}$ subspace of $\mathcal{R}^L$.

- There is no simple way to compute Q^b from Q.

- $Q^b - Q$ is no projection matrix.

- It can be shown that $Q \geq Q^b + Q^a$, with equality only when Q is block diagonal, which happens when $(\Phi^a)^T \Phi^b = 0$. This means that the second alternative gives a smaller test statistic.

The geometrical interpretation is as follows. V^1 is the part of the residual energy $V = \|Y - Y_0\|_Q^2$ that belongs to the subspace generated by the projection Q^a. Similarly, V^2 is the residual energy V subtracted by part of it that belongs to the subspace generated by the projection Q^b. The measures V^1 and V^2 are equal only if these subspaces (Q^a and Q^b) are orthogonal.

Example 6.9 Diagnosis of sensor faults

Suppose that we measure signals from two sensors, where each one can be subject to an offset (fault). After removing the known signal part, a simple model of the offset problem is

$$y_t = \begin{pmatrix} \theta^a \\ \theta^b \end{pmatrix} + e_t.$$

That is, the measurements here take the role of residuals in the general case. Suppose also that the nominal model has no sensor offset, so $\theta_0 = 0$ and $Y_0 = 0$. Consider a change detection and isolation algorithm using a sliding window of size L. First, for detection, the following distance measure should be used

$$Q = I_{2L} \Rightarrow V = Y^T Q Y = \sum_{i=t-L+1}^{t} \sum_{j=1}^{2} (y_i^{(j)})^2.$$

Secondly, for fault isolation, we compute the following two measures for approach 1:

$$\Phi^a = \text{diag}(1,0,1,0,\dots,1,0) \Rightarrow V^a = Y^T Q^a Y = \sum_{i=t-L+1}^{t} (y_i^{(1)})^2$$

$$\Phi^b = \text{diag}(0,1,0,1,\dots,0,1) \Rightarrow V^b = Y^T Q^b Y = \sum_{i=t-L+1}^{t} (y_i^{(2)})^2.$$

Note that Y is a column vector, and the projection Q^a picks out every second entry in Y, that is every sensor 1 sample, and similarly for Q^b. As a remark, approach 2 will coincide in this example, since $Q^a = Q - Q^b$ in this example.

To simplify isolation, we can make a table with possible faults and their influence of the distance measures:

Fault	θ^a	θ^b
V^1	1	0
V^2	0	1

Here 1 should be read 'large' and 0 means 'small' in some measure. Compare with Figure 11.1 and the approaches in Chapter 11.

6.3.2. A general approach

General approaches to detection and isolation using likelihood-based methods can be derived from the formulas in Section 5.B. Consider a general expression for the likelihood of a hypothesis

$$p(Y|H_i) = p(Y|\theta = \theta_i, \lambda = \lambda_i).$$

Several hypotheses may be considered simultaneously, each one considering changes in different subsets of the parameters, which can be divided in subsets as $(\theta^a, \theta^b, \lambda)$. Nuisance parameters can be treated as described in Section 6.3.1, by nullifying or estimating them. A third alternative is marginalization. The likelihoods are computed exactly as described in Section 5.B.4. This section outlines how this can be achieved for the particular case of isolating parametric and variance changes. See also the application in Section 6.5.3.

6.3.3. Diagnosis of parameter and variance changes

The following signal model can be used in order to determine whether an abrupt change in the parameters or noise variance has occurred at time t_0:

$$y_t = \begin{cases} \varphi_t^T \theta^0 + e_t, & \text{Var}(e_t) = \lambda_0 R_t \quad \text{if } t \le t_0 \\ \varphi_t^T \theta^1 + e_t, & \text{Var}(e_t) = \lambda_1 R_t \quad \text{if } t > t_0 \end{cases} \tag{6.11}$$

For generality, a known time-varying noise (co-)variance R_t is introduced. We can think of λ as either a scaling of the noise variance or the variance itself ($R_t = 1$). Neither θ_0, θ_1, λ_0 or λ_1 are known.

The following hypotheses are used:

$$\begin{aligned} H_0 &: \theta_0 = \theta_1 \text{ and } \lambda_0 = \lambda_1 \\ H_1 &: \theta_0 \neq \theta_1 \text{ and } \lambda_0 = \lambda_1 \\ H_2 &: \theta_0 = \theta_1 \text{ and } \lambda_0 \neq \lambda_1. \end{aligned} \tag{6.12}$$

Figure 6.3 illustrates the setup and the sufficient statistics from the filters are given in (6.13).

$$\begin{array}{rcc} \text{Data} & \underbrace{y_1, y_2, \dots, y_{t-L},} & \underbrace{y_{t-L+1}, \dots, y_t} \\ \text{Model} & M_0 & M_1 \\ \text{Time interval} & T_0 & T_1 \\ \text{RLS quantities} & \hat{\theta}_0,\ P_0 & \hat{\theta}_1,\ P_1 \\ \text{Loss function} & V_0 & V_1 \\ \text{Number of data} & n_0 = t - L & n_1 = L, \end{array} \tag{6.13}$$

where P_j, $j = 0, 1$, denotes the covariance of the parameter estimate achieved from the RLS algorithm. The loss functions are defined by

$$V_j(\theta) = \sum_{k \in T_j} \left(y_k - \varphi_k^T \theta\right)^T (\lambda_j R_k)^{-1} \left(y_k - \varphi_k^T \theta\right), \ j = 0, 1. \tag{6.14}$$

Note that it makes sense to compute $V_1(\hat{\theta}_0)$ to test how the first model performs on the new data set. The maximum likelihood approach will here be stated in the slightly more general maximum *a posteriori* approach, where the prior

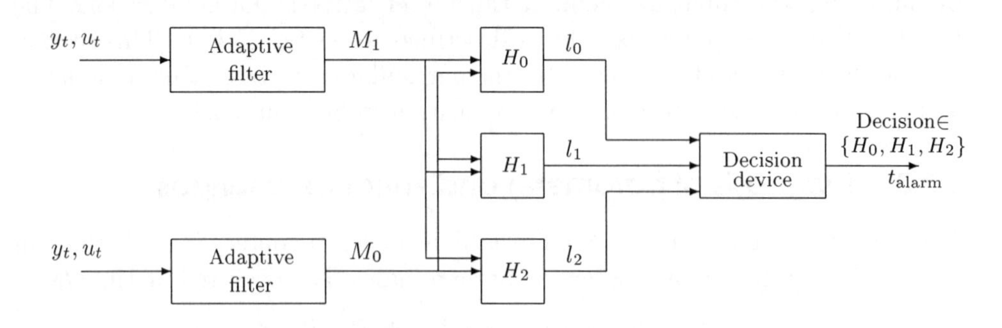

Figure 6.3. The model used to describe the detection scheme.

probabilities q_i for each hypothesis can be incorporated. The exact *a posteriori* probabilities

$$l_i = -2 \log p(H_i | y_1, y_2, \ldots, y_t), \ i = 0, 1, 2 \tag{6.15}$$

are derived below.

Assuming that $H_i, i = 0, 1, 2$, is Bernoulli distributed with probability q_i, i.e.

$$H_i = \begin{cases} \text{does not hold,} & \text{with probability } 1 - q_i \\ \text{holds,} & \text{with probability } q_i, \end{cases} \tag{6.16}$$

$\log p(H_i)$ is given by

$$\begin{aligned} \log p(H_i) &= \log \left(q_i^2 (1 - q_i)^{n_0 + n_1 - 2} \right) \\ &= 2 \log(q_i) + (n_0 + n_1 - 2) \log(1 - q_i), \ i = 0, 1, 2. \end{aligned} \tag{6.17}$$

Consider model (6.11), where $e \in N(0, \lambda)$. For marginalization purposes, the prior distribution on λ can be taken as inverse Wishart. The inverse Wishart distribution has two parameters, m and σ, and is denoted by $W^{-1}(m, \sigma)$. Its probability density function is given by

$$p(\lambda) = \frac{\sigma^{m/2} e^{-\frac{\sigma}{2\lambda}}}{2^{m/2} \Gamma(m/2) \lambda^{(m+2)/2}}. \tag{6.18}$$

The expected mean value of λ is

$$\mathrm{E}(\lambda) = \frac{\sigma}{m - 2} \tag{6.19}$$

and the variance is given by

$$\mathrm{Var}(\lambda) = \frac{2\sigma^2}{(m - 2)^2 (m - 4)}. \tag{6.20}$$

The mean value (6.19) and noise variance (6.20) are design parameters. From these, the Wishart parameter m and σ can be computed.

Algorithm 6.1 Diagnosis of parameter and variance changes

Consider the signal model (6.11) and the hypotheses given in (6.12). Let the prior for λ be as in (6.18) and the prior for the parameter vector be $\theta \in \mathrm{N}(0, P_0)$. With the loss function (6.14) and standard least squares estimation, the *a posteriori* probabilities are approximately given by

$$l_0 \approx (n_0 + n_1 - 2 + m) \log\left(\frac{V_0(\hat{\theta}_0) + V_1(\hat{\theta}_0) + \sigma}{n_0 + n_1 - 4}\right) + \log\det(P_0^{-1} + P_1^{-1}) + 2\log(q_0), \tag{6.21}$$

$$l_1 \approx (n_0 + n_1 - 2 + m) \log\left(\frac{V_0(\hat{\theta}_0) + V_1(\hat{\theta}_1) + \sigma}{n_0 + n_1 - 4}\right) - \log\det P_0 - \log\det P_1 + 2\log(q_1), \tag{6.22}$$

$$l_2 \approx (n_0 - 2 + m) \log\left(\frac{V_0(\hat{\theta}_0) + \sigma}{n_0 - 4}\right) + (n_1 - 2 + m)\log\left(\frac{V_1(\hat{\theta}_0) + \sigma}{n_1 - 4}\right) - 2\log\det P_0 + 2\log(q_2). \tag{6.23}$$

Derivation: Using the same type of calculations as in Section 5.B.4, the following *a posteriori* probabilities can be derived. They are the sum of the negative log likelihood and the prior in (6.17):

$$l_0 = -2\log p(H_0) + 2\log p(\{y_k\}_{k=1}^t) + (n_0 + n_1)\log(\pi) + 2\log(2) + D_0 + D_1 + 2\log\Gamma\left(\frac{m}{2}\right) - m\log(\sigma) - 2\log\Gamma\left(\frac{n_0 + n_1 - 2}{2}\right) + (n_0 + n_1 - 2 + m)\log\left(V_0(\hat{\theta}_0) + V_1(\hat{\theta}_0) + \sigma\right) \tag{6.24}$$

$$l_1 = -2\log p(H_1) + 2\log p(\{y_k\}_{k=1}^t) + (n_0 + n_1)\log(\pi) + 2\log(2) + D_0 + D_1 + 2\log\Gamma\left(\frac{m}{2}\right) - m\log(\sigma) - 2\log\Gamma\left(\frac{n_0 + n_1 - 2}{2}\right) + (n_0 + n_1 - 2 + m)\log\left(V_0(\hat{\theta}_0) + V_1(\hat{\theta}_1) + \sigma\right), \tag{6.25}$$

$$
\begin{aligned}
l_2 = &- 2\log p(H_2) + 2\log p(\{y_k\}_{k=1}^t) + (n_0+n_1)\log(\pi) + 4\log(2) \\
&+ D_0 + (n_0 - 2 + m)\log(V_0(\hat{\theta}_0)+\sigma) - 2\log\Gamma\left(\frac{n_0-2}{2}\right) \\
&+ 2\log\Gamma\left(\frac{m}{2}\right) - m\log(\sigma) + D_1 \\
&+ (n_1 - 2 + m)\log(V_1(\hat{\theta}_0)+\sigma) - 2\log\Gamma\left(\frac{n_1-2}{2}\right).
\end{aligned}
\tag{6.26}
$$

By using

$$
D_i = \log\det P_0 - \log\det P_i + \sum_{t=1}^{n_0+n_1} \log\det R_t, \tag{6.27}
$$

removing terms that are small for large t and L, $t \gg L$, and removing constants that are equal in Equations (6.24), (6.25) and (6.26), and *Stirling's formula* $\Gamma(n+1) \approx \sqrt{2\pi} n^{n+1/2} e^{-n}$, the approximate formulas are achieved.

6.4. Design optimization

It is wellknown that low order models are usually preferred to the full model in change detection. An example is speech signals, where a high order AR model is used for modeling, but a second order AR model is sufficient for segmentation; see Section 11.1.3 in Basseville and Nikiforov (1993), and Example 6.7. One advantage of using low order models is of course lower complexity. Another heuristic argument is that, since the variance on the model is proportional to the model order, a change can be determined faster with a constant significance level for a low order model. The price paid is that certain changes are not visible in the model anymore, and are thus not detectable. A good example is a FIR model for the impulse response of an IIR filter; changes in the impulse response beyond the FIR truncation level are not detectable.

Model order selection for change detection

The smaller the model order, the easier it is to detect changes. A reduced order model implies that there are certain subspaces of the parameter vector that are not detectable. By proper model reduction, these subspaces can be designed so that they would not be detectable in any model order due to the a poor signal-to-noise ratio for that change.

It is customary to fix the significance level, that is the probability for false alarms, and try to maximize the power of the test or, equivalently, minimize the average delay for detection. The power is of course a function of the true change, and using low order models there are always some changes that cannot be detected. However, for large enough model orders the variance contribution will inevitably outweigh the contribution from the system change. These ideas were first presented in Ninness and Goodwin (1991) and Ninness and Goodwin (February, 1992), and then refined in Gustafsson and Ninness (1995).

There is a close link between this work and the identification of transfer functions, where the trade-off between bias and variance is wellknown; see, for instance, Ljung (1985). In the cited work, an expression for the variance term is derived which is asymptotic in model order and number of data. Asymptotically, the variance term is proportional to the model order and since the bias is decreasing in the model order, this shows that the model order cannot be increased indefinitely. That is, a finite model order gives the smallest overall error in the estimate, although the true system might be infinitely dimensional.

Example 6.10 Under-modeling of an FIR system

In Gustafsson and Ninness (1995), an asymptotic analysis is presented for the case of θ being the impulse response of a system, using an FIR model. The data are generated by the so called 'Åström' system under the no change hypothesis

$$y_t = \frac{q^{-1} + 0.5q^{-2}}{1 - 1.5q^{-1} + 0.7q^{-2}} u_t + e_t.$$

The change is a shift in the phase angle of the complex pole pair from 0.46 to 0.4. The corresponding impulse responses are plotted in Figure 6.4(a). Both the input and noise are white Gaussian noise with variance 1 and 0.1, respectively. The number of data in the sliding window is $L = 50$.

Figure 6.4(b) shows a Monte Carlo simulation for 1000 noise realizations. A χ^2 test was designed to give the desired confidence level. The upper plot shows the chosen and obtained confidence level. The lower plot shows the asymptotic power function (which can be pre-computed) and the result from the Monte Carlo simulation. Qualitatively, they are very similar, with local maxima and minima where expected and a large increase between model orders 4 and 8. The power from the Monte Carlo simulation is, however, much smaller, which depends on a crude approximation of a χ^2 distribution that probably could be refined.

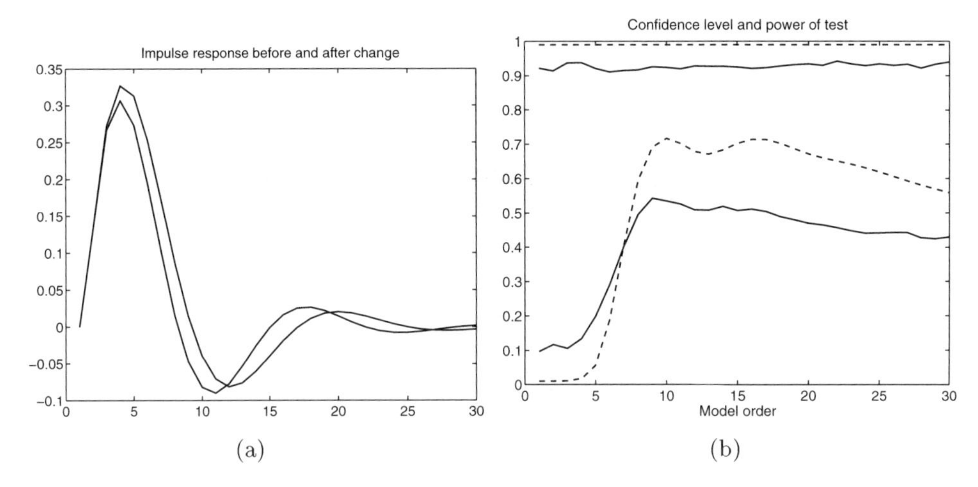

Figure 6.4. Impulse response before and after a change for the system under consideration (a). Significance level and power from asymptotic expression and simulation as a function of the number of parameters in the model (b).

6.5. Applications

6.5.1. Rat EEG

Algorithm 6.1 is applied here to the rat EEG described in Section 2.4.1, using an AR(2) model. The size of the sliding window is chosen to $L = n_1 = 100$ and both types of parameter and variance changes are considered. The result is

Change times	1085	1586	1945	2363	2949	3632	3735
Winning hypothesis	2	2	2	2	2	2	1

That is, the first six changes are due to variance changes, and the last one is a change in dynamics, which is minor, as seen from the parameter plot in Figure 6.5.

6.5.2. Belching sheep

In developing medicine for asthma, sheep are used to study its effects as described in Section 2.5.2. The dynamics between air volume and pressure is believed to depend on the medicine's effect. The dynamics can accurately be adaptively modeled by an ARX model. The main problem is the so-called *outliers*: segments with bad data, caused by belches. One should therefore detect the belch segments, and remove them before modeling. We have here taken a batch of data (shown in Figure 6.6), estimated an ARX(5,5,0) model to all data and applied Algorithm 6.1 as a variance change detector.

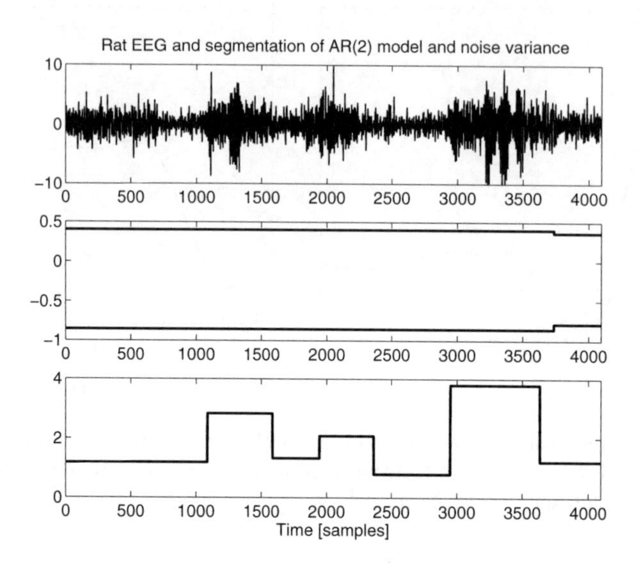

Figure 6.5. EEG for a rat and segmented noise variance from an AR(2) model.

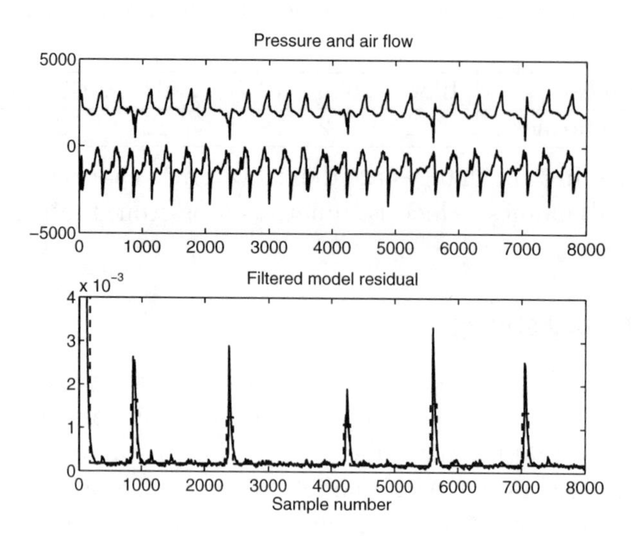

Figure 6.6. ARX residuals. Second: low-pass filtered and segmented noise variance

That is, only hypothesis H_2 is considered. The result is illustrated in the lower plot in Figure 6.6. The solid line is a low-pass filtered version of ε_t^2, and the dashed line the detected belch segments, where the variance is much higher.

6.5.3. Application to digital communication

In this section, it is demonstrated how Algorithm 6.1 can be applied to detect the occurrence of double talk and abrupt changes in the echo path in a communication system.

In a telephone system, it is important to detect a change in the echo path quickly, but not confuse it with double talk, since the echo canceler should react differently for these two phenomena. Figure 6.7 illustrates the echo path of a telephone system. The task for the hybrid is to convert 4-wire to 2-wire connection, which is used by all end users. This device is not perfect, but a part of the far-end talker is reflected and transmitted back again. The total echo path is commonly modeled by a FIR model. The adaptive FIR filter estimates the echo path and subtracts a prediction of the echo. Usually, the adaptation rate is low, but on two events it should be modified:

- After an *echo path change*, caused by a completely different physical connection being used, the adaption rate should be increased.
- During *double talk*, which means that the near-end listener starts to talk

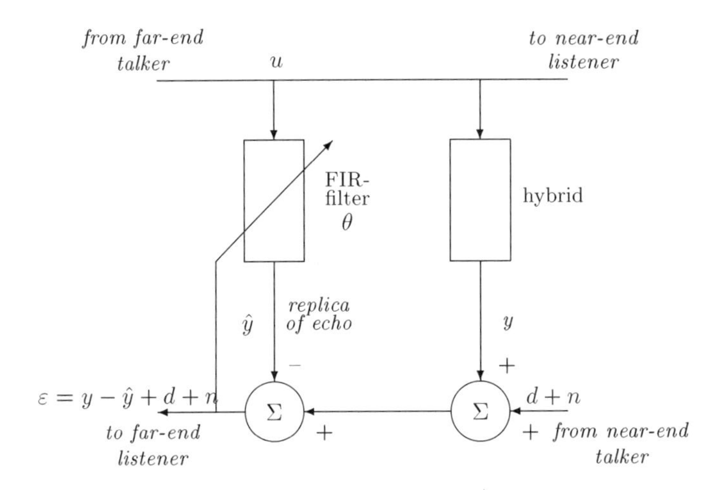

Figure 6.7. The communication system with an echo return path (from Carlemalm and Gustafsson (1998)).

simultaneously, the adaptation rate should be decreased (the residuals are large, implying fast adaptation) and must not be increased.

That is, correct isolation is far more important than detection. It is better to do nothing than to change adaptation rate in the wrong direction. This conclusion is generalized below.

Fundamental adaptive filtering problem

Disturbances and system changes must be isolated. An alarm caused by a system change requires that the adaptation rate should be increased, while an alarm caused by a disturbance (false alarm) implies that the adaptivity should be frozen.

The impulse response of a telephone channel is short, i.e. about 4 ms long (i.e. 32 samples with the standard 8 kHz sampling frequency), but since the delay can be large, the FIR filter is often of a length of between 128–512.

Algorithm 6.1 can be applied directly to the current problem. Hypothesis H_1 corresponds to echo path change and H_2 to double talk. The details of the design and successful numerical evaluation are presented in Carlemalm and Gustafsson (1998).

7 Change detection based on filter banks

7.1. Basics

Let us start with considering change detection in linear regressions as an off-line problem, which will be referred to as *segmentation*. The goal is to find a sequence of time indices $k^n = (k_1, k_2, .., k_n)$, where both the number n and the locations k_i are unknown, such that a linear regression model with piecewise constant parameters,

$$y_t = \varphi_t^T \theta(i) + e_t, \quad \mathrm{E}(e_t^2) = \lambda(i) R_t \text{ when } k_{i-1} < t \le k_i, \tag{7.1}$$

is a good description of the observed signal y_t. In this chapter, the measurements may be vector valued, and the nominal covariance matrix of the noise is R_t, and $\lambda(i)$ is a possibly unknown scaling, which is piecewise constant.

One way to guarantee that the best possible solution is found, is to consider all possible segmentations k^n, estimate one linear regression model in each segment, and then choose the particular k^n that minimizes an optimality criteria,

$$\widehat{k^n} = \arg \min_{n \ge 1, 0 < k_1 < \cdots < k_n = N} V(k^n).$$

The procedure and, as it turns out, sufficient statistics as are defined in (7.6)–(7.8), are shown below:

Data	$\underbrace{y_1, y_2, .., y_{k_1}}$	$\underbrace{y_{k_1+1}, .., y_{k_2}}$	$\cdots$	$\underbrace{y_{k_{n-1}+1}, .., y_{k_n}}$
Segmentation	Segment 1	Segment 2	$\cdots$	Segment n
LS estimates	$\hat{\theta}(1), P(1)$	$\hat{\theta}(2), P(2)$	$\cdots$	$\hat{\theta}(n), P(n)$
Statistics	$V(1), D(1), N(1)$	$V(2), D(2), N(2)$	$\cdots$	$V(n), D(n), N(n)$

What is needed from each data segment is the sufficient statistics V (sum of squared residuals), D ($-\log\det$ of the covariance matrix) and number of data N in each segment, as defined in equations (7.6), (7.7) and (7.8). The segmentation k^n has $n-1$ degrees of freedom.

Two types of optimality criteria have been proposed:

- Statistical criterion: the maximum likelihood or maximum *a posteriori* estimate of k^n is studied.

- Information based criterion: the information of data in each segment is $V(i)$ (the sum of squared residuals), and the total information is the sum of these. Since the total information is minimized for the degenerated solution $k^n = 1, 2, 3, ..., N$, giving $V(i) = 0$, a *penalty term* is needed. Similar problems have been studied in the context of model structure selection, and from this literature Akaike's AIC and BIC criteria have been proposed for segmentation.

The real challenge in segmentation is to cope with the *curse of dimensionality.* The number of segmentations k^n is 2^N (there can be either a change or no change at each time instant). Here, several strategies have been proposed:

- Numerical searches based on dynamic programming or MCMC techniques.
- Recursive local search schemes.

The main part of this chapter is devoted to the second approach, which provides a solution to adaptive filtering, which is an on-line problem.

7.2. Problem setup

7.2.1. The changing regression model

The segmentation model is based on a linear regression with piecewise constant parameters,

$$y_t = \varphi_t^T \theta(i) + e_t, \quad \text{when } k_{i-1} < t \leq k_i. \tag{7.2}$$

Here $\theta(i)$ is the d-dimensional parameter vector in segment i, φ_t is the regressor and k_i denotes the change times. The measurement vector is assumed to have dimension p. The noise e_t in (7.2) is assumed to be Gaussian with variance $\lambda(i)R_t$, where $\lambda(i)$ is a possibly segment dependent scaling of the noise. We will assume R_t to be known and the scaling as a possibly unknown parameter. The problem is now to estimate the number of segments n and the sequence of change times, denoted $k^n = (k_1, k_2, .., k_n)$. Note that both the number n and positions of change times k_i are considered unknown.

Two important special cases of (7.2) are a changing mean model where $\varphi_t = 1$ and an auto-regression, where $\varphi_t = (-y_{t-1}, .., -y_{t-d})^T$.

For the analysis in Section 7.A, and for defining the prior on each segmentation, the following equivalent state space model turns out be be more convenient:

$$\begin{aligned} \theta_{t+1} &= (1 - \delta_t)\theta_t + \delta_t v_t \\ y_t &= \varphi_t^T \theta_t + e_t. \end{aligned} \tag{7.3}$$

Here δ_t is a binary variable, which equals one when the parameter vector changes and is zero otherwise, and v_t is a sequence of unknown parameter vectors. Putting $\delta_t = 0$ into (7.3) gives a standard regression model with constant parameters, but when $\delta_t = 1$ it is assigned a completely new parameter vector v_t taken at random. Thus, models (7.3) and (7.2) are equivalent. For convenience, it is assumed that $k_0 = 0$ and $\delta_0 = 1$, so the first segment begins at time 1. The segmentation problem can be formulated as estimating the number of jumps n and the jump instants k^n, or alternatively the jump parameter sequence $\delta^N = (\delta_1, .., \delta_N)$.

The models (7.3) and (7.2) will be referred to as *changing regressions*, because they change between different regression models. The most important feature with the changing regression model is that the jumps divide the measurements into a number of *independent* segments. This follows, since the parameter vectors in the different segments are independent; they are two different samples of the stochastic process $\{v_t\}$.

A related model studied in Andersson (1985) is a *jumping regression* model. The difference to the approach herein, is that the changes are added to the paremeter vector. In (7.3), this would mean that the parameter variation model is $\theta_{t+1} = \theta_t + \delta_t v_t$. We lose the property of independent segments. The optimal algorithms proposed here are then only sub-optimal.

7.2.2. Notation

Given a segmentation k^n, it will be useful to introduce compact notation $Y(i)$ for the measurements in the ith segment, that is $y_{k_{i-1}+1}, .., y_{k_i} = y_{k_{i-1}+1}^{k_i}$. The least squares estimate and its covariance matrix for the ith segment are denoted:

$$\hat{\theta}(i) = P(i) \sum_{t=k_{i-1}+1}^{k_i} \varphi_t R_t^{-1} y_t, \tag{7.4}$$

$$P(i) = \left(\sum_{t=k_{i-1}+1}^{k_i} \varphi_t R_t^{-1} \varphi_t^T \right)^{-1}. \tag{7.5}$$

Although these are off-line expressions, $\hat{\theta}(i)$ and $P(i)$ can of course be computed recursively using the *Recursive Least Squares* (RLS) scheme.

Finally, the following quantities will be shown to represent sufficient statistics in each segment:

$$V(i) = \sum_{t=k_{i-1}+1}^{k_i} (y_t - \varphi_t^T \hat{\theta}(i))^T R_t^{-1} (y_t - \varphi_t^T \hat{\theta}(i)) \tag{7.6}$$

$$D(i) = -\log\det P(i) \tag{7.7}$$

$$N(i) = k_i - k_{i-1}. \tag{7.8}$$

7.3. Statistical criteria

7.3.1. The MML estimator

Let k^n, θ^n, λ^n denote the sets of jump times, parameter vectors and noise scalings, respectively, needed in the signal model (7.2). The likelihood for data

y^N given all parameters is denoted $p(y^N|k^n, \theta^n, \lambda^n)$. We will assume independent Gaussian noise distributions, so $p(e^N) = \prod_{i=1}^{n} \prod_{t=k_{i-1}+1}^{k_i} (2\pi\lambda(i))^{-p/2} \cdot (\det R_t)^{-1/2} \exp(-e_t^T R_t^{-1} e_t / (2\lambda(i)))$. Then, we have

$$
\begin{aligned}
&-2\log p(y^N|k^n, \theta^n, \lambda^n) \\
&\quad = Np\log(2\pi) + \sum_{t=1}^{N} \log\det R_t + \sum_{i=1}^{n} N(i)\log(\lambda(i)^p) \\
&\quad + \sum_{i=1}^{n} \frac{\sum_{t=k_{i-1}+1}^{k_i} (y_t - \varphi_t^T \theta(i))^T R_t^{-1} (y_t - \varphi_t^T \theta(i))}{\lambda(i)}
\end{aligned} \tag{7.9}
$$

Here and in the sequel, p is the dimension of the measurement vector y_t. There are two ways of eliminating the nuisance parameters θ^n, λ^n, leading to the *marginalized* and *generalized* likelihoods, respectively. The latter is the standard approach where the nuisance parameters are removed by minimization of (7.9). A relation between these is given in Section 7.4. See Wald (1947) for a discussion on generalized and marginalized (or weighted) likelihoods.

We next investigate the use of the marginalized likelihood, where (7.9) is integrated with respect to a prior distribution of the nuisance parameters. The likelihood given only k^n is then given by

$$
p(y^N|k^n) = \int_{\theta^n, \lambda^n} p(y^N|k^n, \theta^n, \lambda^n) p(\theta^n|\lambda^n) p(\lambda^n) d\theta^n d\lambda^n. \tag{7.10}
$$

In this expression, the prior for θ, $p(\theta^n|\lambda^n)$, is technically a function of the noise variance scaling λ, but is usually chosen as an independent function. The maximum likelihood estimator is given by maximization of $p(y^N|k^n)$. Finally, the *a posteriori* probabilities can be computed from Bayes' law,

$$
p(k^n|y^N) = p(y^N|k^n)\frac{p(k^n)}{p(y^N)}, \tag{7.11}
$$

where $p(y^N)$ is just a constant, and the *Maximum A posteriori Probability* (*MAP*) estimate is given by maximization. In this way, a prior on the segmentation can be included. In the sequel, only the more general MAP estimator is considered.

The prior $p(k^n) = p(k^n|n)p(n)$ or, equivalently, $p(\delta^N)$ on the segmentation is a user's choice (in fact the only one). A natural and powerful possibility is to use $p(\delta^N)$ and assume a fixed probability q of jump at each new time instant. That is, consider the jump sequence δ^N as independent *Bernoulli variables* $\delta_t \in \text{Be}(q)$, which means

$$
\delta_t = \begin{cases} 0 & \text{with probability } 1-q \\ 1 & \text{with probability } q. \end{cases}
$$

It might be useful in some applications to tune the jump probability q above, because it controls the number of jumps estimated. Since there is a one-to-one correspondence between k^n and δ^N, both priors are given by

$$p(k^n) = p(\delta^n) = q^n(1-q)^{N-n}. \tag{7.12}$$

A q less than 0.5 penalizes a large number of segments. A non-informative prior $p(k^n) = 0.5^N$ is obtained with $q = 0.5$. In this case, the MAP estimator equals the *Maximum Likelihood* (*ML*) estimator, which follows from (7.11).

7.3.2. The *a posteriori* probabilities

In Appendix 7.B, the *a posteriori* probabilities are derived in three theorems for the three different cases of treating the measurement covariance: completely known, known except for a constant scaling and finally known with an unknown changing scaling. The case of completely unknown covariance matrix is not solved in the literature. These are generalizations and extensions of results for a changing mean models ($\varphi_t = 1$) presented in Chapter 3; see also Smith (1975) and Lee and Hefhinian (1978). Appendix 7.B also contains a discussion and motivation of the particular prior distributions used in marginalization. The different steps in the MAP estimator can be summarized as follows; see also (7.16).

Filter bank segmentation

- Examine every possible segmentation, parameterized in the number of jumps n and jump times k^n, separately.
- For each segmentation, compute the best models in each segment parameterized in the least squares estimates $\hat{\theta}(i)$ and their covariance matrices $P(i)$.
- Compute the sum of squared prediction errors $V(i)$ and $D(i) = -\log\det P(i)$ in each segment.
- The MAP estimate of the model structure for the three different assumptions on noise scaling (known $\lambda(i) = \lambda_o$, unknown but constant $\lambda(i) = \lambda$ and finally unknown and changing $\lambda(i)$) is given in equations (7.13), (7.14) and (7.15), respectively,

$$\widehat{k^n} = \arg\min_{k^n, n} \sum_{i=1}^{n} (D(i) + V(i)) + 2n \log \frac{1-q}{q} \tag{7.13}$$

$$\widehat{k^n} = \arg\min_{k^n, n} \sum_{i=1}^{n} D(i) + (Np - nd - 2) \log \sum_{i=1}^{n} \frac{V(i)}{Np - nd - 4} + 2n \log \frac{1-q}{q} \tag{7.14}$$

$$\widehat{k^n} = \arg\min_{k^n, n} \sum_{i=1}^{n} \left(D(i) + (N(i)p - d - 2) \log \frac{V(i)}{N(i)p - d - 4} \right) + 2n \log \frac{1-q}{q}. \tag{7.15}$$

The last two *a posteriori* probabilities are only approximate, since Stirling's formula has been used to eliminate gamma functions; the exact expressions are found in Appendix 7.B. Equation (7.16) defines the involved statistics.

$$\begin{array}{lcccc}
\text{Data} & \underbrace{y_1, y_2, .., y_{k_1}} & \underbrace{y_{k_1+1}, .., y_{k_2}} & \cdots & \underbrace{y_{k_{n-1}+1}, .., y_{k_n}} \\
\text{Segmentation} & \text{Segment } 1 & \text{Segment } 2 & \cdots & \text{Segment } n \\
\text{LS estimates} & \hat{\theta}(1), P(1) & \hat{\theta}(2), P(2) & \cdots & \hat{\theta}(n), P(n) \\
\text{Statistics} & V(1), D(1) & V(2), D(2) & \cdots & V(n), D(n)
\end{array} \tag{7.16}$$

The required steps in computing the MAP estimated segmentation are as follows. First, every possible segmentation of the data is examined separately. For each segmentation, one model for every segment is estimated and the test statistics are computed. Finally, one of equations (7.13)–(7.15) is evaluated.

In all cases, constants in the *a posteriori* probabilities are omitted. The difference in the three approaches is thus basically only how to treat the sum of squared prediction errors. A prior probability q causes a penalty term increasing linearly in n for $q < 0.5$. As noted before, $q = 0.5$ corresponds to ML estimation.

The derivations of (7.13) to (7.15) are valid only if all terms are well-defined. The condition is that $P(i)$ has full rank for all i, and that the denominator under $V(i)$ is positive. That is, $Np - nd - 4 > 0$ in (7.14) and $N(i)p - d - 4 > 0$ in (7.15). The segments must therefore be forced to be long enough.

7.3.3. On the choice of priors

The Gaussian assumption on the noise is a standard one, partly because it gives analytical expressions and partly because it has proven to work well in practice. Other alternatives are rarely seen. The Laplacian distribution is shown in Wu and Fitzgerald (1995) to also give an analytical solution in the case of unknown mean models. It was there found that it is less sensitive to large measurement errors.

The standard approach used here for marginalization is to consider both Gaussian and non-informative prior in parallel. We often give priority to a non-informative prior on θ, using a flat density function, in our aim to have as few non-intuitive design parameters as possible. That is, $p(\theta^n|\lambda^n) = C$ is an arbitrary constant in (7.10). The use of non-informative priors, and especially improper ones, is sometimes criticized. See Aitken (1991) for an interesting discussion. Specifically, here the flat prior introduces an arbitrary term $n \log C$ in the log likelihood. The idea of using a flat prior, or non-informative prior, in marginalization is perhaps best explained by an example.

Example 7.1 Marginalized likelihood for variance estimation

Suppose we have t observations from a Gaussian distribution; $y_t \in \mathrm{N}(\mu, \lambda)$. Thus the likelihood $p(y^t|\mu, \lambda)$ is Gaussian. We want to compute the likelihood conditioned on just λ using marginalization: $p(y^t|\lambda) = \int p(y^t|\mu, \lambda)p(\mu)d\mu$. Two alternatives of priors are a Gaussian, $\mu \in \mathrm{N}(\mu_0, P_0)$, and a flat prior, $p(\mu) = C$. In both cases, we end up with an inverse Wishart density function (3.54) with maximas

$$\mu \in \mathrm{N}(\mu_0, P_0) \Rightarrow \hat{\lambda} = \frac{1}{t-1}\left(\sum_{k=1}^{t}(y_k - \bar{y})^2 + \frac{(\mu_0 - \bar{y})^2}{P_0}\right)$$

$$p(\mu) = C \Rightarrow \hat{\lambda} = \frac{1}{t-1}\sum_{k=1}^{t}(y_k - \bar{y})^2,$$

where $\bar{y}$ is the sample average. Note the scaling factor $1/(t-1)$, which makes the estimate unbiased. The joint likelihood estimate of both mean and variance gives a variance estimator scaling factor $1/t$. The prior thus induces a bias in the estimate.

Thus, a flat prior eliminates the bias induced by the prior. We remark that the likelihood interpreted as a conditional density function is proper, and it does not depend upon the constant C.

The use of a flat prior can be motivated as follows:

- The data dependent terms in the log likelihood increase like $\log N$. That is, whatever the choice of C, the prior dependent term will be insignificant for a large amount of data.

- The choice $C \approx 1$ can be shown to give approximately the same likelihood as a proper informative Gaussian prior would give if the true parameters were known and used in the prior. See Gustafsson (1996), where an example is given.

More precisely, with the prior $\mathrm{N}(\theta_0, P_0)$, where θ_0 is the true value of $\theta(i)$ the constant should be chosen as $C = \det P_0$. The uncertainty about θ_0 reflected in P_0 should be much larger than the data information in $P(i)$ if one wants the data to speak for themselves. Still, the choice of P_0 is ambiguous. The larger value, the higher is the penalty on a large number of segments. This is exactly *Lindley's paradox* (Lindley, 1957):

Lindley's paradox

The more non-informative prior, the more the zero-hypothesis is favored.

Thus, the prior should be chosen to be as informative as possible without interfering with data. For auto-regressions and other regressions where the parameters are scaled to be around or less than 1, the choice $P_0 = I$ is appropriate. Since the true value θ_0 is not known, this discussion seems to validate the use of a flat prior with the choice $C = 1$, which has also been confirmed to work well by simulations. An unknown noise variance is assigned a flat prior as well with the same pragmatic motivation.

Example 7.2 Lindley's paradox

Consider the hypothesis test

$$\begin{aligned} \mathrm{H}_0 &: y \in \mathrm{N}(0,1) \\ \mathrm{H}_1 &: y \in \mathrm{N}(\theta,1), \end{aligned}$$

and assume that the prior on θ is $\mathrm{N}(\theta_0, P_0)$. Equation (5.98) gives for scalar measurements that

$$p(y|H_1) \sim p(y|\hat{\theta}_N) p_\theta(\hat{\theta}_N) \sqrt{P_N}.$$

Here we have $N = 1$, $P_1 = (P_0^{-1} + 1)^{-1} \to 1$ and $\hat{\theta}_1 = P_1 y \to y$. Then the likelihood ratio is

$$\begin{aligned} \frac{p(y|H_1)}{p(y|H_0)} &= \frac{1}{\sqrt{2\pi P_1}} e^{-\frac{(y-\hat{\theta}_1)^2}{2P_1}} \frac{1}{\sqrt{2\pi P_0}} e^{-\frac{(\hat{\theta}_1)^2}{2P_0}} \sqrt{P_1} \sqrt{2\pi} e^{+\frac{(y)^2}{2}} \\ &\to 0, \quad P_0 \to \infty, \end{aligned}$$

since the whole expression behaves like $1/\sqrt{P_0}$. This fact is not influenced by the number of data or what the true mean is, or what θ_0 is. That is, the more non-informative the prior, the more H_0 is favored!

7.4. Information based criteria

The information based approach of this section can be called a penalized *Maximum Generalized Likelihood* (MGL) approach.

7.4.1. The MGL estimator

It is straightforward to show that the minimum of (7.9) with respect to θ^n, assuming a known $\lambda(i)$, is

$$\begin{aligned}\mathrm{MGL}(k^n) &= \min_{\theta^n} -2\log p(y^N|k^n,\theta^n,\lambda^n) \\ &= Np\log(2\pi) + \sum_{t=1}^{N} \log\det(R_t) \\ &\quad + \sum_{i=1}^{n}\left(\frac{V(i)}{\lambda(i)} + N(i)\log(\lambda(i)^p)\right). \end{aligned} \tag{7.17}$$

Minimizing the right-hand side of (7.17) with respect to a constant unknown noise scaling $\lambda(i) = \lambda$ gives

$$\begin{aligned}\mathrm{MGL}(k^n) &= \min_{\theta^n,\lambda} -2\log p(y^N|k^n,\theta^n,\lambda) \\ &= Np\log(2\pi) + \sum_{t=1}^{N} \log\det(R_t) + Np\left(1 + \log\frac{\sum_{i=1}^{n} V(i)}{Np}\right) \end{aligned} \tag{7.18}$$

and finally, for a changing noise scaling

$$\begin{aligned}\mathrm{MGL}(k^n) &= \min_{\theta^n,\lambda^n} -2\log p(y^N|k^n,\theta^n,\lambda^n) \\ &= Np\log(2\pi) + \sum_{t=1}^{N} \log\det(R_t) \\ &\quad + \sum_{i=1}^{n} N(i)p\left(1 + \log\frac{V(i)}{N(i)p}\right). \end{aligned} \tag{7.19}$$

In summary, the counterparts to the MML estimates (7.13)-(7.15) are given by the MGL estimates

$$\widehat{k^n} = \arg\min_{k^n,n} \sum_{i=1}^{n} \frac{V(i)}{\lambda(i)} + N(i)\log(\lambda(i)^p) \tag{7.20}$$

$$\widehat{k^n} = \arg\min_{k^n,n} \sum_{i=1}^{n} V(i) \tag{7.21}$$

$$\widehat{k^n} = \arg\min_{k^n,n} \sum_{i=1}^{n} N(i)p\left(1+\log\frac{V(i)}{N(i)p}\right). \tag{7.22}$$

It is easily realized that these generalized likelihoods cannot directly be used for estimating k^n, where n is unknown, because

- for any given segmentation, inclusion of one more change time will strictly increase the generalized likelihood ($\sum V(i)$ decreases), and
- the generalized likelihoods (7.18) and (7.19) can be made arbitrarily large, since $\sum V(i) = 0$ and $V(i) = 0$, respectively, if there are enough segments. Note that $n = N$ and $k_i = i$ is one permissible solution.

That is, the *parsimonious principle* is not fulfilled–there is no trade-off between model complexity and data fit.

7.4.2. MGL with penalty term

An attempt to satisfy the parsimonious principle is to add a *penalty term* to the generalized likelihoods (7.17)–(7.19). A general form of suggested penalty terms is $n(d+1)\gamma(N)$, which is proportional to the number of parameters used to describe the signal (here the change time itself is counted as one parameter). Penalty terms occuring in model order selection problems can be used in this application as well, like Akaike's *AIC* (Akaike, 1969) or the equivalent criteria: Akaike's *BIC* (Akaike, 1977), Rissanen's *Minimum Description Length* (*MDL*) approach (Rissanen, 1989) and *Schwartz criterion* (Schwartz, 1978). The penalty term in AIC is $2n(d+1)$ and in BIC $n(d+1)\log N$.

AIC is proposed in Kitagawa and Akaike (1978) for auto-regressive models with a changing noise variance (one more parameter per segment), leading to

$$\widehat{k^n} = \arg\min_{k^n,n} \sum_{i=1}^{n} N(i)p\log\frac{V(i)}{N(i)p} + 2n(d+2), \tag{7.23}$$

and BIC is suggested in Yao (1988) for a changing mean model ($\varphi_t = 1$) and unknown constant noise variance:

$$\widehat{k^n} = \arg\min_{k^n, n} Np \log \frac{\sum_{i=1}^{n} V(i)}{Np} + n(d+1)\log N. \tag{7.24}$$

Both (7.23) and (7.24) are globally maximized for $n = N$ and $k_i = i$. This is solved in Yao (1988) by assuming that an upper bound on n is known, but it is not commented upon in Kitagawa and Akaike (1978).

The MDL theory provides a nice interpretation of the segmentation problem: choose the segments such that the fewest possible data bits are used to describe the signal up to a certain accuracy, given that both the parameter vectors and the prediction errors are stored with finite accuracy.

Both AIC and BIC are based on an assumption on a large number of data, and its use in segmentation where each segment could be quite short is questioned in Kitagawa and Akaike (1978). Simulations in Djuric (1994) indicate that AIC and BIC tend to over-segment data in a simple example where marginalized ML works fine.

7.4.3. Relation to MML

A comparison of the generalized likelihoods (7.17)–(7.19) with the marginalized likelihoods (7.13)–(7.15) (assuming $q = 1/2$), shows that the penalty term introduced by marginalization is $\sum_{i=1}^{n} D(i)$ in all cases. It is therefore interesting to study this term in more detail.

Lemma 5.5 shows

$$-\frac{\log \det P_N}{\log N} \to d, \quad N \to \infty.$$

This implies that $\sum_{i=1}^{n} D(i) = \sum_{i=1}^{n} -\log\det P(i) \approx \sum_{i=1}^{n} d \log N(i)$. The penalty term in MML is thus of the same form asymptotically as BIC. If the segments are roughly of the same length, then $\sum_{i=1}^{n} D(i) \approx nd\log(N/n)$. Note however, that the behavior for short segments might improve using MML.

The BIC criterion in the context of model order selection is known to be a consistent estimate of the model order. It is shown in Yao (1988) that BIC is a weakly consistent estimate of the number of the change times in segmentation of changing mean models. The asymptotic link with BIC supports the use of marginalized likelihoods.

7.5. On-line local search for optimum

Computing the exact likelihood or information based estimate is computationally intractable because of the exponential complexity. This section reviews

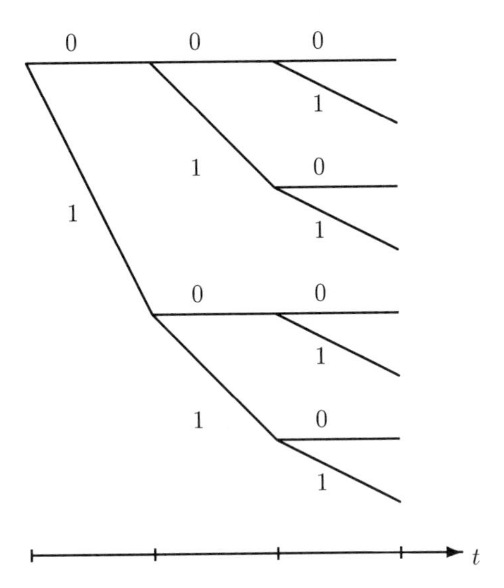

Figure 7.1. The tree of jump sequences. A path marked 0 corresponds to no jump, while 1 in the δ-parameterization of the jump sequence corresponds to a jump.

local search techniques, while the next section comments on numerical methods.

7.5.1. Local tree search

In Section 7.A, an exact pruning possibility having quadratic in time complexity is described. Here a natural recursive (linear in time) approximate algorithm will be given. The complexity of the problem can be compared to the growing tree in Figure 7.1. The algorithm will use terminology from this analogy, like cutting, pruning and merging branches. Generally, the global maximum can be found only by searching through the whole tree. However, the following arguments indicate heuristically how the complexity can be decreased dramatically.

At time t, every branch splits into two branches where one corresponds to a jump. Past data contain no information about what happens after a jump. Therefore, only one sequence among all those with a jump at a given time instant has to be considered, i.e. the most likely one. This is the point in the first step, after which only one new branch in the tree is started at each time instant. That is, there are only N branches left. This exploration of a finite memory property has much in common with the famous Viterbi algorithm in equalization, see Algorithm 5.5 or the articles Viterbi (1967) and Forney (1973).

It seems to be a waste of computational power to keep updating probabilities for sequences which have been unlikely for a long time. However, one still cannot be sure that one of them will not start to grow and become the MAP estimate. The solution offered in Section 7.A, is to compute a common upper bound on the *a posteriori* probabilities. If this bound does not exceed the MAP estimate's probability, which is normally the case, one can be sure that the true MAP estimate is found. The approximation in the following algorithm is to simply reject these sequences.

The following algorithm is a straightforward extension of Algorithm 4.1.

Algorithm 7.1 Recursive parameter segmentation

1. Choose an optimality criterion. The options are the *a posteriori* probabilities as in Theorem 7.3, 7.4 or 7.5, or the information criteria AIC (7.23) or BIC (7.24).
2. Compute recursively the optimality criterion using a bank of least squares estimators, each one matched to a particular segmentation.
3. Use the following rules for maintaining the hypotheses and keeping the number of considered sequences (M) fixed:
   a) Let only the most probable sequence split.
   b) Cut off the least probable sequence, so only M are left.
   c) Assume a minimum segment length: let the most probable sequence split *only if it is not too young*. A suitable default value is 0.
   d) Assure that sequences are not cut off immediately after they are born: cut off the least probable sequences *among those that are older than a certain minimum lifelength*, until only M are left. This should mostly be chosen as large as possible.

The last two restrictions are important for performance. A tuning rule in simulations is to simulate the signal without noise for tuning the local search parameters.

The output of the algorithm at time t is the parameter estimate of the most probable sequence, or possibly a weighted sum of all estimates. However, it should be pointed out that the fixed interval smoothing estimate is readily available by back-tracking the history of the most probable sequence, which can be realized from (7.16). Algorithm 7.1 is similar to the one proposed in Andersson (1985). However, this algorithm is *ad hoc*, and works only for the case of known noise.

Section 4.3 contains some illustrative examples, while Section 7.7 uses the algorithm in a number of applications.

7.5.2. Design parameters

It has already been noted that the segments have to be longer than a minimum segment length, otherwise the derivations in Appendix 7.B of (7.14) and (7.15) are not valid. Consider Theorem 7.5. Since there is a term $\Gamma((N(i)p-d-2)/2)$ and the gamma function $\Gamma(z)$ has poles for $z = 0, -1, -2, \ldots$, the segment lengths must be larger than $(2+d)/p$. This is intuitively logical, since d data points are required to estimate θ and two more to estimate λ.

That it could be wise to use a minimum lifelength of the sequences can be determined as follows. Suppose the model structure on the regression is a third order model. Then at least three measurements are needed to estimate the parameters, and more are needed to judge the fit of the model to data. That is, after at least four samples, something intelligent can be said about the data fit. Thus, the choice of a minimum lifelength is related to the identifiability of the model, and should be chosen larger than $\dim(\theta) + 2$.

It is interesting to point out the possibility of forcing the algorithm to give the exact MAP estimate by specifying the minimum lifelength and the number of sequences to N. In this way, only the first rule is actually performed (which is the first step in Algorithm 7.3). The MAP estimate is, in this way, found in quadratic time.

Finally, the jump probability q is used to tune the number of segments.

7.6. Off-line global search for optimum

Numerical approximations that have been suggested include dynamic programming (Djuric, 1992), batch-wise processing where only a small number of jump times is considered (Kitagawa and Akaike, 1978), and MCMC methods, but it is fairly easy to construct examples where these approaches have shortcomings, as demonstrated in Section 4.4.

Algorithm 4.2 for signal estimation is straightforward to generalize to the parameter estimation problem. This more general form is given in Fitzgerald et al. (1994), and is a combination of Gibbs sampling and the Metropolis algorithm.

Algorithm 7.2 MCMC segmentation

Decide the number of changes n and choose which likelihood to use. The options are the *a posteriori* probabilities in Theorems 7.3, 7.4 or 7.5 with $q = 0.5$:

1. Iterate Monte Carlo run i.
2. Iterate Gibbs sampler for component j in k^n, where a random number from

$$\bar{k}_j \sim p(k_j | k_1^n \text{ except } k_j)$$

   is taken. Denote the new candidate sequence $\overline{k^n}$. The distribution may be taken as flat, or Gaussian centered around the previous estimate.
3. The candidate j is accepted if the likelihood increases, $p(\overline{k^n}) > p(k^n)$. Otherwise, candidate j is accepted (the Metropolis step) if a random number from a uniform distribution is less than the likelihood ratio

$$0 \le \frac{p(\overline{k^n})}{p(k^n)} \le 1.$$

After the *burn-in* (convergence) time, the distribution of change times can be computed by Monte Carlo techniques.

The last step of random rejection sampling defines the Metropolis algorithm. Here the candidate will be rejected with large probability if its value is unlikely.

We refer to Section 4.4 for illustrative examples and Section 7.7.3 for an application.

7.7. Applications

The first application uses segmentation as a means for signal compression, modeling an EKG signal as a piecewise constant polynomial. In the second application, the proposed method is compared to existing segmentation methods. In an attempt to be as fair as possible, first we choose a test signal that has been examined before in the literature. In this way, it is clear that the algorithms under comparison are tuned as well as possible. The last application concerns real time estimation for navigation in a car.

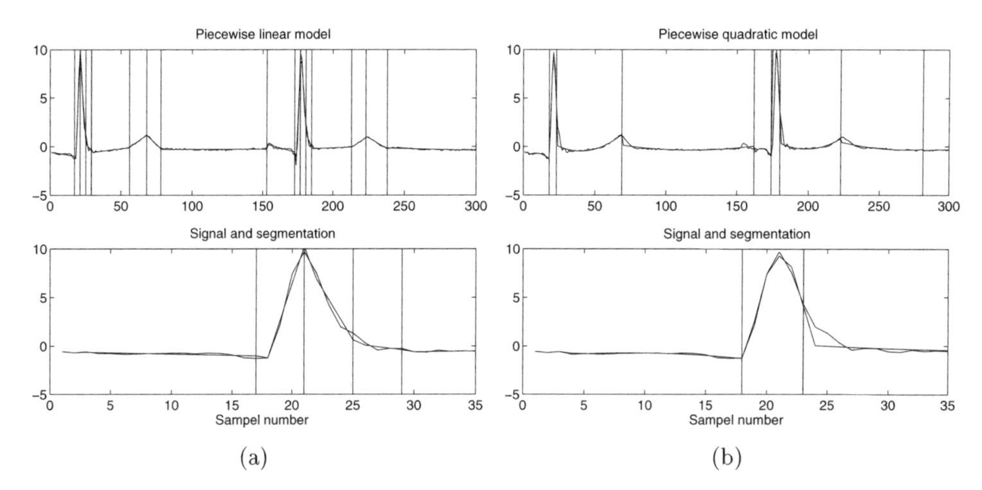

Figure 7.2. An EKG signal and a piecewise constant linear model (a) and quadratic model (b), respectively.

7.7.1. Storing EKG signals

The EKG compression problem defined in Section 2.6.2 is here approached by segmentation. Algorithm 7.1 is used with 10 parallel filters and fixed noise variance $\sigma^2 = 0.01$. The assumption of fixed variance gives us a tool to control the accuracy in the compression, and to trade it off to compression rate. Figure 7.2 shows the EKG signal and a possible segmentation. For evaluation, the following statistics are interesting:

Model type	Linear	Quadratic
Regression	$y_t = (1,\ t)\theta_t + e_t$	$y_t = (1,\ t,\ t^2)\theta_t + e_t$
Loss $V_N(\hat{\theta}) = \frac{1}{N}\sum \varepsilon_t^2$	0.032	0.61
Normalized loss $V_N(\hat{\theta})/V_N(0)$	0.85	15
Number of parameters	30	24
Compression rate (%)	10	8

With this algorithm, the linear model gives far less error and almost the same compression rate. The numerical resolution is the reason for the poor performance of the quadratic model, which includes the linear one as a special case. If the lower value of σ^2 is supplied, then the performance will degrade substantially. The remedy seems to be another basis for the quadratic polynomial.

Table 7.1. Estimated change times for different methods.

Method	n_a	Estimated change times								
Noisy signal										
Divergence	16	451	611	1450	1900	2125		2830		3626
Brandt's GLR	16	451	611	1450	1900	2125		2830		3626
Brandt's GLR	2		593	1450		2125		2830		3626
Approx ML	2	451	593	1608		2116	2741	2822		3626
Pre-filtered signal										
Divergence	16	445	645	1550	1800	2151		2797		3626
Brandt's GLR	16	445	645	1550	1800	2151		2797		3626
Brandt's GLR	2	445	645	1550	1750	2151		2797	3400	3626
Approx ML	2	445	626	1609		2151		2797		3627

7.7.2. Speech segmentation

The speech signal[1] under consideration was recorded inside a car by the French National Agency for Telecommunications, as described in Andre-Obrecht (1988). This example is an continuation of Example 6.7, and the performance of the filter bank will be compared to the consistency tests examined in Chapter 6.

To get a direct comparison with the segmentation result in Basseville and Nikiforov (1993), a second order AR model is used. The approximate ML estimate is derived ($q = 1/2$), using 10 parallel filters where each new segment has a guaranteed lifelength of seven. The following should be stressed:

- The resemblance with the result of Brandt's GLR test and the divergence test presented in Basseville and Nikiforov (1993) is striking.
- No tuning parameters are involved (although $q \neq 1/2$ can be used to influence the number of segments, if not satisfactory). This should be compared with the tricky choice of threshold, window size and drift parameter in the divergence test and Brandt's GLR test–which, furthermore, should be different for voiced and unvoiced zones. Presumably, a considerable tuning effort was required in Andre-Obrecht (1988) to obtain a result similar to that which the proposed method gave in a first try using default parameters tuned on simple test signals.
- The drawback compared to the two previously mentioned methods is a somewhat higher computational complexity. Using the same implementation of the required RLS filters, the number of floating point operations for AR(2) models were $1.6 \cdot 10^6$ for Brandt's GLR test and $5.8 \cdot 10^6$ for the approximate ML method.

[1] The author would like to thank Michele Basseville and Regine Andree-Obrecht for sharing the speech signals in this application.

- The design parameters of the search scheme are not very critical. There is a certain lower bound where the performance drastically deteriorates, but there is no trade-off, as is common for design parameters.
- With the chosen search strategy, the algorithm is recursive and the estimated change points are delivered with a time delay of 10 samples. This is much faster than the other methods due to their sliding window of width 160. For instance, the change at time 2741 for the noisy signal, where the noise variance increases by a factor 3 (see below), is only 80 samples away from a more significant change, and cannot be distinguished with the chosen sliding window.

Much of the power of the algorithm is due to the model with changing noise variance. A speech signal has very large variations in the driving noise. For these two signals, the sequences of noise variances are estimated to

$$10^5 \times (0.035,\ 0.13,\ 1.6,\ 0.37,\ 1.3,\ 0.058,\ 1.7)$$

and

$$10^5 \times (0.038,\ 0.11,\ 1.6,\ 0.38,\ 1.6,\ 0.54,\ 0.055,\ 1.8),$$

respectively. Note, that the noise variance differs as much as a factor 50. No algorithm based on a fixed noise variance can handle that.

Therefore, the proposed algorithm seems to be an efficient tool for getting a quick and reliable result. The lack of design parameters makes it very suitable for general purpose software implementations.

7.7.3. Segmentation of a car's driven path

We will here study the case described in Section 2.6.1.

Signal model

The model is that the heading angle is piecewise constant or piecewise linear, corresponding to straight paths and bends or roundabouts. The changing regression model is here

$$\begin{aligned}\theta_{t+1} &= (1-\delta_t)\theta_t + \delta_t v_t\\ \psi_t &= \theta_t^1 + \theta_t^2 t + e_t\\ \mathrm{E}\, e_t^2 &= \lambda_t.\end{aligned}$$

The approximate MAP estimate of the change times can now be computed in real time (the sampling interval is 100 times larger than needed for computations). The number of parallel filters is 10, the minimum allowed segment length is 0 and each new jump hypothesis is guaranteed to survive at least six samples. The prior probability of a jump is 0.05 at each time.

Local search

Segmentation with a fixed accuracy of the model, using a fixed noise variance $\lambda = 0.05$, gives the result in Figure 7.3. Figure 7.3 also shows the segmentation where the noise variance is unknown and changing over the segments. In both cases, the roundabout is perfectly modeled by one segment and the bends are detected. The seemingly bad performance after the first turn is actually a proof of the power of this approach. Little data are available, and there is no good model for them, so why waste segments? It is more logical to tolerate larger errors here and just use one model. This proves that the adaptive noise variance works for this application as well. In any case, the model with fixed noise scaling seems to be the most appropriate for this application. The main reason is to exclude small, though significant, changes, like lane changes on highways.

Optimal search

The optimal segmentation using Algorithm 7.3 gives almost the same segmentation. The estimated change time sequences are

$$\begin{aligned}
\widehat{k^n}^{rec,fix} &= (20,\ 46,\ 83,\ 111,\ 130,\ 173)\\
\widehat{k^n}^{rec,marg} &= (18,\ 42,\ 66,\ 79,\ 89,\ 121,\ 136,\ 165,\ 173,\ 180)\\
\widehat{k^n}^{opt,marg} &= (18,\ 42,\ 65,\ 79,\ 88,\ 110,\ 133,\ 162,\ 173,\ 180),
\end{aligned}$$

respectively.

Robustness to design parameters

The robustness with respect to the design parameters of the approximation is as follows:

- The exact MAP estimate is almost identical to the approximation. The number of segments is correct, but three jumps differ slightly.
- A number of different values on the jump probability were examined. Any value between $q = 0.001$ and $q = 0.1$ gives the same number of segments. A q between 0.1 and 0.5 gives one more segment, just at the entrance to the roundabout.
- A smaller number of filters than $M = 10$ gives one or two more segments.

That is, a reasonable performance is obtained for almost any choice of design parameters.

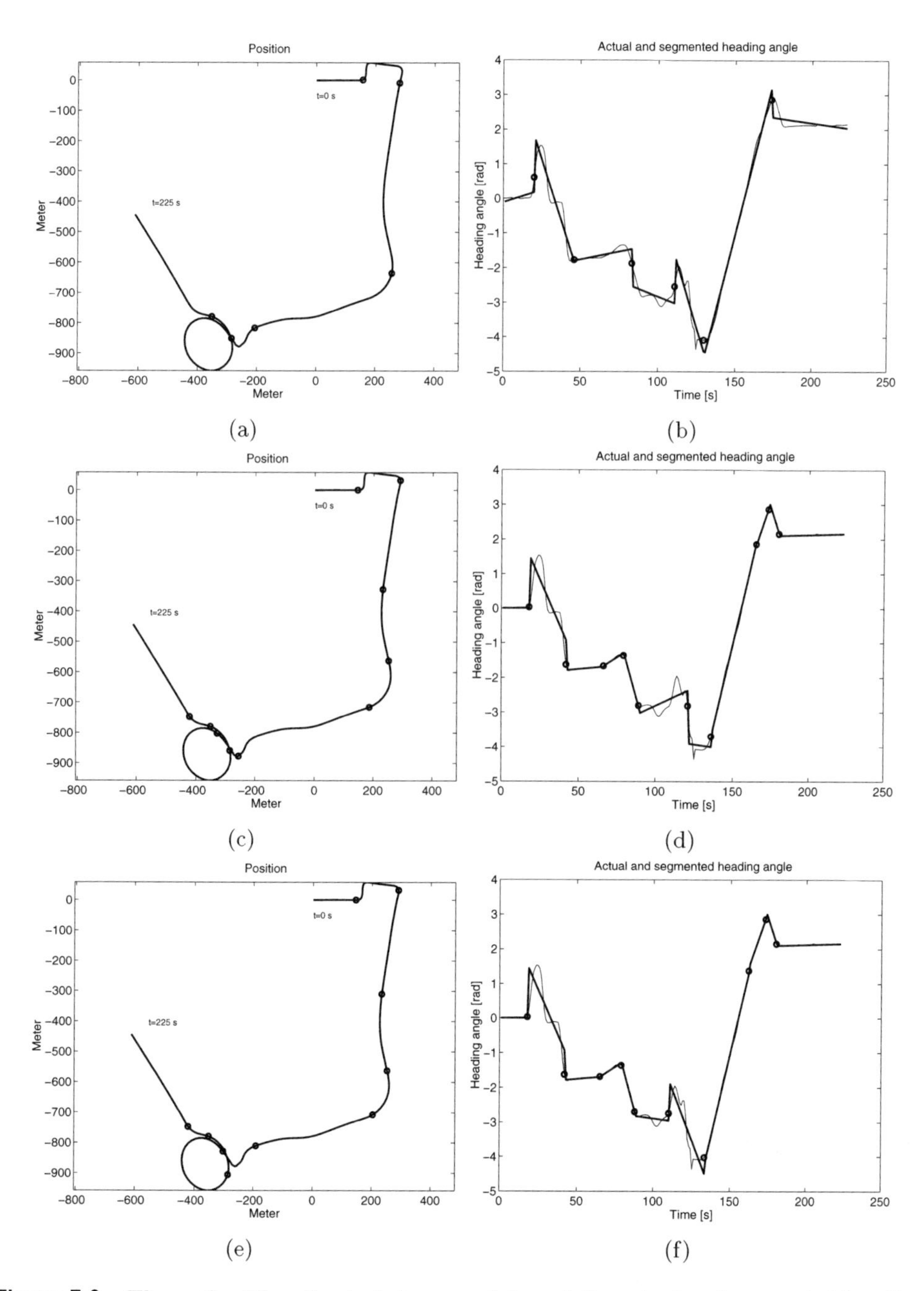

Figure 7.3. The path with estimated change points and the actual and segmented heading angle. First row for fix noise variance 0.05, second row for marginalized noise variance, and finally, optimal ML estimate for marginalized noise variance.

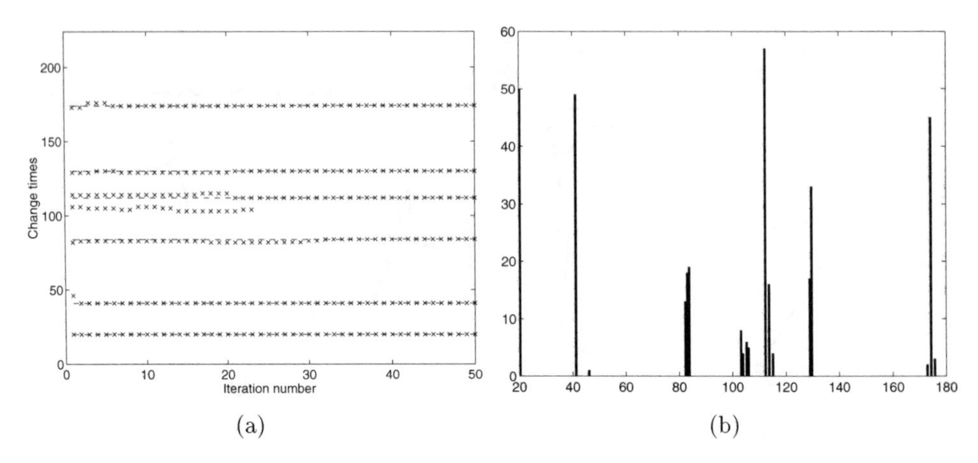

Figure 7.4. Result of the MCMC Algorithm 7.2. The left plot shows the jump sequence examined in each iteration. The encountered sequence with the overall largest likelihood is marked with dashed lines. The right plot shows a histogram over all considered jump sequences. The burn-in time is not excluded.

MCMC search

Figure 7.4 shows the result of the MCMC algorithm in Algorithm 7.2. The algorithm was initiated at the estimated change times from Algorithm 7.1. We note that two change times at the entrance to the roundabout are merged into one after 25 iterations. The histogram shows that some of the changes are easier to locate in time. This follows intuitively when studying the path. For example, the first two turns are sharper than the following ones.

7.A. Two inequalities for likelihoods

In this section, two inequalities for the likelihoods will be derived. They hold for both generalized and marginalized likelihoods, although they will be applied to the latter only.

7.A.1. The first inequality

The following theorems consider the δ-parameterized changing regression model (7.3).

Theorem 7.1
Consider the problem of either maximum likelihood segmentation using (7.13) or (7.15) with $q = 0.5$, or MAP segmentation using (7.13) or (7.15), or information based segmentation using (7.20), (7.21) or (7.22) with an arbitrary

penalty term. Let $\delta_1^{t_0-1}$ *and* $\delta_{t_0+1}^t$ *be two arbitrary sequences of length* $t_0 - 1$ *and* $t - t_0$*, respectively. Then the following inequality holds:*

$$p(\{\delta_1^{t_0-1}, 1, \delta_{t_0-1}^t\}|y^t) \leq p(\{\widehat{\delta_1^{t_0-1}}, 1, \delta_{t_0-1}^t\}|y^t). \tag{7.25}$$

Remarks:

- Similar to the Viterbi algorithm 5.5, there is a finite memory property after a change which implies that we can take the most likely sequence before the change and dismiss with all other possibilities.

- For MAP and ML segmentation, the *a posteriori* probabilities below are given in Theorems 7.3 and 7.5.

- Note that the case (7.14) is not covered by the theorem. The reason is that the segments are not independent for the assumption of a constant and unknown noise scaling.

- The theorem implies that, conditioned on a change at time t_0, the MAP sequences at time t, $\widehat{\delta_{MAP}^t}$, must begin with the MAP sequence at time t_0, $\widehat{\delta_{MAP}^{t_0}}$.

Proof: Given a jump at t_0,

$$\begin{aligned} p(\delta^t|y^t, \delta_{t_0} = 1) &= p(\{\delta^{t_0}, \delta_{t_0+1}^t\}|y^t, \delta_{t_0} = 1) \\ &= p(\delta^{t_0}|y^t, \delta_{t_0} = 1)p(\delta_{t_0+1}^t|y^t, \delta_{t_0} = 1, \delta^{t_0}) \\ &= \underbrace{p(\delta^{t_0}|y^{t_0})}_{\leq p(\widehat{\delta^{t_0}}|y^{t_0})} p(\delta_{t_0+1}^t|y_{t_0+1}^t, \delta_{t_0} = 1). \end{aligned} \tag{7.26}$$

The second equality is Bayes' law and the last one follows since the supposed jump implies that the measurements before the jump are uncorrelated with the jump sequence after the jump, and *vice versa*. We have also used causality at several instances. The theorem now follows from the fact that the first factor is always less than or equal to $p(\widehat{\delta^{t_0}}|y^{t_0})$. □

The reason that the case of unknown and constant noise scaling does not work can be realized form the last sentence of the proof. All measurements contain information about λ, so measurements from all segments influence the model and δ^t.

The Viterbi algorithm proposed in Viterbi (1967) is a powerful tool in equalization; see Forney (1973). There is a close connection between this step and the Viterbi algorithm. Compare with the derivation of the Viterbi Algorithm 5.5.

Theorem 7.1 leaves the following $t+1$ candidates for the MAP estimate of the jump sequence at time t:

$$\begin{aligned} \delta(0) &= (0,0,...,0) \\ \delta(k) &= (\hat{\delta}^{k-1}_{MAP},1,0,...,0) \quad k=1,2,...,t. \end{aligned} \tag{7.27}$$

At time t, t filters are updated which should be compared to 2^t for a straightforward implementation. The total number of updates sums up to $t^2/2$.

7.A.2. The second inequality

Consider the two jump sequences

$$\begin{aligned} \delta^t(t_0-1) &= (\hat{\delta}^{t_0-2}_{MAP},1,\ \ 0\ \ ,0,...,0) \\ \delta^t(t_0) &= (\hat{\delta}^{t_0-1}_{MAP},\ \ 1\ \ ,0,...,0). \end{aligned} \tag{7.28}$$

The second step intuitively works as follows. Given a jump at time t_0 *or* t_0+1, the measurements before t_0, y^{t_0-1}, are independent of the jump sequence after t_0+1 and *vice versa*. Thus, it is only the measurement y_{t_0} that contains any information about which jump sequence is the correct one. If this measurement was not available, then these two sequences would be the same and indistinguishable. One then compensates for the 'deleted' measurement y_{t_0} according to the worst case, and gets an upper bound on the probabilities.

Theorem 7.2
Consider the problem of either maximum likelihood segmentation using (7.13) or (7.15) with $q=0.5$, or MAP segmentation using (7.13) or (7.15), or information based segmentation using (7.20), (7.21) or (7.22) with an arbitrary penalty term. Consider the particular sequences $p(\delta^t(t_0)|y^t)$ and $p(\delta^t(t_0-1)|y^t)$ in (7.28). These are both bounded from above,

$$\begin{aligned} p(\delta^t(t_0-1)|y^t) &< \alpha p(\delta^t_{t_0+1}=0|y^t_{t_0+1}) \\ p(\delta^t(t_0)|y^t) &< \alpha p(\delta^t_{t_0+1}=0|y^t_{t_0+1}), \end{aligned}$$

where

$$\alpha = \frac{\gamma_{max}}{p(y_{t_0}|y^{t_0-1})}\max\Big((1-q)p(\hat{\delta}^{t_0-1}_{MAP}|y^{t_0-1}),\, qp(\hat{\delta}^{t_0-2}_{MAP},\delta_{t_0-1}=1|y^{t_0-1})\Big)$$

and $\gamma_{max}=(2\pi)^{-p/2}\det R^{-1/2}_{t_0}$.

Proof: The *a posteriori* probability $p(\delta(t_0-1)|y^t)$ is expanded by using Bayes' law and the fact that the supposed jump at t_0 divides the measurements into two independent parts:

$$\begin{aligned}
p(\delta^t(t_0-1)|y^t) =& \frac{p(\delta^t(t_0-1))}{p(y^t)} p(y^t|\delta^t(t_0-1)) \\
=& \frac{p(\delta^t(t_0-1))}{p(y^t)} p(y^{t_0-1}|\delta^t(t_0-1)) p(y^t_{t_0}|\delta^t(t_0-1), y^{t_0-1}) \\
=& \frac{p(\hat{\delta}^{t_0-2}_{MAP}, \delta_{t_0-1}=1)}{p(y^{t_0-1})} p(y^{t_0-1}|\hat{\delta}^{t_0-2}_{MAP}, \delta_{t_0-1}=1) \\
& \cdot \frac{p(\delta_{t_0}=0)}{p(y_{t_0}|y^{t_0-1})} p(y_{t_0}|\delta t_0 - 1 = 1, \delta_{t_0}=0, .., \delta_t = 0, y^t_{t_0+1}) \\
& \cdot \frac{p(\delta_{t_0+1}=0, .., \delta_t=0)}{p(y^t|y^{t_0})} p(y^t_{t_0+1}|\delta_{t_0+1}=0, .., \delta_t=0).
\end{aligned}$$

Expanding $p(\delta(t_0)|y^t)$ in a similar way gives

$$\begin{aligned}
p(\delta(t_0)|y^t) =& \frac{1}{p(y^{t_0-1})} p(y^{t_0-1}|\hat{\delta}^{t_0-1}_{MAP}) p(\hat{\delta}^{t_0-1}_{MAP}) \\
& \cdot \frac{p(\delta_{t_0}=1)}{p(y_{t_0}|y^{t_0-1})} p(y_{t_0}|\hat{\delta}^{t_0-1}_{MAP}, y^{t_0-1}) \\
& \cdot \frac{1}{p(y^t|y^{t_0})} p(y_{t_0+1}, .., y_t|\delta_{t_0+1}=0, .., \delta_t=0) p(\delta_{t_0+1}=0, .., \delta_t=0)
\end{aligned}$$

Note that the last factor of each expansion is the same, and the first is known at time t_0. The point is that $p(y_{t_0}|$ whatsoever $)$ is bounded from above by $\gamma_{max} = \gamma(0, R_{t_0})$ and the theorem follows. □

Note that all probabilities needed are already available. Also, note that merged sequences may be merged again, which implies that one upper bound is in common for more than two sequences. An interesting question is how powerful this second inequality is. It was shown in Gustafsson (1992) that this second inequality reduces the complexity from $O(t^2)$ to $O\left(\frac{t^2}{\log t}\right)$ if there is just one segment in the data, and to $O(nt)$ if the number of segments n is large. In simulated examples, the number of sequences is very close to nt. In Section 7.7, the complexity is reduced by a factor of four in an application with real data and imperfect modeling.

7.A.3. The exact pruning algorithm

The two inequalities lead to the following algorithm, which finds the exact MAP estimate with less than t filters.

Algorithm 7.3 Optimal segmentation

Start from the *a posteriori* probabilities given in Theorem 7.3 or Theorem 7.5 for the sequences under consideration at time t:

1. At time $t+1$, let only the most likely sequence jump.
2. Decide whether two or more sequences should be merged. If so, compute a common upper bound on their *a posteriori* probabilities, and consider in the sequel these two merged branches as just one.

If the most probable sequence has larger probability than all upper bounds, then the MAP estimate is found. Otherwise, restart the algorithm or backtrack the history of that upper bound and be more restrictive in merging.

Note that the case of unknown constant noise scaling does not apply here. The first step is the most powerful one: it is trivial to implement and makes it possible to compute the exact ML and MAP estimate for real signals. It is also very useful for evaluating the accuracy of low-complexity approximations.

The first steps in Algorithms 7.3 and 7.1 are the same. The second step in Algorithm 7.1 corresponds to the merging step in Algorithm 7.3. Instead of computing an upper bound, the unlikely sequences are just cut off.

7.B. The posterior probabilities of a jump sequence

7.B.1. Main theorems

Theorem 7.3 (Segmentation with known noise variance)
Consider the changing regression model (7.3). The a posteriori probability of k^n for a known noise scalings λ^n and a flat prior on θ^n is given by

$$-2\log p(k^n|y^N,\lambda^n) = C + 2n\log\frac{1-q}{q} + \sum_{i=1}^{n}(D(i)+V(i)) \tag{7.29}$$

if $P(i)$ is non-singular for all i.

Proof: Without loss of generality, it can be assumed that $\lambda(i)=1$ since it can be included in R_t. Let $Y(i)$ denote the measurements $y_{k_{i-1}+1}^{k_i}$ in segment i. Bayes' law implies the relations $p(A|B)=p(B|A)\frac{p(A)}{p(B)}$ and $p(A_1,A_2,..,A_n)=p(A_1)p(A_2|A_1)..p(A_n|A_1,A_2,..,A_{n-1})$, and this yields

$$\begin{aligned}
p(k^n|y^N) &= p(y^N|k^n)\frac{p(k^n)}{p(y^N)}\\
&= \frac{p(k^n)}{p(y^N)}p(Y(1))p(Y(2)|Y(1))..p(Y(n)|Y(1),Y(2),..,Y(n-1)).
\end{aligned}$$

Since the model parameters are independent between the segments, the measurements from one segment are independent of other segments,

$$p(k^n|y^N) = \frac{p(k^n)}{p(y^N)} \prod_{i=1}^{n} p(Y(i)). \tag{7.30}$$

The law of total probability gives

$$p(Y(i)) = \int_\theta p(Y(i)|\theta)p(\theta)d\theta.$$

As mentioned, a flat and non-informative prior on θ is assumed, that is $p(\theta) \sim 1$, so

$$p(Y(i)) = \int_\theta (2\pi)^{-N(i)p/2}(\det \lambda(i)R)^{-N(i)/2} \times \exp\left(- \frac{\sum_{t=k_{i-1}+1}^{k_i} (y_t - \varphi_t^T\theta)^T R^{-1}(y_t - \varphi_t^T\theta)}{2\lambda(i)} \right) d\theta.$$

To simplify the expressions somewhat, a constant $R_t = R$ is assumed. It is now straightforward to complete the squares for θ, similarly as done to prove (5.103), in order to rewrite the integrand as a Gaussian density function which integrates to one. The result is the counterpart to equation (5.98) for a non-informative prior, and the remaining factor is

$$\begin{aligned} p(Y(i)) =&(2\pi)^{-N(i)p/2}(\det R)^{-N(i)/2}\lambda(i)^{-N(i)p/2}(2\pi)^{d/2}\sqrt{\det \lambda(i)P(i)} \\ &\cdot \exp\left(- \frac{\sum_{t=k_{i-1}+1}^{k_i} (y_t - \varphi_t^T\hat\theta(i))^T R^{-1}(y_t - \varphi_t^T\hat\theta(i))}{2\lambda(i)} \right) \\ =&(2\pi)^{-(N(i)p-d)/2}(\det R)^{-N(i)/2}\lambda(i)^{-\frac{N(i)p-d}{2}} \exp\left(-\frac{D(i)}{2} - \frac{V(i)}{2\lambda(i)}\right), \end{aligned} \tag{7.31}$$

where $V(i)$ and $D(i)$ are as defined in (7.6) and (7.7), respectively. Taking the logarithm, using (7.12) and collecting all terms that do not depend on the jump sequence in the constant C the result follows. □

Theorem 7.4 (Segmentation with constant noise variance)
The a posteriori probability of k^n in the changing regression model (7.3) for an unknown but constant noise covariance scaling $\lambda(i) = \lambda$, with a non-informative prior on both $\theta(i)$ and λ, is given by

$$\begin{aligned} -2\log p(k^n|y^N) = C + 2n\log\frac{1-q}{q} + \sum_{i=1}^{n} D(i) \\ +(Np - nd - 2)\log\frac{1}{2}\sum_{i=1}^{n} V(i) - 2\log\Gamma\left(\frac{Np-nd-2}{2}\right) \end{aligned} \tag{7.32}$$

if $P(i)$ is non-singular for all i and $\sum_{i=1}^{n} V(i) \neq 0$.

Proof: The proof starts from equation (7.31), and the conditioning on λ is written out explicitely: $p(k^n|y^N) = p(k^n|y^N, \lambda)$ in (7.30) and $p(Y^N) = p(Y^N|\lambda)$ in (7.31). The law of total probability gives

$$\begin{aligned}
p(k^n|y^N) &= \frac{p(k^n)}{p(y^N)} \int_\lambda p(y^N|k^n, \lambda) p(\lambda|k^n) d\lambda \\
&= \frac{p(k^n)}{p(y^N)} \int_\lambda \prod_{i=1}^{n} p(Y(i)|\lambda) p(\lambda|k^n) d\lambda \\
&= \frac{p(k^n)}{p(y^N)} \int_\lambda \prod_{i=1}^{n} \lambda(i)^{-\frac{N(i)p-d}{2}} \exp\left(-\frac{D(i)}{2} - \frac{V(i)}{2\lambda(i)}\right) p(\lambda|k^n) d\lambda \\
&= Cp(k^n) \exp\left(-\frac{1}{2}\sum_{i=1}^{n} D(i)\right) \int_\lambda \lambda^{-(Np-nd)/2} e^{-\frac{\sum_{i=1}^{n} V(i)}{2\lambda}} d\lambda.
\end{aligned}$$

Here λ is separated from the definitions (7.6) and (7.7).

To solve the integral, the inverse Wishart *Probability Density Function* (PDF)

$$p(\lambda) = \frac{V^{m/2}}{2^{m/2}\Gamma(m/2)} \frac{e^{-\frac{V}{2\lambda}}}{\lambda^{(m+2)/2}}, \quad \lambda > 0$$

is utilized, where Γ is the gamma-function. A PDF integrates to one, so

$$p(k^n|y^N) = Cp(k^n) e^{-\frac{1}{2}\sum_{i=1}^{n} D(i)} \frac{2^{(Np-nd-2)/2}\Gamma((Np-nd-2)/2)}{(\sum_{i=1}^{n} V(i))^{(Np-nd-2)/2}},$$

and the result follows. □

Theorem 7.5 (Segmentation with changing noise variance)
The a posteriori probability of k^n in the changing regression model (7.3) for an unknown changing noise scaling $\lambda(i)$, with a flat prior on both $\theta(i)$ and $\lambda(i)$, is given by

$$\begin{aligned}
&-2\log p(k^n|y^N) = C + 2n \log \frac{1-q}{q} \qquad (7.33)\\
&+ \sum_{i=1}^{n} \left(D(i) + (N(i)p - d - 2) \log \frac{1}{2} V(i) - 2 \log \Gamma\left(\frac{N(i)p - d - 2}{2}\right)\right)
\end{aligned}$$

if $P(i)$ is non-singular and $V(i) \neq 0$ for all i.

Proof: The proof is almost identical to the one of Theorem 7.4. The difference is that there is one integral for each segment, and again they are solved by identification with the inverse Wishart PDF.

$$
\begin{aligned}
&p(k^n|y^N) \\
&= \frac{p(k^n)}{p(y^N)} \int_{\lambda^n} p(y^N|k^n,\lambda)p(\lambda|k^n)d\lambda^n \\
&= \frac{p(k^n)}{p(y^N)} \prod_{i=1}^{n} \int_{\lambda(i)} p(Y(i)|\lambda(i))p(\lambda(i)|k^n)d\lambda(i) \\
&= \frac{p(k^n)}{p(y^N)} \prod_{i=1}^{n} \int_{\lambda(i)} \lambda(i)^{-\frac{N(i)p-d}{2}} \exp\left(-\frac{D(i)}{2} - \frac{V(i)}{2\lambda(i)}\right)p(\lambda(i)|k^n)d\lambda(i) \\
&= Cp(k^n)\exp\left(-\frac{1}{2}\sum_{i=1}^{n} D(i)\right)\prod_{i=1}^{n}\int_{\lambda(i)} \lambda(i)^{-(N(i)p-d)/2} e^{-\frac{\sum_{i=1}^{n} V(i)}{2\lambda(i)}} d\lambda(i) \\
&= Cp(k^n)e^{-\frac{1}{2}\sum_{i=1}^{n} D(i)} \prod_{i=1}^{n} \frac{2^{(N(i)p-d-2)/2}\Gamma((N(i)p-d-2)/2)}{(V(i))^{(N(i)p-d-2)/2}}
\end{aligned}
$$

and the result follows. □

The results can be somewhat simplified by using $\Gamma(n+1) \approx \sqrt{2\pi}n^{n+1/2}e^{-n}$, which is *Stirling's formula.* It follows that the following expression can be used for reasonably large segment lengths (> 30 roughly):

$$
\begin{aligned}
-2\log\Gamma\left(\frac{N(i)p-d-2}{2}\right) &\approx -\log(2\pi) + (N(i)p-d-4) \\
&\quad - (N(i)p-d-5)\log\frac{N(i)p-d-4}{2} \\
&\approx -(N(i)p-d-2)\log\frac{N(i)p-d-4}{2},
\end{aligned}
$$

and similarly for $\Gamma(\frac{Np-nd-2}{2})$. This relation is used in (7.13)–(7.15).

Part IV: State estimation

8

Kalman filtering

8.1. Basics

The goal in this section is to explain the fundamentals of Kalman filter theory by a few illustrative examples.

The Kalman filter requires a state space model for describing the signal dynamics. To describe its role, we need a concrete example, so let us return to the target tracking example from Chapter 1. Assume that we want a model with the states $x^1 = X$, $x^2 = Y$, $x^3 = \dot{X}$ och $x^4 = \dot{Y}$. This is the simplest possible case of state vector used in practice. Before we derive a model in the next section, a few remarks will be given on what role the different terms in the model has.

Example 8.1 Target tracking: function of Kalman filter

Figure 8.1(a) illustrates how the Kalman filter makes use of the model. Suppose the object in a target tracking problem (exactly the same reasoning holds for navigation as well) is located at the origin with velocity vector $(1,1)$. For simplicity, assume that we have an estimate of the state vector which coincides with the true value

$$\hat{x}_{0|0} = (0,0,1,1)^T = x_0^o.$$

In practice, there will of course be some error or uncertainty in the estimate. The uncertainty can be described by a confidence interval, which in the Kalman filter approach is always shaped as an ellipsoid. In the figure, the uncertainty is larger in the longitudinal direction than in the lateral direction.

Given the initial state, future positions can be predicted by just integrating the velocity vector (like in dead reckoning). This simulation yields a straight line for the position. With a more complex state space model with more states, more complex manoeuvres can be simulated. In the state noise description,

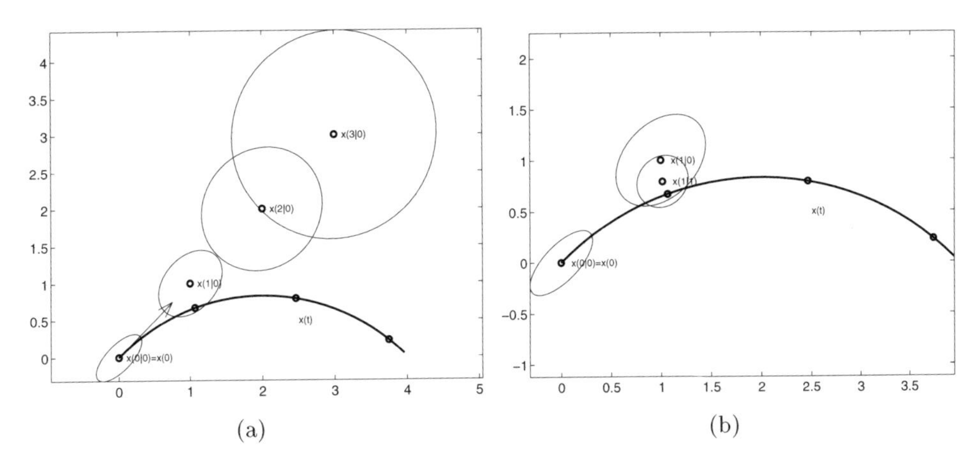

Figure 8.1. The plot in (a) shows how the model predicts future positions of the object, given the state vector at time 0, and how the uncertainty region increases in time. The plot in (b) shows how the filter takes care of the first measurement by correcting the state vector and decreasing the uncertainty region.

we can model what kind of manoeuvres can be performed by the aircraft, for instance, 1 m/s^2 in longitudinal direction and 3 m/s^2 in lateral direction. The possible and unpredictable manoeuvres in the future imply that the uncertainty grows in time, just like illustrated by the ellipsoids in Figure 8.1(a). Note that the object makes a very sharp manoeuvre that makes the actual path x_t going outside the confidence interval.

Figure 8.1(b) shows how the Kalman filter acts when the first measurement x_1 becomes available.

1. First, the estimate is corrected towards the measurement. The velocity state component is adapted likewise.
2. Then the uncertainty is decreased accordingly.

The index rules can now be summarized as:

- For the true trajectory x_t, a single index is used as usual. Time is indeed continuous here.
- For a simulation starting at time 0, as shown in Figure 8.1(a), the sequence $x_{0|0}, x_{1|0}, x_{2|0}, x_{3|0}, \dots$ is used. The rule is that the first index is time and the second one indicates when the simulation is started.
- When the Kalman filter updates the estimate, the second index is increased one unit. The Kalman filter is alternating between a one time step simulation, and updating the state vector, which yields the sequence $x_{0|0}, x_{1|0}, x_{1|1}, x_{2|1}, x_{2|2}, x_{3|2}, \dots$.

- Hats are used on all computed quantities (simulated or estimated), for instance $\hat{x}_{t|t}$.

The state space model used in this chapter is

$$\begin{aligned} x_{t+1} =& Ax_t + B_u u_t + B_v v_t \\ y_t =& Cx_t + Du_t + e_t, \end{aligned} \tag{8.1}$$

Here y_t is a measured signal, A, B, C, D are known matrices and x_t an unknown state vector. There are three inputs to the system: the observable (and controllable) u_t, the non-observable process noise v_t and the measurement noise e_t. The Kalman filter is given by

$$\begin{aligned} \hat{x}_{t+1} =& A\hat{x}_t + K_t \varepsilon_t \\ \varepsilon_t =& y_t - \hat{y}_t = y_t - C\hat{x}_t, \end{aligned} \tag{8.2}$$

where the update gain K_t is computed by the the Kalman filter equations as a function

$$K_t = K_t(A, B_u, B_v, C, D, Q, R). \tag{8.3}$$

Figure 8.2 summarizes the signal flow for the signal model and the Kalman filter.

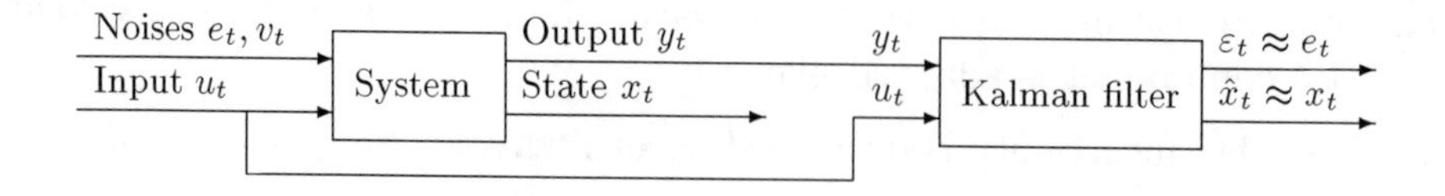

Figure 8.2. Definition of signals for the signal model and Kalman filter.

Example 8.2 Target tracking: function of state space model

The role of the state space model in Example 8.1 is summarized as follows:

- The deterministic model (the A matrix) describes how to simulate the state vector. In the example, the velocity state is used to find future positions.

- The stochastic term in the state equation, $B_v v_t$, decides how the confidence interval grows (the size of the random walk). One extreme case is $Q \to \infty$, when anything can happen in one time step and all old information must be considered as useless, and the ellipsoid becomes infinitely large. The other extreme case is $Q \to 0$, which corresponds to the ellipsoid not growing at all.
- The deterministic part of the measurement equation $y_t = Cx_t$ tells the Kalman filter in what direction one measurement should affect the estimate.
- The covariance R of the measurement noise e_t describes how reliable the measurement information is. Again, the extreme points are important to understand. First, $R = 0$ says that the measurement is exact, and in the example this would imply that $\hat{x}_{1|1}$ would coincide with x_1, and the corresponding ellipsoid must break down to a point. Secondly, $R \to \infty$ means that the measurement is useless and should be discarded.

Much of the advantage of using the Kalman filter compared to more *ad hoc* filters (low-pass) is that the design is moved from an abstract pole-zero placement to a more concrete level of model design.

Literature

The standard reference for all computational aspects of the Kalman filter has for a long time been Anderson and Moore (1979), but from now on it is likely that the complete reference will be Kailath et al. (1998). This is a thorough work covering everything related to state space estimation. These two references are a bit weak regarding applications. A suitable book with navigation applications is Minkler and Minkler (1990). Other monographs are Brown and Hwang (1997) and Chui and Chen (1987).

8.2. State space modeling

Designing a good Kalman filter is indeed more a modeling than a filter design task. All that is needed to derive a good Kalman filter in an application is to understand the modeling and a few tuning issues. The rest is implementation problems that can be hidden in good software. That is, the first sub-goal is to derive a state space model

$$x_{t+1} = A_t x_t + B_{u,t} u_t + B_{v,t} v_t, \quad \text{Cov}(v_t) = Q_t \tag{8.4}$$

$$y_t = C_t x_t + D_t u_t + e_t, \quad \text{Cov}(e_t) = R_t. \tag{8.5}$$

The model is completely specified by the matrices $A, B = [B_u,\ B_v], C, D, Q, R$. System modeling is a subject covered by many textbooks; see for instance Ljung and Glad (1996). The purpose here is to give a few but representative examples to cover the applications in this part. Since many models are derived in continuous time, but we are concerned with discrete time filtering, we start by reviewing some sampling formulas.

8.2.1. Sampling formula

Integration over one sample period T of the continuous time model

$$\dot{x}_t = A^c x_t + B_u^c u_t$$

gives

$$x_{t+T} = x_t + \int_t^{t+T} (A^c x_\tau + B_u^c u_\tau)\, d\tau. \tag{8.6}$$

Assuming that the deterministic input is constant during the sampling interval, which is the case in for instance computer controlled systems, the solution can be written as the discrete time state space model

$$x_{t+T} = Ax_t + B_u u_t,$$

where

$$A = e^{A^c T}$$
$$B_u = \int_0^T e^{A^c \tau} B_u^c d\tau.$$

The same sampling formulas can be used for the stochastic input v_t in (8.4), although it is seldom the case that v_t is constant during the sampling intervals. Other alternatives are surveyed in Section 8.9. There are three main ways to compute the matrix exponential: using the Laplace transform, series expansion or algebraic equations. See, for example, Åström and Wittenmark (1984) for details. Standard computer packages for control theory can be used, like the function `c2d` in Control System Toolbox in MATLAB™ .

8.2.2. Physical modeling

Basic relations from different disciplines in form of differential equations can be used to build up a state space model. The example below illustrates many aspects of how the signal processing problem can be seen as a modeling one, rather than a pure filter design. Much can be gained in performance by designing the model appropriately.

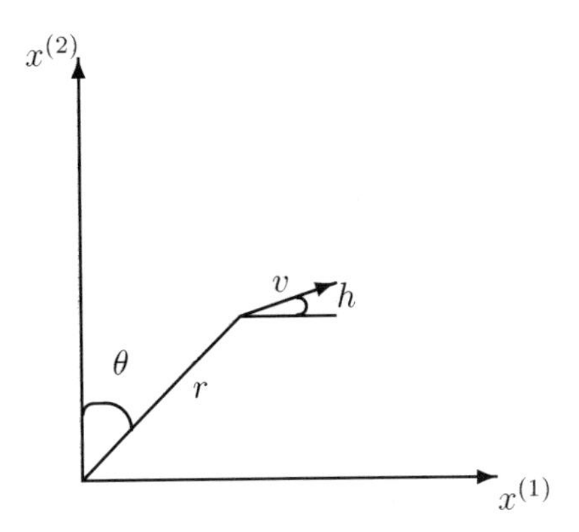

Figure 8.3. Tracking coordinates.

Example 8.3 Target tracking: modeling

Consider an aircraft moving in two dimensions as in Figure 8.3. Using Newton's law $F = ma$ separately in two dimensions gives

$$\ddot{x}_t^1 = \frac{F^1}{m}$$
$$\ddot{x}_t^2 = \frac{F^2}{m}.$$

Here F^i are forces acting on the aircraft, normally due to the pilot's manoeuvres, which are unknown to the tracker. From now on, the right-hand side will be considered as random components w_t^i. In vector form, we get

$$\ddot{x}_t = w_t.$$

The standard form of a state space model only has first order derivatives in the left-hand side. We therefore introduce dummy states, which here has the physical interpretation of velocities:

$$\dot{x}_t^1 = x_t^3$$
$$\dot{x}_t^2 = x_t^4$$
$$\dot{x}_t^3 = w_t^1$$
$$\dot{x}_t^4 = w_t^2.$$

Here the noise is denoted w^i rather than v so as not to confuse it with the velocities $v^i = \dot{x}^i$, $i = 1, 2$. Assuming that the position is measurable, we get a continuous time state space model for the state vector $x = (x^1, x^2, v^1, v^2)$ in standard form:

$$\dot{x}_t = A^c x_t + B^c_w w_t = \begin{pmatrix} 0 & 0 & 1 & 0 \\ 0 & 0 & 0 & 1 \\ 0 & 0 & 0 & 0 \\ 0 & 0 & 0 & 0 \end{pmatrix} x_t + \begin{pmatrix} 0 & 0 \\ 0 & 0 \\ 1 & 0 \\ 0 & 1 \end{pmatrix} w_t,$$

$$y_t = \begin{pmatrix} 1 & 0 & 0 & 0 \\ 0 & 1 & 0 & 0 \end{pmatrix} x_t + e_t.$$

Here superscript c indicates continuous time. The sample formula with sampling interval T gives

$$\begin{aligned} x_{t+T} &= A(T)x_t + B(T)\bar{w}_t \\ &= \begin{pmatrix} 1 & 0 & T & 0 \\ 0 & 1 & 0 & T \\ 0 & 0 & 1 & 0 \\ 0 & 0 & 0 & 1 \end{pmatrix} x_t + \begin{pmatrix} T^2/2 & 0 \\ 0 & T^2/2 \\ T & 0 \\ 0 & T \end{pmatrix} \bar{w}_t. \end{aligned} \tag{8.7}$$

There are two possibilities for the choice of Q. The simplest is a diagonal $Q = qI_2$. An aircraft can accelerate much faster orthogonally to the velocity vector than parallel to it. That is, the forces during turns are much larger than during acceleration and retardation. This implies that we would like to have different adaptation gains in these two directions. The only problem is that the aircraft has its own coordinate system which is not suitable for tracking. However, we can use the heading (velocity direction) estimate

$$\hat{h} = \tan^{-1}(\hat{x}^4/\hat{x}^3),$$

and assume that the velocity changes mainly orthogonally to h. It can be shown that the appropriate state dependent covariance matrix is

$$Q = \begin{pmatrix} q_\omega \sin^2(h) + q_v \cos^2(h) & (q_v - q_\omega)\sin(h)\cos(h) \\ (q_v - q_\omega)\sin(h)\cos(h) & q_\omega \cos^2(h) + q_v \sin^2(h) \end{pmatrix}. \tag{8.8}$$

Here q_v is the force variance along the velocity vector and q_ω is perpendicular to it, where $q_\omega \gg q_v$.

It has been pointed out by many authors that during manoeuvres, better tracking can be achieved by a so-called *jerk model*. Here two more acceleration states are included in the model, and the state noise is now the jerk rather than acceleration. The state space model for the state vector $x = (x^1, x^2, v^1, v^2, a^1, a^2)^T$ is:

$$\dot{x}_t = A^c x_t + B^c w_t = \begin{pmatrix} 0 & 0 & 1 & 0 & 0 & 0 \\ 0 & 0 & 0 & 1 & 0 & 0 \\ 0 & 0 & 0 & 0 & 1 & 0 \\ 0 & 0 & 0 & 0 & 0 & 1 \\ 0 & 0 & 0 & 0 & 0 & 0 \\ 0 & 0 & 0 & 0 & 0 & 0 \end{pmatrix} x_t + \begin{pmatrix} 0 & 0 \\ 0 & 0 \\ 0 & 0 \\ 0 & 0 \\ 1 & 0 \\ 0 & 1 \end{pmatrix} w_t.$$

Sampling gives

$$x_{t+T} = A(T)x_t + B(T)\bar{w}_t \tag{8.9}$$

$$= \begin{pmatrix} 1 & 0 & T & 0 & 0 & 0 \\ 0 & 1 & 0 & T & 0 & 0 \\ 0 & 0 & 1 & 0 & T & 0 \\ 0 & 0 & 0 & 1 & 0 & T \\ 0 & 0 & 0 & 0 & 1 & 0 \\ 0 & 0 & 0 & 0 & 0 & 1 \end{pmatrix} x_t + \begin{pmatrix} T^3/3 & 0 \\ 0 & T^3/3 \\ T^2/2 & 0 \\ 0 & T^2/2 \\ T & 0 \\ 0 & T \end{pmatrix} \bar{w}_t. \tag{8.10}$$

Again, the covariance matrix in (8.8) is a reasonable option.

Another strategy is a compromise between the acceleration and jerk models, inspired by the physical constraints of the motion of the aircraft. Since the acceleration is mainly orthogonal to the velocity vector, one more state can be introduced for this acceleration. The acceleration orthogonal to the velocity is the *turn rate* ω, and the state equations for the state vector $(x^1, x^2, v^1, v^2, \omega)^T$ is given by:

$$\begin{aligned} \dot{x}_t^1 &= v_t^1 \\ \dot{x}_t^2 &= v_t^2 \\ \dot{v}_t^1 &= -\omega_t v_t^2 \\ \dot{v}_t^2 &= \omega_t v_t^1 \\ \dot{\omega}_t &= 0. \end{aligned} \tag{8.11}$$

The main advantage of this model is that circular paths correspond to constant turn rate, so we have a parameter estimation problem rather than a tracking problem in these segments. Paths consisting of straight lines and circle segments are called *coordinated turns* in the literature of *Air Traffic Control (ATC)*.

One further alternative is to use velocities in polar coordinates rather than Cartesian. With the state vector $(x^1, x^2, v, h, \omega)^T$, the state dynamics become:

$$
\begin{aligned}
\dot{x}_t^1 &= v_t \cos(h_t) \\
\dot{x}_t^2 &= v_t \sin(h_t) \\
\dot{v}_t &= 0 \\
\dot{h}_t &= \omega \\
\dot{\omega}_t &= 0.
\end{aligned}
\tag{8.12}
$$

Both (8.11) and (8.12) are non-linear models, and sampling and filtering are not straightforward. Section 8.9 is devoted to this problem, and several solutions to this problem will be described.

As a summary, there are plenty of options for choosing the model, and each one corresponds to one unique Kalman filter. The choice is a compromise between algorithm complexity and performance, but it is not at all certain that a more complex model gives better accuracy.

8.2.3. Using known transfer functions

There is a standard transformation to go from a given transfer function to a state space model which works both for continuous and discrete time models. The *observer companion form* for a transfer function

$$
G(s) = \frac{b_1 s^{n-1} + \cdots + b_{n-1}s + b_n}{s^n + a_1 s^{n-1} + \cdots + a_{n-1}s + a_n}
$$

is

$$
\begin{aligned}
\dot{x}_t &= \begin{pmatrix} -a_1 & 1 & 0 & \cdots & 0 \\ -a_2 & 0 & 1 & 0 & \vdots \\ \vdots & \vdots & & \ddots & 0 \\ -a_{n-1} & 0 & \cdots & 0 & 1 \\ -a_n & 0 & \cdots & 0 & 0 \end{pmatrix} x_t + \begin{pmatrix} b_1 \\ b_2 \\ \vdots \\ b_{n-1} \\ b_n \end{pmatrix} u_t \\
y_t &= \begin{pmatrix} 1 & 0 & \cdots & 0 \end{pmatrix} x_t + e_t.
\end{aligned}
$$

Here the input u_t can be replaced by state noise v_t, fault f_t or disturbances d_t.

It should be mentioned that the state space model is not unique. There are many other transformations.

Example 8.4 DC motor: model

Consider a sampled state space model of a DC motor with continuous time transfer function

$$G(s) = \frac{1}{s(s+1)} = \frac{1}{s^2+s}.$$

The continuous time state space model is in so called observer canonical form,

$$\dot{x}_t = \begin{pmatrix} -1 & 1 \\ 0 & 0 \end{pmatrix} x_t + \begin{pmatrix} 0 \\ 1 \end{pmatrix} u_t$$
$$y_t = \begin{pmatrix} 1 & 0 \end{pmatrix} x_t + e_t.$$

A better choice of state variables for physical interpretations is x^1 being the angle and x^2 the angular velocity of the motor. The derivation of the corresponding state space model is straightforward, and can be found in any textbook in control theory. Sampling with sample interval $T_s = 0.4$ s gives

$$A = \begin{pmatrix} 1 & 0.3297 \\ 0 & 0.6703 \end{pmatrix}, \quad B_u = \begin{pmatrix} 0.0703 \\ 0.3297 \end{pmatrix} \qquad C = \begin{pmatrix} 1 & 0 \end{pmatrix}, \quad D_u = 0.$$

Finally, we revisit the communication channel modeling problem from Chapter 5 in two examples leading to state space models and Kalman filter approaches.

Example 8.5 Fading communication channels

Measurements of the FIR coefficients b^i, $i = 1, 2, \cdots, n_b$, in a fading communication channel can be used to estimate a model of the parameter variability. The spectral content is well modeled by a low order AR model. One possible realization of channel time variations will be shown in Figure 10.6(a). Assume that we have estimated the model $b_t^i = -a_1^i b_{t-1}^i - a_2^i b_{t-2}^i + v_t^i$ for each coefficient b^i. The corresponding state space model is

$$x_{t+1}^i = \underbrace{\begin{pmatrix} -a_1^i & -a_2^i \\ 1 & 0 \end{pmatrix}}_{A^i} x_t^i + \begin{pmatrix} 0 \\ 1 \end{pmatrix} v_t^i$$
$$b_t^i = \begin{pmatrix} 1 & 0 \end{pmatrix} x_t^i.$$

An application on how to utilize the predictive ability of such a model is given in Example 8.6.

Example 8.6 ***Equalization using hyper models***

A so called *hyper model* of parameter variations can be used to improve tracking ability. Consider the digital communication channel in Example 8.5. Equalization incorporating this knowledge of parameter variation can use the following state space model (assuming $n_b = 2$, i.e. an FIR(2) channel model):

$$x_t = \begin{pmatrix} x_t^1 \\ x_t^2 \end{pmatrix}$$

$$x_{t+1} = \begin{pmatrix} A^i & 0 \\ 0 & A^2 \end{pmatrix} x_t + \begin{pmatrix} 0 \\ v_t^1 \\ 0 \\ v_t^2 \end{pmatrix}$$

$$y_t = \begin{pmatrix} u_t & 0 & u_{t-1} & 0 \end{pmatrix} x_t + e_t.$$

See Lindbom (1995) for extensive theory on this matter, and Davis et al. (1997) for one application.

8.2.4. Modeling tricks

A very useful trick in Kalman filtering is to augment the state vector with some auxiliary states x^a, and then to apply the Kalman filter to the augmented state space model. We will denote the augmented state vector $\bar{x} = (x^T, (x^a)^T)^T$. It should be noted here that the Kalman filter is the optimal estimator of any linear combination of the state vector. That is, given the assumptions, there is no better way to estimate the state vector x_t than to estimate the augmented state. This section lists some important cases where this trick is useful.

Colored state noise

Assume the state noise is colored, so we can write $v_t = H(q)\bar{v}_t$, where $\bar{v}_t$ is white noise. Let a state space realization of $H(q)$ be

$$\begin{aligned} x_{t+1}^v &= A^v x_t^v + B^v \bar{v}_t \\ v_t &= C^v x_t^v. \end{aligned}$$

The augmented state space model is

$$\begin{aligned} \bar{x}_{t+1} &= \begin{pmatrix} A & BC^v \\ 0 & A^v \end{pmatrix} \bar{x}_t + \begin{pmatrix} 0 \\ B^v \end{pmatrix} \bar{v}_t \\ y_t &= (C,\ 0)\bar{x}_t + e_t. \end{aligned}$$

The relation is easily verified by expanding each row above separately.

Colored measurement noise

Assume the measurement noise is colored, so we can write $e_t = H(q)\bar{e}_t$, where $\bar{e}_t$ is white noise. We can apply the same trick as in the previous case, but in the case of Markov noise

$$e_{t+1} = A_t^e e_t + w_t, \tag{8.13}$$

there is a way to avoid an increased state dimension.

The obvious approach is to pre-filter the measurements, $\bar{y}_t = H^{-1}(q)y_t$, where $H(q)$ is a stable minimum phase spectral factor of the measurement noise, so that the measurement noise becomes white. The actual way of implementing this in state space form is as follows. Modify the measurements as

$$\begin{aligned}
\bar{y}_{t+1} =& y_{t+1} - A_t^e y_t \\
=& C_{t+1}x_{t+1} - A_t^e C_t x_t + \underbrace{e_{t+1} - A_t^e e_t}_{w_t} \\
=& \underbrace{\left(C_{t+1}A_t - A_t^e C_t\right)}_{\bar{C}_t} x_t + \underbrace{C_{t+1}B_t v_t + w_t}_{\bar{e}_t}.
\end{aligned}$$

Thus, we have a standard state space model again (though the time index of the measurement is non-standard), now with correlated state and measurement noises.

Sensor offset or trend

Most sensors have an unknown offset. This can be solved by off-line calibration, but a more general alternative is given below. Mathematically, this means that we must estimate the mean of the measurements on-line. The straightforward solution is to modify the state and measurement equation as follows:

$$\bar{x}_{t+1} = \begin{pmatrix} A & 0 \\ 0 & I \end{pmatrix} \begin{pmatrix} x_t \\ m \end{pmatrix} + \begin{pmatrix} B \\ 0 \end{pmatrix} v_t \tag{8.14}$$

$$y_t = \begin{pmatrix} C & I \end{pmatrix} \begin{pmatrix} x_t \\ m \end{pmatrix} + e_t. \tag{8.15}$$

Here it should be remarked that in off-line situations, the mean can be estimated directly from the measurements and removed before the Kalman filter is applied. If there is an input u, the deterministic part of y should be subtracted before computing the mean. According to least squares theory, the solution will be the same.

In the same way, drifts are modeled as

$$\bar{x}_{t+1} = \begin{pmatrix} A & 0 & 0 \\ 0 & I & 0 \\ 0 & 0 & I \end{pmatrix} \begin{pmatrix} x_t \\ m \\ c \end{pmatrix} + \begin{pmatrix} B \\ 0 \\ 0 \end{pmatrix} v_t \tag{8.16}$$

$$y_t = \begin{pmatrix} C & I & tI \end{pmatrix} \begin{pmatrix} x_t \\ m \\ c \end{pmatrix} + e_t. \tag{8.17}$$

An alternative is to include the drift in the offset state as follows:

$$\bar{x}_{t+1} = \begin{pmatrix} A & 0 & 0 \\ 0 & I & I \\ 0 & 0 & I \end{pmatrix} \begin{pmatrix} x_t \\ m \\ c \end{pmatrix} + \begin{pmatrix} B \\ 0 \\ 0 \end{pmatrix} v_t \tag{8.18}$$

$$y_t = \begin{pmatrix} C & I & 0 \end{pmatrix} \begin{pmatrix} x_t \\ m \\ c \end{pmatrix} + e_t. \tag{8.19}$$

The advantage of this latter alternative is that the state space model is time-invariant if the original model is time-invariant.

Sensor faults

The sensor offset and drift might be due to a sensor fault. One model of a sensor fault is as a sudden offset or drift. The state space model corresponding to this is

$$\bar{x}_{t+1} = \begin{pmatrix} A & 0 & 0 \\ 0 & I & 0 \\ 0 & 0 & I \end{pmatrix} \begin{pmatrix} x_t \\ m \\ c \end{pmatrix} + \underbrace{\begin{pmatrix} B_v \\ 0 \\ 0 \end{pmatrix}}_{\bar{B}_{v,t}} v_t + \delta_{t-k} \underbrace{\begin{pmatrix} 0 \\ f_m \\ f_c \end{pmatrix}}_{\bar{f}} \tag{8.20}$$

$$y_t = \begin{pmatrix} C & I & tI \end{pmatrix} \begin{pmatrix} x_t \\ m \\ c \end{pmatrix} + e_t, \tag{8.21}$$

where k denotes the time the fault occurs.

Actuator faults

In the same way as a sensor fault, a sudden offset in an actuator is modeled as

$$x_{t+1} = A_t x_t + B_{u,t} u_t + B_{v,t} v_t + \delta_{t-k} B_{u,t} f_u \tag{8.22}$$

$$y_t = C_t x_t + e_t. \tag{8.23}$$

If the offset lasts, rather than being just a pulse, more states $x_{f,t}$ can be included for keeping the fault f_u in a memory

$$\bar{x}_{t+1} = \begin{pmatrix} A_t & B_{u,t} \\ 0 & I \end{pmatrix} \begin{pmatrix} x_t \\ x_{f,t} \end{pmatrix} + \begin{pmatrix} B_{u,t} \\ 0 \end{pmatrix} u_t + \begin{pmatrix} B_{v,t} \\ 0 \end{pmatrix} v_t + \delta_{t-k} \begin{pmatrix} 0 \\ I \end{pmatrix} f_u \tag{8.24}$$

$$y_t = \begin{pmatrix} C & 0 \end{pmatrix} \begin{pmatrix} x_t \\ x_{f,t} \end{pmatrix} + e_t. \tag{8.25}$$

State disturbances

The third kind of additive change in state space model, beside sensor and actuator fault is a state disturbance. Disturbances and, for example, manoeuvres in target tracking are well described by

$$x_{t+1} = A_t x_t + B_{u,t} u_t + B_{v,t} v_t + \delta_{t-k} B_f f \tag{8.26}$$

$$y_t = C_t x_t + e_t. \tag{8.27}$$

Parametric state space models

If unknown system parameters θ enter the state space model linearly, we have a model like

$$x_{t+1} = A_t x_t + B_{u,t} u_t + B_{v,t} v_t + \varphi_t^T \theta \tag{8.28}$$

$$y_t = C_t x_t + D_t u_t + e_t. \tag{8.29}$$

This is like an hybrid of a linear regression and a state space model. Using the state vector $\bar{x}^T = (x^T, \theta^T)$, we get

$$\bar{x}_{t+1} = \begin{pmatrix} A_t & \varphi_t^T \\ 0 & I \end{pmatrix} \begin{pmatrix} x_t \\ \theta \end{pmatrix} + \begin{pmatrix} B_{u,t} \\ 0 \end{pmatrix} u_t + \begin{pmatrix} B_{v,t} \\ 0 \end{pmatrix} v_t \tag{8.30}$$

$$y_t = \begin{pmatrix} C & 0 \end{pmatrix} \begin{pmatrix} x_t \\ \theta \end{pmatrix} + e_t \tag{8.31}$$

The Kalman filter applies, yielding a combined state and parameter estimator. If the parameter vector is time-varying, it can be interpreted as an unknown input. Variants of the Kalman filter as such a kind of an *unknown input observer* are given in for instance Keller and Darouach (1998, 1999).

Smoothing

The method for fixed-lag smoothing given in Section 8.5 is a good example of state augmentation.

8.3. The Kalman filter

Sections 13.1 and 13.2 contain two derivations of the Kalman filter. It is advisable at this stage to study, or recapitulate, the derivation that suits one's background knowledge the best.

8.3.1. Basic formulas

A general time-varying state space model for the Kalman filter is

$$x_{t+1} = A_t x_t + B_{u,t} u_t + B_{v,t} v_t, \quad \mathrm{Cov}(v_t) = Q_t \tag{8.32}$$

$$y_t = C_t x_t + e_t, \quad \mathrm{Cov}(e_t) = R_t. \tag{8.33}$$

The dimensions of the matrices will be denoted n_x, n_u, n_v, n_y, respectively. The time update and measurement update in the Kalman filter are given by:

$$\hat{x}_{t+1|t} = A_t \hat{x}_{t|t} + B_{u,t} u_t \tag{8.34}$$

$$P_{t+1|t} = A_t P_{t|t} A_t^T + B_{v,t} Q_t B_{v,t}^T \tag{8.35}$$

$$\hat{x}_{t|t} = \hat{x}_{t|t-1} + P_{t|t-1} C_t^T (C_t P_{t|t-1} C_t^T + R_t)^{-1} (y_t - C_t \hat{x}_{t|t-1}) \tag{8.36}$$

$$P_{t|t} = P_{t|t-1} - P_{t|t-1} C_t^T (C_t P_{t|t-1} C_t^T + R_t)^{-1} C_t P_{t|t-1}. \tag{8.37}$$

Comments:

- It is convenient to define and compute three auxiliary quantities:

$$\varepsilon_t = y_t - C_t \hat{x}_{t|t-1} \tag{8.38}$$

$$S_t = C_t P_{t|t-1} C_t^T + R_t \tag{8.39}$$

$$K_t = P_{t|t-1} C_t^T (C_t P_{t|t-1} C_t^T + R_t)^{-1} = P_{t|t-1} C_t^T S_t^{-1}. \tag{8.40}$$

  The interpretations are that S_t is the covariance matrix of the innovation ε_t, and K_t is the Kalman gain.

- With the auxiliary quantities above, equation (8.37) can be written more compactly as $P_{t|t} = P_{t|t-1} - K_t S_t K_t^T$.

- It is suitable to introduce a dummy variable $L = P_{t|t-1} C_t^T$ to save computations. Then we compute $S_t = C_t L + R_t$, $K_t = L S_t^{-1}$ and $P_{t|t} = P_{t|t-1} - K_t L^T$.

- The indexing rule is that $\hat{x}_{t|k}$ is the projection of x_t onto the space spanned by the measurements $y_1, y_2, \ldots, y_k$, or in the statistical framework the conditional expectation of x_t, given the set of measurements $y_1, y_2, \ldots, y_k$.

- $P = \mathrm{E}(x-\hat{x})(x-\hat{x})^T$ is the covariance matrix for the state estimate.

- The update equation for P is called the discrete Riccati equation. Even if the state space model is time-invariant, the Kalman filter will be time-varying due to the transient caused by unknown initial conditions. The filter will, however, converge to a time-invariant one. More on this in Section 8.4.

- The Kalman gain and covariance matrix do not depend upon data, and can be pre-computed. This is in a way counter intuitive, because the actual filter can diverge without any noticeable sign except for very large innovations. The explanation is the prior belief in the model. More on this in Section 8.6.1.

- As for adaptive filters, it is very illuminating to analyze the estimation error in terms of bias, variance and tracking error. This will be done in Sections 8.6.3 for bias error, and Section 8.4.1 for the other two errors, respectively.

- The signal model (8.32) is quite general in that all matrices, including the dimensions of the input and output vectors, may change in time. For instance, it is possible to have a time-varying number of sensors, enabling a straightforward solution to the so-called *multi-rate signal processing* problem.

- The covariance matrix of the stochastic contribution to the state equation is $B_{v,t}Q_tB_{v,t}^T$. It is implicitly assumed that the factorization is done so that Q_t is non-singular. (This may imply that its dimension is time-varying as well.) In many cases, this term can be replaced by a (singular) covariance matrix $\bar{Q}_t = B_{v,t}Q_tB_{v,t}^T$ of dimension $n_x \times n_x$. However, certain algorithms require a non-singular Q_t.

There are several ways to rewrite the basic recursions. The most common ones are summarized in the algorithm below.

Algorithm 8.1 Kalman filter

The Kalman filter in its filter form is defined by the recursion

$$\begin{aligned}\hat{x}_{t|t} =& A_{t-1}\hat{x}_{t-1|t-1} + B_{u,t-1}u_{t-1} + K_t(y_t - C_tA_{t-1}\hat{x}_{t-1|t-1} - C_tB_{u,t-1}u_{t-1})\\ =&(A_{t-1} - K_tC_tA_{t-1})\hat{x}_{t-1|t-1} + K_ty_t + (I - K_tC_t)B_{u,t-1}u_{t-1}.\end{aligned} \tag{8.41}$$

The one-step ahead predictor form of the Kalman filter is defined by the recursion

$$\begin{aligned}\hat{x}_{t+1|t} =& A_t\hat{x}_{t|t-1} + B_{u,t}u_t + A_tK_t(y_t - C_t\hat{x}_{t|t-1})\\ =& (A_t - A_tK_tC_t)\hat{x}_{t|t-1} + B_{u,t}u_t + A_tK_ty_t.\end{aligned} \tag{8.42}$$

In both cases, the covariance matrix given by the recursions (8.35) and (8.37) is needed.

8.3.2. Numerical examples

The first numerical example will illustrate the underlying projections in a two-dimensional case.

Example 8.7 Kalman filter: a numerical example

A two-dimensional example is very suitable for graphical illustration. The model under consideration is

$$\begin{aligned} x_{t+1} &= \begin{pmatrix} 0 & -1 \\ 1 & 0 \end{pmatrix} x_t, \qquad x_0 = \begin{pmatrix} 1 \\ 1 \end{pmatrix} \\ y_t &= \begin{pmatrix} 1 & 0 \end{pmatrix} x_t + e_t. \end{aligned}$$

The state equation describes a rotation of the state vector 90° to the left. The absence of state noise facilitates illustration in that there is no random walk in the state process. The measurement equation says that only the first state component is observed. Figure 8.4(a) shows the first three state vectors x_0, x_1, x_2, and the corresponding measurements, which are the projection on the horizontal axis. The measurement noise is assumed negligible. Alternatively, we may assume that the realization of the measurement noise with covariance matrix R is $e_0 = e_1 = 0$.

The Kalman filter is initialized with $\hat{x}_{0|-1} = (0,0)^T$, and covariance matrix

$$P_{0|-1} = \alpha I.$$

Simple and easily checked calculations give

$$\begin{aligned}
K_0 &= \frac{\alpha}{\alpha+R}\begin{pmatrix} 1 \\ 0 \end{pmatrix}, & P_{0|0} &= \begin{pmatrix} \frac{\alpha R}{\alpha+R} & 0 \\ 0 & \alpha \end{pmatrix} \\
P_{1|0} &= \begin{pmatrix} \alpha & 0 \\ 0 & \frac{\alpha R}{\alpha+R} \end{pmatrix}, & K_1 &= \frac{\alpha}{\alpha+R}\begin{pmatrix} 1 \\ 0 \end{pmatrix} \\
\hat{x}_{0|0} &= \frac{\alpha}{\alpha+R}\begin{pmatrix} 1 \\ 0 \end{pmatrix}, & \hat{x}_{1|0} &= \frac{\alpha}{\alpha+R}\begin{pmatrix} 0 \\ 1 \end{pmatrix} \\
\hat{x}_{1|1} &= \frac{\alpha}{\alpha+R}\begin{pmatrix} -1 \\ 1 \end{pmatrix}, & \hat{x}_{2|1} &= \frac{\alpha}{\alpha+R}\begin{pmatrix} -1 \\ -1 \end{pmatrix}.
\end{aligned}$$

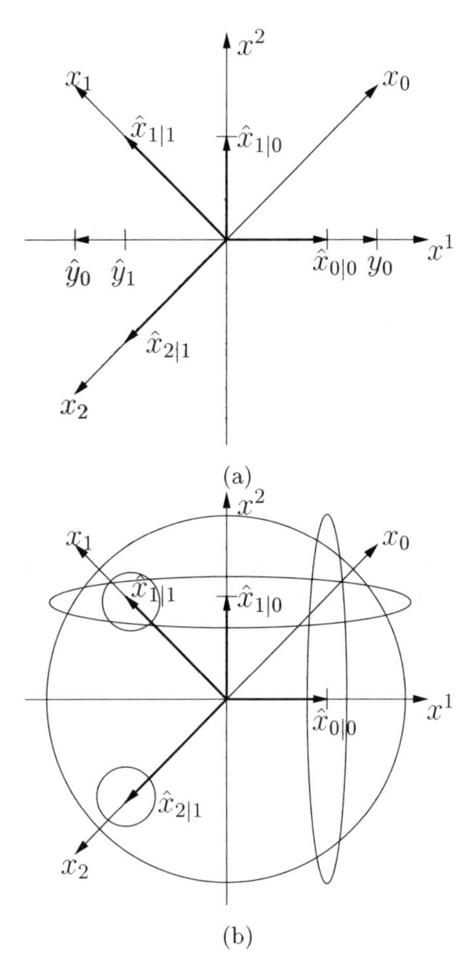

Figure 8.4. True states and estimated estimates from the Kalman filter's two first time recursions. In (a), the measurements are included, and in (b) the covariance ellipses are marked.

Note that $\frac{\alpha}{\alpha+R} < 1$. For large α, or small R, it holds that

$$\hat{x}_{0|0} \approx \begin{pmatrix} 1 \\ 0 \end{pmatrix}, \; \hat{x}_{1|0} \approx \begin{pmatrix} 0 \\ 1 \end{pmatrix}, \; \hat{x}_{1|1} \approx \begin{pmatrix} -1 \\ 1 \end{pmatrix}, \; \hat{x}_{2|1} \approx \begin{pmatrix} -1 \\ -1 \end{pmatrix}.$$

Figure 8.4(b) illustrates the change in shape in the covariance matrix after each update. For each measurement, one dimension is compressed, and the time update is due to the lack of state noise just a rotation.

We make the following observations:

- The time update corresponds to a 90° rotation, just in the same way as for the true state.

- A *model error*, when the true and modeled A matrices are not the same, implies that the rotations are not exactly the same.
- The measurement update is according to the *projection theorem* a projection of the true state onto the measurement. See the derivation in Section 13.1. The plane Π_y is interpreted here as the last measurement. It is only the initial uncertainty reflected in P that prevents the estimate from coinciding exactly with the projection.
- It is the rotation of the state vector that makes the state vector observable! With $A = I$, the true state vector would be constant and x_2 would not be possible to estimate. This can be verified with the observability criterion to be defined in Section 8.6.1

The second example is scalar, and used for illustrating how the Kalman filter equations look like in the simplest possible case.

Example 8.8 Kalman filter: a scalar example

It is often illustrative to rewrite complicated expressions as the simplest possible special case at hand. For a state space model, this corresponds to that all variables are scalars:

$$\begin{aligned} x_{t+1} =& a x_t + b v_t \\ y_t =& c x_t + e_t \\ \mathrm{E}(v_t^2) =& q \\ \mathrm{E}(e_t^2) =& r. \end{aligned}$$

The state equation can be interpreted as an AR(1) process ($|a| < 1$), which is by the measurement equation observed in white noise. Figure 8.5 shows an example with

$$a = 0.9, \quad b = 0.1, \quad c = 3, \quad q = 1, \quad r = 1.$$

The Kalman filter divided into residual generation, measurement update and time update, with the corresponding covariance matrices, is given by

$$\begin{aligned} \varepsilon_t =& y_t - c\hat{x}_{t|t-1} \\ S_t =& c^2 P_{t|t-1} + r \\ \hat{x}_{t|t} =& \hat{x}_{t|t-1} + \frac{P_{t|t-1} c \varepsilon_t}{S_t} \\ P_{t|t} =& P_{t|t-1} - \frac{P_{t|t-1}^2 c^2}{S_t} \\ \hat{x}_{t+1|t} =& a\hat{x}_{t|t} \\ P_{t+1|t} =& a^2 P_{t|t} + b^2 q. \end{aligned} \tag{8.43}$$

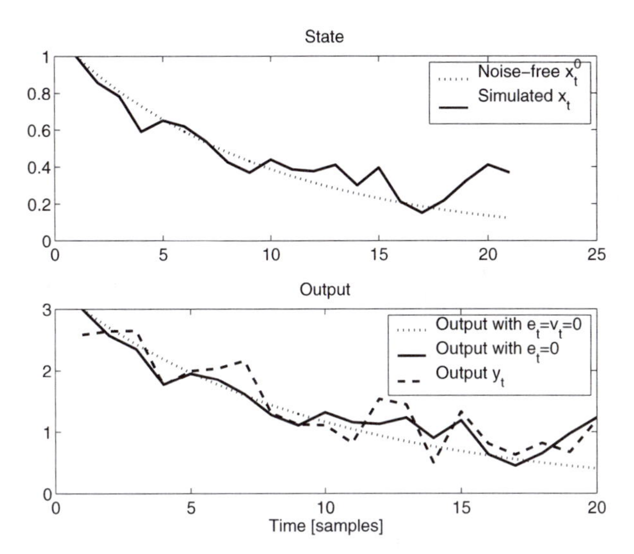

Figure 8.5. Simulation of a scalar state space model. For reference, a state space simulation without state noise is shown, as well as the noise-free output.

Note that the time update is the standard AR predictor. It might be instructive to try to figure out how the structure relates to the general linear filtering formulas in Sections 13.1.3 and 13.2.1, which is a suitable exercise.

Figure 8.6 shows an example for the realization in Figure 8.5. Note how the measurement update improves the prediction from the time update by using the information in the current measurement. Note also that although we are using the best possible filter, we cannot expect to get a perfect estimate of the state.

The 'dummy variables' $P_{t|t}$, $P_{t|t-1}$ and S_t have the interpretation of covariance matrices. We can therefore plot confidence intervals for the estimates in which the true state is expected to be in. In this scalar example, the confidence interval for the filtered estimate is $\hat{x}_{t|t} \pm 2\sqrt{P_{t|t}}$. Since all noises are Gaussian in this example, this corresponds to a 95% confidence interval, which is shown in Figure 8.6(b). We see that the true state is within this interval for all 20 samples (which is better than the average behavior where one sample can be expected to fall outside the interval).

Example 8.9 Kalman filter: DC motor

Consider the state space model of a DC motor in Example 8.4. A simulation of this model is shown in Figure 8.7. It is assumed here that only

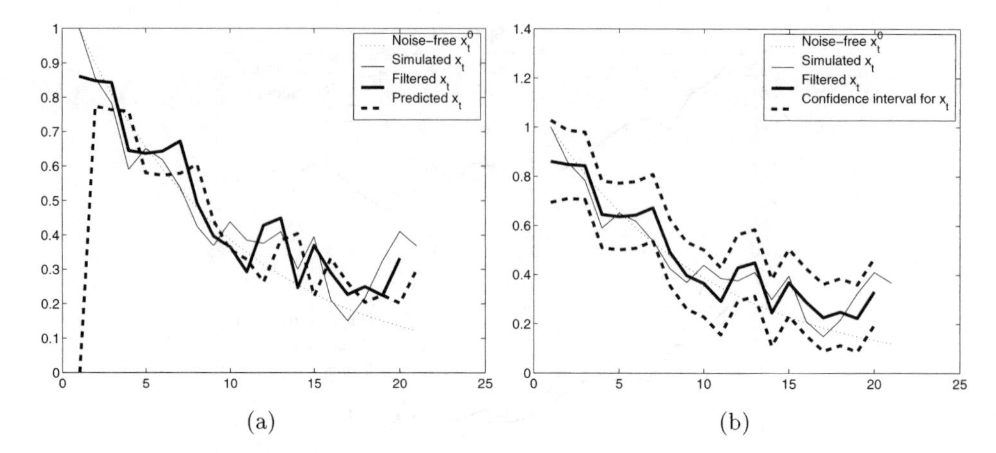

Figure 8.6. The Kalman filter applied to a scalar state space model. In (a), x_t, $\hat{x}_{t|t}$ and $\hat{x}_{t|t-1}$ are compared. In (b), a 95% confidence interval $\hat{x}_{t|t} \pm 2\sqrt{P_{t|t}}$ for x_t is shown. For reference, a state space simulation without state noise is shown. x_t^0 denotes the ensemble average.

the angle is measured, and not the other state which is angular velocity. The Kalman filter estimates of angle and angular velocity are shown in Figure 8.8. As a reference, the measurements and the difference of each two measurements as a velocity measure are plotted. A suboptimal approach to estimate angular velocity would be to take the difference and try to tune a low-pass filter.

8.3.3. Optimality properties

There are several interpretations of what the Kalman filter actually computes:

- The Kalman filter can be interpreted both as an estimator or an algorithm for computing an estimate, cf. Section 13.2.
- If x_0, v_t, e_t are Gaussian variables, then

$$\begin{aligned} x_{t+1}|y_1,\ldots,y_t &\in \mathrm{N}(\hat{x}_{t+1|t}, P_{t+1|t}) \\ x_t|y_1,\ldots,y_t &\in \mathrm{N}(\hat{x}_{t|t}, P_{t|t}) \\ \varepsilon_t &\in \mathrm{N}(0, S_t). \end{aligned}$$

  That is, the Kalman filter provides an algorithm for updating the complete conditional density function.

- If x_0, v_t, e_t are Gaussian variables, then the Kalman filter is the best possible estimator among all *linear and non-linear* ones. Best can here be defined in the mean square sense (conditional or unconditional MSE),

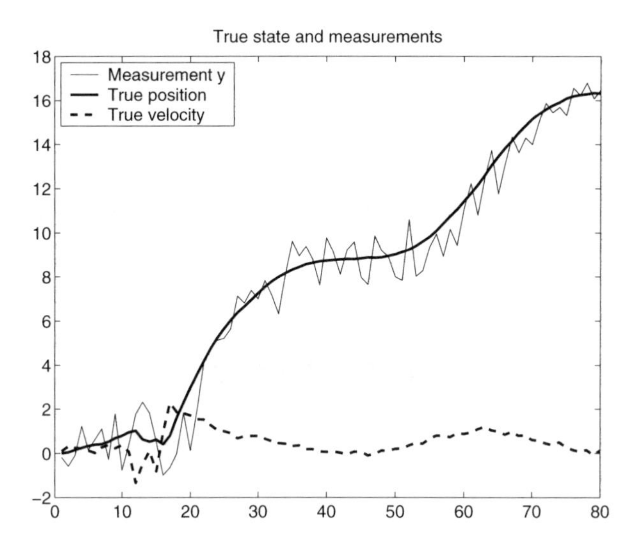

Figure 8.7. Simulation of a DC motor.

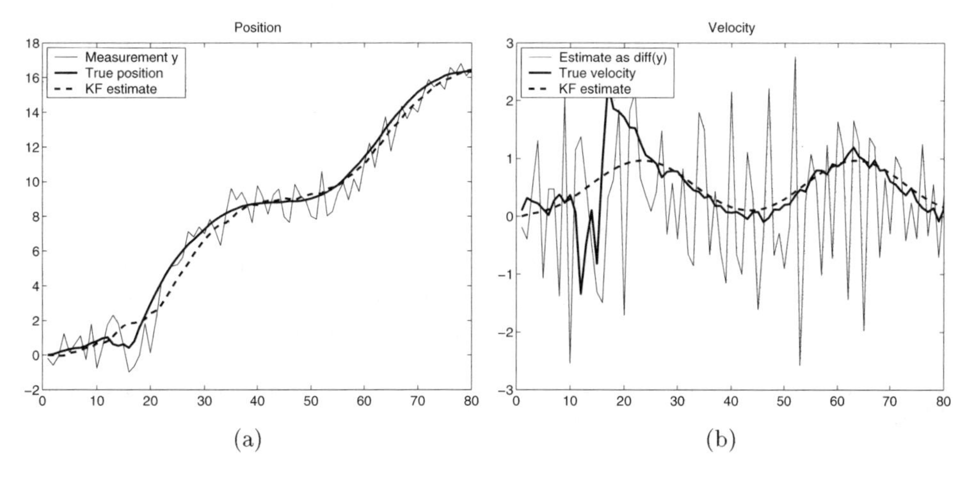

Figure 8.8. Kalman filter estimate of angle (left plot) and angular velocity (right plot).

or in the minimum variance sense, or the maximum *a posteriori* sense, cf. Section 13.2.

- Independently of the distribution of x_0, v_t, e_t, assuming the specified covariance matrices reflect the true second order moments, the Kalman filter is the best possible *linear* filter. What is meant by 'best' is defined as above, but *only* in the unconditional meaning. That is, $P_{t|t-1}$ is the *unconditional* covariance matrix of the MSE. As a counter example, where there are better conditional filters than the Kalman filter, is when the state or measurement noise have frequent outliers. Since the noise is still being white, the Kalman filter is the best linear filter, but then appropriate change detectors in Chapters 9 and 10 often work much better as state estimators.

- All system matrices may depend on data, and the interpretation of $\hat{x}_{t+1|t}$ and $P_{t+1|t}$ as *conditional* estimate and covariance, respectively, still holds. An important case is the linear regression model, where the regression vector C_t contains old measurements.

8.4. Time-invariant signal model

We will here study the special case of a time-invariant state space model

$$x_{t+1} = Ax_t + B_u u_t + B_v v_t, \quad \text{Cov}(v_t) = Q \tag{8.44}$$

$$y_t = Cx_t + e_t, \quad \text{Cov}(e_t) = R. \tag{8.45}$$

in more detail. The reason for this special attention is that we can obtain sharper results for the properties and gain more insight into the Kalman filter.

Assume the Kalman filter is applied at time t_0 for the first time. The Kalman gain K_t will then approach the *steady state Kalman gain* $\bar{K}$ when $t_0 \to -\infty$ or when $t \to \infty$. These interpretations are equivalent. In the same way, $P_t \to \bar{P}$. The existence of such a limit assumes asymptotic time-invariance and stability $|\lambda(A)| < 1$ of the signal model (8.44). To actually prove convergence is quite complicated, and Kalman himself stated it to be his main contribution to the Kalman filter theory. A full proof is found in Anderson and Moore (1979) and Kailath et al. (1998).

Algorithm 8.2 Stationary Kalman filter

The stationary Kalman filter in predictor form is given by

$$\hat{x}_{t+1|t} = (A - A\bar{K}C)\hat{x}_{t|t-1} + A\bar{K}y_t + B_u u_t \tag{8.46}$$

$$\bar{P} = A\bar{P}A^T - A\bar{P}C^T(C\bar{P}C^T + R)^{-1}C\bar{P}A^T + B_v Q B_v^T \tag{8.47}$$

$$\bar{K} = \bar{P}C^T(C\bar{P}C^T + R)^{-1}. \tag{8.48}$$

The stationary covariance matrix, $\bar{P}$, is found by solving the non-linear equation system (8.47), which is referred to as the stationary Riccati equation. Several approaches exist, the simplest one being to iterate the Riccati equation until convergence, which will always work due to stability of the filter. Solving the stationary Riccati equation is the counterpart to spectral factorization in Wiener filtering, cf. Section 13.3.7.

It can be shown from the convergence theory that the Kalman filter is stable,

$$|\lambda(A - A\bar{K}C)| < 1. \tag{8.49}$$

It is interesting to note that stability is ensured even if the signal model in unstable. That is, the Kalman filter is stable even if the system is not. The conditions for stability are that unstable modes in A are excited by the noise $B_v v_t$ and that these modes are observable. In standard terms (Kailath, 1980), these two conditions can be expressed as follows: 1) the pair $[A, C]$ is detectable, and 2) the pair $[A, B_v Q^{1/2}]$ is stabilizable. Here $Q^{1/2}$ denotes any matrix such that $Q^{1/2}(Q^{1/2})^T = Q$ (a square root); see Section 8.7.

8.4.1. Error sources

The time-invariant predictor form is, substituting the measurement equation (8.45) into (8.46),

$$\hat{x}_{t+1|t} = (A - A\bar{K}C)\hat{x}_{t|t-1} + A\bar{K}Cx_t + A\bar{K}e_t. \tag{8.50}$$

In the scalar case, this is an exponential filter with two inputs. The forgetting factor is then $1 - \bar{K}C$, and the adaptation gain is $\bar{K}C$. A small $\bar{K}$ thus implies slow forgetting.

The estimation error $\tilde{x}_{t|t-1} = x_t - \hat{x}_{t|t-1}$ is from (8.50) using the signal model, given by

$$\tilde{x}_{t+1|t} = (A - A\bar{K}C)\tilde{x}_{t|t-1} + A\bar{K}e_t + B_v v_t.$$

Thus, the transient and variance errors are (see Section 8.6.3 for a procedure how to measure the bias error):

$$\tilde{x}_{t+1|t} = \underbrace{(A - A\bar{K}C)^t x_0}_{\text{transient error}} + \underbrace{\sum_{k=1}^{t}(A - A\bar{K}C)^{k-1}\left(A\bar{K}e_k + B_v v_t\right)}_{\text{variance error}}.$$

There are now several philosophies for state estimation:

- The stationary Kalman filter aims at minimizing the variance error, assuming the initial error has faded away.
- An *observer* minimizes the transient, assuming no noise, or assuming deterministic disturbances $B_v v_t$ which cannot be characterized in a stochastic framework (a deterministic disturbance gives a perturbation on the state that can be interpreted as a transient).

- The time-varying Kalman filter minimizes the sum of transient and variance errors.

8.4.2. Observer

We will assume here that there is no process noise. To proceed as for adaptive filters in Section 5.5, we need a further assumption. Assume all eigenvalues of $A - A\bar{K}C$ are distinct. Then we can factorize $T^{-1}DT = A - A\bar{K}C$, where D is a diagonal matrix with diagonal elements being the eigenvalues. In the general case with multiple eigenvalues, we have to deal with so-called Jordan forms (see Kailath (1980)). The transient error can then be written

$$\tilde{x}_{t+1|t} =(A - A\bar{K}C)^t x_0 = T^{-1}D^tTx_0.$$

The transient will decay to zero only if all eigenvalues are strictly less than one, $|\lambda(A - A\bar{K}C)| < 1$. The rate of decay depends upon the largest eigenvalue $\max|\lambda(A - A\bar{K}C)|$. The idea in observer design is to choose $\bar{K}$ such that the eigenvalues get pre-assigned values. This is possible if the pair $[A, C]$ is observable (Kailath, 1980). The only thing that prevents us from making the eigenvalues arbitrarily small is that a disturbance can imply a huge contribution to the error (though it decays quickly after the disturbance has disappeared). Assume a step disturbance entering at time 0, and that the transient error is negligible at this time. Then

$$\tilde{x}_{t+1|t} =\sum_{k=1}^{t}(A - A\bar{K}C)^{k-1}A\bar{K} = \sum_{k=1}^{t-1} T^{-1}D^kTA\bar{K} + A\bar{K}.$$

A fast decay of the transient requires that the gain $\bar{K}$ is large, so the direct term $A\bar{K}$ will be large as well, even if $D = 0$.

Consider the case of a scalar output. We then have n degrees of freedom in designing $\bar{K}$. Since there are n eigenvalues in D, and these can be chosen arbitrarily, we have no degrees of freedom to shape T. This is well known from control theory, since the zeros of the observer are fixed, and the poles can be arbitrarily chosen. The special case of $D = 0$ corresponds to a deadbeat observer (Kailath, 1980). The compromise offered in the Kalman filter is to choose the eigenvalues so that the transient error and variance error are balanced.

8.4.3. Frequency response

The assumption on stationarity allows us to define the frequency response of the Kalman filter. Taking the z-transform of (8.46) gives

$$\hat{X}(z) = \underbrace{(zI - A + A\bar{K}C)^{-1}A\bar{K}}_{H_y(z)} Y(z) + \underbrace{(zI - A + A\bar{K}C)^{-1}B_u}_{H_u(z)} U(z).$$

That is, the transfer functions of the Kalman filter when $t \to \infty$ or $t_0 \to -\infty$ are

$$H_y(e^{i\omega T_s}) = (e^{i\omega T_s} I - A + A\bar{K}C)^{-1}A\bar{K} \quad (8.51)$$

$$H_u(e^{i\omega T_s}) = (e^{i\omega T_s} I - A + A\bar{K}C)^{-1}B_u. \quad (8.52)$$

Here $H_y(e^{i\omega T_s})$ is a $n_x \times n_y$ matrix of transfer functions (or column vector when y is scalar), so there is one scalar transfer function from each measurement to each state, and similarly for $H_u(e^{i\omega T_s})$. Each transfer function has the same n_x poles, and generally $n_x - 1$ zeros (there is at least one time delay).

We now analyze the transfer function to the measurement y instead of to the state x. Assume here that, without loss of generality, there is no input. Since $\hat{Y} = C\hat{X}$ we immediately get

$$H_{KF}^{closed\ loop}(e^{i\omega T_s}) = C(e^{i\omega T_s} I - A + A\bar{K}C)^{-1}A\bar{K}. \quad (8.53)$$

Equation (8.53) gives the closed loop transfer function in the input-output relation $\hat{Y}(z) = H_{KF}^{closed\ loop}(z)Y(z)$ of the Kalman filter. That is, equation (8.53) gives the transfer function from y to $\hat{y}$ in Figure 8.9. An open loop equivalent from innovation ε to y is readily obtained in the same way. We get

$$H_{model} = C(zI - A)^{-1}B_v$$

$$H_{KF}^{open\ loop} = C(zI - A)^{-1}\bar{K}$$

as the transfer functions of the model and Kalman filter, respectively. See Figure 8.9.

8.4.4. Spectral factorization

An interesting side effect of these calculations, is that a spectral factorization can be obtained automatically. The calculations below are based on the *super formula*: if $Y(z) = S(z)U(z)$, then the spectrum is given by $\Phi_{yy}(z) = |S(z)|^2\Phi_{uu}(z) = S(z)S(z^{-1})\Phi_{uu}(z)$ on the unit circle. For matrix valued transfer functions, this formula reads $\Phi_{yy}(z) = S(z)\Phi_{uu}(z)S^T(z^{-1})$.

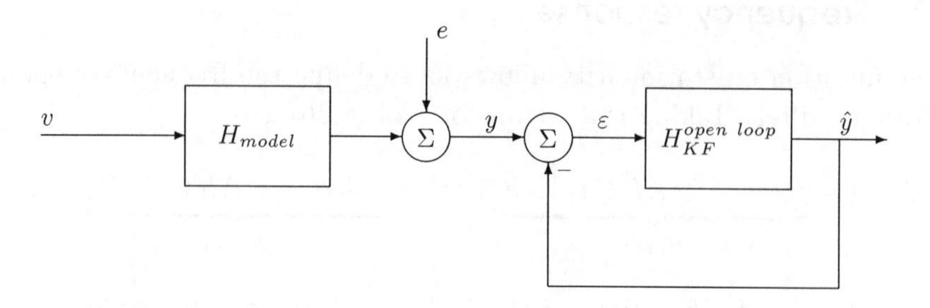

Figure 8.9. Interplay of model and Kalman filter transfer functions.

It follows from Figure 8.9 that the output power spectrum is

$$\Phi_{yy}(z) = R + C(zI - A)^{-1}B_v Q B_v^T (zI - A)^{-T} C^T \tag{8.54}$$

$$= R + H_{model}(z) Q H_{model}(z^{-1}). \tag{8.55}$$

The *spectral factorization* problem is now to find a transfer function that gives the same spectrum $\Phi_{yy}(z)$ for a white noise input with $\Phi_{uu}(z) = 1$. That is, find $S(z)$ such that $\Phi_{yy}(z) = S(z)S^T(z^{-1})$. The result is not immediate from (8.55) because it is the sum of two terms.

Alternatively, we can deduce from Figure 8.9 that $\varepsilon = y - H_{KF}\varepsilon$, so $y = (I + H_{KF})\varepsilon$, and

$$\Phi_{yy}(z) = (I + H_{KF}(z))(R + C\bar{P}C^T)(I + H_{KF}^T(z^{-1})). \tag{8.56}$$

Note that $R + C\bar{P}C^T$ is the covariance matrix of the innovation. That is, the non-minimum phase stable spectral factor of the output spectrum is given by $S(z) = (I + H_{KF}(z))(R + C\bar{P}C^T)^{1/2}$ (the square root is an ordinary matrix one). This expression can be used to compute the Wiener filter for vector valued processes, a task that can hardly be done without computing the stationary Kalman filter. The derivation is based on Anderson and Moore (1979), p. 86.

8.5. Smoothing

8.5.1. Fixed-lag smoothing

Fixed-lag smoothing is to estimate the state vector x_t from measurements $y(s)$, $s \leq t + m$. If m is chosen in the order of two, say, time constants of the system, almost all of the information in the data is utilized. The trick to derive the optimal filter is to augment the state vector with delayed states,

$$x_t^i = x_{t-i}, \quad i = 1, \ldots, m + 1. \tag{8.57}$$

The augmented state vector is (cf. Section 8.2.4)

$$\bar{x}_t = \begin{bmatrix} x_t^0 \\ \vdots \\ x_t^{m+1} \end{bmatrix}. \tag{8.58}$$

For simplicity, we will drop the time indices on the state matrices in this subsection. The full signal model is

$$\bar{x}_{t+1} = \bar{A}\bar{x}_t + \bar{B}_v v_t \tag{8.59}$$

$$y_t = \bar{C}\bar{x}_t + e_t, \tag{8.60}$$

where

$$\bar{A} = \begin{pmatrix} A & 0 & 0 & \cdots & 0 & 0 \\ I & 0 & 0 & \cdots & 0 & 0 \\ 0 & I & 0 & \cdots & 0 & 0 \\ 0 & 0 & I & \cdots & 0 & 0 \\ \vdots & \vdots & \vdots & \ddots & \vdots & \vdots \\ 0 & 0 & 0 & \cdots & I & 0 \end{pmatrix} \qquad \bar{B}_{v,t} = \begin{pmatrix} B_{v,t} \\ 0 \\ \vdots \\ 0 \end{pmatrix} \tag{8.61}$$

$$\bar{C} = \begin{pmatrix} C & 0 & \cdots & 0 \end{pmatrix}. \tag{8.62}$$

Estimate $\hat{x}_{t-m|t}, \hat{x}_{t-m+1|t}, \ldots, \hat{x}_{t|t}$ by applying the Kalman filter on (8.59)–(8.60). The estimate of x_{t-m} given observations of $y_s, 0 \leq s \leq t$ is given by $\hat{x}^{m+1}_{t+1|t}$. This is recognized as the last sub-vector of the Kalman filter prediction $\hat{\bar{x}}_{t+1|t}$.

To simplify the equations somewhat, let us write out the Kalman predictor formulas

$$\hat{\bar{x}}_{t+1|t} = \bar{A}\hat{\bar{x}}_{t|t-1} + \bar{A}\bar{K}_t \left(y_t - \bar{C}^T \hat{\bar{x}}_{t|t-1}\right) \tag{8.63}$$

$$\bar{K}_t = \bar{P}_{t|t-1}\bar{C}^T \left(\bar{C}\bar{P}_{t|t-1}\bar{C}^T + R\right)^{-1} \tag{8.64}$$

$$\bar{P}_{t+1|t} = \bar{A}\bar{P}_{t|t-1}\bar{A}^T + \bar{B}_{v,t}Q\bar{B}_{v,t}^T - \bar{A}\bar{P}_{t|t-1}\bar{C}^T \left(\bar{C}\bar{P}_{t|t-1}\bar{C}^T + R\right)^{-1} \bar{C}\bar{P}_{t|t-1}\bar{A}^T. \tag{8.65}$$

Next, we aim at simplifying (8.63)–(8.65). It follows from (8.62) that the innovation covariance can be simplified to its usual low dimensional form

$$\bar{C}\bar{P}_{t|t-1}\bar{C}^T + R = C P_{t|t-1} C^T + R. \tag{8.66}$$

Split $\bar{P}_{t|t-1}$ into $n \times n$ blocks

$$\bar{P}_{t|t-1} = \begin{pmatrix} \bar{P}^{0,0}_{t|t-1} & \bar{P}^{0,1}_{t|t-1} & \cdots & \bar{P}^{0,m+1}_{t|t-1} \\ \bar{P}^{1,0}_{t|t-1} & \bar{P}^{1,1}_{t|t-1} & \cdots & \bar{P}^{1,m+1}_{t|t-1} \\ \vdots & \vdots & \ddots & \vdots \\ \bar{P}^{m+1,0}_{t|t-1} & \bar{P}^{m+1,1}_{t|t-1} & \cdots & \bar{P}^{m+1,m+1}_{t|t-1} \end{pmatrix}, \tag{8.67}$$

which gives

$$\bar{K}_t = \begin{pmatrix} \bar{K}^0_t \\ \vdots \\ \bar{K}^{m+1}_t \end{pmatrix} = \begin{pmatrix} \bar{P}^{0,0}_{t|t-1} \\ \bar{P}^{1,0}_{t|t-1} \\ \vdots \\ \bar{P}^{m+1,0}_{t|t-1} \end{pmatrix} C^T \left(CP_{t|t-1}C^T + R\right)^{-1}. \tag{8.68}$$

To compute the Kalman gain, we need to know $\bar{P}^{i,0}, \quad i = 0, \dots, m+1$. These are given by a substitution of (8.67) in the Riccati equation (8.65)

$$\bar{P}^{i+1,0}_{t+1|t} = \bar{P}^{i,0}_{t|t-1}(A - AK_tC)^T, \quad i = 0, \dots, m, \tag{8.69}$$

where $K_t = \bar{K}^0_t$. That is, to compute the state update we only need to know the quantities from the standard Kalman predictor and the original state space matrices of size $n \times n$, and the covariance matrices updated in (8.69). To compute the diagonal matrices $P^{i,i}_{t|t-1}, i = 0, \dots, m+1$, corresponding to the smoothed estimate's covariance matrix, a substitution of (8.67) in the Riccati equation (8.65) gives

$$\bar{P}^{i+1,i+1}_{t+1|t} = \bar{P}^{i,i}_{t|t-1} - \bar{K}^i_t C \bar{P}^{0,i}_{t|t-1} \quad i = 0, \dots, m. \tag{8.70}$$

$\bar{P}^{0,0}_{t|t-1}$ is found from the Riccati equation for the original signal model.

8.5.2. Fixed-interval smoothing

In this off-line filtering situation, we have access to all N measurements and want to find the best possible state estimate $\hat{x}_{t|N}$. One rather naive way is to use the fixed-lag smoother from the previous section, let $m = N$ and introduce N fictive measurements (e.g. zeros) with infinite variance. Then we apply the fixed-lag smoother and at time $t = 2N$ we have all smoothed estimates available. The reason for introducing information less measurements is solely to being able to run the Kalman filter until all useful information in the measurements have propagated down to the last sub-vector of the augmented state vector in (8.58).

A better way is to use so-called *forward-backward* filters. We will study two different approaches.

Rauch-Tung-Striebel formulas

Algorithm 8.3 gives the *Rauch–Tung–Striebel formulas* for fixed-interval smoothing, which are given without proof. The notation $\hat{x}_{t|N}$ means as usual the estimate of x_t given measurements up to time N.

Algorithm 8.3 Fixed interval smoothing: Rauch-Tung-Striebel

Available observations are $y_t, t = 0, \ldots, N$. Run the standard Kalman filter and store both the time and measurement updates, $\hat{x}_{t|t}, \hat{x}_{t|t-1}, P_{t|t}, P_{t|t-1}$. Apply the following time recursion backwards in time:

$$\hat{x}_{t-1|N} = \hat{x}_{t-1|t-1} + P_{t-1|t-1}A_{t-1}^T P_{t|t-1}^{-1}\left(\hat{x}_{t|N} - \hat{x}_{t|t-1}\right). \tag{8.71a}$$

The covariance matrix of the estimation error $P_{t|N}$ is

$$P_{t-1|N} = P_{t-1|t-1} + P_{t-1|t-1}A_{t-1}^T P_{t|t-1}^{-1}\left(P_{t|N} - P_{t|t-1}\right)P_{t|t-1}^{-1}A_{t-1}P_{t-1|t-1}. \tag{8.71b}$$

This algorithm is quite simple to implement, but it requires that all state and covariance matrices, both for the prediction and filter errors, are stored. That also means that we must explicitely apply both time and measurement updates.

Two-filter smoothing formulas

We will here derive a *two-filter smoothing formula*, which will turn out to be useful for change detection.

Denote the conditional expectations of past and future data

$$(x_t|y_1^t, x_0) \in \mathrm{N}(\hat{x}_{t|t}^F, P_{t|t}^F)$$
$$(x_t|y_{t+1}^N) \in \mathrm{N}(\hat{x}_{t|t+1}^B, P_{t|t+1}^B),$$

respectively. Here $\hat{x}_{t|t}^F$ and $P_{t|t}^F$ are the estimates from the Kalman filter (these will not be given explicitly). The index F is introduced to stress that the filter runs forwards in time, in contrast to the filter running backwards in time, yielding $\hat{x}_{t|t+1}^B$ and $P_{t|t+1}^B$, to be introduced. Quite logically, $\hat{x}_{t|t+1}^B$ is the estimate of x_t based on measurements $y_{t+1}, y_{t+2}, \ldots, y_N$, which is up to time $t+1$ backwards in time.

The smoothed estimate is, given these distributions, a standard fusion problem, whose solution will be derived in Section 8.8.3. The result is

$$\begin{aligned}(x_t|y_1^N, x_0) \in &\mathrm{N}(\hat{x}_{t|N}, P_{t|N}) \\ P_{t|N} &= \left((P_{t|t}^F)^{-1} + (P_{t|t+1}^B)^{-1}\right)^{-1} \\ \hat{x}_{t|N} &= P_{t|N}\left((P_{t|t}^F)^{-1}\hat{x}_{t|t}^F + (P_{t|t+1}^B)^{-1}\hat{x}_{t|t+1}^B\right).\end{aligned}$$

There is an elegant way to use the Kalman filter backwards on data to compute $\hat{x}_{t|t+1}^B$ and $P_{t|t+1}^B$. What is needed is a so-called backward model.

The following lemma gives the desired backwards Markovian model that is sample path equivalent to (8.32). The Markov property implies that the noise process $\{v_t\}$, and the final value of the state vector x_N, are independent. Not only are the first and second order statistics equal for these two models, but they are indeed sample path equivalent, since they both produce the same state and output vectors.

Lemma 8.1

The following model is sample path equivalent to (8.32) and is its corresponding backward Markovian model

$$\begin{aligned}x_t &= A_t^B x_{t+1} + v_{t+1}^B \\ y_t &= C_t x_t + e_t. \end{aligned} \tag{8.72}$$

Here

$$A_t^B = \Pi_t A_t^T \Pi_{t+1}^{-1} = A_t^{-1} - A_t^{-1} Q_t \Pi_{t+1}^{-1} \tag{8.73}$$

$$Q_t^B = \Pi_t - \Pi_t A_t^T \Pi_{t+1}^{-1} A_t \Pi_t \tag{8.74}$$

$$= A_t^{-1} Q_t A_t^{-T} - A_t^{-1} Q_t \Pi_{t+1}^{-1} Q_t A_t^{-T}, \tag{8.75}$$

where $\Pi_t = E[x_t x_t^T]$ is the a priori *covariance matrix of the state vector, computed recursively by $\Pi_{t+1} = A_t \Pi_t A_t^T + Q_t$. The last equalities of (8.73) and (8.75) hold if A_t is invertible.*

Proof: See Verghese and Kailath (1979). □

Note that no inversion of the state noise covariance matrix is needed, so we do not have to deal with the factorized form here.

The Kalman filter applied to the model (8.72) in reverse time provides the quantities sought:

$$\hat{x}_{t|t+1}^B, \quad P_{t|t+1}^B.$$

The backward model can be simplified by assuming no initial knowledge of the state, and thus $\Pi_0 = P_{1|0} = \infty I$, or more formally, $\Pi_0^{-1} = 0$. Then $\Pi_t^{-1} = 0$ for all t, and the latter expressions assuming invertible A_t give

$$\begin{aligned} A_t^B &= A_t^{-1} \\ Q_t^B &= A_t^{-1} Q_t A_t^{-T}. \end{aligned}$$

These last two formulas are quite intuitive, and the result of a straightforward inversion of the state equation as $x_t = A^{-1} x_{t+1} - A^{-1} v_t$.

Algorithm 8.4 Fixed interval smoothing: forward-backward filtering

Consider the state space model

$$\begin{aligned} x_t &= A_t^B x_{t+1} + v_{t+1}^B \\ y_t &= C_t x_t + e_t. \end{aligned}$$

1. Run the standard Kalman filter forwards in time and store the filter quantities $\hat{x}_{t|t}, P_{t|t}$.
2. Compute the backward model

$$\begin{aligned} A_t^B &= \Pi_t A_t^T \Pi_{t+1}^{-1} = A_t^{-1} - A_t^{-1} Q_t \Pi_{t+1}^{-1} \\ Q_t^B &= \Pi_t - \Pi_t A_t^T \Pi_{t+1}^{-1} A_t \Pi_t \\ &= A_t^{-1} Q_t A_t^{-T} - A_t^{-1} Q_t \Pi_{t+1}^{-1} Q_t A_t^{-T} \\ \Pi_{t+1} &= A_t \Pi_t A_t^T + Q_t, \end{aligned}$$

   and run the standard Kalman filter backwards in time and store both the 'predictions' $\hat{x}_{t|t+1}, P_{t|t+1}$.
3. Merge the information from the forward and backward filters

$$\begin{aligned} P_{t|N} &= \left((P_{t|t}^F)^{-1} + (P_{t|t+1}^B)^{-1} \right)^{-1} \\ \hat{x}_{t|N} &= P_{t|N} \left((P_{t|t}^F)^{-1} \hat{x}_{t|t}^F + (P_{t|t+1}^B)^{-1} \hat{x}_{t|t+1}^B \right). \end{aligned}$$

8.6. Computational aspects

First, some brief comments on how to improve the numerical properties are given before we examine major issues:

- In case the measurement noise covariance is *singular*, the filter may be numerically unstable, although the solution to the estimation problem is well defined.
  - One solution to this is to replace the inverse of S_t in K_t with a pseudo-inverse.
  - Another solution is regularization. That is to add a small identity matrix to R_t. This is a common suggestion, especially when the measurement is scalar.
  - A singular R means that one or more linear combinations of the state vector can be exactly computed from one measurement. A third solution is to compute a reduced order filter (Anderson and Moore, 1979). This is quite similar to the Luenberger observer (Kailath, 1980; Luenberger, 1966). The idea is to transform the state and observation vector to $\bar{x}, \bar{y}$ such that the exactly known part of $\bar{y}$ is a sub-vector of $\bar{x}$, which does not need to be estimated.
- Sometimes we have linearly dependent measurements, which give cause to a rank deficient C_t. Then there is the possibility to reduce the complexity of the filter, by replacing the observation vector with an *equivalent observation*, see Murdin (1998), obtained by solving the least squares problem $y_t = C_t x_t + e_t$. For instance, when there are more measurements than states we get the least squares (snapshot) estimate

$$\hat{x}_t = (C_t^T C_t)^{-1} C_t^T y_t = x_t + (C_t^T C_t)^{-1} C_t^T e_t \triangleq \bar{y}_t.$$

  Then

$$\bar{y}_t = I x_t + \bar{e}_t, \quad \bar{e}_t \in \mathrm{N}(0, (C_t^T C_t)^{-1} C_t^T R_t C_t (C_t^T C_t)^{-1})$$

  can be fed to the Kalman filter without loss of performance.
- *Outliers* should be removed from the filter. Here prior knowledge and data pre-processing can be used, or impulsive disturbances can be incorporated into the model (Niedzwiecki and Cisowski, 1996; Settineri et al., 1996).
- All numerical problems are significantly decreased by using square root algorithms. Section 8.7 is devoted to this issue.

8.6.1. Divergence

Divergence is the case when the covariance matrix does not reflect the true uncertainty in the state estimate. The simplest way to diagnose divergence is to compare the innovation variance S_t with an estimate formed from the Kalman filter innovations, e.g.

$$\hat{S}_t = (1-\lambda)\sum_{k=0}^{\infty} \lambda^k \varepsilon_{t-k}\varepsilon_{t-k}^T.$$

Other whiteness-based innovation tests apply, see Section 8.10. Another indication of divergence is that the filter gain K_t tends to zero. Possible causes of divergence are:

- Poor excitation or signal-to-noise ratio. We have seen the importance of excitation in adaptive filtering (when $A_t = I$ and $C_t = \varphi_t^T$). Basically, we want the input signals to make C_t point in many different directions, in the sense that

$$\sum_{k=t-L+1}^{t} C_k^T R_k^{-1} C_k$$

  should be invertible for small L. This expression is a measure of the *information* in the measurements, as will become clear in the information filter, see Section 8.8.1. For a time-invariant signal model, we must require *observability*, that is

$$\begin{pmatrix} C \\ CA \\ CA^2 \\ \vdots \\ CA^{n-1} \end{pmatrix}$$

  has full column rank; see Kailath (1980). To get good numerical properties, we should also require that the condition number is not too small.

- A small offset in combination with unstable or marginally unstable signal models is another case where divergence may occur. Compare this with navigation applications, where the position estimate may drift away, without any possibility of detecting this.

- Bias errors; see Section 8.6.3.

8.6.2. Cross correlated noise

If there is a correlation between the noise processes $\mathrm{E}(v_t e_t^T) = M_t$, so that

$$\mathrm{E}\left(\begin{pmatrix} v_t \\ e_t \end{pmatrix} (v_k,\ e_k)\right) = \begin{pmatrix} Q_t & M_t \\ M_t^T & R_t \end{pmatrix} \delta_{tk},$$

one trick to get back on track is to replace the state noise with

$$\bar{v}_t = v_t - M_t R_t^{-1} e_t \tag{8.76}$$

$$= v_t - M_t R_t^{-1}(y_t - C_t x_t). \tag{8.77}$$

Equation (8.76) gives

$$\mathrm{E}\left(\begin{pmatrix}\bar{v}_t \\ e_t\end{pmatrix}(\bar{v}_k,\ e_k)\right) = \begin{pmatrix} Q_t - M_t R_t^{-1} M_t & 0 \\ 0 & R_t \end{pmatrix}\delta_{tk}.$$

Substitution with (8.77) implies the equivalent signal model

$$x_{t+1} = (A_t - B_{v,t} M_t R_t^{-1} C_t)x_t + B_{u,t}u_t + B_{v,t}\bar{v}_t + B_{v,t}M_t R_t^{-1} y_t \tag{8.78}$$

$$y_t = C_t x_t + e_t. \tag{8.79}$$

The Kalman filter applies to this model. The last output dependent term in the state equation should be interpreted as an extra input.

8.6.3. Bias error

A *bias error* due to modeling errors can be analyzed by distinguishing the true system and the design model, denoted by super-indices o and d, respectively. The state space model for the true system and the Kalman filter for the design model is

$$\begin{pmatrix} x_{t+1}^o \\ \hat{x}_{t+1|t}^d \end{pmatrix} = \begin{pmatrix} A_t^o & 0 \\ A_t^d K_t^d C_t^o & A_t^d - A_t^d K_t^d C_t^d \end{pmatrix} \begin{pmatrix} x_t^o \\ \hat{x}_{t|t-1}^d \end{pmatrix} + \begin{pmatrix} B_{v,t}^o & 0 \\ 0 & A_t^d K_t^d \end{pmatrix}\begin{pmatrix} v_t \\ e_t \end{pmatrix} + \begin{pmatrix} B_{u,t}^o & 0 \\ 0 & B_{u,t}^d \end{pmatrix}\begin{pmatrix} u_t^0 \\ u_t^d \end{pmatrix}. \tag{8.80}$$

Let $\bar{x}$ denote the augmented state vector with covariance matrix $\bar{P}$. This is a state space model (without measurements), so the covariance matrix time update can be applied,

$$\bar{P}_{t+1} = \begin{pmatrix} A_t^o & 0 \\ A_t^d K_t^d C_t^o & A_t^d - A_t^d K_t^d C_t^d \end{pmatrix} \bar{P}_t \begin{pmatrix} A_t^o & 0 \\ A_t^d K_t^d C_t^o & A_t^d - A_t^d K_t^d C_t^d \end{pmatrix}^T + \begin{pmatrix} B_{v,t}^o & 0 \\ 0 & A_t^d K_t^d \end{pmatrix}\begin{pmatrix} Q_t^o & 0 \\ 0 & R_t^o \end{pmatrix}\begin{pmatrix} B_{v,t}^o & 0 \\ 0 & A_t^d K_t^d \end{pmatrix}^T. \tag{8.81}$$

The total error (including both bias and variance) can be expressed as

$$x_t^o - \hat{x}_{t|t-1}^d = (I,\ -I)\bar{x}_t.$$

and the error correlation matrix (note that P is no covariance matrix when there is a bias error) is

$$P^p_{t|t-1} = (I, \ -I)\bar{P}_{t|t-1}\begin{pmatrix} I \\ -I \end{pmatrix}.$$

This matrix is a measure of the *performance* (hence the superscript p) of the mean square error. This should not be confused with the optimal performance P^o and the measure that comes out from the Kalman filter P^d. Note that $P^p > P^o$, and the difference can be interpreted as the bias. Note also that although the Kalman filter is stable, P^p may not even be bounded. This is the case in navigation using an inertial navigation system, where the position estimate will drift away.

The main use of (8.80) is a simple test procedure for evaluating the influence of parameter variations or faults in the system matrices, and measurement errors or faults in the input signal. This kind of *sensitivity analysis* should always be performed before implementation. That is, vary each single partially unknown parameter, and compare P^p to P^o. Sensitivity analysis can here be interpreted as a Taylor expansion

$$P^p \approx P^o + \frac{dP^o}{d\theta}\Delta\theta.$$

The more sensitive P^o is to a certain parameter, the more important it is to model it correctly.

8.6.4. Sequential processing

For large measurement vectors, the main computational burden lies in the matrix inversion of S_t. This can be avoided by sequential processing, in which the measurements are processed one at a time by several consecutive measurement updates. Another advantage, except for computation time, is when a real-time requirement occasionally makes it impossible to complete the measurement update. Then we can interrupt the measurement updates and only lose some of the information, instead of all of it.

Assume that R_t is block diagonal, $R_t = \text{diag}(R_t^1, \ldots, R_t^m)$, which is often the case in practice. Otherwise, the measurements can be transformed as outlined below. Assume also that the measurement vector is partitioned in the same way, $y_t = ((y_t^1)^T, \ldots, (y_t^m)^T)^T$ and $C_t = ((C_t^1)^T, \ldots, (C_t^m)^T)^T$. Then we apply for each $i = 0, \ldots, m-1$, the measurement updates

$$\hat{x}^{i+1}_{t|t} = \hat{x}^i_{t|t-1} + P^i_{t|t-1}(C^i_t)^T(C^i_t P^i_{t|t-1}(C^i_t)^T + R^i_t)^{-1}(y^i_t - C^i_t\hat{x}^i_{t|t-1}) \quad (8.82)$$

$$P^{i+1}_{t|t} = P^i_{t|t-1} - P^i_{t|t-1}(C^i_t)^T(C^i_t P^i_{t|t-1}(C^i_t)^T + R^i_t)^{-1}C^i_t P^i_{t|t-1}, \quad (8.83)$$

with $\hat{x}^0_{t|t} = \hat{x}_{t|t-1}$ and $\hat{P}^0_{t|t} = \hat{P}_{t|t-1}$. Then we let $\hat{x}_{t|t} = \hat{x}^m_{t|t-1}$ and $\hat{P}_{t|t} = \hat{P}^m_{t|t-1}$.

If R_t is not block-diagonal, or if the blocks are large, the measurement vector can easily be transformed to have a diagonal covariance matrix. That is, let

$$\begin{aligned}\bar{y}_t =& R_t^{-1/2} y_t \\ \bar{C}_t =& R_t^{-1/2} C_t \\ \bar{R}_t =& R_t^{-1/2} R_t R_t^{-T/2} = I.\end{aligned}$$

Here $R_t^{1/2}$ denotes the square root; see Section 8.7. For time-varying R_t this may be of little help, since the factorization requires as many operations as the matrix inversion we wanted to avoid. However, for time invariant R_t this is the method to use.

8.7. Square root implementation

Square root algorithms are motivated by numerical problems when updating the covariance matrix P that often occur in practice:

- P is not *symmetric*. This is easily checked, and one remedy is to replace it by $0.5(P + P^T)$.
- P is not *positive definite*. This is not as easy to check as symmetry, and there is no good remedy. One computationally demanding solution might be to compute the SVD of $P = UDU^T$ after each update, and to replace negative values in the singular values in D with zero or a small positive number.
- Due to large differences in the scalings of the states, there might be numerical problems in representing P. Assume, for instance, that $n_x = 2$ and that the states are rescaled

$$\bar{x} = \begin{pmatrix} 10^{10} & 0 \\ 0 & 1 \end{pmatrix} x \Rightarrow \bar{P} = \begin{pmatrix} 10^{10} & 0 \\ 0 & 1 \end{pmatrix} P \begin{pmatrix} 10^{10} & 0 \\ 0 & 1 \end{pmatrix},$$

  and there will almost surely be numerical difficulties in representing the covariance matrix, while there is no similar problem for the state. The solution is a thoughtful scaling of measurements and states from the beginning.
- Due to a numerically sensitive state space model (e.g an almost singular R matrix), there might be numerical problems in computing P.

The square root implementation resolves all these problems. We define the square root as any matrix $P^{1/2}$ of the same dimension as P, satisfying $P = P^{1/2}P^{T/2}$. The reason for the first requirement is to avoid solutions of the kind $\sqrt{1} = (1/\sqrt{2}, 1/\sqrt{2})$. Note that `sqrtm` in MATLAB™ defines a square root without transpose as $P = AA$. This definition is equivalent to the one used here, since P is symmetric.

The idea of updating a *square root* is very old, and fundamental contributions have been done in Bierman (1977), Park and Kailath (1995) and Potter (1963). The theory now seems to have reached a very mature and unified form, as described in Kailath et al. (1998).

8.7.1. Time and measurement updates

The idea is best described by studying the time update,

$$P_{t+1|t} = A_t P_{t|t} A_t^T + B_{v,t} Q_t B_{v,t}^T. \tag{8.84}$$

A first attempt of factorization

$$P_{t+1|t} = \begin{pmatrix} A_t P_{t|t}^{1/2} & B_{v,t} Q_t^{1/2} \end{pmatrix} \begin{pmatrix} A_t P_{t|t}^{1/2} & B_{v,t} Q_t^{1/2} \end{pmatrix}^T \tag{8.85}$$

fails to the condition that a square root must be quadratic. We can, however, apply a *QR factorization* to each factor in (8.85). A QR factorization is defined as

$$X = QR, \tag{8.86}$$

where Q is a unitary matrix such that $Q^T Q = I$ and R is a upper triangular matrix. The R and Q here should not be confused with R_t and Q_t in the state space model.

Example 8.10 QR factorization

The MATLAB™ function `qr` efficiently computes the factorization

$$\begin{pmatrix} 1 & 2 \\ 3 & 4 \\ 5 & 6 \end{pmatrix} = \begin{pmatrix} -0.1690 & 0.8971 & 0.4082 \\ -0.5071 & 0.2760 & -0.8165 \\ -0.8452 & -0.3450 & 0.4082 \end{pmatrix} \begin{pmatrix} -5.9161 & -7.4374 \\ 0 & 0.8281 \\ 0 & 0 \end{pmatrix}.$$

Applying QR factorization to the transpose of the first factor in (8.85) gives

$$\begin{pmatrix} A_t P_{t|t}^{1/2} & B_{v,t} Q_t^{1/2} \end{pmatrix} = R^T Q^T. \tag{8.87}$$

That is, the time update can be written

$$P_{t+1|t} = R^T Q^T Q R = R^T R. \tag{8.88}$$

It is clear that Q is here instrumental, and does not have to be saved after the factorization is completed. Here R consists of a quadratic part (actually triangular) and a part with only zeros. We can identify the square root as the first part of R,

$$R^T = \begin{pmatrix} P_{t+1|t}^{1/2} & 0 \end{pmatrix}. \tag{8.89}$$

To summarize, the time update is as follows.

Algorithm 8.5 Square root algorithm, time update

1. Form the matrix in the left hand side of (8.87). This involves computing one new square root of Q_t, which can be done by the QR factorization (R is taken as the square root) or `sqrtm` in MATLAB™ .
2. Apply the QR factorization in (8.87).
3. Identify the square root of $P_{t+1|t}$ as in (8.89).

More compactly, the relations are

$$\begin{pmatrix} A_t P_{t|t}^{1/2} & B_{v,t} Q_t^{1/2} \end{pmatrix} = \begin{pmatrix} P_{t+1|t}^{1/2} & 0 \end{pmatrix} Q^T. \tag{8.90}$$

The measurement update can be treated analogously. However, we will give a somewhat more complex form that also provides the Kalman gain and the innovation covariance matrix. Apply the QR factorization to the matrix

$$\begin{pmatrix} R_t^{1/2} & C_t P_{t|t-1}^{1/2} \\ 0 & P_{t|t-1}^{1/2} \end{pmatrix} = R^T Q^T, \tag{8.91}$$

where

$$R^T = \begin{pmatrix} X & 0 \\ Y & Z \end{pmatrix}. \tag{8.92}$$

The Kalman filter interpretations of the matrices X, Y, Z can be found by squaring the QR factorization

$$\begin{aligned} R^T Q^T Q R = R^T R &= \begin{pmatrix} XX^T & XY^T \\ YX^T & YY^T + ZZ^T \end{pmatrix} \\ &= \begin{pmatrix} \underbrace{R_t + C_t P_{t|t-1} C_t^T}_{S_t} & C_t P_{t|t-1} \\ P_{t|t-1} C_t^T & P_{t|t-1} \end{pmatrix}, \end{aligned} \tag{8.93}$$

from which we can identify

$$\begin{aligned} XX^T &= S_t \\ YX^T &= P_{t|t-1} C_t^T \Rightarrow Y = P_{t|t-1} C_t^T S_t^{-1/2} = K_t S_t^{1/2} \\ ZZ^T &= P_{t|t-1} - YY^T = P_{t|t-1} - P_{t|t-1} C_t^T X^{-T} X^{-1} C_t P_{t|t-1} \\ &= P_{t|t-1} - P_{t|t-1} C_t^T S_t^{-1} C_t P_{t|t-1} = P_{t|t}. \end{aligned}$$

This gives the algorithm below.

Algorithm 8.6 Square root algorithm, measurement update

1. Form the matrix in the left-hand side of (8.91).
2. Apply the QR factorization in (8.91).
3. Identify R with (8.92), where

$$\begin{aligned} X &= S_t^{1/2} \\ Y &= K_t S_t^{1/2} \\ Z &= P_{t|t}^{1/2}. \end{aligned}$$

All in one equation:

$$\begin{pmatrix} R_t^{1/2} & C_t P_{t|t-1}^{1/2} \\ 0 & P_{t|t-1}^{1/2} \end{pmatrix} = \begin{pmatrix} S_t^{1/2} & 0 \\ K_t S_t^{1/2} & P_{t|t}^{1/2} \end{pmatrix} Q^T. \tag{8.94}$$

Note that the Y can be interpreted as the Kalman gain on a normalized innovation $S_t^{-1/2} \varepsilon_t \in \mathrm{N}(0, I)$. That is, we have to multiply either the gain Y or the innovation by the inverse of X.

Remarks:

- The only non-trivial part of these algorithms, is to come up with the matrix to be factorized in the first place. Then, in all cases here it is trivial to verify that it works.
- There are many ways to factorize a matrix to get a square root. The QR factorization is recommended here for several reasons. First, there are many efficient implementations available, e.g. MATLAB™ 's `qr`. Secondly, this gives the unique triangular square root, which is useful partly because it is easily checked that it is positive definite and partly because inversion is simple (this is needed in the state update).

8.7.2. Kalman predictor

The predictor form, eliminating the measurement update, can be derived using the matrix (8.95) below. The derivation is done by squaring up in the same way as before, and identifying the blocks.

Algorithm 8.7 Square root Kalman predictor

Apply the QR factorization

$$\begin{pmatrix} R_t^{1/2} & C_t P_{t|t-1}^{1/2} & 0 \\ 0 & A_t P_{t|t-1}^{1/2} & B_{v,t} Q_t^{1/2} \end{pmatrix} = \begin{pmatrix} S_t^{1/2} & 0 & 0 \\ A_t K_t S_t^{1/2} & P_{t+1|t}^{1/2} & 0 \end{pmatrix} Q^T. \tag{8.95}$$

The state update is computed by

$$\hat{x}_{t+1|t} = A_t \hat{x}_{t|t-1} + B_{u,t} u_t + A_t K_t S_t^{1/2} S_t^{-1/2} (y_t - C_t \hat{x}_{t|t-1}). \tag{8.96}$$

We need to multiply the gain vector $K_t S_t^{1/2}$ in the lower left corner with the inverse of the upper left corner element $S_t^{1/2}$. Here the triangular structure can be utilized for matrix inversion. However, this matrix inversion can be avoided, by factorizing a larger matrix; see Kailath et al. (1998) for details.

8.7.3. Kalman filter

Similarly, a square root algorithm for the Kalman filter is given below.

Algorithm 8.8 Square root Kalman filter

Apply the QR factorization

$$\begin{pmatrix} R_{t+1}^{1/2} & C_{t+1}A_tP_{t|t}^{1/2} & C_{t+1}B_{v,t}Q_t^{1/2} \\ 0 & A_tP_{t|t}^{1/2} & B_{v,t}Q_t^{1/2} \end{pmatrix} = \begin{pmatrix} S_{t+1}^{1/2} & 0 & 0 \\ K_{t+1}S_{t+1}^{1/2} & P_{t+1|t+1}^{1/2} & 0 \end{pmatrix} Q^T. \tag{8.97}$$

The state update is computed by

$$\hat{x}_{t|t} = A_{t-1}\hat{x}_{t-1|t-1} + K_tS_t^{1/2}S_t^{-1/2}(y_t - C_tA_{t-1}\hat{x}_{t-1|t-1}). \tag{8.98}$$

Example 8.11 DC motor: square root filtering

Consider the DC motor in Examples 8.4 and 8.9, where

$$A = \begin{pmatrix} 1 & 0.3287 \\ 0 & 0.6703 \end{pmatrix}, \quad B = \begin{pmatrix} 0.0703 \\ 0.3297 \end{pmatrix}, \quad C = \begin{pmatrix} 1 & 0 \end{pmatrix}.$$

Assume that we initially have

$$\hat{x}_{0|0} = \begin{pmatrix} 0 \\ 0 \end{pmatrix}$$

$$P_{0|0} = \begin{pmatrix} 1 & 0 \\ 0 & 1 \end{pmatrix}.$$

The left-hand side (LHS) required for the QR factorization in the filter formulation is given by

$$\text{LHS} = \begin{pmatrix} 1 & 1 & 0.3297 & 0 \\ 0 & 1 & 0.3297 & 0 \\ 0 & 0 & 0.6703 & 0.1000 \end{pmatrix}.$$

The sought factor is

$$R^T = \begin{pmatrix} -1.4521 & 0 & 0 & 0 \\ -0.7635 & -0.7251 & 0 & 0 \\ -0.1522 & -0.1445 & -0.6444 & 0 \end{pmatrix}.$$

From this matrix, we identify the Kalman filter quantities

$$P_{1|1}^{1/2} = \begin{pmatrix} -0.7251 & 0 \\ -0.1445 & -0.6444 \end{pmatrix}$$
$$K_1 S_1^{1/2} = \begin{pmatrix} -0.5012 \\ -0.1300 \end{pmatrix}$$
$$S_1^{1/2} = -1.2815.$$

Note that the QR factorization in this example gives negative signs of the square roots of S_t and K_t. The usual Kalman filter quantities are recovered as

$$P_{1|1} = \begin{pmatrix} 0.3911 & 0.1015 \\ 0.1015 & 0.1891 \end{pmatrix}$$
$$K_1 = \begin{pmatrix} 0.3911 \\ 0.1015 \end{pmatrix}$$
$$S_1 = 1.6423.$$

8.8. Sensor fusion

Sensor fusion is the problem of how to merge information from different sensors. One example is when we have sensor redundancy, so we could solve the filtering problem with either of the sensors. In more complicated cases, each sensor provides unique information. An interesting question is posed in Blair and Bar-Shalom (1996): "Does more data always mean better estimates?". The answer should be yes in most cases.

Example 8.12 Fuel consumption: sensor fusion

Consider the fuel consumption problem treated in Section 3.6.1. Here we used a sensor that provides a measurement of instantaneous fuel consumption. It is quite plausible that this signal has a small offset which is impossible to estimate from just this sensor.

A related parameter that the driver wishes to monitor is the amount of fuel left in the tank. Here a sensor in the tank is used to measure fuel level. Because of slosh excited by accelerations, this sensor has low frequency measurement noise (or disturbance), so a filter with very small gain is currently used by car manufacturers.

That is, we have to sensors:

- One sensor measures the derivative with high accuracy, but with a small offset.
- One sensor measures the absolute value with large uncertainty.

A natural idea, which is probably not used in today's systems, is to merge the information into one filter. The rate sensor can be used to get a quicker response to fuel level changes (for instance, after refueling), and more importantly, the long term information about level changes can be used to estimate the offset.

A similar sensor fusion problem as in Example 8.12 is found in navigation systems, where one sensor is accurate for low frequencies (an integrating sensor) and the other for high frequencies.

Example 8.13 Navigation: sensor fusion

The *Inertial Navigation System* (*INS*) provides very accurate measurements of accelerations (second derivative rather than first derivative as in Example 8.12). The problem is possible offsets in accelerometers and gyros. These must be estimated, otherwise the position will drift away.

As a complement, low quality measurements such as the so-called baro-altitude are used. This sensor uses only the barometric pressure, compares it to the ground level, and computes an altitude. Of course, this sensor is unreliable as the only sensor for altitude, but in combination with the INS it can be used to stabilize the drift in altitude.

An important issue is whether the fusion should be made at a central computer or in a distributed fashion. *Central fusion* means that we have access to all measurements when deriving the optimal filter; see Figure 8.10. In contrast, in *decentralized fusion* a filter is applied to each measurement, and the global fusion process has access only to the estimates and their error covariances. Decentralized filtering has certain advantages in fault detection in that the different sub-systems can apply a 'voting' strategy for fault detection. More on this is given in Chapter 11. An obvious disadvantage is increased signaling, since the state, and in particular its covariance matrix, is often of much larger dimension than the measurement vector. One might argue that the global processor can process the measurements in a decentralized fashion to get the advantages of both alternatives. However, many sensoring systems have built-in filters.

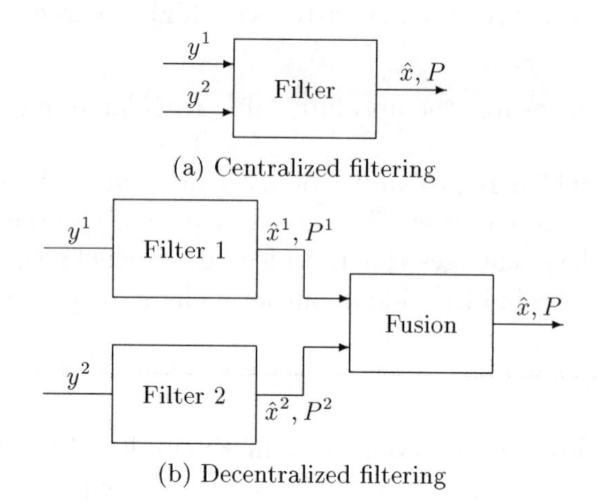

Figure 8.10. The concepts of centralized and decentralized filtering.

8.8.1. The information filter

To understand the relation between centralized and de-centralized filtering, the information filter formulation of the Kalman filter is useful.

Linear regression models are special cases of the signal model (8.32) with $A_t = I$ and $C_t = \varphi_t^T$. When deriving recursive versions of the least squares solution, we started with formulas like (see Section 5.B)

$$\begin{aligned} f_t &= f_{t-1} + C_t^T R_t^{-1} y_t \\ R_t^{\varphi} &= R_{t-1}^{\varphi} + C_t^T R_t^{-1} C_t \\ \hat{x}_{t|t} &= (R_t^{\varphi})^{-1} f_t. \end{aligned}$$

After noticing that the matrix inversion requires many operations, we used the matrix inversion lemma to derive Kalman filter like equations, where no matrix inversion is needed (for scalar y).

Here we will go the other way around. We have a Kalman filter, and we are attracted by the simplicity of the time update of $P_{t|t}^{-1} = R_t^{\varphi}$ above. To keep the efficiency of avoiding a matrix inversion, let us introduce the transformed state vector

$$\begin{aligned} \hat{a}_{t|t} &= P_{t|t}^{-1} \hat{x}_{t|t} \\ \hat{a}_{t|t-1} &= P_{t|t-1}^{-1} \hat{x}_{t|t-1}. \end{aligned}$$

Here, a can be interpreted as the vector f_t in least squares above. That is, the state estimate is never actually computed.

Algorithm 8.9 The information filter

Introduce two auxiliary matrices to simplify notation

$$\begin{aligned} M_t &= A_t^{-1} P_{t|t}^{-1} A_t^{-T} \\ N_t &= A_t B_{v,t} (B_{v,t}^T A_t B_{v,t} + Q_t^{-1})^{-1} \end{aligned}$$

The Kalman filter equations sorted in time and measurement updates for the transformed state vector $\hat{a}_{t|k} = P_{t|k}^{-1} \hat{x}_{t|k}$ is:

$$\begin{aligned} \hat{a}_{t+1|t} &= (I - N_t B_{v,t}^T) A_t^{-1} \hat{a}_{t|t} \\ P_{t+1|t}^{-1} &= (I - A_t B_{v,t} (B_{v,t}^T A_t B_{v,t} + Q_t^{-1})^{-1}) B_{v,t}^T M_t \\ &= (I - N_t B_{v,t}^T) M_t \\ \hat{a}_{t|t} &= \hat{a}_{t|t-1} + C_t^T R_t^{-1} y_t \\ P_{t|t}^{-1} &= P_{t|t-1}^{-1} + C_t^T R_t^{-1} C_t. \end{aligned}$$

The equations for the covariance update follow from the matrix inversion lemma applied to the Kalman filter equations. Note that the measurement update here is trivial, while the main computational burden comes from the time update. This is in contrast to the Kalman filter.
Comments:

- $C_t^T R_t^{-1} C_t$ is the information in a new measurement, and $C_t^T R_t^{-1} y_t$ is a sufficient statistic for updating the estimate. The interpretation is that $P_{t|t}^{-1}$ contains the information contained in all past measurements. The time update implies a forgetting of information.

- Note that this is one occasion where it is important to factorize the state noise covariance, so that Q_t is non-singular. In most other applications, the term $B_{v,t} v_t$ can be replaced with $\bar{v}_t$ with (singular) covariance matrix $\bar{Q}_t = B_{v,t} Q_t B_{v,t}^T$.

Main advantages of the information filter:

- Very vague prior knowledge of the state can now be expressed as $P_0^{-1} = 0$.

- The Kalman filter requires an $n_y \times n_y$ matrix to be inverted ($n_y = \dim y$), while the information filter requires an $n_v \times n_v$ matrix to be inverted. In navigation applications, the number of state noise inputs may be quite small (3, say), while the number of outputs can be large.

- A large measurement vector with block diagonal covariance R_t may be processed sequentially in the Kalman filter, as shown in Section 8.6.4 so that the matrices to be inverted are considerably smaller than n_y. There is a corresponding result for the information filter. If Q_t is block diagonal, the time update can be processed sequentially; see Anderson and Moore (1979) (pp. 146–147).

8.8.2. Centralized fusion

Much of the beauty of the Kalman filter theory is the powerful state space model. A large set of measurements is simply collected in one measurement equation,

$$y_t = \begin{pmatrix} y_1 \\ y_2 \\ \vdots \\ y_m \end{pmatrix} = \begin{pmatrix} C_1 \\ C_2 \\ \vdots \\ C_m \end{pmatrix} x_t + \begin{pmatrix} e_1 \\ e_2 \\ \vdots \\ e_m \end{pmatrix}. \tag{8.99}$$

The only problem that might occur is that the sensors are working in different state coordinates. One example is target tracking, where the sensors might give positions in Cartesian or polar coordinates, or perhaps only bearing. This is indeed an extended Kalman filtering problem; see Section 8.9.

8.8.3. The general fusion formula

If there are two independent state estimates $\hat{x}^1$ and $\hat{x}^2$, with covariance matrices P^1 and P^2 respectively, fusion of these pieces of information is straightforward:

$$\hat{x} = P\left((P^1)^{-1}\hat{x}^1 + (P^2)^{-1}\hat{x}^2\right) \tag{8.100}$$

$$P = \left((P^1)^{-1} + (P^2)^{-1}\right)^{-1}. \tag{8.101}$$

It is not difficult to show that this is the minimum variance estimator. The Kalman filter formulation of this fusion is somewhat awkwardly described by the state space model

$$x_{t+1} = v_t, \qquad \mathrm{Cov}(v_t) = \infty I$$

$$\begin{pmatrix} \hat{x}^1 \\ \hat{x}^2 \end{pmatrix} = \begin{pmatrix} I \\ I \end{pmatrix} x_t + \begin{pmatrix} e^1 \\ e^2 \end{pmatrix}, \quad \mathrm{Cov}(e^i) = P^i.$$

To verify it, use the information filter. The infinite state variance gives $P_{t+1|t}^{-1} = 0$ for all t. The measurement update becomes

$$P_{t|t}^{-1} = 0 + C^T R^{-1} C = (P_{t|t}^1)^{-1} + (P_{t|t}^2)^{-1}.$$

8.8.4. Decentralized fusion

Suppose we get estimates from two Kalman filters working with the same state vector. The total state space model for the *decentralized filters* is

$$\begin{pmatrix} x^1_{t+1} \\ x^2_{t+1} \end{pmatrix} = \begin{pmatrix} A_t & 0 \\ 0 & A_t \end{pmatrix} \begin{pmatrix} x^1_t \\ x^2_t \end{pmatrix} + \begin{pmatrix} B_t & 0 \\ 0 & B_t \end{pmatrix} \begin{pmatrix} v^1_t \\ v^2_t \end{pmatrix}, \quad \text{Cov}(v^i_t) = Q$$
$$y_t = \begin{pmatrix} C^1_t & 0 \\ 0 & C^2_t \end{pmatrix} \begin{pmatrix} x^1_t \\ x^2_t \end{pmatrix} + \begin{pmatrix} e^1_t \\ e^2_t \end{pmatrix}.$$

The diagonal forms of $\bar{A}$, $\bar{B}$ and $\bar{C}$ imply:

1. The updates of the estimates can be done separately, and hence decentralized.

2. The state estimates will be independent.

Thus, because of independence the general *fusion formula* (8.100) applies. The state space model for *decentralized filters* and the *fusion filter* is now

$$\begin{pmatrix} x^1_{t+1} \\ x^2_{t+1} \\ x_{t+1} \end{pmatrix} = \begin{pmatrix} A_t & 0 & 0 \\ 0 & A_t & 0 \\ 0 & 0 & 0 \end{pmatrix} \begin{pmatrix} x^1_t \\ x^2_t \\ x_t \end{pmatrix} + \begin{pmatrix} B_t & 0 & 0 \\ 0 & B_t & 0 \\ 0 & 0 & I \end{pmatrix} \begin{pmatrix} v^1_t \\ v^2_t \\ v^3_t \end{pmatrix}$$
$$\begin{pmatrix} y^1_t \\ y^2_t \\ 0 \\ 0 \end{pmatrix} = \begin{pmatrix} C^1_t & 0 & 0 \\ 0 & C^2_t & 0 \\ I & 0 & -I \\ 0 & I & -I \end{pmatrix} \begin{pmatrix} x^1_t \\ x^2_t \\ x_t \end{pmatrix} + \begin{pmatrix} e^1_t \\ e^2_t \\ 0 \\ 0 \end{pmatrix}$$
$$\text{Cov}(v^1_t) = Q, \ \text{Cov}(v^2_t) = Q, \ \text{Cov}(v^3_t) = \infty I.$$

That is, the Kalman filter applied to this state space model provides the fusioned state estimate as the third sub-vector of the augmented state vector.

The only problem in this formulation is that we have *lost information.* The decentralized filters do not use the fact that they are estimating the same system, which have only one state noise. It can be shown that the covariance matrix will become too small, compared to that which the Kalman filter provides for the model

$$x_{t+1} = A_t x_t + B_t v_t, \quad \text{Cov}(v_t) = Q \tag{8.102}$$
$$y_t = \begin{pmatrix} C^1_t \\ C^2_t \end{pmatrix} x_t + \begin{pmatrix} e^1_t \\ e^2_t \end{pmatrix}. \tag{8.103}$$

To obtain the optimal state estimate, given the measurements and the state space model, we need to invert the Kalman filter and recover the raw *information in the measurements.* One way to do this is to ask all decentralized filters for the Kalman filter transfer functions $\hat{x}_{t|t} = H_t(q)y_t$, and then apply inverse filtering. It should be noted that the inverse Kalman filter is likely to be high-pass, and thus amplifying numerical errors. Perhaps a better alternative is as follows. As indicated, it is the information in the measurements, not the measurements themselves that are needed. Sufficient statistics are provided by the *information filter*; see Section 8.8.1. We recapitulate the measurement update for convenience:

$$\begin{aligned}\hat{a}_{t|t} =& \hat{a}_{t|t-1} + C_t^T R_t^{-1} y_t \\ P_{t|t}^{-1} =& P_{t|t-1}^{-1} + C_t^T R_t^{-1} C_t,\end{aligned}$$

where $\hat{a}_{t|t} = \hat{P}_{t|t}^{-1}\hat{x}_{t|t}$. Since the measurement error covariance in (8.103) is block diagonal, the time update can be split into two parts

$$\begin{aligned}\hat{a}_{t|t} =& \hat{a}_{t|t-1} + (C_t^1)^T (R_t^1)^{-1} y_t^1 + (C_t^2)^T (R_t^2)^{-1} y_t^2, \\ P_{t|t}^{-1} =& P_{t|t-1}^{-1} + (C_t^1)^T (R_t^1)^{-1} C_t^1 + (C_t^2)^T (R_t^2)^{-1} C_t^2.\end{aligned}$$

Each Kalman filter, interpreted as information filters, has computed a measurement update

$$\begin{aligned}\hat{a}_{t|t}^i =& \hat{a}_{t|t-1}^i + (C_t^i)^T (R_t^i)^{-1} y_t^i \\ (P_{t|t}^i)^{-1} =& (P_{t|t-1}^i)^{-1} + (C_t^i)^T (R_t^i)^{-1} C_t^i.\end{aligned}$$

By backward computations, we can now recover the information in each measurement, expressed below in the available Kalman filter quantities:

$$\begin{aligned}(C_t^i)^T (R_t^i)^{-1} y_t^i =& \hat{a}_{t|t}^i - \hat{a}_{t|t-1}^i = (\hat{P}_{t|t}^i)^{-1}\hat{x}_{t|t}^i - (\hat{P}_{t|t-1}^i)^{-1}\hat{x}_{t|t-1}^i \\ (C_t^i)^T (R_t^i)^{-1} C_t^i =& (P_{t|t}^i)^{-1} - (P_{t|t-1}^i)^{-1}.\end{aligned}$$

The findings are summarized and generalized to several sensors in the algorithm below.

Algorithm 8.10 Decentralized filtering

Given the filter and prediction quantities from m Kalman filters working on the same state space model, the optimal sensor fusion is given by the usual time update and the following measurement update:

$$\begin{aligned}\hat{P}_{t|t}^{-1}\hat{x}_{t|t} =& \hat{P}_{t|t-1}^{-1}\hat{x}_{t|t-1} + \sum_{i=1}^{m} \left((\hat{P}_{t|t}^i)^{-1}\hat{x}_{t|t}^i - (\hat{P}_{t|t-1}^i)^{-1}\hat{x}_{t|t-1}^i \right) \\ \hat{P}_{t|t}^{-1} =& \hat{P}_{t|t-1}^{-1} + \sum_{i=1}^{m} \left((\hat{P}_{t|t}^i)^{-1} - (\hat{P}_{t|t-1}^i)^{-1} \right).\end{aligned}$$

The price paid for the decentralized structure is heavy signaling and many matrix inversions. Also for this approach, there might be numerical problems, since differences of terms of the same order are to be computed. These differences correspond to a high-pass filter.

8.9. The extended Kalman filter

A typical non-linear state space model with discrete time observations is:

$$\dot{x}_t = f(x_t) + v_t \tag{8.104}$$
$$y_{kT} = h(x_{kT}) + e_{kT}. \tag{8.105}$$

Here and in the sequel, T will denote the sampling interval when used as an argument, while it means matrix transpose when used as a superscript. The measurement and time update are treated separately. An initialization procedure is outlined in the simulation section.

8.9.1. Measurement update

Linearization

The most straightforward approach to linearize (8.105) is by using a Taylor expansion:

$$\begin{aligned} y_{kT} &= h(x_{kT}) + e_{kT} \\ &\approx h(\hat{x}_{kT|kT-T}) + \underbrace{\left.\frac{dh(x)}{dx}\right|_{x=\hat{x}_{kT|kT-T}}}_{H_{kT}} (x_{kT} - \hat{x}_{kT|kT-T}) + e_{kT}. \end{aligned}$$

With a translation of the observation, we are approximately back to the linear case,

$$\bar{y}_{kT} = y_{kT} - h(\hat{x}_{kT|kT-T}) + H_{kT}\hat{x}_{kT|kT-T} \approx H_{kT}x_{kT} + e_{kT}, \tag{8.106}$$

and the Kalman filter measurement update can be applied to (8.106).

Non-linear transformation

An alternative to making (8.105) linear is to apply a non-linear transformation to the observation. For instance, if $n_y = n_x$ one can try $\bar{y} = h^{-1}(y)$, which gives $H_{kT} = I$. The problem is now to transform the measurement covariance R to $\bar{R}$. This can only be done approximately, and Gaussianity is not preserved. This is illustrated with an example.

Example 8.14 Target tracking: output transformation

In radar applications, range r and bearing θ to the object are measured, while these co-ordinates are not suitable to keep in the state vector. The uncertainty of range and bearing measurements is contained in the variances σ_r^2 and σ_θ^2. The measurements in polar coordinates can be transformed into Cartesian position by

$$x^{(1)} = r\sin(\theta),$$
$$x^{(2)} = r\cos(\theta).$$

The covariance matrix R of the measurement error is neatly expressed in Li and Bar-Shalom (1993) as

$$R = \frac{\sigma_r^2 - r^2\sigma_\theta^2}{2}\begin{pmatrix} b+\cos(2\theta) & \sin(2\theta) \\ \sin(2\theta) & b-\cos(2\theta) \end{pmatrix} \tag{8.107}$$

$$b = \frac{\sigma_r^2 + r^2\sigma_\theta^2}{\sigma_r^2 - r^2\sigma_\theta^2}. \tag{8.108}$$

Indeed, this is an approximation, but it is accurate for $r\sigma_\theta^2/\sigma_r < 0.4$, which is the case normally.

8.9.2. Time update

There are basically two alternatives in passing from a continuous non-linear model to a discrete linear one: first linearizing and then discretizing; or the other way around. An important aspect is to quantify the linearization error, which can be included in the bias error (model mismatch). The conversion of state noise from continuous to discrete time is mostly a philosophical problem, since the alternatives at hand will give about the same performance. In this section, the discrete time noise will be characterized by its assumed covariance matrix $\bar{Q}$, and Section 8.9.4 will discuss different alternatives how to compute it from Q.

Discretized linearization

A Taylor expansion of each component of the state in (8.104) around the current estimate $\hat{x}$ yields

$$\dot{x}_i = f_i(\hat{x}) + f_i'(\hat{x})(x-\hat{x}) + \frac{1}{2}(x-\hat{x})^T f_i''(\xi)(x-\hat{x}) + v_{i,t}, \tag{8.109}$$

where ξ is a point in the neighborhood of x and $\hat{x}$. Here f_i' denotes the derivative of the ith row of f with respect to the state vector x, and f_i'' is similarly the second derivative. We will not investigate transformations of the state noise yet. It will be assumed that the discrete time counterpart has a covariance matrix $\bar{Q}$.

Example 8.15 Target tracking: coordinated turns

In Example 8.3, the dynamics for the state vector $(x^{(1)}, x^{(2)}, v^{(1)}, v^{(2)}, \omega)^T$ is given by:

$$\begin{aligned}
\dot{x}^{(1)} &= v^{(1)} \\
\dot{x}^{(2)} &= v_2 \\
\dot{v}^{(1)} &= -\omega v^{(2)} \\
\dot{v}^{(2)} &= \omega v^{(1)} \\
\dot{\omega} &= 0.
\end{aligned} \tag{8.110}$$

This choice of state is natural when considering so called coordinated turns, where turn rate ω is more or less piecewise constant. The derivative in the Taylor expansion (8.109) is

$$f_{cv}'(x) = \begin{pmatrix} 0 & 0 & 1 & 0 & 0 \\ 0 & 0 & 0 & 1 & 0 \\ 0 & 0 & 0 & -\omega & -v^{(2)} \\ 0 & 0 & \omega & 0 & v^{(1)} \\ 0 & 0 & 0 & 0 & 0 \end{pmatrix}. \tag{8.111}$$

This is the solution of (8.109) found by integration from t to $t+T$, and assuming $x_t = \hat{x}$. With the state vector $(x^{(1)}, x^{(2)}, v, h, \omega)^T$, the state dynamics become:

$$\begin{aligned}
\dot{x}^{(1)} &= v\cos(h) \\
\dot{x}^{(2)} &= v\sin(h) \\
\dot{v} &= 0 \\
\dot{h} &= \omega \\
\dot{\omega} &= 0.
\end{aligned} \tag{8.112}$$

The derivative in the Taylor expansion (8.109) is now

$$f'_{pv}(x) = \begin{pmatrix} 0 & 0 & \cos(h) & -v\sin(h) & 0 \\ 0 & 0 & \sin(h) & v\cos(h) & 0 \\ 0 & 0 & 0 & 0 & 0 \\ 0 & 0 & 0 & 0 & 1 \\ 0 & 0 & 0 & 0 & 0 \end{pmatrix}. \tag{8.113}$$

From standard theory on sampled systems (see for instance Kailath (1980)) the continuous time system (8.109), neglecting the second order rest term, has the discrete time counterpart

$$x_{t+T} = \hat{x} + \left[\int_0^T e^{f'(\hat{x})\tau} d\tau\right] f(\hat{x}). \tag{8.114}$$

The time discrete Kalman filter can now be applied to this and (8.105) to provide a version of one of the wellknown *Extended Kalman Filters* (here called EKF 1) with time update:

$$\hat{x}_{t+T|t} = \hat{x}_{t|t} + \left[\int_0^T e^{f'(\hat{x}_{t|t})\tau} d\tau\right] f(\hat{x}_{t|t}) \tag{8.115}$$

$$P_{t+T|t} = e^{f'(\hat{x}_{t|t})T} P_{t|t} e^{f'(\hat{x}_{t|t})^T T} + \bar{Q}_t. \tag{8.116}$$

This will be referred to as the *discretized linearization* approach.

Linearized discretization

A different and more accurate approach is to first discretize (8.104). In some rare cases, of which tracking with constant turn rate is one example, the state space model can be discretized exactly by solving the sampling formula

$$x_{t+T} = x_t + \int_t^{t+T} f(x(\tau))d\tau \tag{8.117}$$

analytically. The solution can be written

$$x_{t+T} = g(x_t). \tag{8.118}$$

If this is not possible to do analytically, we can get a good approximation of an exact sampling by numerical integration. This can be implemented using the discretized linearization approach in the following way. Suppose we have a function `[x,P]=tu(x,P,T)` for the time update. Then the fast sampling method, in MATLAB™ formalism,

```
for i=1:M;
    [x,P]=tu(x,P,T/M);
end;
```

provides a more accurate time update for the state. This is known as the *iterated Kalman filter*. In the limit, we can expect that the resulting state update converges to $g(x)$. An advantage with numerical integration is that it provides a natural approximation of the state noise.

Example 8.16 Target tracking: coordinated turns and exact sampling

Consider the tracking example with $x_t = (x_t^{(1)}, x_t^{(2)}, v_t, h_t, \omega_t)^T$. The analytical solution of (8.117) using (8.112) is

$$\begin{aligned}
x_{t+T}^{(1)} &= x_t^{(1)} + \frac{2v_t}{\omega_t}\sin(\frac{\omega_t T}{2})\cos(h_t + \frac{\omega_t T}{2})\\
x_{t+T}^{(2)} &= x_t^{(2)} - \frac{2v_t}{\omega_t}\sin(\frac{\omega_t T}{2})\sin(h_t + \frac{\omega_t T}{2})\\
v_{t+T} &= v_t\\
h_{t+T} &= h_t + \omega T\\
\omega_{t+T} &= \omega_t.
\end{aligned} \tag{8.119}$$

The alternate state coordinates $x = (x_t^{(1)}, x_t^{(2)}, v_t^{(1)}, v_t^{(2)}, \omega_t)^T$ give, with (8.110) in (8.117),

$$\begin{aligned}
x_{t+T}^{(1)} &= x_t^{(1)} + \frac{v_t^{(1)}}{\omega_t}\sin(\omega_t T) - \frac{v_t^{(2)}}{\omega_t}(1-\cos(\omega_t T))\\
x_{t+T}^{(2)} &= x_t^{(2)} + \frac{v_t^{(1)}}{\omega_t}(1-\cos(\omega_t T)) + \frac{v_t^{(2)}}{\omega_t}\sin(\omega_t T)\\
v_{t+T}^{(1)} &= v_t^{(1)}\cos(\omega_t T) - v_t^{(2)}\sin(\omega_t T)\\
v_{t+T}^{(2)} &= v_t^{(1)}\sin(\omega_t T) + v_t^{(2)}\cos(\omega_t T)\\
\omega_{t+T} &= \omega_t.
\end{aligned} \tag{8.120}$$

These calculations are quite straightforward to compute using symbolic computation programs, such as *Maple*™ or *Mathematica*™.

A component-wise Taylor expansion around a point $\hat{x}$ gives

$$x_{t+T}^{(i)} = g^{(i)}(\hat{x}) + (g^{(i)})'(\hat{x})(x_t - \hat{x}) + \frac{1}{2}(x_t - \hat{x})^T (g^{(i)})''(\xi)(x_t - \hat{x}). \tag{8.121}$$

Applying the time discrete Kalman filter to (8.121) and (8.106), and neglecting the second order Taylor term in the covariance matrix update, we get the second version of the extended Kalman filter (EKF 2), with time update:

$$\hat{x}_{t+T|t} = g(\hat{x}_{t|t}) \tag{8.122}$$

$$P_{t+T|t} = g'(\hat{x}_{t|t})P_{t|t}g'(\hat{x}_{t|t})^T + \bar{Q}_t. \tag{8.123}$$

This will be referred to as the *linearized discretization* approach, because we have linearized the exact time update of the state for updating its covariance matrix.

One further alternative is to include the second order Taylor term in the state time update. Assuming a Gaussian distribution of $\hat{x}_{t|t}$, a second order covariance update is also possible (see equation (3-33) in Bar-Shalom and Fortmann (1988)). The component-wise time update is given by (EKF 3):

$$\hat{x}^{(i)}_{t+T|t} = (g^{(i)})(\hat{x}_{t|t}) + \mathrm{tr}\left((g^{(i)})''(\hat{x}_{t|t})P_{t|t}\right) \tag{8.124}$$

$$P^{ij}_{t+T|t} = (g^{(i)})'(\hat{x}_{t|t})P_{t|t}(g^{j})'(\hat{x}_{t|t})^T + \bar{Q}^{ij}_t \tag{8.125}$$

$$+ \mathrm{tr}\left((g^{(i)})''(\hat{x}_{t|t})P_{t|t}(g^{j})''(\hat{x}_{t|t})P_{t|t}\right). \tag{8.126}$$

8.9.3. Linearization error

The error we make when going from (8.109) to (8.114) *or* from (8.118) to (8.121) depends only upon the size of $(x - \hat{x})^T(f^{(i)})''(\xi)(x - \hat{x})$ and, for the discrete case, $(x - \hat{x})^T(g^{(i)})''(\xi)(x - \hat{x})$, respectively. This error propagates to the covariance matrix update and for discretized linearization also to the state update. This observation implies that we can use the same analysis for linearized discretization and discretized linearization. The Hessian $(f_i)''$ of component i of $f(x)$ will be used for notational convenience, but the same expressions holds for $(g_i)''$. Let $\tilde{x}_{t+T} = x_{t+T} - \hat{x}_{t+T|t}$ denote the state prediction error. Then we have

$$4\|\dot{\tilde{x}}^{(i)}\|_2^2 = \|(x - \hat{x})^T(f^{(i)})''(\xi)(x - \hat{x})\|_2^2. \tag{8.127}$$

Care must be taken when comparing different state vectors because $\|x - \hat{x}\|_2$ is not an invariant measure of state error in different coordinate systems.

Example 8.17 Target tracking: best state coordinates

In target tracking with coordinated turns, we want to determine which of the following two coordinate systems is likely to give the smallest linearization error:

$$
\begin{aligned}
x =& (x^{(1)}, x^{(2)}, v^{(1)}, v^{(2)}, \omega)^T \\
x =& (x^{(1)}, x^{(2)}, v, h, \omega)^T.
\end{aligned}
$$

For notational simplicity, we will drop the time index in this example. For these two cases, the state dynamics are given by

$$
\begin{aligned}
f_{cv}(x) =& (v^{(1)}, v^{(2)}, -\omega v^{(2)}, \omega v^{(1)}, 0)^T \\
f_{pv}(x) =& (v\cos(h), v\sin(h), 0, \omega, 0)^T,
\end{aligned}
$$

respectively.

Suppose the initial state error is expressed in $\|\tilde{x}_{cv}\|_2^2 = (\tilde{x}^{(1)})^2 + (\tilde{x}^{(2)})^2 + (\tilde{v}^{(1)})^2 + (\tilde{v}^{(2)})^2 + \tilde{\omega}^2$. The problem is that the error in heading angle is numerically much smaller than in velocity. We need to find the scaling matrix such that $\|\tilde{x}_{pv}\|_2^2$ is a comparable error. Using

$$
\begin{aligned}
v^{(1)} =& v\cos(h) \\
v^{(2)} =& v\sin(h),
\end{aligned}
$$

we have

$$
\begin{aligned}
(v^{(1)} - \hat{v}^{(1)})^2 + (v^{(2)} - \hat{v}^{(2)})^2 =& v^2 + \hat{v}^2 - 2v\hat{v}\cos(h - \hat{h}) \\
=& v^2 + \hat{v}^2 - 2v\hat{v} + 2v\hat{v}(1 - \cos(h - \hat{h})) \\
=& (v - \hat{v})^2 + 4v\hat{v}\sin^2\left(\frac{h - \hat{h}}{2}\right) \\
\approx& (v - \hat{v})^2 + v^2(h - \hat{h})^2.
\end{aligned}
$$

The approximation is accurate for angular errors less than, say, 40° and relative velocity errors less than, say, 10%. Thus, a weighting matrix

$$
J_{pv} = \operatorname{diag}(1, 1, 1, v, 1)
$$

should be used in (8.127) for polar velocity and the identity matrix J_{cv} for Cartesian velocity.

The error in linearizing state variable i is

$$
\begin{aligned}
4\|\dot{\tilde{x}}_i\|_2^2 =& \|(x - \hat{x})^T J J^{-1} (f^{(i)})''(\xi) J^{-1} J (x - \hat{x})\|_2^2 \\
\leq& \|J(x - \hat{x})\|_2^4 \|J^{-1}(f^{(i)})''(\xi)J^{-1}\|_2^2.
\end{aligned}
$$

The Frobenius norm defined by $\|X\|_F^2 = \sum(x^{(ij)})^2$, where the sum is taken over all elements $x^{(ij)}$ of X, will be used to bound the 2-norm. Using the inequality $\frac{1}{n}\|X\|_F^2 \leq \|X\|_2^2 \leq \|X\|_F^2$, the linearization error can be bounded by

$$4\|\dot{\tilde{x}}^{(i)}\|_2^2 \leq \|J(x-\hat{x})\|_2^4 \|J^{-1}(f^{(i)})''(\xi)J^{-1}\|_F^2. \tag{8.128}$$

The total error is then

$$4\|\dot{\tilde{x}}\|_2^2 = \sum_{i=1}^{5} 4\|\dot{\tilde{x}}^{(i)}\|_2^2. \tag{8.129}$$

The above calculations show that if we start with the same initial error in two different state coordinate systems, (8.128) quantizes the linearization error, which can be upper bounded by taking the Frobenius norm instead of two-norm, on how large the error in the time update of the extended Kalman filter can be.

The remaining task is to give explicit and exact expressions of the rest terms in the Taylor expansion. The following results are taken from Gustafsson and Isaksson (1996). For discretized linearization, the Frobenius norm of the Hessian for the state transition function $f(x)$ is

$$\sum_{i=1}^{5} \|(f_{cv}^{(i)})''(x)\|_F^2 = 4T^2$$

for coordinates $(x^{(1)}, x^{(2)}, \dot{x}^{(1)}, \dot{x}^{(2)}, \omega)^T$, and

$$\sum_{i=1}^{5} \|J_{pv}^{-1}(f_{pv}^{(i)})''(x)J_{pv}^{-1}\|_F^2 = (1+\frac{2}{v})T^2,$$

for coordinates $(x^{(1)}, x^{(2)}, v, h, \omega)^T$. Here we have scaled the result with T^2 to get the integrated error during one sampling period, so as to be able to compare it with the results below. For linearized discretization, the corresponding norms are

$$\sum_{i=1}^{5} \|(g_{cv}^{(i)})''\|_F^2 \approx 4T^2 + T^4(1+v^2) \tag{8.130}$$

$$+v^2T^6\left(\frac{1}{9}(\omega T)^0 - \frac{1}{240}(\omega T)^2 + \frac{1}{12600}(\omega T)^4 - \frac{1}{1088640}(\omega T)^6 + O\left((\omega T)^8\right)\right)$$

and

$$\sum_{i=1}^{5} \|J_{pv}^{-1}(g_{pv}^{(i)})'' J_{pv}^{-1}\|_F^2 \approx T^2\left(1+\frac{2}{v}\right) + T^4\left(\frac{1}{2}+\frac{v}{2}\right) \tag{8.131}$$

$$+v^2T^6\left(\frac{1}{9}(\omega T)^0 - \frac{1}{240}(\omega T)^2 + \frac{1}{12600}(\omega T)^4 - \frac{1}{1088640}(\omega T)^6 + O\left((\omega T)^8\right)\right).$$

The approximation here is that ω^2 is neglected compared to $4v$. The proof is straightforward, but the use of symbolic computation programs as *Maple* or *Mathematica* is recommended.

Note first that linearized discretization gives an additive extra term that grows with the sampling interval, but vanishes for small sampling intervals, compared to discretized linearization. Note also that

$$\frac{\sum_{i=1}^{5} \|J_{pv}^{-1}(f_{pv}^{(i)})''(x) J_{pv}^{-1}\|_F^2}{\sum_{i=1}^{5} \|(f_{cv}^{(i)})''(x)\|_F^2} = \frac{1}{4} + \frac{1}{2v},$$

$$\lim_{T\to 0} \frac{\sum_{i=1}^{5} \|J_{pv}^{-1}(g_{pv}^{(i)})''(x) J_{pv}^{-1}\|_F^2}{\sum_{i=1}^{5} \|(g_{cv}^{(i)})''(x)\|_F^2} = \frac{1}{4} + \frac{1}{2v}.$$

That is, the continuous time result is consistent with the discrete time result. We can thus conclude the following:

- The formulas imply an upper bound on the rest term that is neglected in the EKF. Other weighting matrices can be used in the derivation to obtain explicit expressions for how e.g. an initial position error influences the time update. This case, where $J = \text{diag}(1,1,0,0,0)$, is particularly interesting, because now both weighting matrices are exact, and no approximation will be involved.
- The Frobenius error for Cartesian velocity has an extra term v^2T^4 compared to the one for polar velocity. For $T = 3$ and $\omega = 0$ this implies twice as large a bound. For $\omega = 0$, the bounds converge as $T \to \infty$.
- We can summarize the results in the plot in Figure 8.11 illustrating the ratio of (8.130) and (8.131) for $\omega = 0$ and three different velocities. As noted above, the asymptotes are 4 and 1, respectively. Note the huge peak around $T \approx 0.1$. For $T = 5$, Cartesian velocity is only 50% worse.

8.9.4. Discretization of state noise

The state noise in (8.104) has hitherto been neglected. In this section, we discuss different ideas on how to define the state noise $\bar{Q}$ in the EKF.

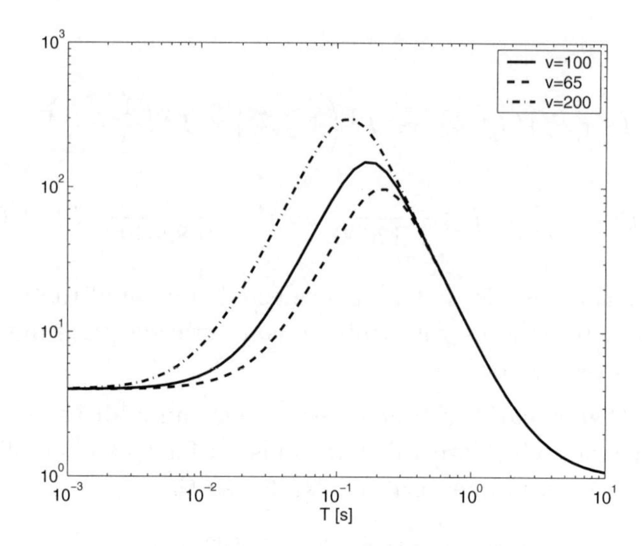

Figure 8.11. Ratio of the upper bounds (8.130) and (8.131) versus sampling interval. Cartesian velocity has a larger upper bound for all sampling intervals.

There are basically five different alternatives for computing a time discrete state noise covariance from a time continuous one. Using one of the alternate models

$$\begin{aligned} \dot{x}_t &= f(x_t) + v_t, \quad \operatorname{Cov}(v_t) = Q, \\ x_{t+T} &= g(x_t) + v_t, \quad \operatorname{Cov}(\bar{v}_t) = \bar{Q}, \end{aligned}$$

we might try

$$\bar{Q}_a = \int_0^T e^{f'\tau} Q e^{(f')^T \tau} \, d\tau \tag{8.132}$$

$$\bar{Q}_b = \frac{1}{T} \int_0^T e^{f'\tau} \, d\tau Q \int_0^T e^{(f')^T \tau} \, d\tau \tag{8.133}$$

$$\bar{Q}_c = T e^{f'T} Q e^{(f')^T T} \tag{8.134}$$

$$\bar{Q}_d = TQ \tag{8.135}$$

$$\bar{Q}_e = T g'(x) Q g'(x)^T. \tag{8.136}$$

All expressions are normalized with T, so that one and the same Q can be used for all of the sampling intervals.

These methods correspond to more or less *ad hoc* assumptions on the state noise for modeling the manoeuvres:

a. v_t is continuous white noise with variance Q.

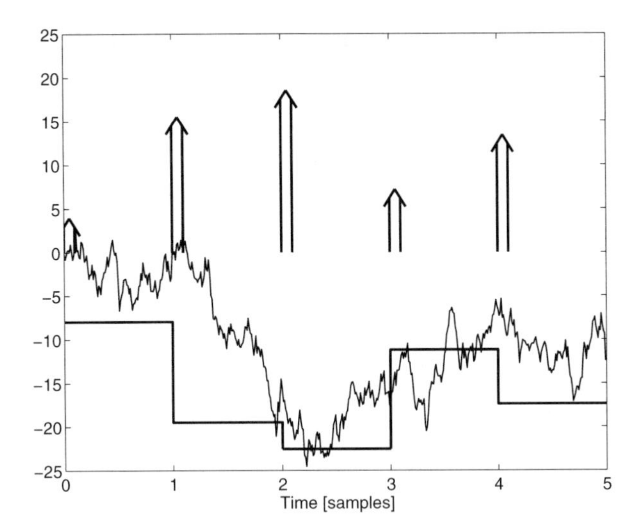

Figure 8.12. Examples of assumptions on the state noise. The arrow denotes impulses in continuous time and pulses in discrete time.

b. $v_t = v_k$ is a stochastic variable which is constant in each sample interval with variance Q/T. That is, each manoeuvre is distributed over the whole sample interval.

c. v_t is a sequence of Dirac impulses active immediately after a sample is taken. Loosely speaking, we assume $\dot{x} = f(x) + \sum_k v_k \delta_{kT-t}$ where v_k is discrete white noise with variance TQ.

d. v_t is white noise such that its total influence during one sample interval is TQ.

e. v_t is a discrete white noise sequence with variance TQ. That is, we assume that all manoeuvres occur suddenly immediately after a sample time, so $x_{t+1} = g(x_t + v_t)$.

The first two approaches require a linear time invariant model for the state noise propagation to be exact. Figure 8.12 shows examples of noise realizations corresponding to assumptions a–c.

It is impossible to say *a priori* which assumption is the most logical one. Instead, one should investigate the alternatives (8.132)–(8.136) by Monte Carlo simulations and determine the importance of this choice for a certain trajectory.

8.10. Whiteness based change detection using the Kalman filter

The simplest form of change detection for state space model is to apply one of the distance measures and stopping rules suggested in Chapter 3, to the normalized innovation, which takes the place as a signal. The principle is illustrated in Figure 8.13.

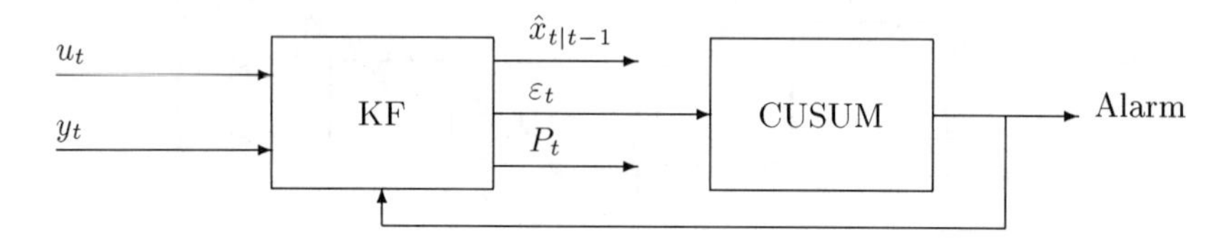

Figure 8.13. Change detection as a whiteness innovation test (here CUSUM) for the Kalman filter, where the alarm feedback controls the adaptation gain.

The *distance measure* s_t is one of the following:

- Normalized innovations

$$s_t = S_t^{-1/2} \varepsilon_t. \tag{8.137}$$

  For vector valued measurements and residuals, we use the sum

$$s_t = \frac{1}{\sqrt{n_y}} \mathbf{1}_{n_y}^T S_t^{-1/2} \varepsilon_t, \tag{8.138}$$

  which will be $\mathrm{N}(0,1)$ distributed under H_0. Here $\mathbf{1}_{n_y}$ is a vector with n_y unit elements.

- The squared normalized innovations

$$s_t = \varepsilon_t^T S_t^{-1} \varepsilon_t - n_y. \tag{8.139}$$

The interpretation here is that we compare the observed error sizes with the predicted ones (S_t) from prior information only. A result along these lines is presented in Bakhache and Nikiforov (1999) for GPS navigation. This idea is used explicitely for comparing confidence intervals of predictions and prior model knowledge in Zolghadri (1996) to detect *filter divergence*. Generally, an important role of the whiteness test is to monitor divergence, and this might be a stand-alone application.

The advantage of (8.139) is that changes which increase the variance without affecting the mean can be detected as well. The disadvantage of (8.139)

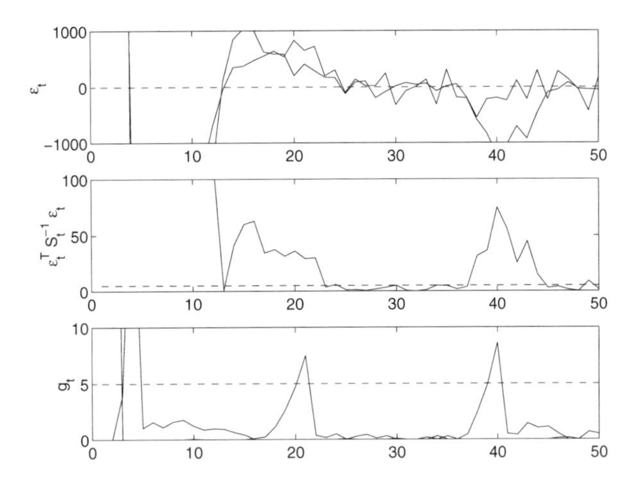

Figure 8.14. Different ways to monitor the innovations from a Kalman filter. First, the two innovations are shown. Second, the normalized square (8.139) (without subtraction of n_y). Third, the test statistics in the CUSUM test using the sum of innovations as input.

is the sensitivity to noise scalings. It is well known that the state estimates are not sensitive to scalings, but the covariances are:

$$\bar{R} = \lambda R, \quad \bar{Q} = \lambda Q, \quad \bar{P}_0 = \lambda P_0, \quad \Rightarrow \bar{P}_{t|t} = \lambda P_{t|t}, \quad \bar{S}_t = \lambda S_t.$$

The usual requirement that the distance measure s_t is zero mean during non-faulty operation is thus sensitive to scalings.

After an alarm, an appropriate action is to increase the process noise covariance matrix momentarily, for instance by multiplying it with a scalar factor α.

Example 8.18 Target tracking: whiteness test

Consider the target tracking example. Figure 8.15(a) shows the trajectory and estimate from a Kalman filter. Figure 8.14 visualizes the size in different ways. Clearly, the means of the innovations (upper plot) are indicators of manoeuvre detection, at least after the transient has faded away around sample 25. The normalized versions from (8.139) also become large in the manoeuvre. The test statistics from the CUSUM test using the innovations as input are shown in the lower plot. It is quite easy here to tune the threshold (and the drift) to get reliable change detection. Manoeuvre detection enables improved tracking, as seen in Figure 8.15(b).

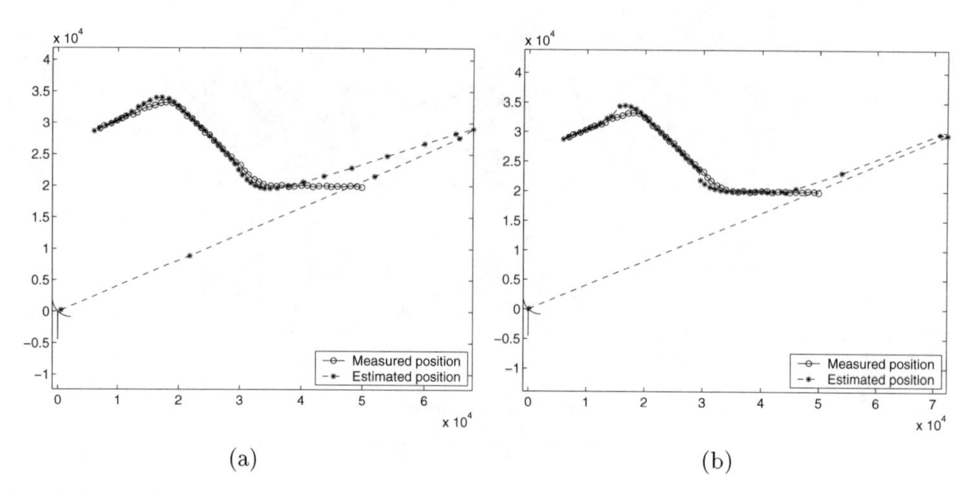

(a) (b)

Figure 8.15. (a) Tracking using a Kalman filter, and (b) where a residual whiteness test is used. The CUSUM parameters are a threshold of 5 and a drift of 0.25. The test statistic is shown as the last subplot in Figure 8.14.

8.11. Estimation of covariances in state space models

A successful design hinges on good tuning of the process noise covariance matrix Q. Several attempts have been presented to estimate Q (Bohlin, 1976; Gutman and Velger, 1990; Isaksson, 1988; Valappila and Georgakisa, 2000; Waller and Saxén, 2000), but as we shall see, there is a fundamental observability problem. At least if both Q and R are to be estimated.

A state space model corresponds to a filter (or differential equation). For a model with scalar process and measurement noise, we can formulate an estimation problem from

$$y_t = F(q)\sigma_v v_t + G(q)\sigma_e e_t,$$

where v_t and e_t are independent white noises with unit variances. Form the vectors $Y = (y_1, y_2, \dots, y_N)^T$, $V = (v_1, v_2, \dots, v_N)^T$, $E = (e_1, e_2, \dots, e_N)^T$ and use the impulse response matrices of the state space model (or filter model):

$$Y = H_v V + H_e E.$$

Assuming Gaussian noise, the distribution for the measurements is given by

$$Y \in \mathrm{N}(0, \underbrace{H_v H_v^T \sigma_v^2 + H_e H_e^T \sigma_e^2}_{P(\sigma_v, \sigma_e)}).$$

The maximum likelihood estimate of the parameters is thus

$$\arg\min_{\sigma_v, \sigma_e} Y^T P^{-1}(\sigma_v, \sigma_e) Y + \log\det P(\sigma_v, \sigma_e).$$

For Kalman filtering, only the ratio σ_v/σ_e is of importance, and we can use, for instance, $\sigma_e = 1$.

Example 8.19 Covariance estimation in state space models

Consider a one-dimensional motion model

$$x_{t+1} = \begin{pmatrix} 1 & 1 \\ 0 & 1 \end{pmatrix} x_t + \sigma_v \begin{pmatrix} 0 \\ 1 \end{pmatrix} v_t$$

$$y_t = \begin{pmatrix} 1 \\ 0 \end{pmatrix} x_t + \sigma_e e_t.$$

The true parameters in the simulation are $N = 36$, $\sigma_v^o = 1$ and $\sigma_e^o = 3$.

Figure 8.16(a) shows that the likelihood gets a well specified minimum. Note, however, that the likelihood will not become convex (compare to Figure 1.21(a)), so the most efficient standard tools for optimization cannot be used. Here we used a global search with a point grid approach.

For change detection, we need good values of both σ_v and σ_e. The joint likelihood for data given both σ_v, σ_e is illustrated in Figure 8.16(b). Note here that the influence of σ_e is minor, and optimization is not well conditioned. This means that the estimates are far from the true values in both approaches.

As a final remark, the numerical efficiency of evaluating the likelihood given here is awkward for large number of data, since large impulse response matrices of dimension $N \times N$ has to be formed, and inversions and determinants of their squares have to be computed. More efficient implementation can be derived in the frequency domain by applying Parseval's formula to the likelihood.

8.12. Applications

8.12.1. DC motor

Consider the DC motor lab experiment described in Sections 2.5.1 and 2.7.1. It was examined with respect to system changes in Section 5.10.2. Here we

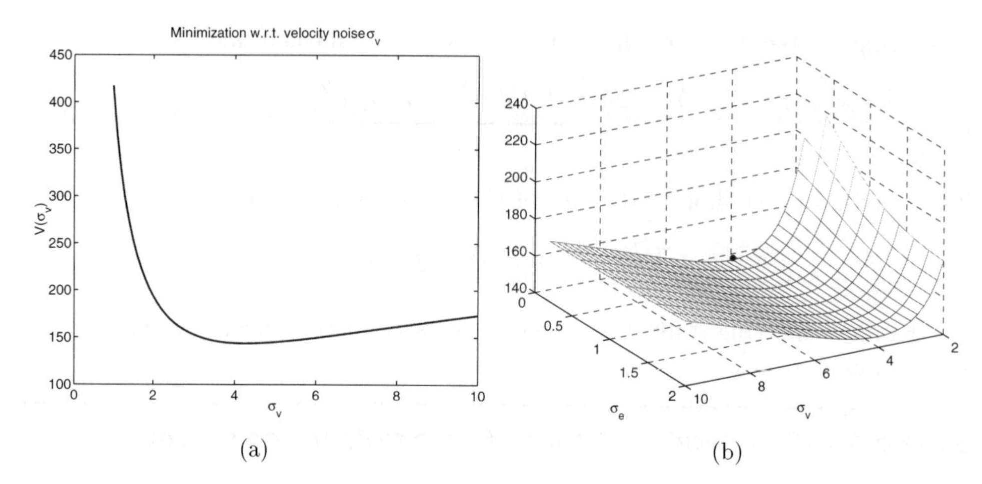

Figure 8.16. Likelihood for σ_v when $\sigma_e = 1$ is assumed (a), and likelihood for σ_v, σ_e (b).

apply the Kalman filter with a whiteness residual test, where the goal is to detect the torque disturbances while being insensitive to system changes. The test cases in Table 2.1 are considered. The squared residuals are fed to the CUSUM test with $h = 10$ and $\nu = 2$. The state space model is given in (2.2).

From Figure 8.17 we conclude the following:

- The design of the CUSUM test is much simpler here than in the parametric case. The basic reason is that state changes are simpler to detect than parameter changes.
- We get alarms at all faults, both system changes and disturbances.
- It does not seem possible to solve the fault isolation problem reliably with this method.

8.12.2. Target tracking

A typical application of target tracking is to estimate the position of aircraft. A civil application is *Air Traffic Control* (*ATC*), where the traffic surveillor at each airport wants to get the position and predicted positions of all aircraft at a certain distance from the airport. There are plenty of military applications where, for several reasons, the position and predicted position of hostile aircraft are crucial information. There are also a few other applications where the target is not an aircraft. See, for instance, the highway surveillance problem below.

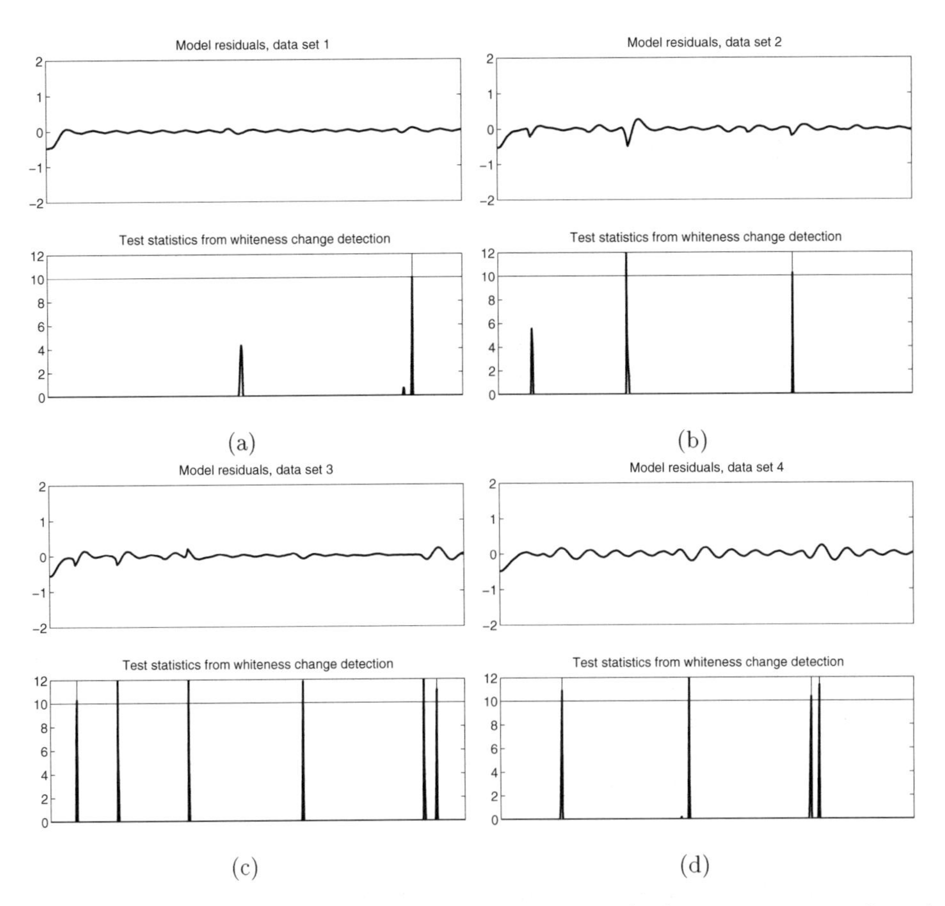

Figure 8.17. Simulation error using the state space model in (2.2) and test statistics from the CUSUM test. Nominal system (a), with torque disturbances (b), with change in dynamics (c) and with both disturbance and change (d).

The classical sensor for target tracking is radar, that at regular sweeps measures range and bearing to the object. Measurements from one flight test are shown in Figure 8.18.

Sensors

The most characteristic features of such a target tracking problem are the available sensors and the dynamical model of the target. Possible sensors include:

- A *radar* where bearing and range to the object are measured.
- *Bearing only sensors* including Infra-Red (IR) and radar warning systems. The advantage of these in military applications is that they are

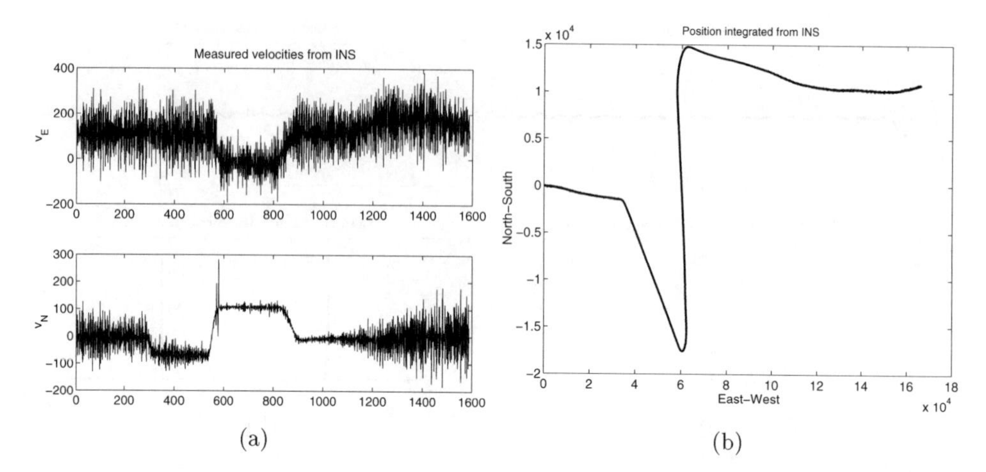

Figure 8.18. Measurements and trajectory from dead reckoning.

passive and do not reveal the position of the tracker.

- The range only information in a GPS should also be mentioned here, although the tracker and target are the same objects in navigation applications as in a GPS.
- Camera and computer vision algorithms.
- Tracking information from other trackers.

Possible models are listed in the ATC application below.

Air Traffic Control (ATC)

There are many sub-problems in target tracking before the Kalman filter can be applied:

- First, one Kalman filter is needed for each aircraft.
- The aircraft is navigating in open-loop in altitude, because altitude is not included in the state vector. The aircraft are supposed to follow their given altitude reference, where a large navigation uncertainty is assumed.
- The *association* problem: each measurement is to be associated with an aircraft trajectory. If this is not working, the Kalman filters are fed with incorrect measurement, belonging to another aircraft.
- The radar detects false echoes that are referred to as *clutter*. These are a problem for the association algorithm.

See Bar-Shalom and Fortmann (1988) and Blackman (1986) for more information. The application of the IMM algorithm from Chapter 10 to ATC is reported in Yeddanapudi et al. (1997).

Choice of state vector and model

The most common model for target tracking by far is a motion model of the kinds listed below. The alternative is to use the aircraft flight dynamics, where the pilot input included in u_t considered as being too computationally intensive. Their are a few investigations of this approach; see, for instance, Koifman and Bar-Itzhack (1999).

Proposed dynamical models for the target that will be examined here include:

1. A four state linear model with $x = (x^{(1)}, x^{(2)}, v^{(1)}, v^{(2)})^T$.
2. A four state linear model with $x = (x^{(1)}, x^{(2)}, v^{(1)}, v^{(2)})^T$ and a time-varying and state dependent covariance matrix $Q(x_t)$ which is matched to the assumption that longitudinal acceleration is only, say, 1% of the lateral acceleration (Bauschlicher et al., 1989; Efe and Atherton, 1998).
3. A six state linear model with $x = (x^{(1)}, x^{(2)}, v^{(1)}, v^{(2)}, a^{(1)}, a^{(2)})^T$.
4. A six state linear model with $x = (x^{(1)}, x^{(2)}, v^{(1)}, v^{(2)}, a^{(1)}, a^{(2)})^T$ and a time-varying and state dependent covariance matrix $Q(x_t)$.
5. A five state non-linear model with $x = (x^{(1)}, x^{(2)}, v^{(1)}, v^{(2)}, \omega)^T$, together with an extended Kalman filter from Section 8.9.

In detail, the coordinated turn assumption in 2 and 4 implies that Q should be computed by using $q_v = 0.01 q_w$ below:

$$\theta = \arctan(x^{(4)}/x^{(3)})$$

$$Q = \begin{pmatrix} q_v \cos^2(\theta) + q_w \sin^2(\theta) & (q_v - q_w)\sin(\theta)\cos(\theta) \\ (q_v - q_w)\sin(\theta)\cos(\theta) & q_v \sin^2(\theta) + q_w \cos^2(\theta) \end{pmatrix}$$

$$B_v = \begin{pmatrix} T^2/2 & 0 \\ 0 & T^2/2 \\ T & 0 \\ 0 & T \end{pmatrix} \quad \text{or} \quad B_v = \begin{pmatrix} T^3/3 & 0 \\ 0 & T^3/3 \\ T^2/2 & 0 \\ 0 & T^2/2 \\ T & 0 \\ 0 & T \end{pmatrix}.$$

The expressions for B_v hold for the assumption of piecewise constant noise between the sampling instants. Compare to equations (8.132)–(8.136).

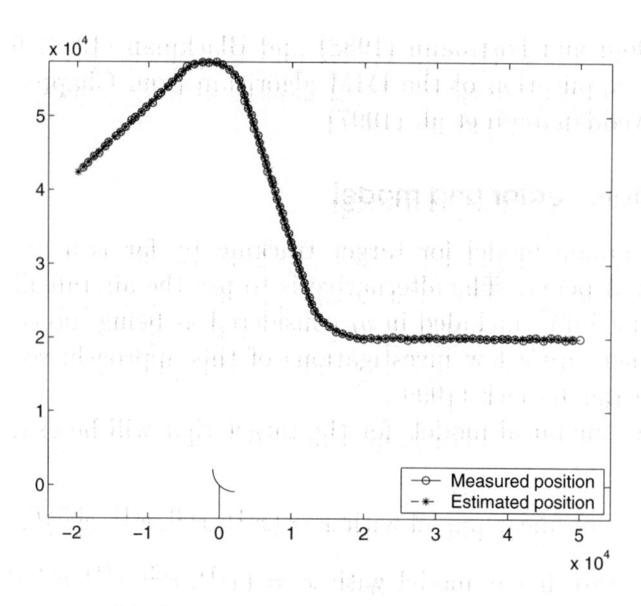

Figure 8.19. Target trajectory and radar measurements.

Example 8.20 Target tracking: choice of state vector

We will compare the five different motion models above on the simulated target in Figure 8.19. The difference in performance is not dramatic by only studying the Mean Square Error (MSE), as seen in Figure 8.20(a). In fact, it is hard to visually see any systematic difference. A very careful study of the plots and general experience in tuning this kind of filters gives the following well-known rules of thumb:

- During a straight course, a four state linear model works fine.
- During manoeuvres (as coordinated turns), the six state linear model works better than using four states.
- The five state non-linear model has the potential of tracking coordinated turns particularly well.

The total RMSE and the corresponding adaptation gain q used to scale Q is given below:

Method	1	2	3	4	5
RMSE	243	330	241	301	260
q	10	100	0.1	1	0.0001

The adaptation gains are not optimized. Several authors have proposed the use of two parallel filters, one with four and one with six states, and

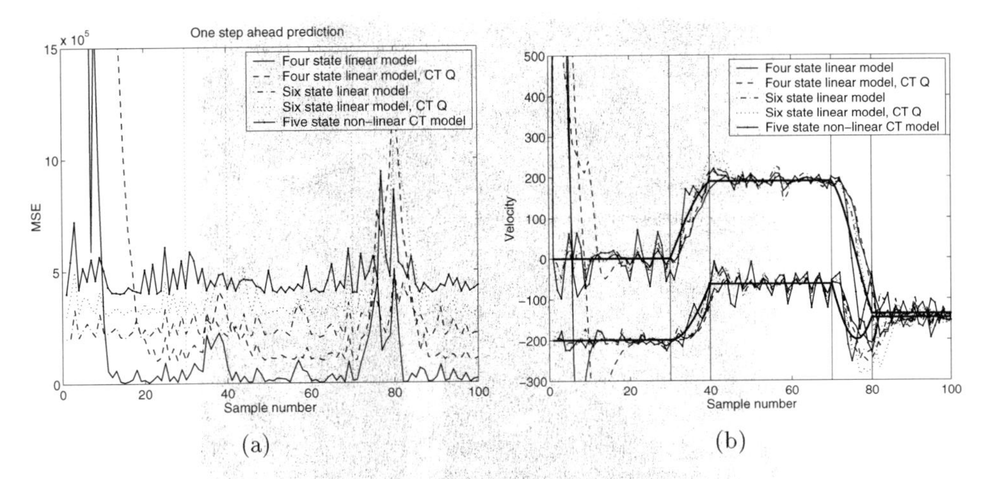

Figure 8.20. Performance of different state models for tracking the object in Figure 8.19. In (a) the mean square error for the different methods are plotted (separated by an offset), and in (b) the velocity estimate is compared to the true value.

a manoeuvre detection algorithm to switch between them. The turn rate model with five states is often used for ATC. Note however, that according to the examples in Section 8.9, there are at least four different alternatives for choosing state coordinates and sampling for the turn rate model.

An intuitive explanation to the rules of thumb above can be derived from general adaptive filtering properties:

- Use as parsimonious a model as possible! That is, four states during a straight course and more states otherwise. Using too many states results in an increased variance error.
- With sufficiently many states, all manoeuvres can be seen as piecewise constant states, and we are facing a segmentation problem rather than a tracking problem. The five state model models coordinated turns as a piecewise constant turn rate ω.

Highway traffic surveillance

The traffic surveillance system described here aims at estimating the position of all cars using data from a video camera. A snapshot from the camera is shown in Figure 8.21. As an application in itself, camera stabilization can be done with the use of Kalman filters (Kasprzak et al., 1994).

Figure 8.22(a) shows the measured position from one track, together with Kalman filtered estimate using a four state linear model. Figure 8.22(b) shows

Figure 8.21. A helicopter hovering over a highway measures the position of all cars for surveillance purposes. Data and picture provided by Centre for Traffic Simulation Research, Royal Institute of Technology, Stockholm.

the time difference of consecutive samples in the east-west direction before and after filtering. Clearly, the filtered estimates have much less noise. It is interesting to note that the uncertainty caused by the bridge can also be attenuated by the filter.

The histogram of the estimation errors in Figure 8.23, where the transient before sample number 20 is excluded, shows a quite typical distribution for applications. Basically, the estimation errors are Gaussian, but the tails are heavier. This can be explained by data outliers. Another more reasonable explanation is that the estimation error is the sum of the measurement error and manoeuvres, and the latter gives rise to outliers. The conclusion from this example, and many others, is that the measurement errors can in many applications be considered as Gaussian. This is important for the optimality property of the Kalman filter. It is also an important conclusion for the change detection algorithms to follow, since Gaussianity is a cornerstone of many of these algorithms.

Bearings only target tracking

The use of passive bearings only sensors is so important that it has become its own research area (Farina, 1998). The basic problem is identifiability. At each measurement time, the target can be anywhere along a line. That is, knowledge of maximal velocity and acceleration is necessary information for tracking to be able to exclude impossible or unrealistic trajectories. It turns

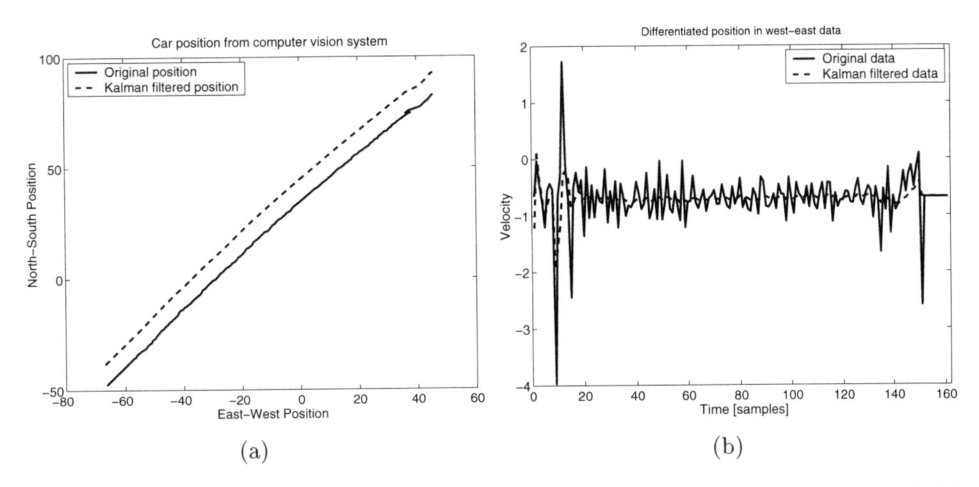

Figure 8.22. (a) Position from computer vision system and Kalman filter estimate (with an offset 10 added to north-south direction); (b) differentiated data in east-west direction shows that the noise level is significantly decreased after filtering.

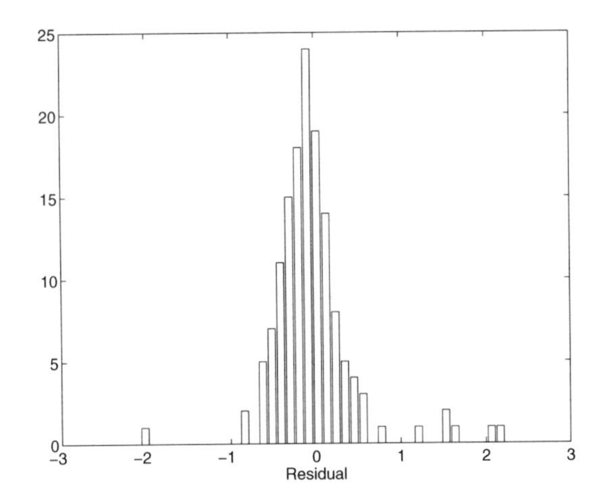

Figure 8.23. Distribution of the estimation errors in Figure 8.22 after sample number 20.

out that even then identifiability is not guaranteed. It is helpful if the own platform is moving. There are papers discussing how the platform should be moving to optimize tracking properties.

It is well known that for a single sensor, the use of spherical state coordinates has an important advantage: the observable subspace is decoupled from the non-observable. The disadvantage is that the motion model must be non-linear. We point here at a special application where two or more bearing only sensors are exchanging information. Since there is no problem with observability here, we keep the Cartesian coordinates. If they are synchronized,

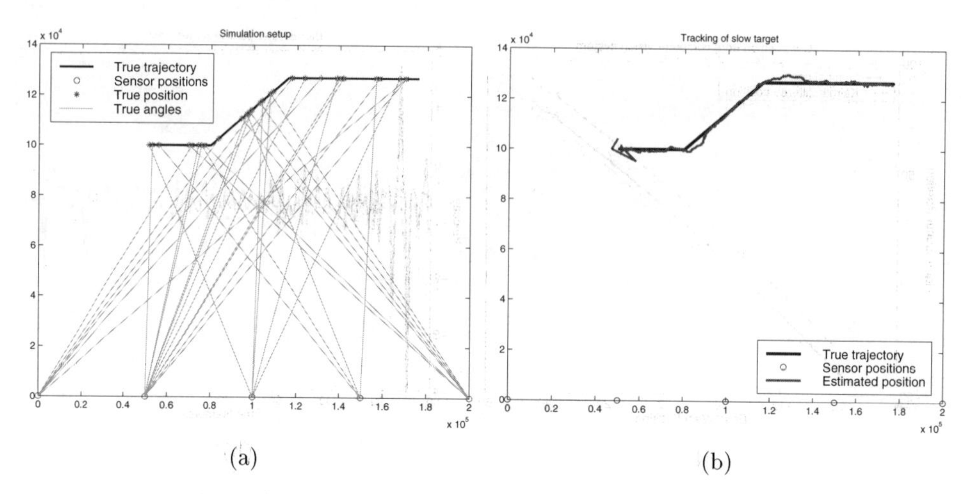

Figure 8.24. Bearing measurements and target trajectory (a) and tracking performance (b).

then the best approach is probably to transform the angle measurements to one Cartesian measurement. If they are not synchronized, then one has to use an extended Kalman filter. The measurement equation is then linearized as follows:

$$
\begin{aligned}
y_t =& \arctan\left(\frac{x_t^{(1)} - \bar{x}_t^{(1)}}{x_t^{(2)} - \bar{x}_t^{(2)}}\right) + e_t \\
\approx& \frac{1}{R^2}((\hat{x}_t^{(2)} - \bar{x}_t^{(2)}), -(\hat{x}_t^{(1)} - \bar{x}_t^{(1)}), 0, 0)x_t + e_t \\
=& C_t x_t + e_t \\
R^2 =& (\hat{x}_t^{(2)} - \bar{x}_t^{(2)})^2 + (\hat{x}_t^{(1)} - \bar{x}_t^{(1)})^2.
\end{aligned}
$$

Here R is the estimated range to the target and C the estimated derivative of the arctangent function. The sensor location at sampling time t is denoted $\bar{x}_t$.

Example 8.21 Bearings only tracking

Figure 8.24(a) shows a trajectory (going from left to right) and how five sensors measure the bearing (angle) to the object. The sensors are not synchronized at all in this simulation. Figure 8.24(b) shows the Kalman filter estimated position.

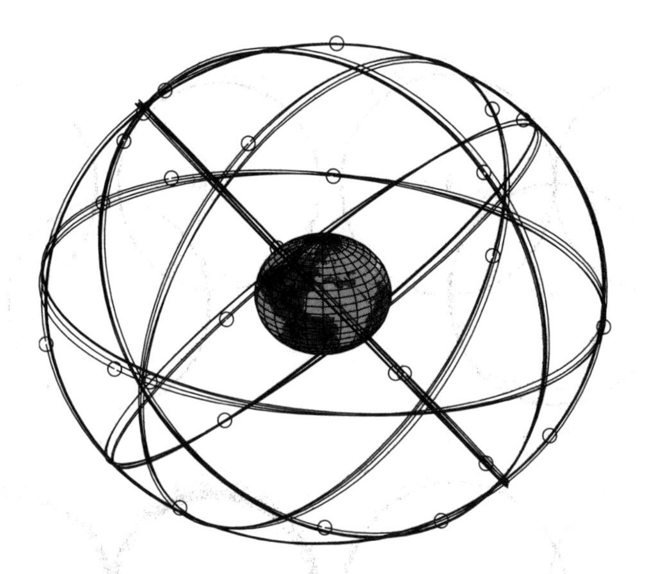

Figure 8.25. GPS satellites in their orbits around the earth.

8.12.3. GPS

Satellite based navigation systems have a long history. It all started with the TRANSIT project 1959–1964 which included seven satellites. The GPS system used today was already initiated 1973 and completed 1994. The American GPS and Russian alternative GLONASS both consist of 24 satellites. The GPS satellites cover six planes, each separated 55° at a distance of 22200 km from the earth, as illustrated in Figure 8.25. The orbit time is 12 hours.

The key to the accuracy is a very accurate atom clock in each satellite, which has a drift of less than 10^{-13}. The clocks are, furthermore, calibrated every 12 hours when they pass the USA. The satellites transmit a message (10.23 MHz) at regular and known time instants. The message includes an identity number and the position of the satellite.

Figure 8.26 illustrates the principles in two dimensions. The result in (a) would be obtained if we knew the time exactly, and there is one unique point where all circles intersect. If the clock is behind true time, we will compute too short a distance to the satellites, and we get the plot in (b). Conversely, plot (c) is obtained if the clock is ahead of true time. It is obvious that by trial and error, we can find the correct time and the 2D position when three satellites are available.

If there are measurement errors and other uncertainties, we get more realistically thicker circles as in Figure 8.26(d), and the intersection becomes diffuse. That is, the position estimate is inaccurate.

The GPS receiver has a clock, but it is not of the same accuracy as the

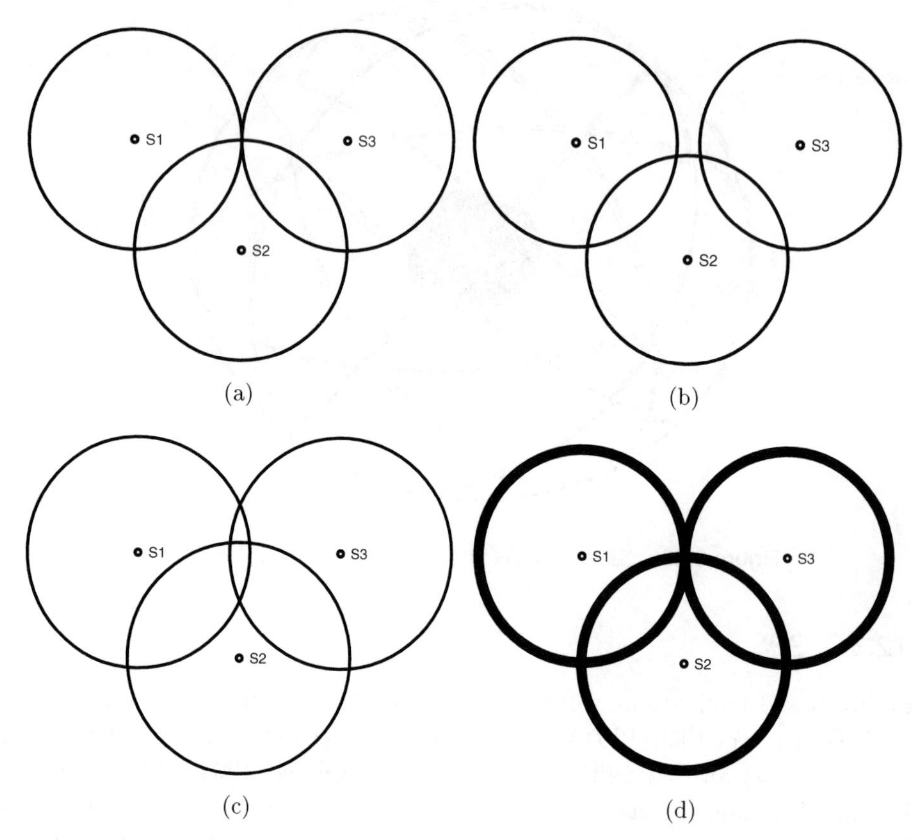

Figure 8.26. Computation of position in 2D from distance computations in case of a perfect clock (a) and where the clock bias gives rise to a common distance error to all satellites as in (b) and (c), and finally, in the case of measurement uncertainty (d).

satellites. We therefore have to include its error, the clock bias, into the state vector, whose elements of primary interest are:

$x^{(1)}$	East-west position [m]
$x^{(2)}$	North-south position [m]
$x^{(3)}$	Up-down position [m]
$x^{(4)}$	Clock bias times velocity of light [m]

Note the scaling of the clock bias to improve the numerical properties (all states in the same order of magnitude).

Non-linear least squares approach

The time delay Δt of the transmitted signal to each satellite j is computed, which can be expressed as

$$\begin{aligned}
(\Delta t)^{(j)} &= \frac{R^{(j)}(x)}{c} + \frac{x^{(4)}}{c} + \frac{e_t^{(j)}}{c} \\
R^{(j)}(x) &= \sqrt{(X^{(1,j)} - x^{(1)})^2 + (X^{(2,j)} - x^{(2)})^2 + (X^{(3,j)} - x^{(3)})^2},
\end{aligned}$$

where $R^{(j)}$ is the distance to and $X^{(j)}$ the position of satellite j, respectively, and c is the velocity of light. We have four unknowns, so we need at least four equations. Four visible satellites give a non-linear equation system and more than four gives an over-determined equation system. These can be solved by iterative methods, but that is not how GPS systems are implemented usually.

Linearized least squares approach

Instead, we use a previous position estimate and linearize the equations. We have

$$\begin{aligned}
y_t^{(j)} &\triangleq c(\Delta t)^{(j)} = R^{(j)}(x_t) + x_t^{(4)} + e_t^{(j)} \\
&\approx R^{(j)}(\hat{x}_t) + \left.\frac{dR^{(j)}(x)}{dx}\right|_{x=\hat{x}_t} (x_t - \hat{x}_t) + x_t^{(4)} + e_t^{(j)}.
\end{aligned}$$

The linearized measurement equation is thus

$$\begin{aligned}
\bar{y}_t^{(j)} &\triangleq y_t^{(j)} - R^{(j)}(\hat{x}_t) + \left.\frac{dR^{(j)}(x)}{dx}\right|_{x=\hat{x}_t} \hat{x}_t \\
&= \underbrace{\left.\frac{dR^{(j)}(x)}{dx}\right|_{x=\hat{x}_t}}_{C^{(j,1:3)}(\hat{x}_t)} x_t + \underbrace{1}_{C^{(j,4)}} x_t^{(4)} + e_t^{(j)}.
\end{aligned}$$

The first part of the regression vector $C_t^{(j)}$, which will be used in a Kalman filter, is given by

$$\begin{aligned}
C^{(j,1:3)}(x) &= \frac{1}{R} \left(-(X^{(1,j)} - x^{(1)}) \quad -(X^{(2,j)} - x^{(2)}) \quad -(X^{(3,j)} - x^{(3)}) \right) \\
R &= \sqrt{(X^{(1,j)} - x^{(1)})^2 + (X^{(2,j)} - x^{(2)})^2 + (X^{(3,j)} - x^{(3)})^2}.
\end{aligned}$$

Collecting all satellite measurements (≤ 4) in one vector $\bar{y}_t$ and the regression vectors in a matrix $C(x)$ gives the least squares solution

$$\hat{x}_t = \left(C(\hat{x}_{t-1})^T C(\hat{x}_{t-1})\right) C(\hat{x}_{t-1})^T \bar{y}_t,$$

where $\hat{x}_{t-1}$ is the estimate from the previous measurements. Note that a very accurate clock compensation is obtained as a by-product in $\hat{x}_t^{(4)}$.

This principle is used in some GPS algorithms. The advantage of this snapshot method is that no dynamical model of the GPS movement is needed, so that no prior information about the application where it is used is needed. The drawback is that no information is obtained when less than four satellites are visible.

Linearized Kalman filter approach

The other alternative is to use an extended Kalman filter. The following state space model can be used:

$$\begin{aligned}
x_{t+1} =& A x_t + B_v v_t \\
A =& \begin{pmatrix} 1 & 0 & 0 & 0 & T_s & 0 & 0 & 0 \\ 0 & 1 & 0 & 0 & 0 & T_s & 0 & 0 \\ 0 & 0 & 1 & 0 & 0 & 0 & T_s & 0 \\ 0 & 0 & 0 & 1 & 0 & 0 & 0 & T_s \\ 0 & 0 & 0 & 0 & 1 & 0 & 0 & 0 \\ 0 & 0 & 0 & 0 & 0 & 1 & 0 & 0 \\ 0 & 0 & 0 & 0 & 0 & 0 & 1 & 0 \\ 0 & 0 & 0 & 0 & 0 & 0 & 0 & 1 \end{pmatrix} \\
B_v =& \begin{pmatrix} 0_{4,4} \\ I_{4,4} \end{pmatrix} \\
Q =& \operatorname{diag}(q^{(1)}, q^{(2)}, q^{(3)}, q^{(4)}).
\end{aligned}$$

The introduced extra states are

$x^{(5)}$	East-west velocity [m/s]
$x^{(6)}$	North-south velocity [m/s]
$x^{(7)}$	Up-down velocity [m/s]
$x^{(8)}$	Clock drift times velocity of light [m/s]

Here the movement of the GPS platform is modeled as velocity random walk, and the clock drift is explicitely modeled as random walk.

The use of a whiteness-based change detector in conjunction with the EKF is reported in Bakhache and Nikiforov (1999). An orbit tracking filter similar to this problem is found in Chaer et al. (1998). Positioning in GSM cellular phone systems using range information of the same kind as in the GPS is reported in Pent et al. (1997).

Performance

The errors included in the noise e_t and their magnitude in range error are listed below:

- Satellite trajectory error (1 m).
- Satellite clock drift (3 m).
- Selected availability (SA) (60 m). Up to the year of 2000, the US military added a disturbance term to the clock of each satellite. This term was transmitted separately and coded so it was not available to civil users. The disturbance was generated as filtered white noise, where the time variations were designed so it takes hours of measurements before the variance of average becomes significantly smaller. One solution to that has been developed to circumvent this problem is to use *Differential GPS* (*DGPS*). Here fixed transmitters in base stations on the ground send out their estimate of the clock bias term, which can be estimated very accurately due to the position of the base station being known exactly. Such base stations will probably be used even in the future to improve positioning accuracy.
- Refraction in the troposphere (2 m) and ionosphere (2 m), and reflections (1 m).
- Measurement noise in the receiver (2 m).

All in all, these error sources add up to a horizontal error of 10 m (100 m with the SA code). Using differential GPS, the accuracy is a couple of meters. ATC systems usually have at least one base station at each airport. It is probably a good guess that the next generation mobile phones will have built-in GPS, where the operators' base stations act as GPS base stations as well.

Long term averaging can bring down the accuracy to decimeters, and by using phase information of the carrier between two measurement points we come down to accuracy in terms of millimeters.

9

Change detection based on likelihood ratios

9.1. Basics

This chapter is devoted to the problem of detecting additive abrupt changes in linear state space models. Sensor and actuator faults as a sudden offset or drift can all be modeled as additive changes. In addition, disturbances are traditionally modeled as additive state changes. The likelihood ratio formulation provides a general framework for detecting such changes, and to isolate the fault/disturbance.

The state space model studied in this chapter is

$$x_{t+1} = A_t x_t + B_{u,t} u_t + B_{v,t} v_t + \sigma_{t-k} B_{\theta,t} \nu \tag{9.1}$$

$$y_t = C_t x_t + e_t + D_{u,t} u_t + \sigma_{t-k} D_{\theta,t} \nu. \tag{9.2}$$

The additive change (fault) ν enters at time k as a step (σ_t denotes the step function). Here v_t, e_t and x_0 are assumed to be independent Gaussian variables:

$$\begin{aligned} v_t &\in \mathrm{N}(0, Q_t) \\ e_t &\in \mathrm{N}(0, R_t) \\ x_0 &\in \mathrm{N}(0, \Pi_0). \end{aligned}$$

Furthermore, they are assumed to be mutually independent. The state change ν occurs at the unknown time instant k, and $\delta(j)$ is the pulse function that is one if $j = 0$ and zero otherwise. The set of measurements $y_1, y_2, \ldots, y_N$, each of dimension p, will be denoted y^N and y_t^N denotes the set $y_t, y_{t+1}, \ldots, y_N$.

This formulation of the change detection problem can be interpreted as an *input observer* or *input estimator* approach. A similar model is used in Chapter 11.

To motivate the ideas of this chapter, let us consider the augmented state space model, assuming a change at time $t = k$ (compare with Examples 6.6 and 8.2.4):

$$\begin{aligned} \bar{x}_{t+1} &= \begin{pmatrix} x_{t+1} \\ \theta_{t+1} \end{pmatrix} = \begin{pmatrix} A_t & B_{\theta,t} \\ 0 & I \end{pmatrix} \bar{x}_t + \begin{pmatrix} B_{u,t} \\ 0 \end{pmatrix} u_t + \begin{pmatrix} B_{v,t} & B_{\theta,k} \\ 0 & I \end{pmatrix} \begin{pmatrix} v_t \\ \delta_{t-k}\nu \end{pmatrix} \\ y_t &= \begin{pmatrix} C_t & D_{\theta,t} \end{pmatrix} \bar{x}_t + e_t + D_{u,t} u_t \\ \bar{x}_{0|0} &= \begin{pmatrix} x_0 \\ 0 \end{pmatrix} \\ \bar{P}_{0|0} &= \begin{pmatrix} P_0 & 0 \\ 0 & 0 \end{pmatrix}. \end{aligned} \tag{9.3}$$

That is, at time $t = k$ the parameter value is changed as a step from $\theta_t = 0$ for $t < k$ to $\theta_t = \nu$ for $t \geq k$. It should be noted that ν and θ both denote the magnitude of the additive change, but the former is seen as an input and the latter as a state, or parameter.

The advantage of the re-parameterization is that we can apply the Kalman filter directly, with or without change detection, and we have an explicit fault state that can be used for fault isolation. The Kalman filter applied to the augmented state space model gives a parameter estimator

$$\hat{\theta}_{t+1|t} = \hat{\theta}_{t|t-1} + K_t^{\theta}(y_t - C_t \hat{x}_{t|t-1} - D_{\theta,t}\hat{\theta}_{t|t-1} - D_{u,t} u_t),$$

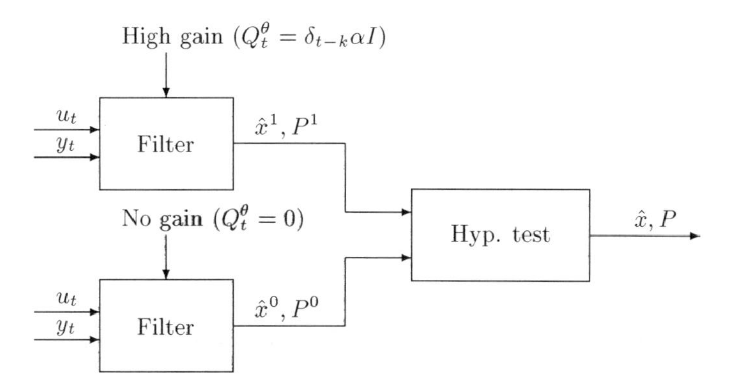

Figure 9.1. Two parallel filters. One is based on the hypothesis no change (H_0) and the other on a change at time k ($H_1(k)$). By including more hypothesis for k, a filter bank is obtained.

Here we have split the Kalman filter quantities as

$$K_t = \begin{pmatrix} K_t^x \\ K_t^\theta \end{pmatrix}, \quad P_t = \begin{pmatrix} P_t^{xx} & P_t^{x\theta} \\ P_t^{\theta x} & P_t^{\theta\theta} \end{pmatrix},$$

so the covariance matrix of the change (fault component) is $P_t^{\theta\theta}$. Note that $K_t^\theta = 0$ before the change. The following alternatives directly appear:

- Kalman filter-based adaptive filtering, where the state noise covariance

$$\bar{Q}_t = \begin{pmatrix} Q_t & 0 \\ 0 & Q_t^\theta \end{pmatrix}$$

  is used to track θ.

- Whiteness-based residual test, where the Kalman filter innovations are used and the state covariance block Q_t^θ is momentarily increased when a change is detected.

- A parallel filter structure as in Figure 9.1. The hypothesis test can be accomplished by one of the distance measures in Chapter 6.

This chapter is devoted to customized approaches for detecting and explicitely estimating the change ν. The approach is based on *Likelihood Ratio* (*LR*) tests using *Generalized Likelihood Ratio* (*GLR*) or *Marginalized Likelihood Ratio* (*MLR*). The derivation of LR is straightforward from (9.3), but the special structure of the state space model can be used to derive lower order filters. The basic idea is that the residuals from a Kalman filter, assuming no change, can be expressed as a linear regression.

Linear regression formulation

The nominal Kalman filter, assuming no abrupt change, is applied, and the additive change is expressed as a linear regression with the innovations as measurements with the following notation:

Kalman filter →	$\hat{x}_{t\|t-1},\ \varepsilon_t$
Auxiliary recursion →	$\varphi_t,\ \mu_t$
Residual regression	$\varepsilon_t = \varphi_t^T \nu + e_t$
Compensation	$x_t \approx \hat{x}_{t\|t-1} + \mu_t \nu.$

The third equation indicates that we can use RLS to estimate the change ν, and the fourth equation shows how to solve the *compensation* problem after detection of change and estimation (isolation) of ν.

Chapter 10 gives an alternative approach to this problem, where the change is not explicitely parameterized.

9.2. The likelihood approach

Some modifications of the Kalman filter equations are given, and the likelihood ratio is defined for the problem at hand.

9.2.1. Notation

The Kalman filter equations for a change $\nu \in \mathrm{N}(0, P_\nu)$ at a given time k follows directly from (8.34)–(8.37) by considering ν as an extra state noise component $\bar{v}_t = \delta_{t-k}\nu$, with $\bar{Q}_t = \delta_{t-k}P_\nu$.

$$\begin{aligned}
\hat{x}_{1|0}(k) &= x_0 \\
P_{1|0}(k) &= P_0 \\
\varepsilon_t(k) &= y_t - C_t \hat{x}_{t|t-1}(k) - D_{u,t} u_t \\
S_t(k) &= C_t P_{t|t-1}(k) C_t^T + R_t \\
K_t(k) &= P_{t|t-1}(k) C_t^T S_t^{-1}(k) \\
\hat{x}_{t+1|t}(k) &= A_t \hat{x}_{t|t-1}(k) + A_t K_t(k) \varepsilon_t(k) + B_{u,t} u_t \\
P_{t+1|t}(k) &= A_t \left(P_{t|t-1}(k) - K_t(k) C_t P_{t|t-1}(k) \right) A_t^T \\
&\quad + B_{v,t} Q_t B_{v,t}^T + \delta_{t-k} B_{\nu,t} P_\nu B_{\nu,t}^T.
\end{aligned} \tag{9.4}$$

The addressed problem is to modify these equations to the case where k and ν are unknown. The change instant k is of primary interest, but good state estimates may also be desired.

In GLR, ν is an unknown constant, while it is considered as a stochastic variable in the MLR test. To start with, the change will be assumed to have a Gaussian prior. Later on, a non-informative prior will be used which is sometimes called a prior of ignorance; see Lehmann (1991). This prior is characterized by a constant density function, $p(\nu) = C$.

Example 9.1 Modeling a change in the mean

We can use Eq. (9.1) to detect abrupt changes in the mean of a sequence of stochastic variables by letting $A_t = 1, C_t = 1, Q_t = 0, B_{u,t} = 0$. Furthermore, if the mean before the change is supposed to be 0, a case often considered in the literature (see Basseville and Nikiforov (1993)), we have $x_0 = 0$ and $\Pi_0 = 0$.

It is worth mentioning, that parametric models from Part III can fit this framework as well.

Example 9.2 Modeling a change in an ARX model

By letting $A_t = I$ and $C_t = (y_{t-1}, y_{t-2}, \ldots, u_{t-1}, u_{t-2}, \ldots)$, a special case of equation (9.1) is obtained. We then have a linear regression description of an ARX model, where x_t is the (time-varying) parameter vector and C_t the regressors. In this way, we can detect abrupt changes in the transfer function of ARX models. Note that the change occurs in the *dynamics* of the system in this case, and not in the system's *state*.

9.2.2. Likelihood

The likelihood for the measurements up to time N given the change ν at time k is denoted $p(y^N|k, \nu)$. The same notation is used for the conditional density function for y^N, given k, ν. For simplicity, $k = N$ is agreed to mean no change. There are two principally different possibilities to estimate the change time k:

- Joint ML estimate of k and ν,

$$\widehat{(k, \nu)} = \arg \max_{k \in [1,N], \nu} p(y^N|k, \nu). \tag{9.5}$$

  Here $\arg\max_{k\in[1,N],\nu} p(y^N|k,\nu)$ means the maximizing arguments of the likelihood $p(y^N|k, \nu)$ where k is restricted to $[1, N]$.

- The ML estimate of just k using marginalization of the conditional density function $p(y^N|k,\nu)$:

$$p(y^N|k) = \int p(y^N|k,\nu)p(\nu)d\nu \tag{9.6}$$

$$\hat{k} = \arg \max_{k \in [1,N]} p(y^N|k). \tag{9.7}$$

The likelihood for data given just k in (9.6) is the starting point in this approach.

A tool in the derivations is the so-called flat prior, of the form $p(\nu) = C$, which is not a proper density function. See Section 7.3.3 for a discussion and two examples for the parametric case, whose conclusions are applicable here as well.

9.2.3. Likelihood ratio

In the context of hypothesis testing, the likelihood ratios rather than the likelihoods are used. The LR test is a multiple hypotheses test, where the different change hypotheses are compared to the no change hypothesis pairwise. In the LR test, the change magnitude is assumed to be known. The hypotheses under consideration are

$$\begin{aligned} H_0 : &\quad \text{no change} \\ H_1(k,\nu) : &\quad \text{a change of magnitude } \nu \text{ at time } k. \end{aligned}$$

The test is as follows. Introduce the log likelihood ratio for the hypotheses as the test statistic:

$$l_N(k,\nu) \triangleq 2\log \frac{p(y^N|H_1(k,\nu))}{p(y^N|H_0)} = 2\log \frac{p(y^N|k,\nu)}{p(y^N|k=N)}. \tag{9.8}$$

The factor 2 is just for notational convenience. We use the convention that $H_1(N,\nu) = H_0$, so again, $k = N$ means no change. Then the LR estimate can be expressed as

$$\hat{k}^{ML} = \arg \max_k l_N(k,\nu), \tag{9.9}$$

when ν is known. Exactly as in (9.5) and (9.7), we have two possibilities of how to eliminate the unknown nuisance parameter ν. Double maximization gives the GLR test, proposed for change detection in Willsky and Jones (1976), and marginalization the MLR test, proposed in Gustafsson (1996).

9.3. The GLR test

Why not just use the augmented state space model (9.3) and the Kalman filter equations in (9.4)? It would be straightforward to evaluate the likelihood ratios in (9.8) for each possible k. The answer is as follows:

> The GLR *algorithm* is mainly a computational tool that splits the Kalman filter for the full order model (9.3) into a low order Kalman filter (which is perhaps already designed and running) and a cascade coupled filter bank with least squares filters.

The GLR test proposed in Willsky and Jones (1976) utilizes this approach. GLR's general applicability has contributed to it now being a standard tool in change detection. As summarized in Kerr (1987), GLR has an appealing analytic framework, is widely understood by many researchers and is readily applicable to systems already utilizing a Kalman filter. Another advantage with GLR is that it partially solves the *isolation* problem in fault detection, i.e. to locate the physical cause of the change. In Kerr (1987), a number of drawbacks with GLR is pointed out as well. Among these, we mention problems with choosing decision thresholds, and for some applications an untenable computational burden.

The use of likelihood ratios in hypothesis testing is motivated by the *Neyman–Pearson Lemma*; see, for instance, Theorem 3.1 in Lehmann (1991). In the application considered here, it says that the likelihood ratio is the optimal test statistic when the change magnitude is known and just one change time is considered. This is not the case here, but a sub-optimal extension is immediate: the test is computed for each possible change time, or a restriction to a sliding window, and if several tests indicate a change the most significant is taken as the estimated change time. In GLR, the actual change in the state of a linear system is estimated from data and then used in the likelihood ratio.

Starting with the likelihood ratio in (9.8), the GLR test is a double maximization over k and ν,

$$\hat{\nu}(k) = \arg\max_{\nu} 2\log\frac{p(y^N|k,\nu)}{p(y^N|k=N)}$$
$$\hat{k} = \arg\max_{k} 2\log\frac{p(y^N|k,\hat{\nu}(k))}{p(y^N|k=N)},$$

where $\hat{\nu}(k)$ is the maximum likelihood estimate of ν, given a change at time k. The change candidate $\hat{k}$ in the GLR test is accepted if

$$l_N(\hat{k},\hat{\nu}(\hat{k})) > h. \tag{9.10}$$

The threshold h characterizes a hypothesis test and distinguishes the GLR test from the ML method (9.5). Note that (9.5) is a special case of (9.10), where $h = 0$. If the zero-change hypothesis is rejected, the state estimate can easily be compensated for the detected change.

The idea in the implementation of GLR in Willsky and Jones (1976) is to make the dependence on ν explicit. This task is solved in Appendix 9.A. The key point is that the innovations from the Kalman filter (9.4) with $k = N$ can be expressed as a linear regression in ν,

$$\varepsilon_t(k) = \varphi_t^T(k)\nu + \varepsilon_t,$$

where $\varepsilon_t(k)$ are the innovations from the Kalman filter if ν and k were known. Here and in the sequel, non-indexed quantities as ε_t are the output from the nominal Kalman filter, assuming no change. The GLR algorithm can be implemented as follows.

Algorithm 9.1 GLR

Given the signal model (9.1):

- Calculate the innovations from the Kalman filter (9.4) assuming no change.
- Compute the regressors $\varphi_t(k)$ using

$$\varphi_{t+1}^T(k) = C_{t+1}\left(\prod_{i=k}^{t} A_i - A_t\mu_t(k)\right)$$
$$\mu_{t+1}(k) = A_t\mu_t(k) + K_{t+1}\varphi_{t+1}^T(k),$$

  initialized by zeros at time $t = k$; see Lemma 9.7. Here φ_t is $n_x \times 1$ and μ_t is $n_x \times n_x$.
- Compute the linear regression quantities

$$R_t(k) = \sum_{i=1}^{t} \varphi_i(k)S_i^{-1}\varphi_i^T(k)$$
$$f_t(k) = \sum_{i=1}^{t} \varphi_i(k)S_i^{-1}\varepsilon_i$$

  for each k, $1 \le k \le t$.
- At time $t = N$, the test statistic is given by

$$l_N(k, \hat{\nu}(k)) = f_N^T(k)R_N^{-1}(k)f_N(k).$$

- A change candidate is given by $\hat{k} = \arg\max l_N(k, \hat{\nu}(k))$. It is accepted if $l_N(\hat{k}, \hat{\nu}(\hat{k}))$ is greater than some threshold h (otherwise $\hat{k} = N$) and the corresponding estimate of the change magnitude is given by $\hat{\nu}_N(\hat{k}) = R_N^{-1}(\hat{k}) f_N(\hat{k})$.

We now make some comments on the algorithm:

- It can be shown that the test statistic $l_N(k, \nu(k))$ under the null hypothesis is χ^2 distributed. Thus, given the confidence level on the test, the threshold h can be found from standard statistical tables. Note that this is a multiple hypothesis test performed for each $k = 1, 2, \ldots, N-1$, so nothing can be said about the total confidence level.

- The regressor $\varphi_t(k)$ is called a *failure signature matrix* in Willsky and Jones (1976).

- The regressors are pre-computable. Furthermore, if the system and the Kalman filter are time-invariant, the regressor is only a function of $t - k$, which simplifies the calculations.

- The formulation in Algorithm 9.1 is off-line. Since the test statistic involves a matrix inversion of R_N, a more efficient on-line method is as follows. From (9.34) and (9.37) we get

$$l_t(k, \hat{\nu}(k)) = f_t^T(k) \hat{\nu}_t(k),$$

  where t is used as time index instead of N. The *Recursive Least Squares* (RLS) scheme (see Algorithm 5.3), can now be used to update $\hat{\nu}_t(k)$ recursively, eliminating the matrix inversion of $R_t(k)$. Thus, the best implementation requires t parallel RLS schemes and one Kalman filter.

The choice of threshold is difficult. It depends not only upon the system's signal-to-noise ratio, but also on the actual noise levels, as will be pointed out in Section 9.4.3.

Example 9.3 DC motor: the GLR test

Consider the DC motor in Example 8.4. Assume impulsive additive state changes at times 60, 80, 100 and 120. First the angle is increased by five units, and then decreased again. Then the same fault is simulated on angular velocity. That is,

$$\nu_1 = \begin{pmatrix} 5 \\ 0 \end{pmatrix}, \nu_2 = \begin{pmatrix} -5 \\ 0 \end{pmatrix}, \nu_3 = \begin{pmatrix} 0 \\ 5 \end{pmatrix}, \nu_4 = \begin{pmatrix} 0 \\ -5 \end{pmatrix}.$$

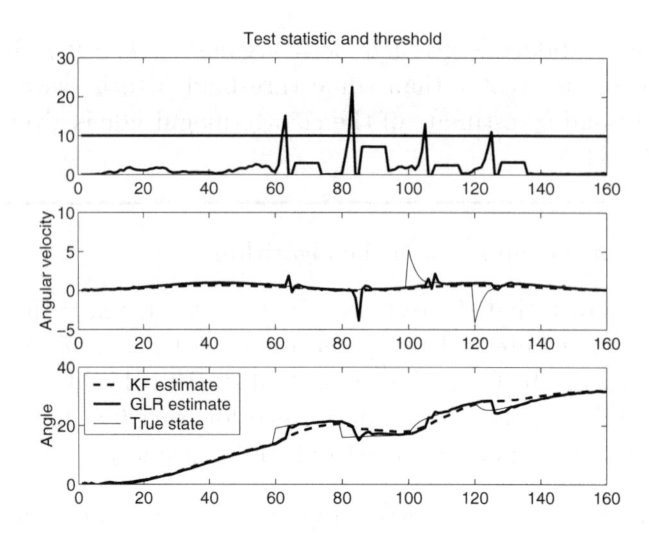

Figure 9.2. Test statistic $\max_{t-L\leq k<t} l_t(k)$ for change detection on a simulated DC motor (first) and state estimates from GLR test and the Kalman filter (second and third). Note, in particular, the improved angle tracking of GLR.

Figure 9.2(a) shows how the maximum value $\max_{t-L\leq k<t} l_t(k)$ of the test statistics evolves in time, and how it exceeds the threshold level $h = 10$ four times. The delay for detection is three samples for angular change and five samples for velocity change.

The GLR state estimate adapts to the true state as shown in Figure 9.2(b). The Kalman filter also comes back to the true state, but much more slowly. The change identification is not very reliable:

$$\hat{\nu}_1 = \begin{pmatrix} 1.81 \\ 2.35 \end{pmatrix}, \hat{\nu}_2 = \begin{pmatrix} -2.53 \\ -2.47 \end{pmatrix}, \hat{\nu}_3 = \begin{pmatrix} 2.32 \\ 0.96 \end{pmatrix}, \hat{\nu}_4 = \begin{pmatrix} -1.25 \\ -1.74 \end{pmatrix}.$$

Compared to the simulated changes, these look like random numbers. The explanation is that detection is so fast that there are too few data for fault estimation. To get good isolations, we have to wait and get considerably more data. The incorrect compensation explains the short transients we can see in the angular velocity estimate.

Navigation examples and references to such are presented in Kerr (1987). As a non-standard application, GLR is applied to noise suppression in image processing in Hong and Brzakovic (1980).

9.4. The MLR test

Another alternative is to consider the change magnitude as a stochastic nuisance parameter. This is then eliminated not by estimation, but by *marginalization*. Marginalization is wellknown in estimation theory, and is also used in other detection problems; see, for instance, Wald (1950). The resulting test will be called the *Marginalized Likelihood Ratio* (*MLR*) test. The MLR test applies to all cases where GLR does, but we point out three advantages with using the former:

- *Tuning.* Unlike GLR, there is no sensitive threshold to choose in MLR. One interpretation is that a reasonable threshold in GLR is chosen automatically.

- *Robustness to modeling errors.* The performance of GLR deteriorates in the case of incorrectly chosen noise variances. The noise level in MLR is allowed to be considered as another unknown nuisance parameter. This approach increases the robustness of MLR.

- *Complexity.* GLR requires a linearly increasing number of parallel filters. An approximation involving a sliding window technique is proposed in Willsky and Jones (1976) to obtain a constant number of filters, typically equivalent to 10–20 parallel filters. For off-line processing, the MLR test can be computed exactly from only two filters. This implementation is of particularly great impact in the design step. Here the false alarm rate, robustness properties and detectability of different changes can be evaluated quickly using Monte-Carlo simulations. In fact, the computation of one single exact GLR test for a realistic data size (> 1000) is already far from inter-active.

9.4.1. Relation between GLR and MLR

In Appendix 9.B the MLR test is derived using the quantities from the GLR test in Algorithm 9.1. This derivation gives a nice relationship between GLR and MLR. In fact, they coincide for a certain choice of threshold.

Lemma 9.1
If (9.1) is time invariant and ν is unknown, then the GLR test in Algorithm 9.1 gives the same estimated change time as the MLR test in Theorem 9.8 as $N-k\to\infty$ and $k\to\infty$ if the threshold is chosen as

$$h = p\log(2\pi) + \log\det \bar{R}_N(k) - p_\nu(\hat{\nu})$$

when the prior of the jump is $\nu \in \mathrm{N}(\nu_0, P_\nu)$, *and*

$$h = \log\det \bar{R}_N(k)$$

for a flat prior. Here $\bar{R}_N(k) = \lim_{N-k\to\infty, k\to\infty} R_N(k)$, *and* $R_N(k)$ *is defined in Algorithm 9.1.*

Proof: In the MLR test a change k is detected if $l_N(k) > l_N(N) = 0$ and in the GLR if $l_N(k, \nu(k)) > h$. From Theorem 9.8 we have $l_N(k) = l_N(k, \nu(k)) + 2\log p_\nu(\hat{\nu}) - \log\det R_N(k) - p\log(2\pi)$. Lemma 9.9 shows that $R_N(k)$ converges as $N \to \infty$, and so does $\log\det R_N(k)$. Since (9.1) is restricted to be time invariant the terms of $R_N(k)$ that depend on the system matrices and the Kalman gain are the same independently of k as $k \to \infty$ according to (9.28). □

Note that (3.48) follows with $p = 1$ and $R(k) = \sum_{t=k+1}^{N} 1^T R^{-1} 1 = (t-k)/R$.

The threshold is thus automatically included in the MLR test. If we want MLR to mimic a GLR test, we can of course include an external threshold $h_{MLR} = h_{GLR} + 2\log p_\nu(\hat{\nu}) - p\log(2\pi) - \log\det \bar{R}_N(k)$. In that case, we accept the change hypothesis only if $l_N(k) > h_{MLR}$. The external threshold can also be included in an *ad hoc* manner to tune the false alarm rate versus probability of correct detection.

We now make a new derivation of the MLR test in a direct way using a linearly increasing number of Kalman filters. This derivation enables first the efficient implementation in the Section 9.4.2, and secondly, the elimination of noise scalings in Section 9.4.3. Since the magnitudes of the likelihoods turn out to be of completely different orders, the log likelihood will be used in order to avoid possible numerical problems.

Theorem 9.2
Consider the signal model (9.1), where the covariance matrix of the Gaussian distributed jump magnitude is P_ν. *For each* $k = 1, 2, \ldots, t$, *update the* k*'th Kalman filter in (9.4). The log likelihood, conditioned on a jump at time* k, *can be recursively computed by*

$$\begin{aligned}\log p(y^t|k) =& \log p(y^{t-1}|k) - \frac{p}{2}\log 2\pi \\ &- \frac{1}{2}\log\det S_t(k) - \frac{1}{2}\varepsilon_t^T(k) S_t^{-1}(k)\varepsilon_t(k).\end{aligned} \tag{9.11}$$

Proof: By Bayes' rule we have

$$p(y^t) = p(y_t|y^{t-1})p(y^{t-1}).$$

It is a well-known property of the Kalman filter; see, for instance, Anderson and Moore (1979), that

$$y_t|k \in \mathrm{N}(C_t\hat{x}_{t|t-1}(k), C_tP_{t|t-1}(k)C_t^T + R_t),$$

and the result follows from the definition of the Gaussian density function. □

This approach requires a linearly growing number with N Kalman filters.

9.4.2. A two-filter implementation

To compute the likelihood ratios efficiently, two statistical tricks are needed:

- Use a flat prior on the jump magnitude ν.
- Use some of the last observations for calculating proper distributions.

The point with the former is that the measurements after the jump are independent of the measurements before the jump, and the likelihood can be computed as a product of the likelihoods before and after the jump. However, this leads to a problem. The likelihood is not uniquely defined immediately after a jump of infinite variance. Therefore, a small part of the data is used for initialization. We also have to assume that A_t in (9.1) is invertible.

The key point in the derivation is the backward model presented in Chapter 8 when discussing smoothing algorithms. The problem here, which is not apparent in smoothing, is that the 'prior' $\Pi_N = \mathrm{E}[x_N x_N^T]$ in the backward recursion generally depends upon k, so we must be careful in using a common Kalman filter for all hypotheses. For this reason, the assumption on infinite variance of the jump magnitude is needed, so Π_N is infinite for all k as well. By infinite we mean that $\Pi_N^{-1} = 0$. The recursion $\Pi_{t+1} = F\Pi_t F^T + Q$ gives $\Pi_{k+1}^{-1} = 0$. The backward model for non-singular A_t becomes

$$\begin{aligned} x_t &= A_t^{-1}x_{t+1} - A_t^{-1}v_t = A_t^{-1}x_{t+1} + v_t^B \\ y_t &= C_t x_t + e_t. \end{aligned} \tag{9.12}$$

Here $Q_t^B = \mathrm{E}[v_t^B(v_t^B)^T] = A_t^{-1}Q_tA_t^{-T}$ and $\Pi_N^{-1} = 0$, where $\Pi_N = \mathrm{E}[x_N x_N^T]$. We now have the backward model and can simply apply the Kalman filter for the estimate $x_{t|t+1}^B$ and its covariance matrix $P_{t|t+1}^B$.

The likelihoods rather than likelihood ratios will be derived. The last L measurements are used for normalization, which means that jumps after time $N-L$ are not considered. This is not a serious restriction, since it suffices to choose $L = \dim x$, and jumps supported by so little data cannot be detected with any significance in any case.

We are now ready for the main result of this section.

Theorem 9.3
Consider the signal model (9.1) for the case of an invertible A_t. The likelihood for the measurements conditioned on a jump at time k and the last L measurements, can be computed by two Kalman filters as follows. First, the likelihoods are separated,

$$p(y^{N-L}|y_{N-L+1}^N, k) = \begin{cases} \frac{p(y^N)}{p(y_{N-L+1}^N)} & \text{if } k = N \\ p(y^k)p(y_{k+1}^{N-L}|y_{N-L+1}^N, k) & \text{if } k < N - L \end{cases} \tag{9.13}$$

The likelihoods involved are computed by

$$p(y^k) = \prod_{t=1}^{k} \gamma(y_t - C_t\hat{x}_{t|t-1}^F, C_t P_{t|t-1}^F C_t^T + R_t) \tag{9.14}$$

$$p(y_{N-L+1}^N) = \prod_{t=N-L+1}^{N} \gamma(y_t - C_t\hat{x}_{t|t-1}^N, C_t P_{t|t-1}^N C_t^T + R_t) \tag{9.15}$$

$$p(y_{k+1}^{N-L}|y_{N-L+1}^N, k) = \prod_{t=k+1}^{N-L} \gamma(y_t - C_t\hat{x}_{t|t+1}^B, C_t P_{t|t+1}^B C_t^T + R_t). \tag{9.16}$$

Here $\gamma(x - \mu, P)$ is the Gaussian probability density function. The quantities $\hat{x}_{t|t-1}^F$ and $P_{t|t-1}^F$ are given by the Kalman filter applied to the forward model and $\hat{x}_{t|t+1}^B$ and $P_{t|t+1}^B$ are given by the Kalman filter applied on the backward model (9.12). The quantities $\hat{x}_{t|t-1}^N$ and $P_{t|t-1}^N$ used for normalization are given by the Kalman filter applied on the forward model initiated at time $t = N - L + 1$ with $P_{N-L+1|N-L} = \Pi_{N-L+1}$.

Proof: Bayes' law gives

$$p(y^{N-L}|y_{N-L+1}^N, k = N) = \frac{p(y^{N-L} \cap y_{N-L+1}^N | k = N)}{p(y_{N-L+1}^N | k = N)} \tag{9.17}$$

$$= \frac{p(y^N | k = N)}{p(y_{N-L+1}^N | k = N)} \tag{9.18}$$

$$\cdot\, p(y^{N-L}|y_{N-L+1}^N, k < N - L)$$

$$= p(y^k | k < N - L) p(y_{k+1}^{N-L} | y^k, y_{N-L+1}^N, k) \tag{9.19}$$

$$= p(y^k) p(y_{k+1}^{N-L} | y_{N-L+1}^N, k). \tag{9.20}$$

The fact that the jump at time k does not affect the measurements before time k (by causality) is used in the last equality, so $p(y^k|k) = p(y^k)$. Here, the infinite variance jump makes the measurements after the jump independent of those before.

The likelihood for a set y_m^n can be expanded either forwards or backwards using Bayes' chain rule:

$$p(y_m^n) = \prod_{t=m}^{n} p(y_t|y_m^{t-1}) \tag{9.21}$$

$$p(y_m^n) = \prod_{t=m}^{n} p(y_t|y_{t+1}^n). \tag{9.22}$$

Now $p(y^N|k = N)$ and $p(y^k)$ are computed using the forward recursion (9.21), and since x_t is Gaussian, it follows immediately that $y_t|y^{t-1}$ is Gaussian with mean $C_t\hat{x}^F_{t|t-1}$ and covariance $C_t P^F_{t|t-1} C_t^T + R_t$, and (9.14) follows.

Also, $p(y^N_{N-L+1}|k = N)$ is computed in the same way; the difference is that the Kalman filter is initiated at time $N - L + 1$. Finally, $p(y_{k+1}^{N-L}|y^N_{N-L+1}, k)$ is computed using (9.22) where $y_t|y^N_{t+1}$ is Gaussian with mean $C_t\hat{x}^B_{t|t+1}$ and covariance $C_t P^B_{t|t+1} C_t^T + R_t$ and (9.16) follows. $\square$

As can be seen, all that is needed to compute the likelihoods are one Kalman filter running backwards in time, one running forwards in time, and one processing the normalizing data at the end. The resulting algorithm is as follows, where the log likelihoods are used because of possible numerical problems caused by very large differences in the magnitude of the likelihoods. The notation introduced here will be used in the sequel.

Algorithm 9.2 Two-filter detection

The likelihood given in Theorem 9.3 of a jump at time k, $k = 1, 2, \dots, N$, is computed with two filters as follows.

Forward filter for $t = 1, 1, \dots, N$:

$$
\begin{aligned}
\hat{x}^F_{1|0} =& x_0 \\
P^F_{1|0} =& \Pi_0 \\
V^F(0) =& 0 \\
D^F(0) =& 0 \\
\varepsilon^F_t =& y_t - C_t \hat{x}^F_{t|t-1} - D_{u,t} u_t \\
S^F_t =& C_t P^F_{t|t-1} C_t^T + R_t \\
\hat{x}^F_{t+1|t} =& A_t \hat{x}^F_{t|t-1} + A_t P^F_{t|t-1} C_t (S^F_t)^{-1} \varepsilon^F_t + B_{u,t} u_t \\
P^F_{t+1|t} =& A_t \left(P^F_{t|t-1} - P^F_{t|t-1} C_t^T (S^F_t)^{-1} C_t P^F_{t|t-1} \right) A_t^T + B_{v,t} Q_t B_{v,t}^T \\
\Pi_{t+1} =& A_t \Pi_t A_t^T + B_{v,t} Q_t B_{v,t}^T \\
V^F(t) =& V^F(t-1) + (\varepsilon^F_t)^T (S^F_t)^{-1} \varepsilon^F_t \\
D^F(t) =& D^F(t-1) + \log\det S^F_t
\end{aligned}
$$

Normalization filter for $t = N - L + 1, N - L + 2, \dots, N$:

$$
\begin{aligned}
\hat{x}^N_{N-L+1|N-L} =& \left(\prod_{t=1}^{N-L} A_t \right) x_0 \\
P^N_{N-L+1|N-L} =& \Pi_{N-L} \\
V^N(N-L) =& 0 \\
D^N(N-L) =& 0 \\
\varepsilon^N_t =& y_t - C_t \hat{x}^N_{t|t-1} - D_{u,t} u_t \\
S^N_t =& C_t P^N_{t|t-1} C_t^T + R_t \\
\hat{x}^N_{t+1|t} =& A_t \hat{x}^N_{t|t-1} + A_t P^N_{t|t-1} C_t (S^N_t)^{-1} \varepsilon^N_t + B_{u,t} u_t \\
P^N_{t+1|t} =& A_t \left(P^N_{t|t-1} - P^N_{t|t-1} C_t^T (S^N_t)^{-1} C_t P^N_{t|t-1} \right) A_t^T + B_{v,t} Q_t B_{v,t}^T \\
V^N(t) =& V^N(t-1) + (\varepsilon^N_t)^T (S^N_t)^{-1} \varepsilon^N_t \\
D^N(t) =& D^N(t-1) + \log\det(S^N_t)
\end{aligned}
$$

Backward information filter for $t = N, N-1, \ldots, N-L+1$:

$$\begin{aligned}
\hat{a}^B_{N|N+1} =& 0 \\
(P^B_{N|N+1})^{-1} =& 0 \\
\hat{a}^B_{t|t} =& \hat{a}^B_{t|t+1} + C_t^T R_t^{-1}(y_t - D_{u,t}u_t) \\
(P^B_{t|t})^{-1} =& (P^B_{t|t+1})^{-1} + C_t^T R_t^{-1} C_t \\
F =& A_t^T (P^B_{t|t})^{-1} A_t \\
G =& F A_t^{-1} B_{v,t} \left(B^T_{v,t} A_t^{-T} F A_t^{-1} B_{v,t} + Q_t^{-1} \right)^{-1} \\
\hat{a}^B_{t-1|t} =& \left(I - G B^T_{u,t} A_t^{-T} \right) A_t^T \hat{a}^B_{t|t} \\
(P^B_{t-1|t})^{-1} =& \left(I - G B^T_{u,t} A_t^{-T} \right) F
\end{aligned}$$

Backward Kalman filter for $t = N-L, N-L-1, \ldots, 1$:

$$\begin{aligned}
P^B_{N-L|N-L+1} & \text{from backward information filter} \\
\hat{x}^B_{N-L|N-L+1} =& P^B_{N-L|N-L+1} \hat{a}^B_{N-L|N-L+1} \\
V^B(N-L+1) =& 0 \\
D^B(N-L+1) =& 0 \\
\varepsilon^B_t =& y_t - C_t \hat{x}^B_{t|t+1} - D_{u,t} u_t \\
S^B_t =& C_t P^B_{t|t+1} C_t^T + R_t \\
\hat{x}^B_{t-1|t} =& A_t^{-1} \hat{x}^B_{t|t+1} + A_t^{-1} P^B_{t|t+1} C_t (S^B_t)^{-1} \varepsilon^B_t - B_{u,t-1} u_{t-1} \\
P^B_{t-1|t} =& A_t^{-1} \left(P^B_{t|t+1} - P^B_{t|t+1} C_t^T (S^B_t)^{-1} C_t P^B_{t|t+1} \right) A_t^{-T} \\
& + A_t^{-1} Q_t A_t^{-T} \\
V^B(t) =& V^B(t+1) + (\varepsilon^B_t)^T (S^B_t)^{-1} \varepsilon^B_t \\
D^B(t) =& D^B(t+1) + \log\det S^B_t
\end{aligned}$$

Compute the likelihood ratios

$$\begin{aligned}
l_N(k) =& V^F(N) - V^N(N) - V^F(k) - V^B(k+1) + \\
& D^F(N) - D^N(N) - D^F(k) - D^B(k+1)
\end{aligned}$$

We make some remarks on the algorithm:

- The normalization filter and the backward information filter play a minor role for the likelihood, and might be omitted in an approximate algorithm.
- The jump magnitude, conditioned on the jump time, can be estimated from the available information using the signal model (9.1) and the filter state estimates:

$$B_{\nu,k}\hat{\nu}(k) = \hat{x}^B(k+1|k+1) - A_k\hat{x}^F(k|k) - B_{u,k}u_k.$$

  This is an over-determined system of equations.
- The *a posteriori* probability of k is easily computed by using Bayes' law. Assuming $p(k) = C$,

$$p(k|y^N) = \frac{p(k)p(y^N|k)}{p(y^N)} = \frac{p(y^N|k)}{\sum_{i=1}^N p(y^N|i)}.$$

  This means that the *a posteriori* probability of a wrong decision can be computed as $1 - p(\hat{k}|y^N)$.
- The relation to fixed-interval smoothing is as follows. The smoothed estimates under the no jump hypothesis can be computed by

$$P_{t|N} = \left((P^F_{t|t})^{-1} + (P^B_{t|t+1})^{-1}\right)^{-1}$$
$$\hat{x}_{t|N} = P_{t|N}\left((P^F_{t|t})^{-1}\hat{x}^F_{t|t} + (P^B_{t|t+1})^{-1}\hat{x}^B_{t|t+1}\right).$$

  Here $\hat{x}^F_{t|t}$ and $P^F_{t|t}$ are the filtered estimates from the forward filter (these are not given explicitly above).
- If the data are collected in batches, the two-filter algorithm can be applied after each batch saving computation time.

It should be stressed for the theorem that it is necessary for this two-filter implementation that the jump is considered as stochastic with infinite variance, which implies the important separability possibility (9.13). If not, the theorem will provide a sub-optimal algorithm with good properties in practice.

A related two-filter idea is found in Niedzwiecki (1991), where a sub-optimal two-filter detection algorithm is proposed for detecting changes in the parameters of a finite impulse response model.

9.4.3. Marginalization of the noise level

Introduction

Knowledge of the covariance matrices is crucial for the performance of model-based detectors. The amount of prior information can be substantially relaxed, by eliminating unknown scalings from the covariance matrices in the state space model (9.1):

$$\overline{R} = \lambda R, \quad \overline{P}_0 = \lambda P_0, \quad \overline{Q} = \lambda Q. \tag{9.23}$$

Here the matrices without a bar are chosen by the user, and those with bar are the 'true' ones or at least give good performance. This means that one chooses the tracking ability of the filter, which is known to be insensitive to scalings; see Anderson and Moore (1979). The estimator then estimates the actual level, which is decisive for the likelihoods. The assumption (9.23) implies $\overline{P}_t = \lambda P_t$ and $\overline{S}_t = \lambda S_t$, and from (9.36) it follows that

$$\bar{l}_N(k) = l_N(k)/\lambda.$$

For the GLR test, this implies that if all covariance matrices are scaled a factor λ, then the optimal threshold should be scaled a factor λ as well. Thus, it is the ratio between the noise variance and the threshold that determines the detection ability in GLR. In this sense, the problem formulation is over-parameterized, since both λ and the threshold have to be chosen by the user.

Equation (9.23) is an interesting assumption from the practical point of view. Scaling does not influence the filtered state estimates. It is relatively easy for the user to tune the tracking ability, but the sizes of the covariances are harder to judge. The robustness to unknown scalings is one of the advantages with the MLR test, as will be shown.

Another point is that a changing measurement noise variance is known to cause problems to many proposed detection algorithms. This is treated here by allowing the noise covariance scaling to be abruptly changing.

A summary of the MLRs is given in Section 9.4.5.

State jump

In this section, the two filter detector in Theorem 9.3 will be derived for the unknown scaling assumption in (9.23). If A_t is not invertible as assumed in Theorem 9.3, the direct implementation of MLR in Theorem 9.2 can also be modified in the same way. The following theorem is the counterpart to the two filter detection method in Algorithm 9.2, for the case of an unknown λ.

Theorem 9.4
Consider the signal model (9.1) in the case of an unknown noise variance, and suppose that (9.23) holds. With notation as in the two filter detector in Algorithm 9.2, the log likelihood ratios are given by

$$\begin{aligned} l_N(k) =&(Np-2)\left(\log(V^F(N)-V^N(N))\right.\\ &\left.-\log(V^F(k)+V^B(k+1))\right)\\ &+D^F(N)-D^N(N)-D^F(k)-D^B(k+1)\end{aligned}$$

with a flat prior on λ.

Proof: The flat prior on λ here means $p(\lambda|y_{N-L+1}^N)=C$, corresponding to being completely unknown. By marginalization, if $k<N-L$ we get

$$\begin{aligned} &p(y^{N-L}|y_{N-L+1}^N,k)\\ =&\int_0^\infty p(y^{N-L}|y_{N-L+1}^N,k,\lambda)p(\lambda|y_{N-L+1}^N)d\lambda\\ =&C\int_0^\infty p(y^k|\lambda)p(y_{k+1}^{N-L}|y_{N-L+1}^N,k,\lambda)d\lambda\\ =&C\int_0^\infty (2\pi)^{-Np/2}\exp\left(-\frac{1}{2}(D^F(k)+D^B(k))\right)\lambda^{-Np/2}\\ &\cdot\exp\left(-\frac{V^F(k)+V^B(k)}{2\lambda}\right)d\lambda\\ =&C(2\pi)^{-Np/2}\exp\left(-\frac{1}{2}(D^F(k)+D^B(k))\right)2^{\frac{Np-2}{2}}\\ &\cdot\Gamma\left(\frac{Np-2}{2}\right)(V^F(k)+V^B(k))^{-\frac{Np-2}{2}}\\ &\cdot\int_0^\infty\frac{(V^F(k)+V^B(k))^{\frac{Np-2}{2}}}{2^{(Np-2)/2}\Gamma\left(\frac{Np-2}{2}\right)}\frac{\exp\left(-\frac{V^F(k)+V^B(k)}{2\lambda}\right)}{\lambda^{(Np-2+2)/2}}d\lambda\\ =&C(2\pi)^{-Np/2}\exp\left(-\frac{1}{2}(D^F(k)+D^B(k))\right)2^{\frac{Np-2}{2}}\\ &\cdot\Gamma\left(\frac{Np-2}{2}\right)(V^F(k)+V^B(k))^{-\frac{Np-2}{2}}.\end{aligned}$$

The gamma function is defined by $\Gamma(a)=\int_0^\infty e^{-t}t^{a-1}dt$. The last equality follows by recognizing the integrand as a density function, namely the inverse Wishart distribution

$$f(\lambda)=\frac{\sigma^{m/2}}{2^{m/2}\Gamma(m/2)}\frac{\exp\left(-\frac{\sigma}{2\lambda}\right)}{\lambda^{(m+2)/2}} \tag{9.24}$$

which integrates to one. In the same way, we have for $k = N$

$$p(y^{N-L}|y_{N-L+1}^N, k=N) = C(2\pi)^{-Np/2} \exp\left(-\frac{1}{2}(D^F(N) - D^N(N))\right) \cdot 2^{\frac{Np-2}{2}} \Gamma\left(\frac{Np-2}{2}\right) (V^F(N) - V^N(N))^{-\frac{Np-2}{2}},$$

and the result follows. □

Here we note that the *a posteriori* distribution for λ, given the jump instant k, is $W^{-1}(Np, V^F(k) + V^B(k))$, where W^{-1} denotes the inverse Wishart distribution (9.24).

9.4.4. State and variance jump

In this section, the likelihood is given for the case when the noise variance is different before and after the jump. This result is of great practical relevance, since variance changes are very common in real signals.

Theorem 9.5
Consider the same detection problem as in Theorem 9.4, but with a noise variance changing at the jump instant,

$$\lambda = \begin{cases} \lambda_1 & , \text{if } t \le k \\ \lambda_2 & , \text{if } k < t \le N. \end{cases}$$

With notation as in Algorithm 9.2, the log likelihood ratios for $2/p < k < N - 2/p$ are given by

$$\begin{aligned} l_N(k) = &(Np-2)\log(V^F(N) - V^N(N)) \\ &- (kp-2)\log V^F(k) - (Np - kp - 2)\log V^B(k+1)) \\ &+ D^F(N) - D^N(N) - D^F(k) - D^B(k+1) \end{aligned}$$

if the prior on λ is flat.

Proof: The proof resembles that of Theorem 9.4. The difference is that the integral over λ is split into two integrals,

$$\begin{aligned} p(y^{N-L}|y_{N-L+1}^N, k) = &p(y^k)p(y_{k+1}^N|y_{N-L+1}^N, k) \\ = &\int p(y^k|\lambda_1)p(\lambda_1|y_{N-L+1}^N)d\lambda_1 \\ &\cdot \int p(y_{k+1}^N|y_{N-L+1}^N, k, \lambda_2)p(\lambda_2|y_{N-L+1}^N)d\lambda_2. \end{aligned} \quad (9.25)$$

Each integral is evaluated exactly as in Theorem 9.4. □

Remark 9.6
For this particular prior on λ, the integrals in (9.25), and thus also the likelihood, are not defined for $kp \leq 2$ and $Np - kp \leq 2$. This is logical, because too little data are available to evaluate the noise variance.

In this case the *a posteriori* distribution for λ_1, given the jump instant k, is $W^{-1}(kp, V^F(k))$ and for λ_2 it is $W^{-1}((N-k)p, V^B(k))$.

9.4.5. Summary

We can conveniently summarize the results in the three different cases as follows: The MLR's in Theorems 9.4, 9.5 and Algorithm 9.2 are given by

$$
l_N(k) = D^F(N) - D^N(N) - D^F(k) - D^B(k+1)
\begin{cases}
+V^F(N) - V^N(N) - V^B(k) - V^B(k+1) & (i) \\
\\
+(Np-2)\log(V^F(N) - V^N(N)) \\
-(Np-2)\log(V^F(k) - V^B(k+1)) & (ii) \\
\\
+(Np-2)\log(V^F(N) - V^N(N)) \\
-(kp-2)\log V^F(k) - (Np-kp-2)V^B(k+1) & (iii)
\end{cases}
$$

in the cases known lambda (i), unknown constant lambda (ii) and unknown changing lambda (iii), respectively.

- The model and data dependent quantities $V(k)$ and $D(k)$ are all given by the two filter Algorithm 9.2. The decision for which likelihood is to be used can be deferred until after these quantities are computed. In particular, all three possibilities can be examined without much extra computations.
- The dominating terms are $V^F(k)$ and $V^B(k)$. When the noise is unknown, $(V^F(k) + V^B(k))/\lambda$ is essentially replaced by $N\log(V^F(k) + V^B(k))$ and $k\log V^F(k) + (N-k)\log V^B(k)$, respectively. This leads to a more cautious estimator that diminishes the influence of the innovations.
- The term $D^F(N) - D^N(N) - D^F(k) - D^B(k+1)$ appears in all likelihood ratios. It is positive and does not vary much for different jump instants $k = 1, 2, \ldots, N-1$. This term corresponds to the threshold in the GLR test.

These methods are compared in the following section.

9.5. Simulation study

The simulation study investigates performance, robustness and sensitivity for GLR and the three different MLR variants for a first order motion model.

The applicability of GLR is wellknown, as mentioned in the introduction. The MLR uses the same model and, thus, can be applied to the same problems as GLR. Therefore, the purpose of the current section is to show what can be gained in robustness and computational burden by using MLR instead of GLR illustrated by a quite simple example. A quite short data length will be used, which allows us to compute the exact GLR test.

A sampled double integrator will be examined in this comparative study of the different methods. For instance, it can be thought of as a model for the position of an object influenced by a random force. The state space model for sample interval 1 is

$$\begin{aligned} x_{t+1} &= \begin{pmatrix} 1 & 1 \\ 0 & 1 \end{pmatrix} x_t + \begin{pmatrix} 0.5 \\ 1 \end{pmatrix} v_t + \delta_{t-k}\nu \\ y_t &= (1\ 0)x_t + e_t \end{aligned}$$

where

$$\begin{aligned} v_t &\in \mathrm{N}(0, \lambda) \\ e_t &\in \mathrm{N}(0, \lambda R). \end{aligned}$$

The jump change corresponds to a sudden force disturbance, caused by a manoeuvre, for example. All filters are initialized assuming $\hat{x}_{1|0} \in \mathrm{N}(0, 1000\lambda I)$ and the number of measurements is N. Default values are given by

$$\begin{aligned} \nu &= (5,\ 10)^T \\ k &= 25 \\ N &= 50 \\ \lambda &= 1 \\ R &= 1. \end{aligned}$$

The default values on λ and R are used to compute the change detectors. The detectors are kept the same in all cases, and the data generation is varied in order to examine the robustness properties.

9.5.1. A Monte Carlo simulation

The following detection methods are compared:

- GLR in Algorithm 9.1 and using a sliding window of size 10 (referred to as GLR(10)).
- MLR in Algorithm 9.2 for an assumed known scaling $\lambda = 1$ (MLR 1), Theorem 9.4 for unknown scaling (MLR 2) and Theorem 9.5 for unknown and changing scaling (MLR 3), respectively.

For the two-filter methods MLR 1–3, five extra data points were simulated and used for initialization. Table 9.1 shows the alarm rates for no jump and jump, respectively, while Table 9.2 shows the estimated jump time in the cases of a jump. The left column indicates what has been changed from the perfect modeling case.

We note that the sliding window approximation of GLR is indistinguishable from the exact implementation. The alarm rates of GLR for the perfect modeling case are slightly larger than for MLR with the chosen threshold. Taking this fact into account, there is no significant difference between GLR and MLR 1. MLR 2 and 3 give somewhat larger false alarm rates and smaller detection probabilities in the perfect modeling case. This is no surprise as less prior information is used, which becomes apparent in these short data sets. The cases where $\lambda < 1$ or $R < 1$, both implying that the measurement noise variance is smaller than expected, cause no problems for GLR and MLR 1. Note, however, how MLR 2 and 3 take advantage of this situation, and here can detect smaller changes than can GLR and MLR. In the case $\lambda = 0.01$ and $\nu = [2;4]$ the probability of detection is 50%, while MLR 1 only detects 2% of these changes.

The real problem is, of course, the cases where the measurement noise variance is larger than modeled. Here the false alarm rate for GLR and MLR 1 is close to one. On the other hand, MLR 2 has a very small and MLR 3 a fairly small false alarm rate. Of course, it becomes harder to detect the fixed size change that is hidden in large noise.

The cases of a suddenly increasing noise scaling is excellently handled by MLR 2 and 3. The former gives no alarm, because this kind of change is not included in the model, and the latter quite correctly estimates a change at time 25.

We can also illustrate the difference in performance by plotting the average log likelihood ratios $2\log p(y^N|k,\hat{\nu})\,/p(y^N|k=N)$ and $2\log p(y^N|k)/p(y^N|k=N)$, respectively, as a function of change time $k = 1, 2, \ldots, N$. This is done in Figure 9.3 for GLR and MLR 1,2,3. A change is detected if the peak value of the log likelihood ratio is larger than zero for MLR and larger than $h = 6$ for GLR. Remember that the GLR log likelihood ratio is always positive.

The first plot shows the perfect modeling case, and the peak values are well above the respective thresholds. Note that MLR 1 and GLR are very

Table 9.1. Alarm rate for 1000 Monte Carlo simulations for different cases of modeling errors. In each case, a state change [5;10] and no change are compared. Sensitivity to incorrect assumptions on the noise variances is investigated, where λ denotes the scaling of Q and R.

Case	GLR	GLR(10)	MLR 1	MLR 2	MLR 3
Perfect modeling	0.08	0.08	0.01	0.01	0.04
Perfect modeling, change	0.99	0.99	0.97	0.92	0.95
Perfect modeling, change [1;2]	0.10	0.10	0.02	0.01	0.04
Perfect modeling, change at $t = 40$	1	1	0.94	0.88	0.91
Perfect modeling, change at $t = 10$	1	1	0.96	0.91	0.97
10 times increase in λ	1	1	0.97	0.10	0.99
10 times increase in λ, change	1	1	0.99	0.14	1
100 times increase in λ	1	1	1	0.14	1
100 times increase in λ, change	1	1	1	0.15	1
$\lambda = 100$	1	1	1	0.01	1
$\lambda = 100$, change	1	1	1	0.01	1
$\lambda = 10$	1	1	1	0.01	0.43
$\lambda = 10$, change	1	1	1	0.02	0.49
$\lambda = 2$	0.50	0.50	0.17	0.01	0.08
$\lambda = 2$, change	0.99	0.99	0.96	0.61	0.79
$\lambda = 0.5$	0.02	0.02	0	0.03	0.05
$\lambda = 0.5$, change	1	1	0.96	0.98	0.99
$\lambda = 0.1$	0.02	0.02	0	0.12	0.10
$\lambda = 0.1$, change	0.99	0.99	0.96	1	1
$\lambda = 0.01$	0	0	0	0.23	0.20
$\lambda = 0.01$, change	0.99	0.99	0.97	1	1
$\lambda = 0.01$, change [2;4]	0.12	0.12	0.02	0.50	0.48
$R = 0.1$	0	0	0	0.01	0
$R = 0.1$, change	1	1	1	1	1
$R = 0.5$	0	0	0	0.01	0.01
$R = 0.5$, change	1	1	0.99	1	1
$R = 2$	0.77	0.77	0.35	0.01	0.10
$R = 2$, change	0.99	0.99	0.91	0.45	0.66
$R = 10$	1	1	0.99	0.02	0.59
$R = 10$, change	1	1	0.99	0.03	0.63

similar in their shape except for a constant offset as stated already in Lemma 9.1.

The second plot illustrates what happens after an abrupt change in noise scaling. GLR and MLR 1 become large for all $t > 25$ and the estimated change times are distributed over the interval $[25, 50]$. MLR 2, which assumes an unknown and constant scaling, handles this case excellently without any peak, while MLR 3 quite correctly has a peak at $t = 25$ where a change in scaling is accurately estimated.

Table 9.2. Estimated change time for 1000 Monte Carlo simulations for different cases of modeling errors.

Case	GLR	GLR(10)	MLR 1	MLR 2	MLR 3
Standard change	25	25	25	25	25
Change [1;2]	29	29	25	23	22
Change at $t = 40$	40	40	40	40	40
Change at $t = 10$	10	10	10	10	10
10 times increase in λ	31	31	33	30	25
100 times increase in λ	37	37	38	39	25
$\lambda = 100$	25	25	27	25	21
$\lambda = 10$	26	26	26	27	23
$\lambda = 2$	25	25	25	25	25
$\lambda = 0.5$	25	25	25	25	25
$\lambda = 0.1$	25	25	25	25	25
$\lambda = 0.01$, change [2; 4]	25	25	27	22	20
$R = 0.1$	25	25	25	25	25
$R = 0.5$	25	25	25	25	25
$R = 2$	25	25	25	25	25
$R = 10$	25	25	27	22	20

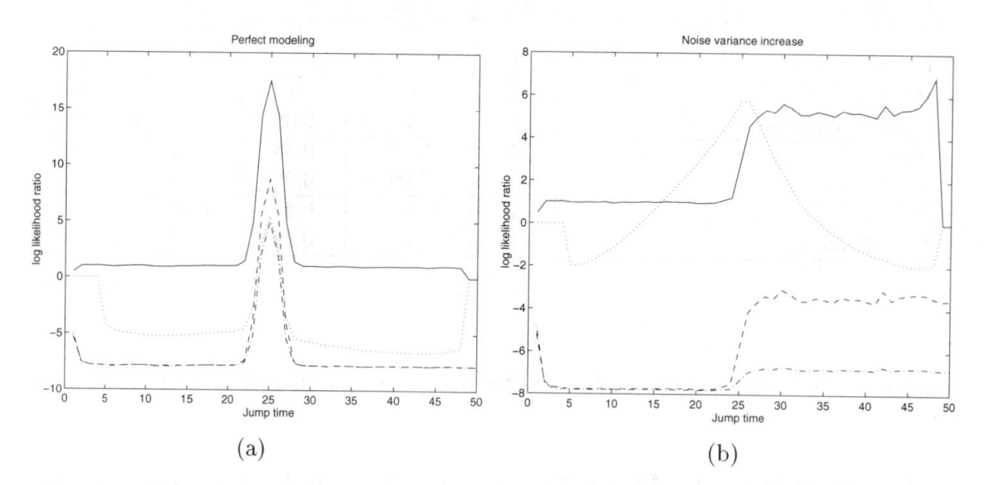

Figure 9.3. Log likelihood ratios for GLR (solid) and MLR 1,2,3 (dashed, dashed-dotted and dotted, respectively) averaged over 100 realizations. Plot for perfect modeling and a change (a), and plot for the case of no change but the noise variance changes abruptly from 1 to 100 at time 25 (b).

9.5.2. Complexity

Figure 9.4 shows the the complexity as a function of the number of observations N, counted in the number of used flops, for the following methods: GLR (M1) and GLR with sliding window (M7), MLR using GLR quantities as in Theorem 9.8 (M2), the direct implementation of MLR in Theorem 9.2 (M3), MLR 1 (M4), MLR 2 (M5) and MLR 3 (M6). It should be noted that the algorithms are intended to be implemented as efficiently as possible, and identical Kalman filter implementations are used.

The implementation in Willsky and Jones (1976) with matched filters is not very efficient, since the direct implementation with Kalman filters instead of RLS schemes is actually faster. This is due to the computation of the regressors. However, both algorithms have a quadratic increase in the number of measurements. The big difference is for the two filter implementation. As expected, it shows only a linear increase in the computational complexity.

The time consumption for GLR with a sliding window of size 10 increases linearly with time, and it is about five times slower than the two filter approach.

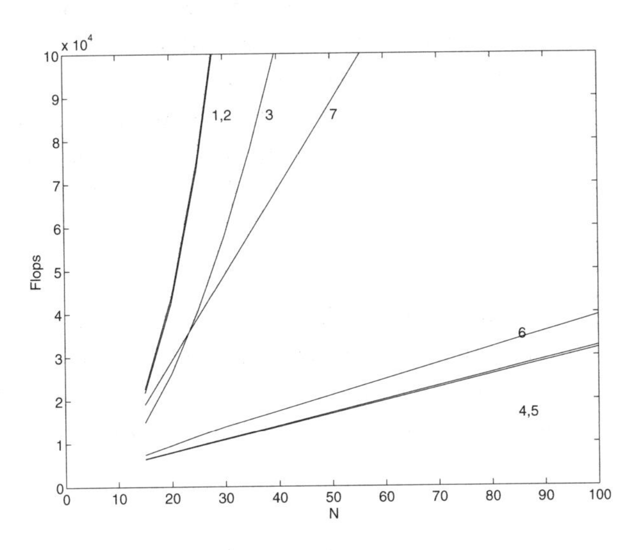

Figure 9.4. Complexity in flops for different implementations of GLR (M1 and M7) and MLR (M2–M6) as a function of the number of measurements.

9.A. Derivation of the GLR test

9.A.1. Regression model for the jump

First, process the measurements through a Kalman filter acting under the hypothesis of no jump. With notation as in (9.4), denote its state estimate, gain, prediction error and prediction error covariance by

$$\hat{x}_{t|t}, \quad K_t = P_{t|t-1}C_tS_t^{-1}, \quad \varepsilon_t = y_t - C_t\hat{x}_{t|t-1},$$

$$S_t = R_t + C_tP_{t|t-1}C_t^T.$$

Next, suppose that there was a jump ν at time k. Due to the linear model, the dependence of ν on the state estimates and innovations will be linear as well. That is, we can postulate the following model:

$$\hat{x}_{t|t}(k) = \hat{x}_{t|t} + \mu_t(k)\nu \tag{9.26}$$

$$\varepsilon_t(k) = \varepsilon_t + \varphi_t^T(k)\nu. \tag{9.27}$$

Here $\hat{x}_{t|t}(k)$ and $\varepsilon_t(k)$ are the quantities one would have obtained from a Kalman filter applied under the assumption that we have a jump at time k. These Kalman filters will, however, not be used explicitly, and that is the key point. Again, we follow the convention that k is equal to the final time, here t, means no jump. We have the following update formulas for the $n_x \times n_x$ matrix $\mu_t(k)$ and the $n_x \times n_y$ matrix $\varphi_t(k)$, first given in Willsky and Jones (1976).

Lemma 9.7

Consider the residuals from the Kalman filter applied to the model (9.1), assuming no jump. The relation between the residuals from a Kalman filter conditioned on a jump of magnitude ν at time k, and from a Kalman filter conditioned on no jump is given by the linear regression

$$\varepsilon_t(k) = \varepsilon_t + \varphi_t^T(k)B_{\theta,t}\nu,$$

where $\varepsilon_t(k)$ is a white noise sequence with variance S_t. Here $\varphi_t(k)$ is computed recursively by

$$\varphi_{t+1}^T(k) = C_{t+1}\left(\prod_{i=k}^{t} A_i - A_t\mu_t(k)\right) \tag{9.28}$$

$$\mu_{t+1}(k) = A_t\mu_t(k) + K_{t+1}\varphi_{t+1}^T(k), \tag{9.29}$$

with the initial conditions $\mu_k(k) = 0$ and $\varphi_k(k) = 0$. Here K_t is the Kalman gain at time t.

Proof: The result is proved by induction. The initial condition is a direct consequence of the signal model (9.1). The induction step follows from the Kalman equations, (9.26) and (9.27). First we have the relation $x_t(k) = x_t + \prod_{i=k}^{t-1} A_i B_{\theta,t}\nu$. This and equations (9.26) and (9.27) give

$$\begin{aligned}\varphi_{t+1}^T B_{\theta,t}\nu =& \varepsilon_{t+1}(k) - \varepsilon_{t+1}\\ =& C_{t+1}\left(x_{t+1}(k) - x_{t+1}\right) - C_{t+1}A_t\left(\hat{x}_{t|t}(k) - \hat{x}_{t|t}\right)\\ =& C_{t+1}\left(\prod_{i=k}^{t} A_i - A_t\mu_t(k)\right) B_{\theta,t}\nu,\end{aligned}$$

and

$$\begin{aligned}\mu_{t+1}(k)B_{\theta,t}\nu =& \hat{x}_{t+1|t+1}(k) - \hat{x}_{t+1|t+1}\\ =& A_t\left(\hat{x}_{t|t}(k) - \hat{x}_{t|t}\right) + K_{t+1}\left(\varepsilon_{t+1} + \varphi_{t+1}^T B_{\theta,t}\nu - \varepsilon_{t+1}\right)\\ =& A_t\mu_t(k)B_{\theta,t}\nu + K_{t+1}\varphi_{t+1}^T B_{\theta,t}\nu.\end{aligned}$$

Since this holds for all $B_{\theta,t}\nu$, the result follows by induction. □

The detection problem is thus moved from a state space to a linear regression framework.

9.A.2. The GLR test

We begin by deriving the classical GLR test, first given in Willsky and Jones (1976), where the jump magnitude is considered as deterministic. Lemma 5.2 gives

$$p(y^N) = p(\varepsilon^N).$$

That is, the measurements and the residuals have the same probability density function. Given a jump ν at time k, the residuals get a bias,

$$\varepsilon_t \in \mathrm{N}(\varphi_t^T(k)\nu, S_t). \tag{9.30}$$

Thus, the test statistic (9.8) can be written as

$$l_N(k,\hat{\nu}(k)) = 2\log\frac{p(\varepsilon_1^N|k,\hat{\nu}(k))}{p(\varepsilon_1^N|k=N)}. \tag{9.31}$$

Introduce the well known compact quantities of the LS estimator

$$f_N(k) = \sum_{t=1}^{N}\varphi_t(k)S_t^{-1}\varepsilon_t \tag{9.32}$$

$$R_N^{\varphi}(k) = \sum_{t=1}^{N}\varphi_t(k)S_t^{-1}\varphi_t^T(k). \tag{9.33}$$

Then the ML estimate of ν, given the jump instant k, can be written

$$\hat{\nu}(k) = (R_N^{\varphi})^{-1}(k) f_N(k). \tag{9.34}$$

We get

$$\begin{aligned} l_N(k, \hat{\nu}(k)) =& 2 \log \frac{p(\varepsilon_1^N | k, \hat{\nu}(k))}{p(\varepsilon_1^N | k = N)} & (9.35) \\ =& \sum_{t=k+1}^{N} \varepsilon_t^T S_t^{-1} \varepsilon_t \\ & - (\varepsilon_t - \varphi_t^T(k)\hat{\nu}(k))^T S_t^{-1} (\varepsilon_t - \varphi_t^T(k)\hat{\nu}(k)) & (9.36) \\ =& f_N^T(k)(R_N^{\varphi})^{-1}(k) f_N(k), & (9.37) \end{aligned}$$

where the second equality follows from (9.30) and the Gaussian probability density function. The third equality follows from straightforward calculations using (9.34), (9.32) and (9.33). This simple expression for the test statistic is an appealing property of the GLR test.

9.B. LS-based derivation of the MLR test

The idea here is to consider the jump magnitude as a stochastic variable. In this way, the maximization over ν in $l_N(k, \nu)$ (9.8) can be avoided. Instead, the jump magnitude is eliminated by integration,

$$p(\varepsilon_1^N) = \int p(\varepsilon_1^N | \nu) p(\nu) d\nu.$$

In the context of likelihood ratio tests, the possibility of integrating out the nuisance parameter ν is also discussed in Wald (1950). Here, there are two choices. Either ν is assumed to have a Gaussian prior, or it is considered to have infinite variance, so the (improper) prior is constant. The log likelihood ratio is given in the following theorem.

Theorem 9.8

Consider the GLR test in Algorithm 9.1 as an estimator of a change of magnitude ν at time k, $k = 1, 2, \ldots, N-1$, for the signal model (9.1), given the measurements y^N. The LR, corresponding to the estimator of k alone, is given by

$$\begin{aligned} l_N(k) =& 2 \log \frac{p(y^N | k)}{p(y^N | k = N)} \\ =& \arg\max_{\nu} 2 \log \frac{p(y^N | k, \nu)}{p(y^N | k = N)} - \log\det R_N^{\varphi}(k) + C_{prior}(k) \\ =& l_N(k, \hat{\nu}(k)) - \log\det R_N^{\varphi}(k) + C_{prior}(k), \end{aligned}$$

where $l_N(k, \hat{\nu}(k)) = f_N^T (R_N^{\varphi})^{-1} f_N$ is given in Algorithm 9.1. Here $C_{prior}(k)$ is a prior dependent constant that equals

$$C_{prior}(k) = 0 \tag{9.38}$$

if the prior is chosen to be non-informative, and

$$C_{prior}(k)Y = 2 \log p_{\nu}(\hat{\nu}) - p \log 2\pi \tag{9.39}$$

if the prior is chosen to be Gaussian, $\nu \in N(\nu_0, P_{\nu})$. Here $R_N^{\varphi}(k)$ and $\hat{\nu}_N(k)$ are given by (9.33) and (9.34), respectively.

Proof: We begin with the Gaussian case,

$$\begin{aligned}
2 \log \frac{p(y^N|k)}{p(y^N)} =& 2 \log \frac{p(y^k)p(y_{k+1}^N|y^k, k)}{p(y^k)p(y_{k+1}^N|y^k)} \\
=& \sum_{t=k+1}^{N} \varepsilon_t^T S_t^{-1} \varepsilon_t \\
& - (\varepsilon_t - \varphi_t^T(k)\hat{\nu}_N(k))^T S_t^{-1} (\varepsilon_t - \varphi_t^T(k)\hat{\nu}_N(k)) \\
& + \log \det P_N(k) - \log \det P_{\nu} \\
& - (\hat{\nu}_N(k) - \nu_0)^T P_{\nu}^{-1} (\hat{\nu}_N(k) - \nu_0) \\
=& l_N(k, \hat{\nu}(k)) - \log \det R_N^{\varphi}(k) + C_{prior}(k).
\end{aligned}$$

In the second equality the off-line expressions (5.96) and (5.97) are used.

The case of non-informative prior ($P_0^{-1} = 0$) is proved as above. Again, (5.96) and (5.97) are used:

$$\begin{aligned}
2 \log \frac{p(y^N|k)}{p(y^N)} =& 2 \log \frac{p(y^k)p(y_{k+1}^N|y^k, k)}{p(y^k)p(y_{k+1}^N|y^k)} \\
=& \sum_{t=k+1}^{N} \varepsilon_t^T S_t^{-1} \varepsilon_t - (\varepsilon_t - \varphi_t^T(k)\hat{\nu}_N(k))^T S_t^{-1} (\varepsilon_t - \varphi_t^T(k)\hat{\nu}_N(k)) \\
& + \log \det P_N(k) \\
=& l_N(k, \hat{\nu}(k)) - \log \det R_N^{\varphi}(k).
\end{aligned}$$

Here the fact $P_N(k) = (R_N^{\varphi})^{-1}(k)$ is used. □

The conclusion is that the ML estimates $\widehat{(k, \nu)}$ and $\hat{k}$ are closely related. In fact, the likelihood ratios are asymptotically equivalent except for a constant. This constant can be interpreted as different thresholds, as done in Lemma 9.1.

In this constant, the term $(\hat{\nu}_N - \nu_0)^T P_\nu^{-1} (\hat{\nu}_N - \nu_0)$ is negligible if the prior uncertainty P_ν is large or, since $\hat{\nu}_N(k)$ converges to ν_0, if $N-k$ is large. Here $\hat{\nu}_N(k) \to \nu_0$ as $N-k \to \infty$, because the Kalman filter eventually tracks the abrupt change in the state vector. As demonstrated in in the simulation section, the term $\log\det R_N^\varphi(k)$ does not alter the likelihood significantly. In fact, $\log\det R_N^\varphi(k)$ is asymptotically constant, and this is formally proved in the following lemma.

Lemma 9.9
Let $\varphi_t(k)$ be as given in Result 9.7. Then, if the signal model (9.1) is stable and time-invariant,

$$R_N^\varphi(k) = \sum_{t=k+1}^{N} \varphi_t(k)\varphi_t(k)^T$$

converges as $N-k$ tends to infinity.

Proof: Rewriting (9.29) using (9.28) gives

$$\mu_{t+1}(k) = A(I - \overline{K}C)\mu_t(k) + \overline{K}CA^{t-k},$$

where $\overline{K}$ is the stationary Kalman gain in the measurement update. Now by assumption $\lambda_1 = \max \lambda_i(A) < 1$ so the Kalman filter theory gives that $\lambda_2 = \max \lambda_i(A - A\overline{K}C) < 1$ as well; see Anderson and Moore (1979) p. 77 (they define $\overline{K}$ as $F\overline{K}$ here). Thus, $\overline{K}CA^{t-k}$ is bounded and

$$\|\mu_t(k)\| < C_1 \lambda_2^{t-k}.$$

This implies that

$$\|\varphi_t(k)\| < C_2 \lambda^{t-k},$$

where $\lambda = \max(\lambda_1, \lambda_2)$. Now if $l > m$

$$\begin{aligned}\|R_m(k) - R_l(k)\| &= \left\| \sum_{t=m+1}^{l} \varphi_t(k)\varphi_t^T(k) \right\| \leq \sum_{t=m+1}^{l} \|\varphi_t(k)\varphi_t^T(k)\| \\ &\leq \sum_{t=m+1}^{l} C_2^2 \lambda^{2(t-k)} \\ &= C_2^2 \lambda^{2(m+1-k)} \frac{1-\lambda^{2l}}{1-\lambda} \to 0 \text{ as } m, l \to \infty.\end{aligned}$$

Thus, we have proved that $R_N^\varphi(k)$ is a Cauchy sequence and since it belongs to a complete space it converges. □

We have now proved that the log likelihoods for the two variants of parameter vectors (k, ν) and k are approximately equal for $k = 1, 2, \ldots, N-1$, except for an unknown constant. Thus,

$$l_N(k, \hat{\nu}_N(k)) \approx l_N(k) + C.$$

Therefore, they are likely to give the same ML estimate. Note, however, that this result does not hold for $k = N$ (that is no jump), since $l_N(N, \hat{\nu}_N(N)) = l_N(N) = 0$. To get equivalence for this case as well, the threshold in the GLR test has to be chosen to this unknown constant C.

10

Change detection based on multiple models

10.1. Basics

This chapter addresses the most general problem formulation of detection in linear systems. Basically, all problem formulations that have been discussed so far are included in the framework considered. The main purpose is to survey multiple model algorithms, and a secondary purpose is to overview and compare the state of the art in different application areas for reducing complexity, where similar algorithms have been developed independently.

The goal is to detect abrupt changes in the state space model

$$\begin{aligned} x_{t+1} =& A_t(\delta_t)x_t + B_{u,t}(\delta_t)u_t + B_{v,t}(\delta_t)v_t \\ y_t =& C_t(\delta_t)x_t + D_{u,t}(\delta_t)u_t + e_t \\ v_t \in& \mathrm{N}(m_{v,t}(\delta_t), Q_t(\delta_t)) \\ e_t \in& \mathrm{N}(m_{e,t}(\delta_t), R_t(\delta_t)). \end{aligned} \tag{10.1}$$

Here δ_t is a discrete parameter representing the *mode* of the system (linearized mode, faulty mode etc.), and it takes on one of S different values (mostly we have the case $S = 2$). This model incorporates all previously discussed problems in this book, and is therefore the most general formulation of the estimation and detection problem. Section 10.2 gives a number of applications, including change detection and segmentation, but also model structure selection, blind and standard equalization, missing data and outliers. The common theme in these examples is that there is an unknown discrete parameter, mode, in a linear system.

One natural strategy for choosing a δ is the following:

- For each possible δ, filter the data through a Kalman filter for the (conditional) known state space model (10.1).
- Choose the particular value of δ, whose Kalman filter gives the smallest prediction errors.

In fact, this is basically how the MAP estimator

$$\hat{\delta}_{MAP} = \arg\max_{\delta} p(\delta|y^N) \tag{10.2}$$

works, as will be proven in Theorem 10.1. The structure is illustrated in Figure 10.1.

The key tool in this chapter is a repeated application of Bayes' law to compute *a posteriori* probabilities:

$$p(\delta^t|y^t) = \frac{p(\delta^t)}{p(y^t)} p(y^t|\delta^t) \tag{10.3}$$

$$= \frac{p(\delta^t)}{p(y^t)} \prod_{k=1}^{t} \mathrm{N}(y_k - C_k(\delta_k)\hat{x}_{k|k-1}(\delta^k), R_k(\delta_k) + C_k(\delta_k)P_{k|k-1}(\delta_k)C_k^T(\delta_k))$$

$$= p(\delta^{t-1}|y^{t-1}) \frac{p(\delta_t|\delta^{t-1})}{p(y_t|y^{t-1})}$$

$$\cdot \mathrm{N}(y_t - C_t(\delta_t)\hat{x}_{t|t-1}(\delta^t), R_t(\delta_t) + C_t(\delta_t)P_{t|t-1}(\delta_t)C_t^T(\delta_t)).$$

A proof is given in Section 10.A. The latter equation is recursive and suitable for implementation. This recursion immediately leads to a multiple model algorithm summarized in Table 10.1. This table also serves as a summary of the chapter.

10.2. Examples of applications

A classical signal processing problem is to find a sinusoid in noise, where the phase, amplitude and frequency may change in time. Multiple model approaches are found in Caciotta and Carbone (1996) and Spanjaard and White

Table 10.1. A generic multiple model algorithm.

1. **Kalman filtering:** conditioned on a particular sequence δ^t, the state estimation problem in (10.1) is solved by a Kalman filter. This will be called the conditional Kalman filter, and its outputs are

$$\hat{x}_{t|t-1}(\delta^t), \quad P_{t|t-1}(\delta^t), \quad \hat{x}_{t|t}(\delta^t), \quad P_{t|t}(\delta^t).$$

2. **Mode evaluation:** for each sequence, we can compute, up to an unknown scaling factor, the posterior probability given the measurements,

$$p(\delta^t|y^t) = \frac{p(\delta^t)}{p(y^t)} p(y^t|\delta^t)$$

using (10.3).

3. **Distribution:** at time t, there are S^t different sequences δ^t, which will be labeled $\delta^t(i)$, $i = 1, 2, \ldots, S^t$. It follows from the theorem of total probability that the exact posterior density of the state vector is

$$p(x_t|y^t) = \frac{1}{\sum_{i=1}^{S^t} p(\delta^t(i)|y^t)} \sum_{i=1}^{S^t} p(\delta^t(i)|y^t) \, \mathrm{N}\left(\hat{x}_{t|t}(\delta^t(i)), P_{t|t}(\delta^t(i))\right).$$

This distribution is a *Gaussian mixture* with S^t modes.

4. **Pruning and merging (on-line):** for *on-line* applications, there are two approaches to approximate the Gaussian mixture, both aiming at removing modes so only a fixed number of modes in the Gaussian mixture are kept. The exponential growth can be interpreted as a tree, and the approximation strategies are *merging* and *pruning*. Pruning is simply to cut off modes in the mixture with low probability. In merging, two or more modes are replaced by one new Gaussian distribution.

5. **Numerical search (off-line):** for off-line analysis, there are numerical approaches based on the EM algorithm or MCMC methods. We will detail some suggestions for how to generate sequences of δ^t which will theoretically belong to the true posterior distribution.

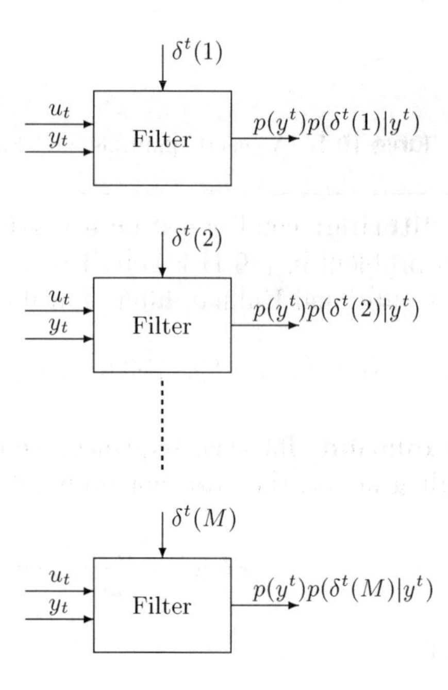

Figure 10.1. The multiple model approach.

(1995). In Daumera and Falka (1998), multiple models are used to find the change points in biomedical time series. In Caputi (1995), the multiple model is used to model the input to a linear system as a switching Gaussian process. Actuator and sensor faults are modeled by multiple models in Maybeck and Hanlon (1995). Wheaton and Maybeck (1995) used the multiple model approach for acceleration modeling in target tracking, and Yeddanapudi et al. (1997) applied the framework to target tracking in ATC. These are just a few examples, more references can be found in Section 10.3.4. Below, important special cases of the general model are listed as examples. It should be stressed that the general algorithm and its approximations can be applied to all of them.

Example 10.1 Detection in changing mean model

Consider the case of an unknown constant in white noise. Suppose that we want to test the hypothesis that the 'constant' has been changed at some unknown time instant. We can then model the signal by

$$y_t = \theta_1 + \sigma(t - \delta + 1)\theta_2 + e_t,$$

where $\sigma(t)$ is the step function. If all possible change instants are to be considered, the variable δ takes its value from the set $\{1, 2, \dots, t-1, t\}$, where $\delta = t$

should be interpreted as no change (yet). This example can be interpreted as a special case of (10.1), where

$$\delta = \{1, 2, \ldots, t\}, \quad x_t = (\theta_1, \theta_2)^T, \quad A_t(\delta) = \begin{pmatrix} 1 & 0 \\ 0 & \sigma(t-\delta+1) \end{pmatrix}$$
$$C_t(\delta) = (1, 1), \quad Q_t(\delta) = 0. \quad R_t(\delta) = \lambda.$$

The *detection* problem is to estimate δ.

Example 10.2 Segmentation in changing mean model

Suppose in Example 10.1 that there can be arbitrarily many changes in the mean. The model used can be extended by including more step functions, but such a description would be rather inconvenient. A better alternative to model the signal is

$$\begin{aligned} \theta_{t+1} &= \theta_t + \delta_t v_t \\ y_t &= \theta_t + e_t \\ \delta_t &\in \{0, 1\}. \end{aligned}$$

Here the changes are modeled as the noise v_t, and the discrete parameter δ_t is 1 if a change occurs at time t and 0 otherwise. Obviously, this is a special case of (10.1) where the discrete variable is $\delta^N = (\delta_1, \delta_2, \ldots, \delta_N)$ and

$$\delta^N = \{0, 1\}^N, \quad x_t = \theta_t, \quad A_t(\delta) = 1, \quad C_t = 1, \quad Q_t(\delta) = \delta_t Q_t, \quad R_t = \lambda.$$

Here $\{0, 1\}^N$ denotes all possible sequences of zeros and ones of length N. The problem of estimating the sequence δ^N is called *segmentation*.

Example 10.3 Model structure selection

Suppose that there are two possible model structures for describing a measured signal, namely two auto-regressions with one or two parameters,

$$\begin{aligned} \delta = 1: &\quad y_t = -a_1 y_{t-1} + e_t \\ \delta = 2: &\quad y_t = -a_1 y_{t-1} - a_2 y_{t-2} + e_t. \end{aligned}$$

Here, e_t is white Gaussian noise with variance λ. We want to determine from a given data set which model is the most suitable. One solution is to refer to the general problem with discrete parameters in (10.1). Here we can take

$$A_t(\delta) = I, \quad Q_t(\delta) = 0, \quad R_t(\delta) = \lambda$$

and

$$\begin{aligned} x_t &= a_1, & C_t(\delta) &= -y_{t-1} & \text{if } \delta = 1 \\ x_t &= (a_1, a_2)^T, & C_t(\delta) &= (-y_{t-1}, -y_{t-2}) & \text{if } \delta = 2. \end{aligned}$$

The problem of estimating δ is called *model structure selection.*

Example 10.4 Equalization

A typical digital communication problem is to estimate a binary signal, u_t, transmitted through a channel with a known characteristic and measured at the output. A simple example is

$$\begin{aligned} y_t =& u_t - 0.5u_{t-1} + e_t \\ u_t \in& \{-1, +1\}. \end{aligned}$$

We can bring back this model to (10.1) by letting

$$\delta_t = u_t, \quad \delta^N = \{-1, +1\}^N, \quad x_t = (1, -0.5)^T, \quad A_t(\delta) = I,$$
$$C_t(\delta) = (\delta_t, \delta_{t-1}), \quad Q_t(\delta) = 0, \quad R_t(\delta) = \lambda.$$

We refer to the problem of estimating the input sequence with a known channel as *equalization.*

Example 10.5 Blind equalization

Consider again the communication problem in Example 10.4, but assume now that both the channel model and the binary signal are unknown *a priori.* We can try to estimate the channel parameters as well by using the model

$$\begin{aligned} y_t =& b_0 u_t + b_1 u_{t-1} + e_t \\ u_t \in& \{-1, +1\}. \end{aligned}$$

Compared to (10.1), we have here

$$\delta_t = u_t, \quad \delta^N = \{-1, +1\}^N, \quad x_t = (b_0, b_1)^T, \quad A_t(\delta) = I, \quad C_t(\delta) = (\delta_t, \delta_{t-1}),$$
$$Q_t(\delta) = 0, \quad R_t(\delta) = \lambda.$$

The problem of estimating the input sequence with an unknown channel is called *blind equalization.*

Example 10.6 Outliers

In practice it is not uncommon that some of the measurements are much worse then the others. These are usually called *outliers*. See Huber (1981) for a thorough treatment of this problem. Consider, for instance, a state space model

$$\begin{aligned} x_{t+1} &= A_t x_t + v_t \\ y_t &= C_t x_t + e_t, \end{aligned} \tag{10.4}$$

where some of the measurements are known to be bad. One possible approach to this problem is to model the measurement noise as a Gaussian mixture,

$$e_t \in \sum_{i=1}^{M} \alpha_i \mathrm{N}(\mu_i, Q_i),$$

where $\sum \alpha_i = 1$. With this notation we mean that the density function for e_t is

$$f_{e_t}(x) \in \sum_{i=1}^{M} \alpha_i \gamma(x - \mu_i, Q_i).$$

In this way, any density function can be approximated arbitrarily well, including heavy-tailed distributions describing the outliers. To put a Gaussian mixture in our framework, express it as

$$e_t \in \begin{cases} \mathrm{N}(\mu_1, Q_1) & \text{with probability } \alpha_1 \\ \mathrm{N}(\mu_2, Q_2) & \text{with probability } \alpha_2 \\ \vdots & \\ \mathrm{N}(\mu_M, Q_M) & \text{with probability } \alpha_M. \end{cases}$$

Hence, the noise distribution can be written

$$e_t \in \mathrm{N}(\mu(\delta_t), Q(\delta_t))$$

where $\delta_t \in \{1, 2, \ldots, M\}$ and the prior is chosen as $p(\delta_t = i) = \alpha_i$.

The simplest way to describe possible outliers is to take $\mu_1 = \mu_2 = 0$, Q_1 equal to the nominal noise variance, Q_2 as much larger than Q_1 and $\alpha_2 = 1 - \alpha_1$ equal to a small number. This models the fraction α_2 of all measurements as outliers with a very large variance. The Kalman filter will then ignore these measurements, and the *a posteriori* probabilities are almost unchanged.

Example 10.7 Missing data

In some applications it frequently happens that measurements are missing, typically due to sensor failure. A suitable model for this situation is

$$\begin{aligned} x_{t+1} =& A_t x_t + v_t \\ y_t =& (1-\delta_t) C_t x_t + e_t. \end{aligned} \tag{10.5}$$

This model is used in Lainiotis (1971). The model (10.5) corresponds to the choices

$$\begin{aligned} &\delta^t = \{0,1\}^N, \quad A_t(\delta) = A_t, \\ &C_t(\delta) = (1-\delta_t)C_t, \quad Q_t(\delta) = Q_t, \quad R_t(\delta) = R_t \end{aligned}$$

in the general formulation (10.1). For a thorough treatment of missing data, see Tanaka and Katayama (1990) and Parzen (1984).

Example 10.8 Markov models

Consider again the case of missing data, modeled by (10.5). In applications, one can expect that a very low fraction, say p_{11}, of the data is missing. On the other hand, if one measurement is missing, there is a fairly high probability, say p_{22}, that the next one is missing as well. This is nothing but a prior assumption on δ_t corresponding to a Markov chain. Such a state space model is commonly referred to as a *jump linear model*. A Markov chain is completely specified by its transition probabilities

$$p(\delta_t = j | \delta_{t-1} = i) = p_{ij},$$

and the initial probabilities $p(\delta_t = i) = p_i$. Here we must have $p_{12} = 1 - p_{22}$ and $p_{21} = 1 - p_{11}$. In our framework, this is only a recursive description of the prior probability of each sequence,

$$p(\delta^t) = p(\delta_t | \delta^{t-1}) p(\delta^{t-1}) = p_{\delta_1} \prod_{k=2}^{t} p_{\delta_{k-1}\delta_k}.$$

For outliers, and especially missing data, the assumption of an underlying Markov chain is particularly logical. It is used, for instance, in MacGarty (1975).

10.3. On-line algorithms

10.3.1. General ideas

Interpret the exponentially increasing number of discrete sequences δ^t as a growing tree, as illustrated in Figure 10.2. It is inevitable that we either *prune* or *merge* this tree.

In this section, we examine how one can discard elements in δ by cutting off branches in the tree, and lump sequences into subsets of δ by merging branches.

Thus, the basic possibilities for pruning the tree are to *cut off* branches and to *merge* two or more branches into one. That is, two state sequences are merged and in the following treated as just one. There is also a timing question: at what instant in the time recursion should the pruning be performed? To understand this, the main steps in updating the *a posteriori* probabilities can be divided into a *time update* and a *measurement update* as follows:

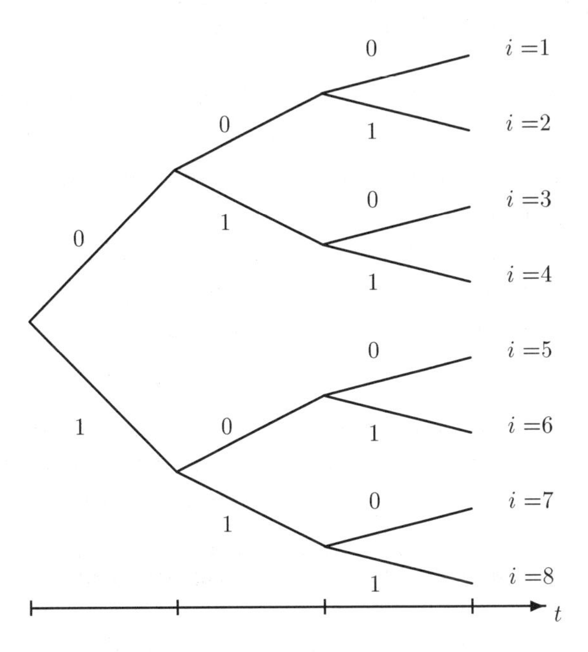

Figure 10.2. A growing tree of discrete state sequences. In GPB(2) the sequences (1,5), (2,6), (3,7) and (4,8), respectively, are merged. In GPB(1) the sequences (1,3,5,7) and (2,4,6,8), respectively, are merged.

- Time update:

$$p(\delta^{t-1}|y^{t-1}) \longrightarrow p(\delta^t|y^{t-1}) \tag{10.6}$$
$$p(x_{t-1}|\delta^{t-1}, y^{t-1}) \longrightarrow p(x_t|\delta^t, y^{t-1}). \tag{10.7}$$

- Measurement update:

$$p(\delta^t|y^{t-1}) \longrightarrow p(\delta^t|y^t) \tag{10.8}$$
$$p(x_t|\delta^t, y^{t-1}) \longrightarrow p(x_t|\delta^t, y^t). \tag{10.9}$$

Here, the splitting of each branch into S branches is performed in (10.7). We define the *most probable branch* as the sequence δ^t, with the largest *a posteriori* probability $p(\delta^t|y^t)$ in (10.8).

A survey on proposed search strategies in the different applications in Section 10.2 is presented in Section 10.3.4.

10.3.2. Pruning algorithms

First, a quite general pruning algorithm is given.

Algorithm 10.1 Multiple model pruning

1. Compute recursively the conditional Kalman filter for a bank of M sequences $\delta^t(i) = (\delta_1(i), \delta_2(i), \ldots, \delta_t(i))^T$, $i = 1, 2, \ldots, M$.
2. After the measurement update at time t, prune all but the M/S most probable branches $\delta^t(i)$.
3. At time $t+1$: let the M/S considered branches split into $S \cdot M/S = M$ branches, $\delta^{t+1}(j) = (\delta^t(i), \delta_{t+1})$ for all $\delta^t(i)$ and δ_{t+1}. Update their *a posteriori* probabilities according to Theorem 10.1.

For change detection purposes, where $\delta_t = 0$ is the normal outcome and $\delta_t \neq 0$ corresponds to different fault modes, we can save a lot of filters in the filter bank by using a local search scheme similar to that in Algorithm 7.1.

Algorithm 10.2 Local pruning for multiple models

1. Compute recursively the conditional Kalman filter for a bank of M hypotheses of $\delta^t(i) = (\delta_1(i), \delta_2(i), \ldots, \delta_t(i))^T$, $i = 1, 2, \ldots, M$.
2. After the measurement update at time t, prune the $S-1$ least probable branches δ^t.

3. At time $t+1$: let only the most probable branch split into S branches, $\delta^{t+1}(j) = (\delta^t(i), \delta_{t+1})$.
4. Update their posterior probabilities according to Theorem 10.1.

Some restrictions on the rules above can sometimes be useful:

- Assume a minimum segment length: let the most probable sequence split *only if it is not too young.*
- Assure that sequences are not cut off immediately after they are born: cut off the least probable sequences *among those that are older than a certain minimum life-length,* until only M ones are left.

10.3.3. Merging strategies

A general merging formula

The exact posterior density of the state vector is a mixture of St Gaussian distributions. The key point in merging is to replace, or approximate, a number of Gaussian distributions by one single Gaussian distribution in such a way that the first and second moments are matched. That is, a sum of L Gaussian distributions

$$p(x) = \sum_{i=1}^{L} \alpha(i)\,\mathrm{N}(\hat{x}^j, P^j)$$

is approximated by

$$p(x) = \alpha\,\mathrm{N}(\hat{x}, P),$$

where

$$\begin{aligned}
\alpha =& \sum_{i=1}^{L} \alpha(i)\\
\hat{x} =& \frac{1}{\alpha}\sum_{i=1}^{L} \alpha(i)\hat{x}(i)\\
P =& \frac{1}{\alpha}\sum_{i=1}^{L} \alpha(i)\left(P(i) + (\hat{x}(i)-\hat{x})(\hat{x}(i)-\hat{x})^T\right).
\end{aligned}$$

The second term in P is the *spread of the mean* (see (10.21)). It is easy to verify that the expectation and covariance are unchanged under the distribution approximation. When merging, all discrete information of the history is lost. That is, merging is less useful for fault detection and isolation than pruning.

The GPB algorithm

The idea of the *Generalized Pseudo-Bayesian* (*GPB*) approach is to merge the mixture after the measurement update.

Algorithm 10.3 GPB

The mode parameter δ is an independent sequence with S outcomes used to switch modes in a linear state space model. Decide on the sliding window memory L. Represent the posterior distribution of the state at time t with a Gaussian mixture of $M = S^{L-1}$ distributions,

$$p(x_t) = \sum_{i=1}^{S^{L-1}} \alpha(i)\,\mathrm{N}(\hat{x}^j, P^j).$$

Repeat the following recursion:

1. Let these split into S^L sequences by considering all S new branches at time $t+1$.
2. For each i, apply the conditional Kalman filter measurement and time update giving $\hat{x}_{t+1|t}(i)$, $\hat{x}_{t+1|t+1}(i)$, $P_{t+1|t}(i)$, $P_{t+1|t+1}(i)$, $\varepsilon_{t+1}(i)$ and $S_{t+1}(i)$.
3. Time update the weight factors $\alpha(i)$ according to

$$\alpha_{t+1|t}(i) = \alpha_{t|t}(i)p(\delta_{t+1}(i)). \tag{10.10}$$

4. Measurement update the weight factors $\alpha(i)$ according to

$$\alpha_{t+1|t+1}(i) = \alpha_{t+1|t}(i)\frac{1}{\sqrt{\det S_{t+1}(i)}}e^{-\frac{1}{2}\epsilon^T_{t+1}(i)S^{-1}_{t+1}(i)\epsilon_{t+1}(i)}. \tag{10.11}$$

5. Merge S sequences corresponding to the same history up to time $t-L$. This requires S^{L-1} separate merging steps using the formula

$$\begin{aligned}
\alpha &= \sum_{i=1}^{S} \alpha(i)\\
\hat{x} &= \frac{1}{\alpha}\sum_{i=1}^{S} \alpha(i)\hat{x}(i)\\
P &= \frac{1}{\alpha}\sum_{i=1}^{S} \alpha(i)\left(P(i) + (\hat{x}(i)-\hat{x})(\hat{x}(i)-\hat{x})^T\right).
\end{aligned}$$

The hypotheses that are merged are identical up to time $t - L$. That is, we do a complete search in a sliding window of size L. In the extreme case of $L = 0$, all hypotheses are merged at the end of the measurement update. This leaves us with S time and measurement updates. Figure 10.2 illustrates how the memory L influences the search strategy.

Note that we prefer to call α_i weight factors rather than posterior probabilities, as in Theorem 10.1. First, we do not bother to compute the appropriate scaling factors (which are never needed), and secondly, these are probabilities of merged sequences that are not easy to interprete afterwards.

The IMM algorithm

The IMM algorithm is very similar to GPB. The only difference is that merging is applied after the time update of the weights rather than after the measurement update. In this way, a lot of time updates are omitted, which usually do not contribute to performance. Computationally, IMM should be seen as an improvement over GPB.

Algorithm 10.4 IMM

As the GPB Algorithm 10.3, but change the order of steps 4 and 5.

For target tracking, IMM has become a standard method (Bar-Shalom and Fortmann, 1988; Bar-Shalom and Li, 1993). Here there is an ambiguity in how the mode parameter should be utilized in the model. A survey of alternatives is given in Efe and Atherton (1998).

10.3.4. A literature survey

Detection

In *detection*, the number of branches increases linearly in time: one branch for each possible time instant for the change. An approximation suggested in Willsky and Jones (1976) is to restrict the jump to a window, so that only jumps in the, say, L last time instants are considered. This is an example of global search, and it would have been the optimal thing to do if there really was a finite memory in the process, so new measurements contain no information about possible jumps L time instants ago. This leaves L branches in the tree.

Segmentation

A common approach to *segmentation* is to apply a recursive detection method, which is restarted each time a jump is decided. This is clearly also a sort of global search.

A pruning strategy is proposed in Andersson (1985). The method is called *Adaptive Forgetting through Multiple Models* (*AFMM*), and basically is a variant of Algorithm 10.2.

Equalization

For *equalization* there is an optimal search algorithm for a finite impulse response channel, namely the *Viterbi algorithm* 5.5. The Viterbi algorithm is in its simplicity indeed the most powerful result for search strategies. The assumption of finite memory is, however, not very often satisfied. Equalization of FIR channels is one exception. In our terminology, one can say that the Viterbi algorithm uses a sliding window, where all possible sequences are examined.

Despite the optimality and finite dimensionality of the Viterbi algorithm, the memory, and accordingly also the number of branches, is sometimes too high. Therefore, a number of approximate search algorithms have been suggested. The simplest example is *Decision-directed Feedback* (DF), where only the most probable branch is saved at each time instant.

An example of a global search is *Reduced State Space Estimation* (RSSE), proposed in Eyuboglu and Qureshi (1988). Similar algorithms are independently developed in Duel-Hallen and Heegard (1989) and Chevillat and Eleftheriou (1989). Here, the possible state sequences are merged to classes of sequences. One example is when the size of the optimal Viterbi window is decreased to less than L. A different and more complicated merging scheme, called *State Space Partitioning* (SSP), appears in Larsson (1991).

A pruning approach is used in Aulin (1991), and it is there called *Search Algorithm* (SA). Apparently, the same algorithm is used in Anderson and Mohan (1984), where it is called the *M-algorithm.* In both algorithms, the M locally best sequences survive.

In Aulin (1991) and Seshadri and Sundberg (1989), a search algorithm is proposed, called the HA(B, L) and *generalized Viterbi algorithm* (GVA), respectively. Here, the B most probable sequences preceding all combinations of sequences in a sliding window of size L are saved, making a total of BS^L branches. The HA(B, L) thus contains DF, RSSE, SA and even the Viterbi algorithm as special cases.

Blind equalization

In *blind equalization*, the approach of examining each input sequence in a tree structure is quite new. However, the DF algorithm, see Sato (1975), can be considered as a local search where only the most probable branch is saved. In Sato (1975), an approximation is proposed, where the possible inputs are merged into two classes: one for positive and one for negative values of the input. The most probable branch defined in these two classes is saved. This algorithm is, however, not an approximation of the optimal algorithm; rather of a suboptimal one where the LMS (Least Mean Squares, see Ljung and Söderström (1983)) algorithm is used for updating the parameter estimate.

Markov models

The search strategy problem is perhaps best developed in the context of *Markov models*; see the excellent survey Tugnait (1982) and also Blom and Bar-Shalom (1988).

The earliest reference on this subject is Ackerson and Fu (1970). In their global algorithm the Gaussian mixture at time t – remember that the *a posteriori* distribution from a Gaussian prior is a Gaussian mixture – is approximated by one Gaussian distribution. That is, all branches are merged into one. This approach is also used in Wernersson (1975), Kazakov (1979) and Segal (1979).

An extension on this algorithm is given in Jaffer and Gupta (1971). Here, all possible sequences over a sliding window are considered, and the preceding sequences are merged by using one Gaussian distribution just as above. Two special cases of this algorithm are given in Bruckner et al. (1973) and Chang and Athans (1978), where the window size is one. The difference is just the model complexity. The most general algorithm in Jaffer and Gupta (1971) is the *Generalized Pseudo Bayes* (GPB), which got this name in Blom and Bar-Shalom (1988), see Algorithm 10.3. The *Interacting Multiple Model* (IMM) algorithm was proposed in Blom and Bar-Shalom (1988). As stated before, the difference of GPB and IMM is the timing when the merging is performed. Merging in (10.7) gives the IMM and after (10.9) the GPB algorithm.

An unconventional approach to global search is presented in Akashi and Kumamoto (1977). Here the small number of considered sequences are chosen *at random*. This was before genetic algorithms and MCMC methods become popular.

A pruning scheme, given in Tugnait (1979), is the *Detection-Estimation Algorithm* (DEA). Here, the M most likely sequences are saved.

10.4. Off-line algorithms

10.4.1. The EM algorithm

The *Expectation Maximization* (*EM*) algorithm (see Baum et al. (1970)), alternates between estimating the state vector by a conditional mean, given a sequence δ^N, and maximizing the posterior probability $p(\delta^N|x^N)$. Application to state space models and some recursive implementations are surveyed in Krishnamurthy and Moore (1993). The MCMC Algorithm 10.6 can be interpreted as a stochastic version of EM.

10.4.2. MCMC algorithms

MCMC algorithms are off-line, though related simulation-based methods have been proposed to find recursive implementations (Bergman, 1999; de Freitas et al., 2000; Doucet, 1998).

Algorithm 7.2 proposed in Fitzgerald et al. (1994) for segmentation, can be generalized to the general detection problem. It is here formulated for the case of binary δ_t, which can be represented by n change times $k_1, k_2, \ldots, k_n$.

Algorithm 10.5 Gibbs-Metropolis MCMC detection

Decide the number of changes n:

1. Iterate Monte Carlo run i
2. Iterate *Gibbs sampler* for component j in k^n, where a random number from
$$\bar{k}_j \sim p(k_j | k_1, k_2, \ldots, k_{j-1}, k_{j+1}, k_n)$$
is taken. Denote the new candidate sequence $\overline{k^n}$. The distribution may be taken as flat, or Gaussian centered around k_j. If independent jump instants are assumed, this task simplifies to taking random numbers $\bar{k}_j \sim p(k_j)$.
3. Run the conditional Kalman filter using the sequence $\overline{k^n}$, and save the innovations ε_t and its covariances S_t.
4. The candidate j is accepted with probability
$$\min\left(1, \frac{p(\varepsilon^N(\overline{k^n}))}{p(\varepsilon^N(k^n))}\right).$$
That is, if the likelihood increases we always keep the new candidate. Otherwise we keep it with a certain probability which depends on its likeliness. This random rejection step is the *Metropolis step.*

After the *burn-in* (convergence) time, the distribution of change times can be computed by Monte Carlo techniques.

Example 10.9 Gibbs-Metropolis change detection

Consider the tracking example described in Section 8.12.2. Figure 10.3(a) shows the trajectory and the result from the Kalman filter. The jump hypothesis is that the state covariance is ten times larger $Q(1) = 10\ Q(0)$. The

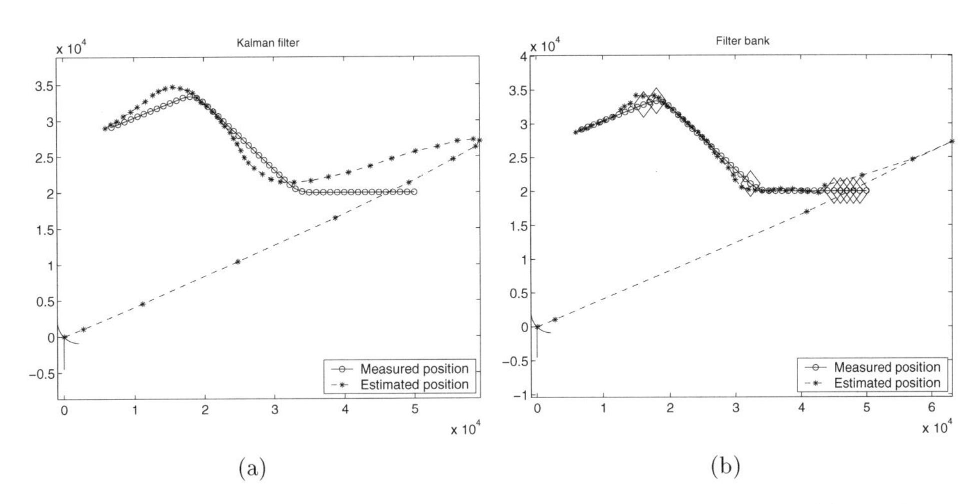

(a) (b)

Figure 10.3. Trajectory and filtered position estimate from Kalman filter (a) and Gibbs-Metropolis algorithm with $n = 5$ (b).

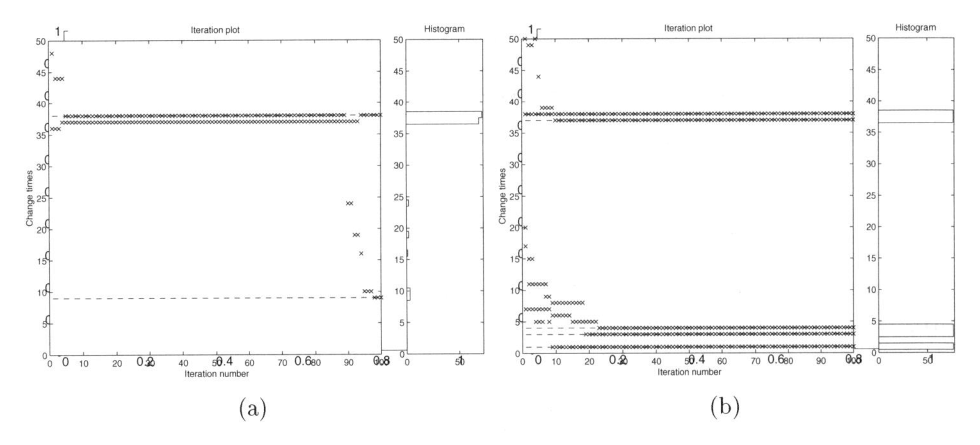

(a) (b)

Figure 10.4. Convergence of Gibbs sequence of change times. To the left for $n = 2$ and to the right for $n = 5$. For each iteration, there are n sub-iterations, in which each change time in the current sequence k^n is replaced by a random one. The accepted sequences k^n are marked with 'x'.

distribution in step 5 is Gaussian N(0, 5). The jump sequences as a function of iterations and sub-iterations of the Gibbs sampler are shown in Figure 10.4. The iteration scheme converges to two change points at 20 and 36. For the case of overestimating the change points, these are placed at one border of the data sequence (here at the end). The improvement in tracking performance is shown in Figure 10.3(b). The distribution we would like to have in practice is the one from Monte Carlo simulations. Figure 10.5 shows the result from Kalman filter whiteness test and a Kalman filter bank. See Bergman and Gustafsson (1999) for more information.

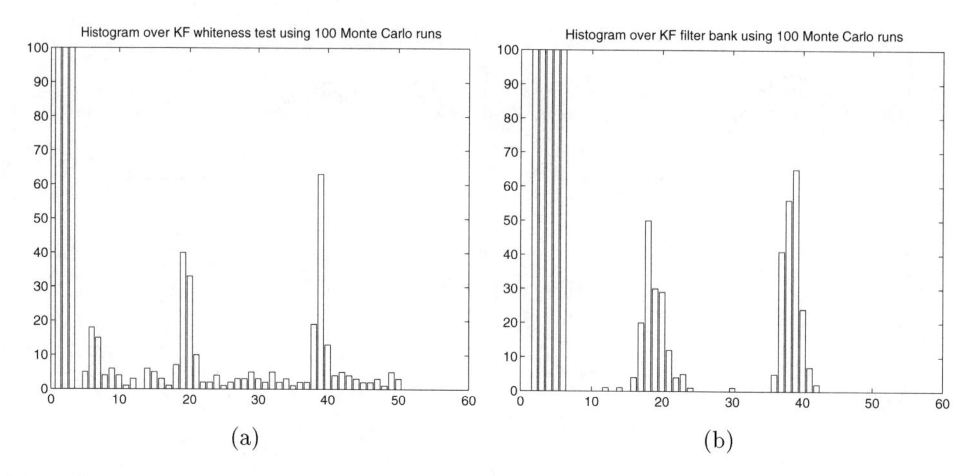

Figure 10.5. Histograms of estimated change times from 100 Monte Carlo simulations using Kalman filter whiteness test and a Kalman filter bank.

We can also generalize the MCMC approach given in Chapter 4.

Algorithm 10.6 MCMC change detection

Assume Gaussian noise in the state space model (10.1). Denote the sequence of mode parameters $\delta^N(i) = (\delta_1, \ldots, \delta_N)^T$, and the stacked vector of states for $X = (x_1^T, \ldots, x_N^T)^T$. The Gibbs sequence of change times is generated by alternating taking random samples from

$$\begin{aligned}(x^N)^{(i+1)} &\sim p(x^N | y^N, (\delta^N)^{(i)}) \\ (\delta^N)^{(i+1)} &\sim p(\delta^N | y^N, (x^N)^{(i+1)}).\end{aligned}$$

The first distribution is given by the conditional Kalman smoother, since

$$(x_t | \delta^N, y^N) \in \mathrm{N}(\hat{x}_{t|N}(\delta^N), P_{t|N}(\delta^N)).$$

The second distribution is

$$\delta_t | x^N \sim \frac{1}{\sqrt{(2\pi)^{n_v} \det(Q_t)}} e^{-\frac{1}{2}(B_{v,t}^{\dagger}(x_{t+1} - A_t(\delta_t)x_t))^T Q_t^{-1}(\delta_t)(B_{v,t}^{\dagger}(x_{t+1} - A_t(\delta_t)x_t))}.$$

Here $A^{\dagger}$ denotes the Moore-Penrose pseudo-inverse.

The interpretation of the last step is the following: If the smoothed sequence x_t is changing rapidly (large derivative), then a change is associated with that time instant with high probability.

A computational advantage of assuming independence is that the random δ_t variables can be generated independently. An alternative is to assume a hidden Markov model for the sequence δ^N.

10.5. Local pruning in blind equalization

As an application, we will study blind equalization. The algorithm is a straightforward application of Algorithm 10.1. It was proposed in Gustafsson and Wahlberg (1995), and a similar algorithm is called *MAPSD* (*Maximum A Posteriori Sequence Detection*) (Giridhar et al., 1996).

10.5.1. Algorithm

A time-varying ARX model is used for the channel:

$$\begin{aligned} \theta_{t+1} =& \theta_t + v_t \\ y_t =& \varphi_t^T \theta_t + e_t, \end{aligned}$$

where

$$\varphi_t =(-y_{t-1}, \cdots, -y_{t-n_a}, \delta_t, \cdots, \delta_{t-n_b+1})^T \tag{10.12}$$

$$\theta_t =(a_1(t), \cdots, a_{n_a}(t), b_1(t), \cdots, b_{n_b}(t))^T. \tag{10.13}$$

The unknown input δ_t belongs to a finite alphabet of size S. With mainly notational changes, this model can encompass multi-variable and complex channels as well. The pruning Algorithm 10.1 now becomes Algorithm 10.7.

Algorithm 10.7 Blind equalization

Assume there are M sequences $\delta^{t-1}(i)$ given at time $t-1$, and that their relative *a posteriori* probabilities $p(\delta^{t-1}(i)|y^{t-1})$ have been computed. At time t, compute the following:

1. **Evaluation:** Update $p(\delta^t(i)|y^t)$ by

$$\begin{aligned} p(\delta^t(i)|y^t) =& C_t p(\delta^{t-1}(i)|y^{t-1}) p(\delta_t(i)|\delta^{t-1}(i)) \\ & \cdot \gamma(y_t - \varphi_t^T(i)\hat{\theta}_{t-1}(i), R_t + \varphi_t^T(i) P_{t-1}(i)\varphi_t(i)) \end{aligned}$$

2. **Pruning:** Reject all but the M most probable sequences – that is, those which have the largest $p(\delta^{t+1}(i)|y^{t+1})$.

3. **Estimation:** For each sequence i apply the Kalman filter

$$\hat{\theta}_t(i) = \theta_{t-1}(i) + P_{t-1}(i)\varphi_t(i)\left((\varphi_t(i))^T P_{t-1}(i)\varphi_t(i) + R_t\right)^{-1} \tag{10.14}$$

$$\cdot\left(y_t - (\varphi_t(i))^T\hat{\theta}_{t-1}(i)\right), \tag{10.15}$$

$$P_t(i) = P_{t-1}(i) + Q_t + P_{t-1}(i)\varphi_t(i)\left((\varphi_t(i))^T P_{t-1}(i)\varphi_t(i) + R_t\right)^{-1} \tag{10.16}$$

$$\cdot(\varphi_t(i))^T P_{t-1}(i). \tag{10.17}$$

Here $\varphi_t(i) = \varphi_t(\delta^t(i))$ and $\hat{\theta}_t(i)$ are conditional of the input sequence $\delta^t(i)$. This gives SM sequences, by considering all S expansions of each sequence at time t.

4. Repeat from step 1.

We conclude with a numerical evaluation.

Example 10.10 Blind equalization using multiple models

In this example we will examine how Algorithm 10.7 performs in the case of a Rayleigh fading communication channel. Rayleigh fading is an important problem in mobile communication. The motion of the receiver causes a time-varying channel characteristics. The Rayleigh fading channel is simulated using the following premises: The frequency of the carrier wave is 900 MHz, and the baseband sampling frequency is 25 kHz. The receiver is moving with the velocity 83 km/h so the maximum Doppler frequency can be shown to be approximately 70 Hz. A channel with two time-varying taps, corresponding to this maximum Doppler frequency, will be used.[1] An example of a tap is shown in Figure 10.6. For more details and a thorough treatment of fading in mobile communication, see Lee (1982).

In Figure 10.6(a), a typical parameter convergence is shown. The true FIR parameter values are here compared to the least squares estimates conditioned on the estimated input sequence at time t. The convergence to the true parameter settings is quite fast (only a few samples are needed), and the tracking ability very good.

To test the sensitivity to noise, the measurement noise variance is varied over a wide range. Figure 10.6(b) shows *Bit Error Rate* (BER) as a function of

[1] The taps are simulated by filtering white Gaussian noise with unit variance by a second order resonance filter, with the resonance frequency equal to 70/25000 Hz, followed by a seventh order Butterworth low-pass filter with cut-off frequency $\pi/2 \cdot 70/25000$.

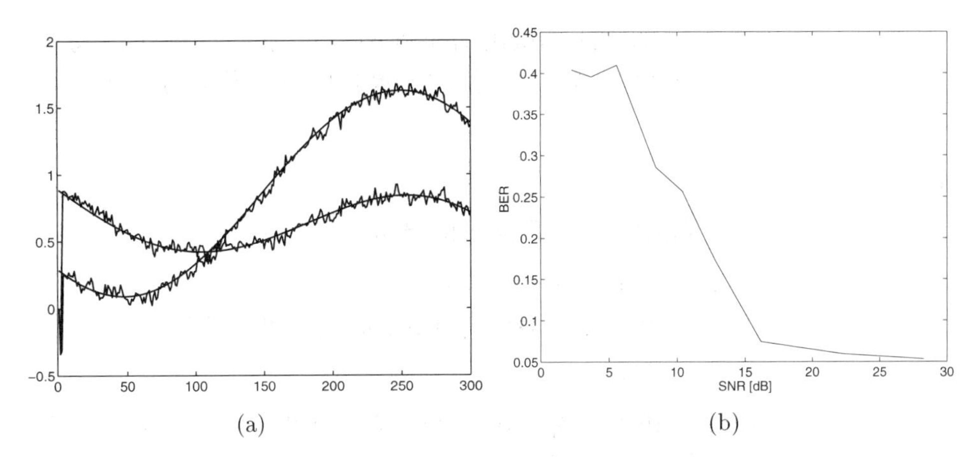

Figure 10.6. Example of estimated and true parameters in a Rayleigh fading channel (a). Bit-error as a function of SNR with 64 parallel filters, using Monte Carlo average over 10 simulations (b).

Signal-to-Noise Ratio (SNR). Rather surprisingly, the mean performance is not much worse compared to the Viterbi algorithm in Example 5.19. Compared to the result in Figure 5.29, the parameters are here rapidly changing in time, and there is no training sequence available.

10.A. Posterior distribution

Theorem 10.1 (MAP estimate of δ^N)
Consider the signal model (10.1). The MAP estimate of δ^N is given by minimizing its negative a posteriori probability

$$-2\log p(\delta^N|y^N) = -2\log p(\delta^N) + 2\log p(y^N) + Np\log 2\pi + \sum_{t=1}^{N}\left\{\log\det S_t(\delta^N) + \epsilon_t^T(\delta^N)S_t^{-1}(\delta^N)\epsilon_t(\delta^N)\right\}. \quad (10.18)$$

Here $\epsilon_t(\delta^N)$ and $S_t(\delta^N)$ are the prediction error and its covariance matrix conditioned on δ^N, computed recursively by the Kalman filter equations

$$\begin{aligned}
\hat{x}_{1|0}(\delta^1) =& x_0 \\
P_{1|0}(\delta^1) =& P_0 \\
\epsilon_t(\delta^t) =& y_t - C_t(\delta_t)\hat{x}_{t|t-1}(\delta^t) - D_{u,t}(\delta_t)u_t + m_{e,t}(\delta_t) \\
S_t(\delta^t) =& C_t(\delta_t)P_{t|t-1}(\delta^t))C_t^T(\delta_t) + R_t(\delta_t) \\
\hat{x}_{t+1|t}(\delta^t) =& A_t(\delta_t)\hat{x}_{t|t-1}(\delta^t) + A_t(\delta_t)P_{t|t-1}(\delta^t)C_t(\delta_t)S_t^{-1}(\delta^t)\epsilon_t(\delta^t) + \\
& B_{u,t}(\delta_t)u_t + m_{v,t}(\delta_t) \\
P_{t+1|t}(\delta^t) =& A_t(\delta_t)\left(P_{t|t-1}(\delta^t) - P_{t|t-1}(\delta^t)C_t^T(\delta_t)S_t^{-1}(\delta^t)C_t(\delta_t)P_{t|t-1}(\delta^t)\right)A_t(\delta_t) \\
& + B_{v,t}(\delta_t)Q_t(\delta_t)B_{v,t}^T(\delta_t).
\end{aligned}$$

Proof: By Bayes' rule we have

$$\begin{aligned}
p(\delta^t|y^t) =& \frac{p(\delta^t)}{p(y^t)}p(y^t|\delta^t) \\
=& \frac{p(\delta^t)}{p(y^t)}p(y_t|y^{t-1},\delta^t)p(y^{t-1}|\delta^t).
\end{aligned}$$

It is a wellknown property of the Kalman filter that

$$(y_t|y^{t-1},\delta^t) \in \mathrm{N}\left(C_t(\delta^t)\hat{x}_{t|t-1}(\delta^t), C_t(\delta_t)P_{t|t-1}(\delta^t)C_t^T(\delta_t) + R_t(\delta_t)\right).$$

Thus, we get

$$p(y_t|y^{t-1},\delta^t) = \gamma\left(y_t - C_t(\delta_t)\hat{x}_{t|t-1}(\delta^t), C_t(\delta_t)P_{t|t-1}(\delta^t)C_t^T(\delta_t) + R_t(\delta_t)\right)$$

which gives the desired result by taking the logarithm. Here $\gamma(x-\mu, P)$ denotes the Gaussian probability density function, with mean μ and covariance matrix P, evaluated at x. □

Note from the proof that (10.18) actually holds if δ^t is continuous as well. The problem is how to minimize the *a posteriori* probability. From (10.18) it follows that the MAP estimator essentially minimizes the sum of squared prediction errors, weighted by their inverse covariance matrices. The first term in the sum is a counterweight to the second. It prevents the covariance matrix from being too large, because that would make the weighted sum of prediction errors very small.

10.A.1. Posterior distribution of the continuous state

Sometimes the distribution of the continuous state is interesting. It follows easily from Theorem 10.1.

Corollary 10.2
Under the same assumptions and notation as in Theorem 10.1, the a posteriori distribution of the continuous state vector is a Gaussian mixture

$$p(x_t|y^N) = \sum_{i=1}^{S^N} p(\delta_i|y^N)\gamma(x_t - \hat{x}_{t|N}(\delta_i), P_{t|N}(\delta_i)). \tag{10.19}$$

Here S^N is the finite number of different δ's, arbitrarily enumerated by the index i, and $\gamma(\cdot,\cdot)$ is the Gaussian density function. If $t < N$, $\hat{x}_{t|N}$ and $P_{t|N}$ denote the smoothed state estimate and its covariance matrix, respectively. For $t = N$ and $t = N+1$, we take the Kalman filter and predictor quantities, respectively.

Proof: The law of total probability gives, since the different δ's are mutually exclusive,

$$p(x_t|y^N) = \sum_{\delta} p(\delta|y^N)p(x_t|y^N, \delta).$$

For Gaussian noise, $p(x_t|y^N, \delta)$ is Gaussian with mean and covariance as given by the Kalman filter. □

The MAP estimator of x_t is $\arg\max_{x_t} p(x_t|y^N)$. Another possible estimate is the *conditional expectation*, which coincides with the minimum variance estimate (see Section 13.2), and it is given by

$$\hat{x}_t^{CE} = \hat{x}_t^{MV} = \sum_{i=1}^{S^N} p(\delta_i|y^N)\hat{x}_{t|N}(\delta_i). \tag{10.20}$$

It is often interesting to compute or plot confidence regions for the state. This is quite complicated for Gaussian mixtures. What one can do is compute the conditional covariance matrix $P_{t|N}$ for x_t, given the measurements y^N. This can then be used for giving approximate confidence regions for the conditional mean. Using the *spread of the mean* formula

$$\begin{aligned} \mathrm{Cov}[X] =& \mathrm{E}_k\mathrm{Cov}[X|k] + \mathrm{Cov}_k\mathrm{E}[X|k] \qquad (10.21)\\ =& \mathrm{E}_k\mathrm{Cov}[X|k] + \mathrm{E}_k\left(\mathrm{E}[X|k]\mathrm{E}^T[X|k]\right) - \left(\mathrm{E}_k\mathrm{E}[X|k]\right)\left(\mathrm{E}_k\mathrm{E}[X|k]\right)^T \end{aligned}$$

where index k means with respect to the probability distribution for k, the conditional covariance matrix follows, as

$$P_{t|N} = \sum_{i=1}^{S^N} p(\delta_i|y^N)\left\{P_{t|N}(\delta_i) + \hat{x}_{t|N}(\delta_i)\hat{x}_{t|N}^T(\delta_i)\right\} - \hat{x}_t^{MV}\left(\hat{x}_t^{MV}\right)^T.$$

Thus, the *a posteriori* covariance matrix for x_t, given measurements y^N is given by three terms. The first is a weighted mean of the covariance matrices for each δ. The last two take the variation in the estimates themselves into consideration. If the estimate of x_t is approximately the same for all δ the first term is dominating, otherwise the variations in estimate might make the covariance matrices negligible.

10.A.2. Unknown noise level

It is wellknown that the properties of the Kalman filter is scale invariant with respect to the noise covariance matrices Q and R. Suppose that all prior covariance matrices are scaled a factor λ. Marking the new quantities with bars we have,

$$\overline{P}_0 = \lambda P_0, \quad \overline{Q} = \lambda Q, \quad \overline{R} = \lambda R. \tag{10.22}$$

It is easily checked from the Kalman filter equations that the estimates $\hat{x}_{t|t-1}$ are still the same. The only difference is that all covariance matrices are scaled a factor λ,

$$\begin{aligned} \overline{P}_{t|t-1} =& \lambda P_{t|t-1} \\ \overline{S}_t =& \lambda S_t. \end{aligned}$$

Thus, for pure filtering the actual level of the covariance matrices does not need to be known.

However, from equation (10.18) in Theorem 10.1 it can easily be checked that the scaling changes the *a posteriori* probability of δ as

$$\begin{aligned} -2\log p(\delta|y^N) =& -2\log p(\delta|y^N, \lambda = 1) \\ &+ Np\log\lambda - (1 - \frac{1}{\lambda})\sum_{t=1}^{N} \epsilon_t^T(\delta) S_t^{-1}(\delta)\epsilon_t(\delta), \end{aligned}$$

and this can have quite serious effects. Consider, for instance, the case when (10.1) describes a single output linear regression. Then, λ scales the measurement noise variance. Over-estimating it gives non-informative results, since the influence of the prediction errors on (10.18) is almost negligible in this case. The MAP estimator then chooses the δ which gives the smallest prediction error covariance matrices, which is independent of the prediction errors.

Thus, there is a need for robustifying the MAP estimate with respect to such unknown levels. This can be done by considering λ as a stochastic variable and using *marginalization*.

Theorem 10.3
Consider the model (10.1) under the same assumptions as in Theorem 10.1, but with an unknown level λ in (10.22). Then if λ is considered as a stochastic variable with no prior information, the a posteriori probability for δ^N is given by

$$\begin{aligned} -2\log p(\delta^N|y^N) =& C - 2\log p(\delta^N) + \sum_{t=1}^{N} \log\det S_t(\delta^t) \\ &+ (Np-2)\log \sum_{t=1}^{N} \epsilon_t^T(\delta^t) S_t^{-1}(\delta^t)\epsilon_t(\delta^t), \end{aligned} \tag{10.23}$$

where

$$C = 2\log p(y^N) + Np\log\pi - 2\log\Gamma\left(\frac{Np-2}{2}\right) + 2\log 2.$$

Here $\Gamma(n)$ is the gamma-function.

Proof: The proof is a minor modification of Theorem 10.1, which is quite similar to Theorem 7.4. □

At this stage, it might be interesting to compare the expressions (10.23) and (10.18) for known and stochastic noise level, respectively. Let

$$D_N = \sum_{t=1}^{N} \log\det S_t(\delta^t) \quad \text{and} \quad V_N = \sum_{t=1}^{N} \epsilon_t^T(\delta^t) S_t^{-1} \epsilon_t(\delta^t).$$

Clearly, the estimator here minimizes $D_N + N\log V_N$ rather than $D_N + V_N$,

$$\begin{aligned} \hat{\delta}^t_{ML} &= \arg\min D_N + V_N, && \text{if } \lambda \text{ is known} \\ \hat{\delta}^t_{ML} &= \arg\min D_N + (Np-2)\log V_N, && \text{if } \lambda \text{ is stochastic.} \end{aligned}$$

Exactly the same statistics are involved; the difference being only how the normalized sum of residuals appears.

11

Change detection based on algebraical consistency tests

11.1. Basics

Consider a batch of data over a sliding window, collected in a measurement vector Y and input vector U. As in Chapter 6, the idea of a consistency test is to apply a linear transformation to a batch of data, $A_iY + B_iU + c_i$. The matrices A_i, B_i and vector c_i are chosen so that the norm of the linear transformation is small when there is no change/fault according to hypothesis H_i, and large when fault H_i has appeared. The approach in this chapter measures the size of

$$\|A_iY + B_iU + c_i\|$$

as a *distance function* in a algebraic meaning (in contrast to the statistical meaning in Chapter 6). The distance measure becomes exactly 'zero' in the non-faulty case, and any deviation from zero is explained by modeling errors and unmodeled disturbances rather than noise.

That is, we will study a noise free state space model

$$x_{t+1} = A_t x_t + B_{u,t} u_t + B_{d,t} d_t + B_{f,t} f_t \tag{11.1}$$
$$y_t = C_t x_t + D_{u,t} u_t + D_{d,t} d_t + D_{f,t} f_t. \tag{11.2}$$

The differences to, for example, the state space model (9.1) are the following:

- The measurement noise is removed.
- The state noise is replaced by a *deterministic disturbance* d_t, which may have a direct term to y_t.
- The state change may have a dynamic profile f_t. This is more general than the step change in (9.1), which is a special case with

$$f_t = \sigma_{t-k}\nu.$$

- There is an ambiguity over how to split the influence from one fault between B_f, D_f and f_t. The convention here is that the *fault direction* is included in B_f, D_f (which are typically time-invariant). That is, $B_f^i f_t^i$ is the influence of fault i, and the scalar *fault profile* f_t^i is the time-varying size of the fault.

The algebraic approach uses the batch model

$$Y_t = \mathcal{O} x_{t-L+1} + H_u U_t + H_d D_t + H_f F_t, \tag{11.3}$$

where

$$Y_t = \begin{pmatrix} y_{t-L+1} \\ y_{t-L+2} \\ \vdots \\ y_t \end{pmatrix}, \quad U_t = \begin{pmatrix} u_{t-L+1} \\ u_{t-L+2} \\ \vdots \\ u_t \end{pmatrix}, \quad D_t = \begin{pmatrix} d_{t-L+1} \\ d_{t-L+2} \\ \vdots \\ d_t \end{pmatrix}, \tag{11.4}$$

$$F_t = \begin{pmatrix} f_{t-L+1} \\ f_{t-L+2} \\ \vdots \\ f_t \end{pmatrix}, \quad \mathcal{O} = \begin{pmatrix} C \\ CA \\ \vdots \\ CA^{L-1} \end{pmatrix}, \quad H = \begin{pmatrix} D & 0 & \cdots & 0 \\ CB & D & & 0 \\ & \ddots & \ddots & 0 \\ CA^{L-2}B & \cdots & CB & D \end{pmatrix}.$$

For simplicity, time-invariant matrices are assumed here. The *Hankel matrix* H is defined identically for all three input signals u, d and f (the subscript defines which one is meant).

The idea now is to compute a linear combination of data, which is usually referred to as a *residual*

$$r_t \triangleq w^T (Y_t - H_u U_t) = w^T (\mathcal{O} x_{t-L+1} + H_d D_t + H_f F_t). \tag{11.5}$$

This equation assumes a stable system (Kinnaert et al., 1995). The equation $r_t = 0$ is called a *parity equation*. For the residual to be useful, we have to impose the following constraints on the choice of the vector or matrix w:

1. Insensitive to the value of the state x_t and disturbances:
$$w^T \begin{pmatrix} \mathcal{O}, & H_d \end{pmatrix} = 0. \tag{11.6}$$
That is, $r_t = w^T(Y_t - H_u U_t) = w^T(\mathcal{O}x_{t-L+1} + H_d D_t) = 0$ when there is no fault. The columns of w^T are therefore vectors in the null space of the matrix $(\mathcal{O},\ H_d)$. This can be referred to as *decoupling* of the residuals from the state vector and disturbances.

2. Sensitive to faults:
$$w^T H_f \neq 0. \tag{11.7}$$
Together with condition 1, this implies $r_t = w^T(Y_t - H_u U_t) = w^T H_f F_t \neq 0$ whenever $F_t \neq 0$.

3. For *isolation*, we would like the residual to react differently to the different faults. That is, the residual vectors from different faults $f^1, f^2, \ldots, f^{n_f}$ should form a certain pattern, called a *residual structure* R. There are two possible approaches:

   a) Transformation of the residuals;
   $$T r_t = T w^T H_f = R^i. \tag{11.8}$$
   This design assumes stationarity in the fault. That is, its magnitude is constant within the sliding window. This implies that there will be a transient in the residual of length L. See Table 11.1 for two common examples on structures R.

   b) For *fault decoupling*, a slight modification of the null space above is needed. Let H_f^i be the fault matrix from fault f^i. There are now n_f such matrices. Replace (11.6) and (11.7) with the iterative scheme
   $$\begin{aligned} w^T \begin{pmatrix} \mathcal{O}, & H_d & H_f^1 & \cdots & H_f^{i-1} & H_f^{i+1} & \cdots & H_f^{n_f} \end{pmatrix} &= 0 \\ w^T H_f^i &\neq 0. \end{aligned}$$
   In this way, the transients in, for instance, Figure 11.4 will disappear. A risk with the previous design is that the time profile f_t^i might excite other residuals causing incorrect isolation, a risk which is eliminated here. On the other hand, it should be remarked that detection should be done faster than isolation, which should be done after transient effects have passed away.

Table 11.1. The residual vector $r_t = w^T(Y_t - H_u U_t)$ is one of the columns of the residual structure matrix R, of which two common examples are given.

Fault	f^1	f^2	f^3
$r_t^{(1)}$	1	0	0
$r_t^{(2)}$	0	1	0
$r_t^{(3)}$	0	0	1

Fault	f^1	f^2	f^3
$r_t^{(1)}$	0	1	1
$r_t^{(2)}$	1	0	1
$r_t^{(3)}$	1	1	0

The derivation based on the three first conditions is given in Section 11.2.

The main disadvantage of this approach is *sensitivity* and *robustness*. That is, the residuals become quite noisy even for small levels of the measurement noise, or when the model used in the design deviates from the true system. This problem is not well treated in the literature, and there are no design rules to be found. However, it should be clear from the previously described design approaches that it is the window size L which is the main design parameter to trade off sensitivity (and robustness) to decreased detection performance.

Equation (11.5) can be expressed in filter form as

$$r_t = A(q)y_t - B(q)u_t,$$

where $A(q)$ and $B(q)$ are polynomials of order L. There are approaches described in Section 11.4, that design the FIR filters in the frequency domain. There is also a close link to observer design presented in Section 11.3. Basically, (11.5) is a dead-beat observer of the faults. Other observer designs correspond to pole placements different from origin, which can be achieved by filtering the residuals in (11.5) by an observer polynomial $C(q)$.

Another approach is also based on observers. The idea is to run a bank of filters, each one using only one output. This is called an *dedicated observer* in Clark (1979). The observer outputs are then compared and by a simple *voting* strategy faulty sensors are detected. The residual structure is the left one in Table 11.1. A variant of this is to include all but one of the measurements in the observer. This is an efficient solution for, for example, navigation systems, since the recovery after a detected change is simple; use the output from the observer not using the faulty sensor. A fault in one sensor will then affect all but one observer, and voting can be applied according to the right-hand structure in Table 11.1. An extension of this idea is to design observers that use all but a subset of inputs, corresponding to the hypothesized faulty actuators. This is called unknown *input observer* (Wünnenberg, 1990). A further alternative is the *generalized observer*, see Patton (1994) and Wünnenberg (1990), which is outlined in Section 11.3. Finally, it will be argued that all observer and frequency domain approaches are equivalent to, or special cases of, the parity space approach detailed here.

Literature

For ten years, the collection by Patton et al. (1989) was the main reference in this area. Now, there are three single-authored monographs in Gertler (1998), Chen and Patton (1999) and Mangoubi (1998). There are also several survey papers, of which we mention Isermann and Balle (1997), Isermann (1997) and Gertler (1997).

The approaches to design suitable residuals are: parity space design (Chow and Willsky, 1984; Ding et al., 1999; Gertler, 1997), unknown input observer (Hong et al., 1997; Hou and Patton, 1998; Wang and Daley, 1996; Wünnenberg, 1990; Yang and Saif, 1998) and in the frequency domain (Frank and Ding, 1994b; Sauter and Hamelin, 1999). A completely different approach is based on reasoning and computer science, and examples here are Årzén (1996), Blanke et al. (1997), Larsson (1994) and Larsson (1999). In the latter approach, Boolean logics and object-orientation are keywords.

A logical approach to merge the deterministic modeling of this chapter with the stochastic models used by the Kalman filter appears in Keller (1999).

11.2. Parity space change detection

11.2.1. Algorithm derivation

The three conditions (11.6)–(11.8) are used to derive Algorithm 11.1 below. Condition (11.6) implies that w belongs to the null space of $(\mathcal{O} \quad H_d)$. This can be computed by a *Singular Value Decomposition (SVD)*:

$$(\mathcal{O} \quad H_d) = UDV^T.$$

These matrices are written in standard notation and should not be confused with U_t and D_t, etc. in the model (11.3). Here D is a diagonal matrix with elements equal to the singular values of $(\mathcal{O} \quad H_d)$, and its left eigenvectors are the rows of U. The null space $\mathcal{N}$ of $(\mathcal{O} \quad H_d)$ is spanned by the last columns of U, corresponding to eigenvalues zero. In the following, we will use the same notation for the *null space* $\mathcal{N}$, as for a basis represented by the rows of a *matrix* $\mathcal{N}$. In MATLAB™ notation, we can take

```
[U,D,V]=svd([O Hd]);
n=rank(D);
N=U(:,n+1:end)';
```

The MATLAB™ function `null` computes the null space directly and gives a slightly different basis. Condition (11.6) is satisfied for any linear combination of $\mathcal{N}$,

$$w^T = T\mathcal{N},$$

where T is an arbitrary (square or thick) matrix. To satisfy condition (11.7), we just have to check that no rows of $\mathcal{N}$ are orthogonal to H_f. If this is the case, these are then deleted and we save $\bar{\mathcal{N}}$. In MATLAB™ notation, we take

```
ind=[];
for i=1:size(N,1);
  a=N(i,:)*Hf;
  if all(a==0);
    ind=[ind i];
  end
end
N(ind,:)=[];
```

That is, the number of rows in $\bar{\mathcal{N}}$, say $n_{\bar{\mathcal{N}}}$, determines how many residuals that can be computed, which is an upper bound on the number of faults that can be detected. This step can be skipped if the isolation design described next is included.

The last thing to do is to choose T to facilitate isolation. Assume there are $n_f \leq n_{\bar{\mathcal{N}}}$ faults, in directions $f^1, f^2, \ldots, f^{n_f}$. Isolation design is done by first choosing a residual structure R. The two popular choices in Table 11.1 are

```
R=eye(nf);
R=ones(nf)-eye(nf);
```

The transformation matrix is then chosen as the solution to the equation system

$$\begin{aligned} T\mathcal{N}H_f\left(\mathbf{1}_L \otimes \begin{pmatrix} f^1 & f^2 & \cdots & f^{n_f}\end{pmatrix}\right) =& R \\ w^T =& T\mathcal{N}. \end{aligned}$$

Here $\mathbf{1}_L$ is a vector of L ones and $\otimes$ denotes the Kronecker product. That is, $\mathbf{1}_L \otimes f^i$ is another way of writing F_t^i when the fault magnitude is constant during the sliding window. In MATLAB™ notation for $n_f = 3$, this is done by

```
T = R / (N*Hf*kron(ones(L,1),[f1 f2 f3]));
w = (T*N)';
```

To summarize, we have the following algorithm:

Algorithm 11.1 Parity space change detection

Given: a state space model (11.1).
Design parameters: sliding window size L and residual structure R.
Compute recursively:

1. The data vectors Y_t and U_t in (11.3) and the model matrices $\mathcal{O}, H_d, H_f$ in (11.4).
2. The null space $\mathcal{N}$ of $\begin{pmatrix}\mathcal{O} & H_d\end{pmatrix}$ is spanned by the last columns of U, corresponding to eigenvalues zero. In MATLAB™ formalism, the transformation matrix giving residual structure R is computed by:

```
[U,D,V]=svd([O Hd]);
n=rank(D);
N=U(:,n+1:end);
T = R / (N*Hf*kron(ones(L,1),[f1 f2 f3]));
w = (T*N)';
```

3. Compute the residual $r = w^T(Y_t - H_u U_t)$,

```
r=w'*(Y-Hu*U);
```

4. Change detection if $r^T r > 0$, or $r^T r > h$ considering model uncertainties.
5. Change isolation. Fault i in direction f^i where $i = \arg\max_i r^T R_i$. R_i denotes column i of R.

```
[dum,i]=max(r'*R);
```

It should be noted that the residual structure R is no real design parameter, but rather a tool for interpreting and illustrating the result.

11.2.2. Some rank results

Rank of null space

When does a parity equation exist? That is, when is the size of the null space $\mathcal{N}$ different from 0? We can do some quick calculations to find the rank of $\mathcal{N}$. We have

$$\text{rank}(\mathcal{N}) = n_y L - \text{rank}(\mathcal{O}) - \text{rank}(H_d) \tag{11.9}$$

$$\text{rank}(\mathcal{O}) = n_x \tag{11.10}$$

$$\text{rank}(H_d) = n_d L \tag{11.11}$$

$$\Rightarrow \text{rank}(\mathcal{N}) = L(n_y - n_d) - n_x. \tag{11.12}$$

It is assumed that the ranks are determined by the column spaces, which is the case if the number of outputs is larger than the number of disturbances (otherwise condition (11.6) can never be satisfied). In (11.9) it is assumed that the column space of $\mathcal{O}$ and H_d do not overlap, otherwise the rank of $\mathcal{N}$ will be larger. That is, a lower bound on the rank of the null space is $\text{rank}(\mathcal{N}) \geq L(n_y - n_d) - n_x$.

The calculations above give a condition for detectability. For isolability, compute $\mathcal{N}_i$ for each fault f_i. Isolability is implied by the two conditions $\mathcal{N}_i \neq 0$ for all i and that $\mathcal{N}_i$ is not parallel to $\mathcal{N}_j$ for all $i \neq j$.

Requirement for isolation

Let

$$\tilde{H}_f = H_f \left(\mathbf{1}_L \otimes \begin{pmatrix} f^1 & f^2 & \cdots & f^{n_f} \end{pmatrix} \right).$$

Now we can write the residuals as

$$r_t = T\mathcal{N}\tilde{H}_f F_t.$$

The faults can be isolated using the residuals if and only if the matrix $\mathcal{N}\tilde{H}_f$ is of rank n_f,

$$\text{rank}(\mathcal{N}\tilde{H}_f) = n_f.$$

In the case of the same number of residuals as faults, we simply take $f_t = (T\mathcal{N}\tilde{H}_f)^{-1} r_t$. The design of the transformation matrix to get the required fault structure then also has a unique solution,

$$T = (\mathcal{N}\tilde{H}_f)^{-1} R.$$

It should be remarked, however, that the residual structure is cosmetics which may be useful for monitoring purposes only. The information is available in the residuals with or without transformation.

Minimal order residual filters

We discuss why window sizes larger than $L = n_x$ do not need to be considered. The simplest explanation is that a state observer does not need to be of a higher dimension (the state comprises all of the information about the system, thus also detection and isolation information). Now so-called *Luenberger observers* can be used to further decrease the order of the observer. The idea here is that the known part in the state from each measurement can be updated exactly.

A similar result exists here as well, of course. A simple derivation goes as follows.

Take a QR factorization of the basis for the null space,

$$\mathcal{N} = QR.$$

It is clear that by using $T = Q^T$ we get residuals

$$r_t = T\mathcal{N}(Y_t - H_u U_t) = Q^T QR(Y_t - H_u U_t) = R(Y_t - H_u U_t).$$

The matrix R looks like the following:

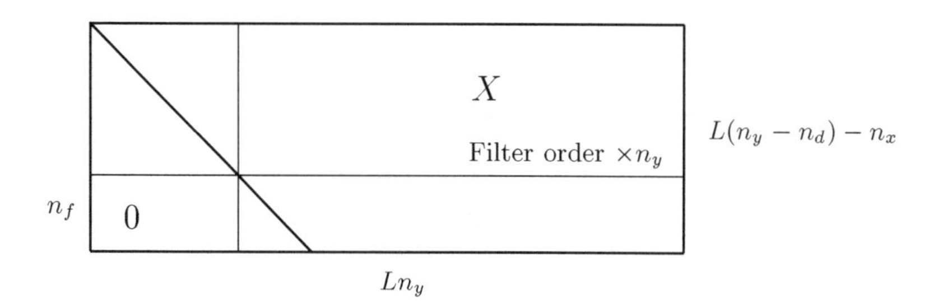

The matrix has zeros below the diagonal, and the numbers indicate dimensions. We just need to use n_f residuals for isolation, which can be taken as the last n_f rows of r_t above. When forming these n_f residuals, a number of elements in the last rows in R are zero. Using geometry in the figure above, the filter order is given by

$$\frac{Ln_y - L(n_y - n_d) + n_x + n_f}{n_y} = \frac{Ln_d + n_x + n_f}{n_y}. \tag{11.13}$$

This number must be rounded upwards to get an integer number of measurement vectors. See Section 11.5.3 for an example.

11.2.3. Sensitivity and robustness

The design methods presented here are based on purely algebraic relations. It turns out, from examples, that the residual filters are extremely sensitive to measurement noise and lack robustness to modeling errors. Consider, for example, the case of measurement noise only, and the case of no fault:

$$r_t = w^T(Y_t + E_t - H_u U_t) = w^T E_t.$$

Here E_t is the stacked vector of measurement noises, which has covariance matrix

$$\text{Cov}(E_t) = \text{E}(E_t E_t^T) = I_L \otimes R.$$

Here $\otimes$ denotes Kronecker product. The covariance matrix of the residuals is given by

$$\text{Cov}(r_t) = w^T \text{Cov}(E_t) w = w^T (I_L \otimes R) w. \tag{11.14}$$

The reason this might blow up is that the components of w might become several order of magnitudes larger than one, and thus magnifying the noise. See Section 11.5.3 for an example. In this section, examples of both measurement noise and modeling error are given, which show that small measurement noise or a small system change can give very 'noisy' residuals, due to large elements in w.

11.2.4. Uncontrollable fault states

A generalization of this approach in the case the state space model has states ('fault states') that are not controllable from the input u_t and disturbance d_t is presented in Nyberg and Nielsen (1997). Suppose the state space model can be written

$$\begin{pmatrix} x^1_{t+1} \\ x^2_{t+1} \end{pmatrix} = \begin{pmatrix} A^{11} & A^{12} \\ 0 & A^{22} \end{pmatrix} \begin{pmatrix} x^1_t \\ x^2_t \end{pmatrix} + \begin{pmatrix} B^1_u \\ 0 \end{pmatrix} u_t + \begin{pmatrix} B^1_d \\ 0 \end{pmatrix} d_t + \begin{pmatrix} B^1_f \\ B^2_f \end{pmatrix} f_t \tag{11.15}$$

$$y_t = \begin{pmatrix} C^1 & C^2 \end{pmatrix} x_t + D_{u,t} u_t + D_{d,t} d_t + D_{f,t} f_t. \tag{11.16}$$

Here x^2 is not controllable from the input and disturbance. Now we can split

$$\mathcal{O} = \begin{pmatrix} \mathcal{O}^1 & \mathcal{O}^2 \end{pmatrix}$$

in an obvious manner. The steps (11.6) and (11.7) can now be replaced by

$$w^T \begin{pmatrix} \mathcal{O}^1 & H_d \end{pmatrix} = 0$$
$$w^T \begin{pmatrix} \mathcal{O}^2 & H_f \end{pmatrix} \neq 0.$$

The reason is that we do not need to decouple the part of the state vector which is not excited by u, d. In this way, the number of candidate residuals – that is, the number of rows in w^T – is much larger than for the original design method. First, the null space of $\begin{pmatrix} \mathcal{O}^1 & H_d \end{pmatrix}$ is larger than for $\begin{pmatrix} \mathcal{O}^1 & \mathcal{O}^2 & H_d \end{pmatrix}$, and secondly, it is less likely that the null space is orthogonal to $\begin{pmatrix} \mathcal{O}^2 & H_f \end{pmatrix}$ than to H_f in the original design. Intuitively, the 'fault states' x^2 are unaffected by input and disturbance decoupling and are observable from the output, which facilitates detection and isolation.

11.2.5. Open problems

The freedoms in the design are the sliding window size L and the transformation matrix T in $w^T = T\mathcal{N}$. The trade-off in L in statistical methods does not appear in this noiseless setting. However, as soon as small measurement noise is introduced, longer window sizes seem to be preferred from a noise rejection point of view.

- In some applications, it might be interesting to force certain columns in w^T to zero, and in that way remove influence from specific sensors or actuators. This is the idea of unknown input observers and dedicated observers.

- From a measurement noise sensitivity viewpoint, a good design gives elements in w of the same order. The examples in the next section show that the elements of w might differ several order of magnitudes from a straightforward design. When L is further increased, the average size of the elements will decrease, and the central limit theorem indicates that the noise attenuation will improve.

11.3. An observer approach

Another approach to residual generation is based on observers. The following facts are important here:

- We know that the state of a system per definition contains all information about the system, and thus also faults.

- There is no better (linear) filter than the observer to compute the states.

- The observer does not have to be of higher order than the system order n_x. The so called *dead-beat observer* can be written as

$$\hat{x}_t = C_y(q)y_t + C_u(q)u_t,$$

  where $C_y(q)$ and $C_u(q)$ are polynomials of order n_x. An arbitrary observer polynomial can be introduced to attenuate disturbances.

This line of arguments indicates that any required residual can be computed by linear combinations of the states estimated by a dead-beat observer,

$$r_t = L\hat{x}_t \triangleq A(q)y_t - B(q)u_t.$$

That is, the residual can be generated by an FIR filter of order n_x. This also indicates that the largest sliding window which needs to be considered is $L = n_x$. This fact is formally proved in Nyberg and Nielsen (1997). The dead-beat observer is sensitive to disturbances. The linear combination L can be designed to decouple the disturbances, and we end up at essentially the same residual as from a design using parity spaces in Section 11.2.

More specifically, let

$$z_t = \begin{pmatrix} \hat{x}_{t+1} - A\hat{x}_t - B_u u_t \\ y_t - C\hat{x}_t - D_u u_t \end{pmatrix}$$

be our candidate residual. Clearly, this is zero when there is no disturbance or fault and when the initial transient has passed away. If the observer states were replaced by the true states, then we would have

$$z_t^0 = \begin{pmatrix} B_d d_t + B_f f_t \\ D_d d_t + D_f f_t \end{pmatrix}.$$

In that case, we could choose L such that the disturbance is decoupled by requiring $L(B_d^T,\ D_d^T)^T = 0$. However, the observer dynamics must be included and the design becomes a bit involved. The bottom line is that the residuals can be expressed as a function of the observer state estimates, which are given by a filter of order n_x.

11.4. An input-output approach

An input-output approach (see Frisk and Nyberg (1999)) to residual generation, is as follows. Let the undisturbed fault-free system be

$$y_t = H_u(q)u_t + H_d(q)d_t + H_f(q)f_t.$$

A residual generator may be taken as

$$r_t = W^T(q)\begin{pmatrix} y_t \\ u_t \end{pmatrix} = W^T(q)\underbrace{\begin{pmatrix} H_u(q) & H_d(q) \\ I & 0 \end{pmatrix}}_{M(q)}\begin{pmatrix} u_t \\ d_t \end{pmatrix} + W^T(q)\begin{pmatrix} H_f(q) \\ 0 \end{pmatrix} f_t,$$

which should be zero when no fault or disturbance is present. $M(q)$ must belong to the left null space of $W^T(q)$. The dimension of this null space and thus the dimension of r_t is n_y if there is no disturbances ($n_d = 0$) and less otherwise.

The order of the residual filter is L, so this design should give exactly the same degrees of freedom as using the parity space.

11.5. Applications

One of most important applications for fault diagnosis is in automotive engines (Dinca et al., 1999; Nyberg, 1999; Soliman et al., 1999). An application to an unmanned underwater vehicle is presented in Alessandri et al. (1999). Here we will consider systems well approximated with a linear model, in contrast to an engine, for instance. This enables a straightforward design and facilitates evaluation.

11.5.1. Simulated DC motor

Consider a sampled state space model of a DC motor with continuous time transfer function

$$G(s) = \frac{1}{s(s+1)}$$

sampled with a sample interval $T_s = 0.4$s. This is the same example used throughout Chapter 8, and the fault detection setup is the same as in Example 9.3.

The state space matrices with x^1 being the angle and x^2 the angular velocity are

$$A = \begin{pmatrix} 1 & 0.3297 \\ 0 & 0.6703 \end{pmatrix}, \quad B_u = \begin{pmatrix} 0.0703 \\ 0.3297 \end{pmatrix}, \quad B_d = \begin{pmatrix} 0 \\ 0 \end{pmatrix}, \quad B_f = \begin{pmatrix} 1 & 0 \\ 0 & 1 \end{pmatrix},$$
$$C = \begin{pmatrix} 1 & 0 \\ 0 & 1 \end{pmatrix}, \quad D_u = \begin{pmatrix} 0 \\ 0 \end{pmatrix}, \quad D_d = \begin{pmatrix} 0 \\ 0 \end{pmatrix}, \quad D_f = \begin{pmatrix} 0 & 0 \\ 0 & 0 \end{pmatrix}.$$

It is assumed that both x_1 and x_2 are measured. Here we have assumed that a fault enters as either an angular or velocity change in the model. The matrices in the sliding window model (11.3) become for $L = 2$:

$$\mathcal{O} = \begin{pmatrix} 1 & 0 \\ 0 & 1 \\ 1 & 0.3297 \\ 0 & 0.6703 \end{pmatrix}, \quad H_u = \begin{pmatrix} 0 & 0 \\ 0 & 0 \\ 0.0703 & 0 \\ 0.3297 & 0 \end{pmatrix}, \quad H_f = \begin{pmatrix} 0 & 0 & 0 & 0 \\ 0 & 0 & 0 & 0 \\ 1 & 0 & 0 & 0 \\ 0 & 1 & 0 & 0 \end{pmatrix},$$
$$\mathcal{N} = \begin{pmatrix} -0.6930 & -0.1901 & 0.6930 & -0.0572 \\ 0.0405 & -0.5466 & -0.0405 & 0.8354 \end{pmatrix}.$$

State faults

A simulation study is performed, where the first angle fault is simulated followed by the second kind of fault in angular velocity. The residuals using $w^T = \mathcal{N}$ are shown in the upper plot in Figure 11.1. The null space is trans-

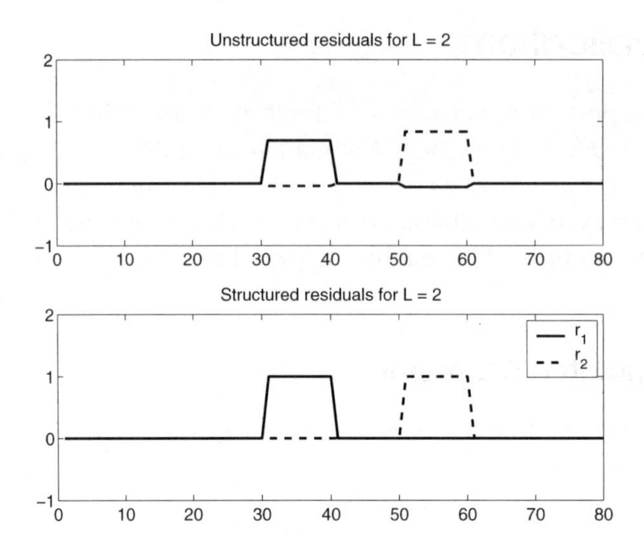

Figure 11.1. Residuals for sliding window $L = 2$. Upper plot shows unstructured residuals and lower plot structured residuals according to the left table in Table 11.1.

formed to get structured residuals according to the pattern in Table 11.1. That is, a fault in angle will show up as a non-zero residual r_t^1, but the second residual r_t^2 remains zero. The data projection matrix becomes

$$w^T = \begin{pmatrix} -1 & -0.3297 & 1 & 0 \\ 0 & -0.6703 & 0 & 1 \end{pmatrix}.$$

One question is what happens if the sliding window size is increased. Figure 11.2 shows the structured and unstructured residuals. One conclusion is that the null space increases linearly as Ln_y, but the rank of H_f also increases as Ln_y, so the number of residual candidates does not change. That is, a larger window size than $L = n_x$ does not help us in the diagnosis.

Actuator and sensor faults

A more realistic fault model is to assume actuator and (for instance angular) sensor offsets. This means that we should use

$$B_f = \begin{pmatrix} B_u & 0 \end{pmatrix}$$
$$D_f = \begin{pmatrix} 0 & 1 \\ 0 & 0 \end{pmatrix}.$$

In this way, $f_t = (\alpha_t,\ 0)^T$ means actuator offset of size α_t, and $f_t = (0,\ \alpha_t)^T$ means angular sensor offset. However, these two faults are not possible to

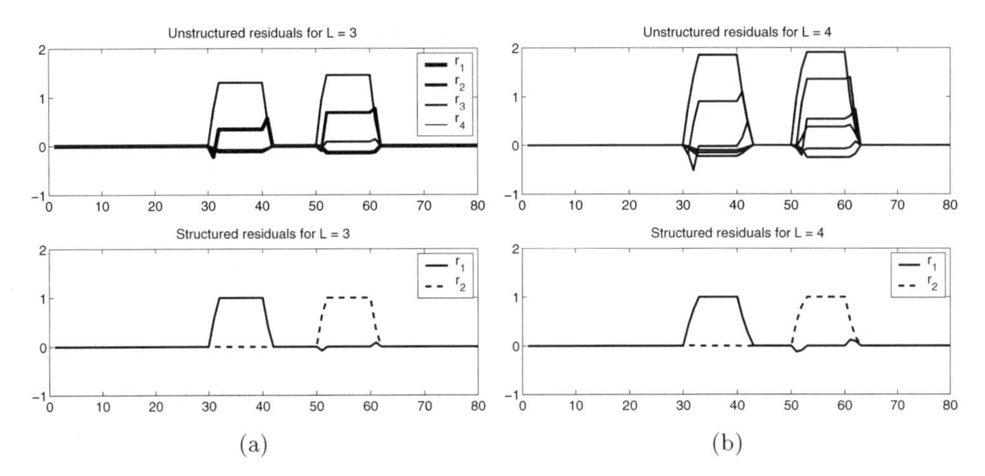

Figure 11.2. Residuals for sliding window $L = 3$ and $L = 4$, respectively. Upper plots show unstructured residuals and lower plots structured residuals according to the left table in Table 11.1

isolate, though they are possible to detect. The only matrix that is influenced by the changed fault assumptions is

$$H_f = \begin{pmatrix} 0 & 1 & 0 & 0 \\ 0 & 0 & 0 & 0 \\ 0.0703 & 0 & 0 & 1 \\ 0.3297 & 0 & 0 & 0 \end{pmatrix}.$$

Here we get a problem, because the second fault (columns 2 and 4 in H_f) are parallel with the first column in $\mathcal{O}$. That is, an angular disturbance cannot be distinguished from the influence of the initial filter states on the angle.

Increasing the window to $L = 3$ does not help:

$$H_f = \begin{pmatrix} 0 & 1 & 0 & 0 & 0 & 0 \\ 0 & 0 & 0 & 0 & 0 & 0 \\ 0.0703 & 0 & 0 & 1 & 0 & 0 \\ 0.3297 & 0 & 0 & 0 & 0 & 0 \\ 0.1790 & 0 & 0.0703 & 0 & 0 & 1 \\ 0.2210 & 0 & 0.3297 & 0 & 0 & 0 \end{pmatrix}.$$

11.5.2. DC motor

Consider the DC motor lab experiment described in Sections 2.5.1 and 2.7.1. It was examined with respect to system changes in Section 5.10.2, and a statistical approach to disturbance detection was presented in Section 8.12.1. Here we apply the parity space residual generator to detect the torque disturbances while being insensitive to system changes. The test cases in Table 2.1 are considered. The state space model is given in (2.2).

From Figure 11.3, we conclude the following:

- The residual gets a significant injection at the time of system changes and disturbances.
- These times coincide with the alarms from the Kalman filter residual whiteness test. The latter method seems to be easier to use and more robust, due to the clear peaks of the test statistics.
- It does not seem possible to solve the fault isolation problem reliably with this method. See Gustafsson and Graebe (1998) for a change detection method that works for this example.

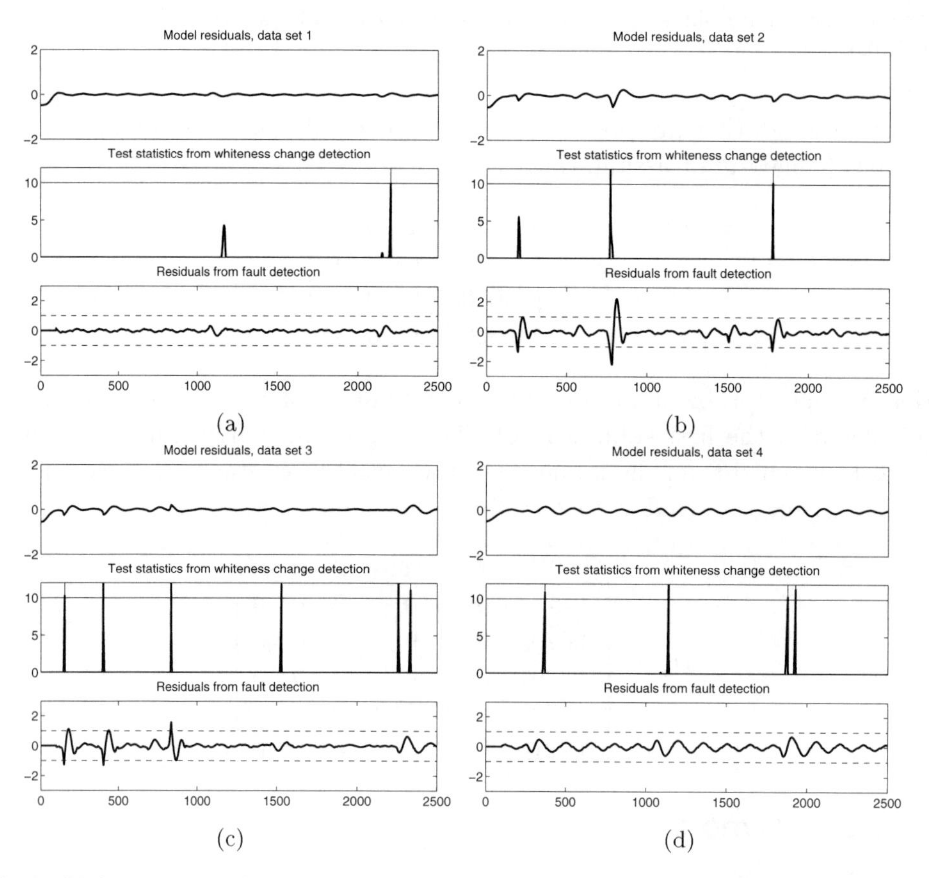

Figure 11.3. Simulation error using the state space model in (2.2), test statistics from the CUSUM whiteness test of the Kalman filter for comparison and parity space residuals. Nominal system (a), with torque disturbances (b), with change in dynamics (c) and both disturbance and change (d).

11.5.3. Vertical aircraft dynamics

We investigate here fault detection in the vertical-plane dynamics of an F-16 aircraft. The F-16 aircraft is, due to accurate public models, used in many simulation studies. For instance, Eide and Maybeck (1995, 1996) use a full scale non-linear model for fault detection. Related studies are Mehra et al. (1995), where fault detection is compared to a Kalman filter, and Maybeck and Hanlon (1995).

The dynamics can be described by a transfer function $y_t = G(q)u_t$ or by a state space model. These models are useful for different purposes, just as the DC motor in Sections 2.5.1 and 2.7.1. One difference here is that the dynamics generally depend upon the operating point.

The state, inputs and outputs for this application are:

Inputs	Outputs	States
u_1: spoiler angle $[0.1deg]$	y_1: relative altitude $[m]$	x_1: altitude $[m]$
u_2: forward acceleration $[m/s^2]$	y_2: forward speed $[m/s]$	x_2: forward speed $[m/s]$
u_3: elevator angle $[deg]$	y_3: pitch angle $[deg]$	x_3: pitch angle $[deg]$
		x_4: pitch rate $[deg/s]$
		x_5: vertical speed $[deg/s]$

The numerical values below are taken from Maciejowski (1989) (given in continuous time) sampled with 10 Hz:

$$A = \begin{pmatrix} 1 & 0.0014 & 0.1133 & 0.0004 & -0.0997 \\ 0 & 0.9945 & -0.0171 & -0.0005 & 0.0070 \\ 0 & 0.0003 & 1.0000 & 0.0957 & -0.0049 \\ 0 & 0.0061 & -0.0000 & 0.9130 & -0.0966 \\ 0 & -0.0286 & 0.0002 & 0.1004 & 0.9879 \end{pmatrix},$$

$$B_u = \begin{pmatrix} -0.0078 & 0.0000 & 0.0003 \\ -0.0115 & 0.0997 & 0.0000 \\ 0.0212 & 0.0000 & -0.0081 \\ 0.4150 & 0.0003 & -0.1589 \\ 0.1794 & -0.0014 & -0.0158 \end{pmatrix},$$

$$B_d = \begin{pmatrix} 0 \\ 1 \\ 0 \\ 0 \\ 0 \end{pmatrix}, \quad B_f = \begin{pmatrix} -0.0078 & 0.0000 & 0.0003 & 0 & 0 & 0 \\ -0.0115 & 0.0997 & 0.0000 & 0 & 0 & 0 \\ 0.0212 & 0.0000 & -0.0081 & 0 & 0 & 0 \\ 0.4150 & 0.0003 & -0.1589 & 0 & 0 & 0 \\ 0.1794 & -0.0014 & -0.0158 & 0 & 0 & 0 \end{pmatrix},$$

$$C = \begin{pmatrix} 1 & 0 & 0 & 0 & 0 \\ 0 & 1 & 0 & 0 & 0 \\ 0 & 0 & 1 & 0 & 0 \end{pmatrix}, \quad D_f = \begin{pmatrix} 0 & 0 & 0 & 1 & 0 & 0 \\ 0 & 0 & 0 & 0 & 1 & 0 \\ 0 & 0 & 0 & 0 & 0 & 1 \end{pmatrix}.$$

The disturbance is assumed to act as an additive term to the forward speed.

Fault detection

The fault model is one component f^i for each possible actuator and sensor fault. In other words, in the input-output domain we have

$$y_t = G(q)\left(u_t + \begin{pmatrix} f_t^1 \\ f_t^2 \\ f_t^3 \end{pmatrix}\right) + \begin{pmatrix} f_t^4 \\ f_t^5 \\ f_t^6 \end{pmatrix}.$$

Using $L = 5$ gives a null space $\mathcal{N}$ being a 6×15 matrix. A QR-factorization of $\mathcal{N} = QR$ gives an upper-diagonal matrix R, whose last row has nine zeros and six non-zero elements,

$$R^{10,\cdot} = (0^{1,9},\ -0.0497,\ -0.7035,\ 0.0032,\ 0.0497,\ 0.7072,\ 0.033).$$

Since the orthogonal matrix Q cannot make any non-zero vector zero, we can use $w^T = R$. More specifically, we can take the last row of R to compute a residual useful for change detection. This residual will only use the current and past measurements of y, u, so the *minimal order residual filter* is thus of order two.

Isolation

Suppose we want to detect and isolate faults in f^1 and f^6. We use the structural matrices

$$F = \begin{pmatrix} 1 & 0 \\ 0 & 0 \\ 0 & 0 \\ 0 & 0 \\ 0 & 0 \\ 0 & 1 \end{pmatrix},$$

$$R = \begin{pmatrix} 1 & 0 \\ 0 & 1 \end{pmatrix}.$$

The projection matrix becomes

$$w^T = \begin{pmatrix} -32 & 0.05 & 0.76 & 30 & 0.10 & -7.6 & 24 & 0.07 & \dots \\ & \dots & 13 & -11 & 0.02 & -5.6 & -12 & 0 & -0.18 \\ 132 & -0.20 & 21 & -135 & -0.39 & 15 & -79 & -0.26 & \dots \\ & \dots & -69 & 41 & -0.067 & 6.73 & 40 & 0 & 28 \end{pmatrix}.$$

The largest element is 132 (compare with (11.14)). One can also note that the ratio of the largest and smallest elements is

$$\frac{\max |w|}{\min |w|} = 8023.$$

Thus, the filter coefficients in w has a large dynamic range, which might indicate numerical difficulties.

A simulation gives the residuals in Figure 11.4. The first plot shows unstructured residuals and the second the structured ones. The transient caused by the filter looks nastier here, largely due to the higher order filters.

Isolation using minimum order filters

The transient problem can be improved on by using minimum order residual filters. Using the QR factorization of $\mathcal{N}$ above, and only using the last two rows of R, we get

$$w^T = \begin{pmatrix} 0 & 0 & 0 & 0 & -0.0032 & 16 & -76 & 0.11 & \dots \\ & \dots & -55 & 153 & 0.12 & 62 & -76 & 0 & -23 \\ 0 & 0 & 0 & 0 & 0.0064 & -32 & 301 & -0.46 & \dots \\ & \dots & 202 & -593 & -0.46 & -315 & 292 & 0 & 145 \end{pmatrix}.$$

Here the largest element is 593 and the ratio of the largest and smallest element is 183666. That is, the price paid for the minimal filter is much larger elements in w, and thus according to (11.14) a much larger residual variance. This explains the sensitivity to noise that will be pointed out.

The order of the filter is 3, since the last 7 components of Y_t corresponds to three measurements (here only the third component of y_{t-2} is used). The result is consistent with the formula (11.13),

$$\frac{Ln_d + n_x + n_f}{n_y} = \frac{0+5+2}{3}.$$

The third plot in Figure 11.4 shows how the transients now only last for two samples, which will allow quicker diagnosis.

Sensitivity to noisy measurements

Suppose we add very small Gaussian noise to the measurements,

$$y_t^m = y_t + e_t, \quad \text{Cov}(e_t) = R.$$

The residuals in Figure 11.4 are now shown in Figure 11.5 for $R = 10^{-6}I$. This level on the noise gives an SNR of $5 \cdot 10^6$. We see that the residual filter is very sensitive to noise, especially the minimal order filter.

Exactly as for other change detection approaches based on sliding windows, the noise attenuation improves when the sliding window increases at the cost of longer delay for detection and isolation. Figure 11.5 also shows the residuals in the case when the window length is increased to $L = 10$.

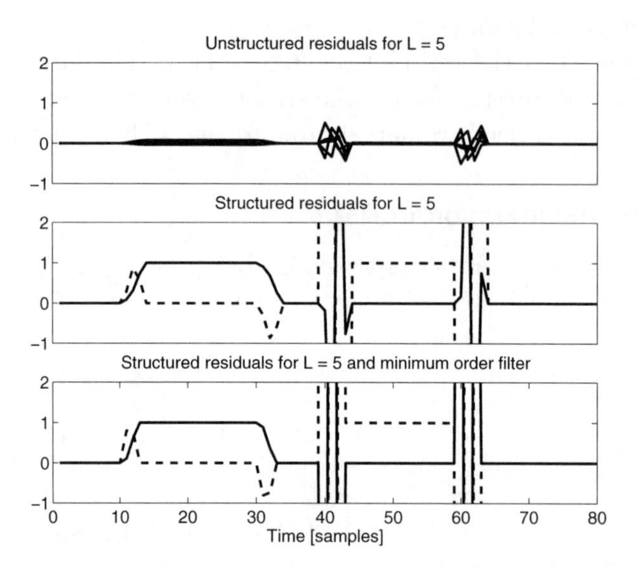

Figure 11.4. Residuals for sliding window $L = 5$ for the aircraft model. The upper plot shows unstructured residuals, and the middle plot structured residuals according to the left table in Table 11.1. The lower plot shows minimal order residuals.

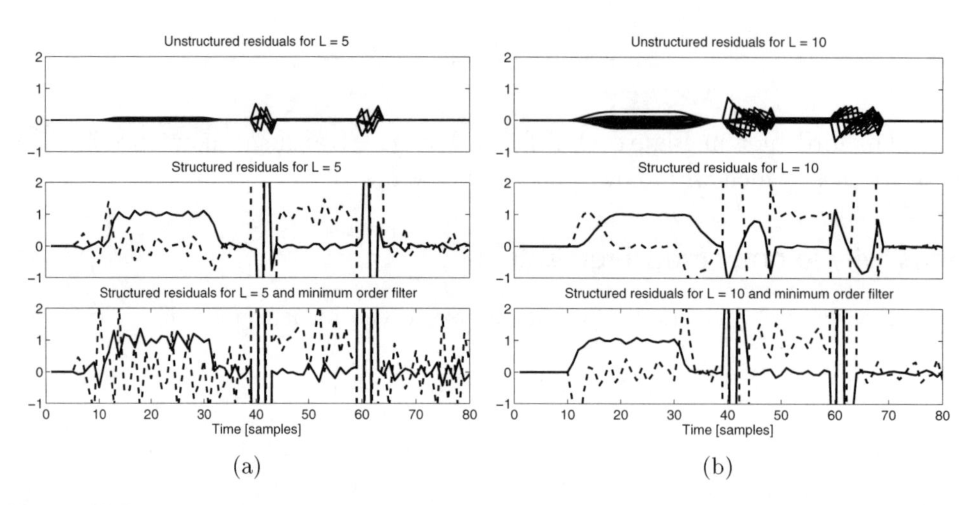

Figure 11.5. Illustration of the sensitivity to measurement noise. Residuals for sliding window $L = 5$ and $L = 10$, respectively, for the aircraft model. The upper plot shows unstructured residuals, and the middle plot structured residuals according to the left table in Table 11.1. The lower plot shows minimal order residuals.

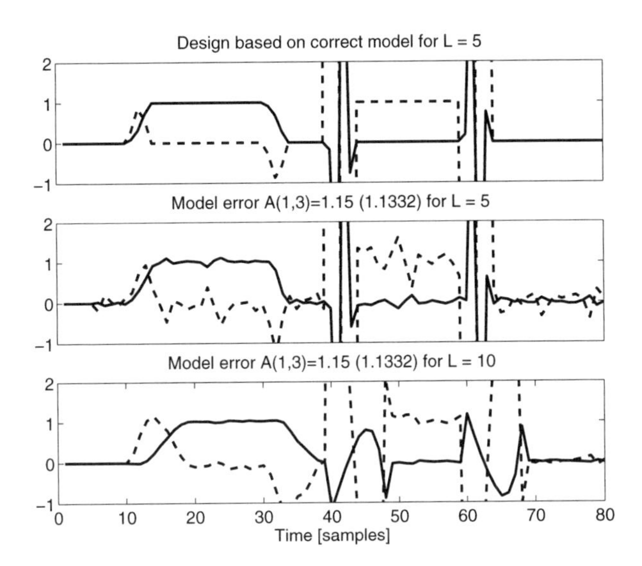

Figure 11.6. Illustration of the robustness to model errors. A design based on the correct model used in the simulation gives structured residuals as in the first subplot. The other two shows the structured residuals when the $A(1,2)$ element is increased 1%, for sliding window sizes $L = 5$ and $L = 10$, respectively.

Robustness to incorrect model

The noise sensitivity indicates that there might be robustness problems as well. This is illustrated by changing one of the parameters in the example by 1%. This is done by letting $A(1,2) = 1.15$ in the design instead of $A(1,2) = 1.1332$ as used when simulating the data. The result in Figure 11.6 shows that small model errors can lead to 'noisy' residuals, and decision making becomes hazardous. Again, improved performance is obtained by increasing the window size at the cost of increased delay for detection.

Part V: Theory

12

Evaluation theory

12.1. Filter evaluation

12.1.1. Filter definitions

Consider a general linear filter (or estimator)

$$\hat{x}_t = \sum_{i=t_1}^{t_2} \alpha_{t,i} z_{t-i}, \tag{12.1}$$

as illustrated in Figure 12.1. Typically, the estimated quantity x is either the parameter vector θ in a parametric model, or the state x in a state space model. It might also be the signal component s_t of the measurement $y_t = s_t + v_t$. The measurements z_t consist of the measured outputs y_t and, when appropriate, the inputs u_t.

The basic definitions that will be used for a *linear filter* are:

- A *time-invariant filter* has $\alpha_{t,i} = \alpha_i$ for all t and i.
- A *non-causal filter* has $t_1 < 0$. This is used in *smoothing*. If $t_1 \geq 0$, the filter is *causal*.

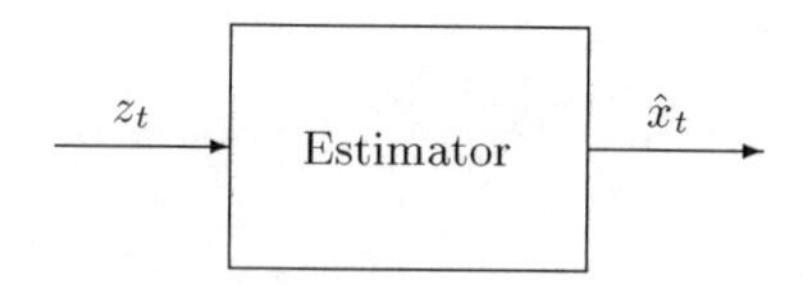

Figure 12.1. An estimator takes the observed signal z_t and transforms it to estimates $\hat{x}_t$.

- A causal filter is *IIR* (*Infinite Impulse Response*) if $t_2 = \infty$, otherwise it is *FIR* (*Finite Impulse Response*).
- $\hat{x}_t$ is a *k-step ahead prediction* if $t_1 = k > 0$.

Most filters use all past data, so $t_2 = \infty$, and the notation $\hat{x}_{t|t-t_1}$ is sometimes useful for highlighting that the estimator is a predictor ($t_1 > 0$), a smoother ($t_1 < 0$) or a filter ($t_1 = 0$).

12.1.2. Performance measures

The estimator gives a so called point estimate of x. For evaluation, we are interested in the variability between different realizations:

- One such measure is the covariance matrix

$$P(t) = \mathrm{E}(\hat{x}_t - x_t^o)(\hat{x}_t - x_t^o)^T. \tag{12.2}$$

  Here and in the sequel, the super-index ^{o} means the true value. Such a covariance matrix is provided by the Kalman filter, for instance, but it reflects the true variability only if the underlying signal model is correct, including its stochastic assumptions.

- A scalar measure of performance is often to prefer when evaluating different filters, such as the square root of the mean value of the norm of the estimation error

$$\sqrt{\mathrm{tr}(P(t))} = \left(\mathrm{E}(\hat{x}_t - x_t^o)^T(\hat{x}_t - x_t^o)\right)^{1/2} = \sqrt{\mathrm{E}(\|\hat{x}_t - x_t^o\|_2^2)}, \tag{12.3}$$

  where the subindex 2 stands for the 2-norm and tr for trace, which is the sum of diagonal elements of the matrix argument. This is a measure of the length of the estimation error. One can think of it as the standard deviation of the estimation error.

- Sometimes the second order properties are not enough, and the complete *Probability Density Function* (PDF) needs to be estimated.

- Confidence intervals for the parameters and hypothesis tests are other related applications of variability measures.

Monte Carlo simulations offer a means for estimating these measures. Before proceeding, let us consider the causes for variability in more detail:

- *Variance error* is caused by the variability of different noise realizations, which gives a variability in $\hat{x}_t - \mathrm{E}(\hat{x}_t)$.

- *Bias* and *tracking error* are caused by an error in the signal model (bias) and inability of the filter to track fast variations in x_t^o (tracking error), respectively. The combined effect is a deviation in $x_t^o - \mathrm{E}(\hat{x}_t)$. Determination of bias error is in general a difficult task and will not be discussed further in this section. We will assume that there is no bias, only tracking error.

Hence, the covariance matrix can be seen as a sum of two terms,

$$P(t) = \mathrm{E}(\|\hat{x}_t - x_t^o\|_2^2) = \underbrace{\mathrm{E}(\|\hat{x}_t - \mathrm{E}(\hat{x}_t)\|_2^2)}_{\text{variance error}} + \underbrace{\mathrm{E}(\|\mathrm{E}(\hat{x}_t) - x_t^o\|_2^2)}_{\text{bias and tracking error}}. \tag{12.4}$$

12.1.3. Monte Carlo simulations

Suppose we can generate M realizations of the data z_t, by means of simulation or data acquisition under the same conditions. Denote them by $z_t^{(j)}$, $j = 1, 2, \ldots, M$. Apply the same estimator (12.1) to all of them,

$$\hat{x}_t^{(j)} = \sum_{i=t_1}^{t_2} \alpha_{t,i} z_{t-i}^{(j)}. \tag{12.5}$$

From these M sets of estimates, we can estimate all kind of statistics. For instance, if the dimension of x is only one, we can make a histogram over $\hat{x}_t$, which graphically approximates the PDF.

The scalar measure (12.3) is estimated by the *Root Mean Square Error* (*RMSE*)

$$\mathbf{RMSE}(t) = \left(\frac{1}{M}\sum_{j=1}^{M} \|x_t^o - \hat{x}_t^{(j)}\|_2^2\right)^{1/2}. \tag{12.6}$$

This equation comprises both tracking and variance errors. If the true value x_t^o is not known for some reason, or if one wants to measure only the variance error, x_t^o can be replaced by its Monte Carlo mean, while at the same time changing M to $M-1$ in the normalization (to get an unbiased estimate of the standard deviation), and we get

$$\mathbf{RMSE}(t) = \left(\frac{1}{M-1} \sum_{j=1}^{M} \left| \left(\frac{1}{M} \sum_{i=1}^{M} \hat{x}_t^{(i)} \right) - \hat{x}_t^{(j)} \right|_2^2 \right)^{1/2}. \tag{12.7}$$

Equation (12.6) is an estimate of the standard deviation of the estimation error norm at each time instant. A scalar measure for the whole data sequence is

$$\mathbf{RMSE} = \left(\frac{1}{t} \sum_{i=1}^{t} \frac{1}{M} \sum_{j=1}^{M} \|x_t^o - \hat{x}_t^{(j)}\|_2^2 \right)^{1/2}. \tag{12.8}$$

Note that the time average is placed inside the square root, in order to get an unbiased estimate of standard deviation. Some authors propose to time average the RMSE(t), but this is not an estimate of standard deviation of the error anymore.[1]

Example 12.1 Signal estimation

The first subplot in Figure 12.2 shows a signal, which includes a ramp change, and a noisy measurement $y_t = s_t + e_t$ of it. To recover the signal from the measurements, a low-pass filter of Butterworth-type is applied. Using many different realization of the measurements, the Monte Carlo mean is illustrated in the last two subplots, where an estimated confidence bound is also marked. This bound is the '$\pm\sigma$' level, where σ is replaced by RMSE(t).

In some applications, including safety critical ones, it is the *peak error* that is crucial,

$$\mathbf{PEAK}(t) = \max_{j \in \{1,M\}} \|x_t^o - \hat{x}_t^{(j)}\|_2. \tag{12.9}$$

From this, a scalar peak measure can be defined as the total peak (max over time) or a time RMSE (time average peak value). Note that this measure is non-decreasing with the number of Monte Carlo runs, so the absolute value should be interpreted with a little care.

[1] According to *Jensen's inequality*, $\mathrm{E}(\sqrt{v_t}) < \sqrt{\mathrm{E}(v_t)}$, so this incorrect procedure would produce an under-biased estimate of standard deviation. In general, standard deviations should not be averaged. Compare the outcomes from `mean(std(randn(10,10000)))` with `sqrt(mean(std(randn(10,10000)).^2))` !

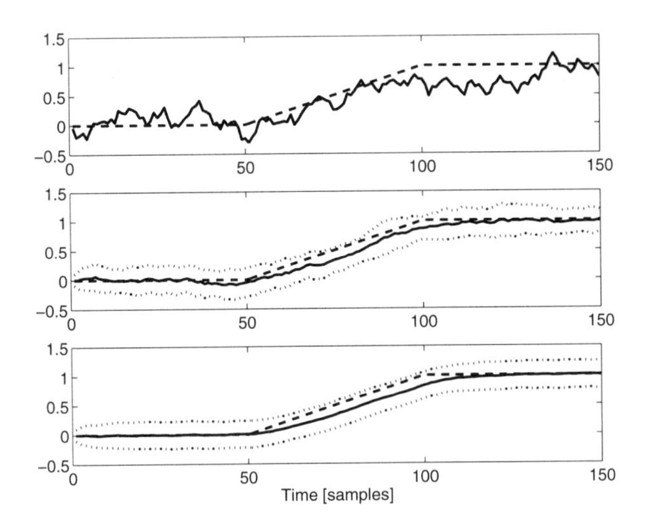

Figure 12.2. First plot shows the signal (dashed line) and the measurements (solid line) from one realization. The second and third plot show the Monte Carlo mean estimate and the one sigma confidence interval using the Monte Carlo standard deviation for 50 and 1000 Monte Carlo runs, respectively.

The scalar performance measures can be used for *auto-tuning*. Suppose the filter is parameterized in a scalar design parameter. As argued in Chapter 1, all linear filters have such a scalar to trade-off variance and tracking errors. Then we can optimize the filter design with respect to this measure, for instance by using a line search. The procedure can be generalized to non-scalar design parameters, at the cost of using more sophisticated and computer intensive optimization routines.

Example 12.2 Target tracking

Consider the target tracking example in Example 1.1. For a particular filter (better tuned than the one in Example 1.1), the filter estimates lie on top of the measurements, so Figure 12.3(a) is not very practical for filter evaluation, because it is hard to see any error at all.

The RMSE position error in 12.3(b) is a much better tool for evaluation. Here we can clearly see the three important phases in adaptive filtering: *transient* (for sample numbers 1–7), the *variance error* (8–15 and 27–32) and *tracking error* (16–26 and 35–45).

All in all, as long as it is possible to generate several data sets under the same premises, Monte Carlo techniques offer a solution to filter evaluation and design. However, this might not be possible, for instance when collecting

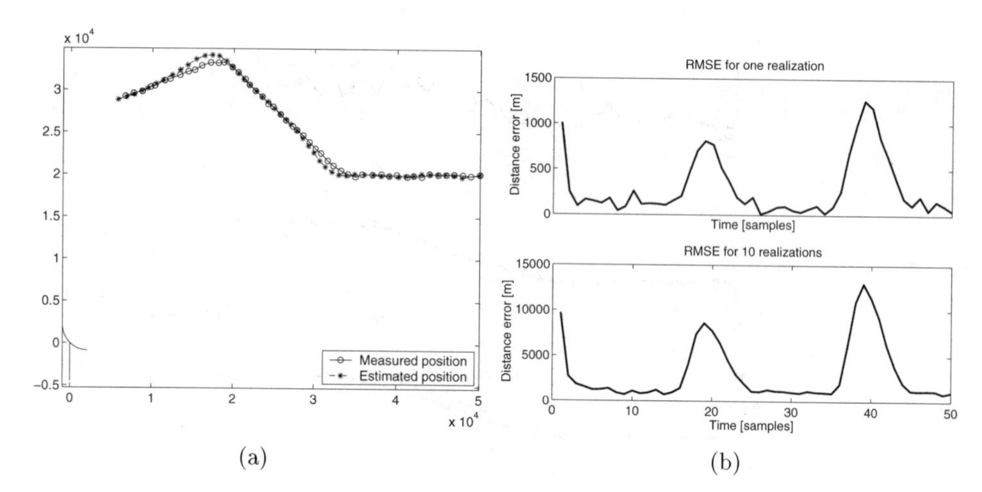

Figure 12.3. Radar trajectory (circles) and filter estimated positions (crosses) in (a). The RMSE position error in (b).

data during one single test trial, where it is either impossible to repeat the same experiment or too expensive. One can then try *resampling techniques*, as described in the next subsections.

- Bootstrap offers a way to reorder the filter residuals and make artificial simulations that are similar to Monte Carlo simulations. The procedure includes iterative filtering and simulation.
- Gibbs resampling is useful when certain marginal distributions can be formulated mathematically. The solution consists in iterative filtering and random number generation.

12.1.4. Bootstrap

Static case

As a non-dynamic example of the bootstrap technique, consider the case of estimating distribution parameters in a sequence of independent identically distributed (i.i.d.) stochastic variables,

$$y_t \in p_y(\theta),$$

where θ are parameters in the distribution. For instance, consider the location θ and scale σ_y parameters in a scalar Gaussian distribution. We know from Section 3.5.2 how to estimate them, and there are also quite simple theoretical expressions for uncertainty in the estimates. In more general cases, it might be difficult, if not impossible, to compute the sample variance of the point estimate, since most known results are asymptotic. As detailed in Section 12.1.1, the standard approach in a simulation environment is to perform Monte Carlo simulations. However, real data sets are often impossible or too costly to reproduce under indentical conditions, and there could be too few data points for the asymptotic expressions to hold.

The key idea in bootstrap is to produce new artificial data sets by picking samples at random from the set y^N *with replacement.* That is, from the measured values $4, 2, 3, 5, 1$ we might generate $2, 2, 5, 1, 5$. Denote the new data set by $(y^N)^{(i)}$. Each new data set is used to compute a point estimate $\theta^{(i)}$. Finally, these estimates $\theta^{(i)}$ are treated as independent outcomes from Monte Carlo simulations, and we can obtain different variability measures as standard deviation and confidence intervals.

Example 12.3 Bootstrap

Following the example in Zoubir and Boushash (1998), we generate 10 samples from $\mathrm{N}(10, 25)$. We want to estimate the mean θ and its variance P in the unknown distribution for the measurements y_t. The point estimate of θ is

$$\hat{\theta} = \frac{1}{10}\sum_{t=1}^{10} y_t$$

and a point estimate of its variance $P = \mathrm{Var}(\hat{\theta})$ is

$$\begin{aligned}\hat{P} &= \frac{\hat{\lambda}}{10}\\ \hat{\lambda} &= \frac{1}{9}\sum_{t=1}^{10}(y_t - \hat{\theta})^2.\end{aligned}$$

We can repeat this experiment a number of times (here 20) in a Monte Carlo simulation, and compute the variability of $\hat{\theta}$ in the different experiments as the Monte Carlo estimate of the variance, or can use the 10 data we have and use bootstrap to generate Monte Carlo like data. Note that its theoretical value is known in this simulation to be $25/10 = 2.5$. The result is as follows:

Statistics	Point estimate	Monte Carlo	bootstrap	Theoretical
$\mathrm{E}(\hat{\theta})$	9.2527	10.0082	9.3104	10
$\mathrm{Var}(\hat{\theta})$	2.338	2.1314	2.3253	2.5

The result is encouraging, in that a good estimate of variability is obtained (pretend that the theoretical value was not available). A new realization of random numbers gives a much worse measure of variability:

Statistics	Point estimate	Monte Carlo	bootstrap	Theoretical
$\mathrm{E}(\hat{\theta})$	8.7932	9.9806	8.8114	10
$\mathrm{Var}(\hat{\theta})$	1.1991	2.6124	1.0693	2.5

Finally, one might ask what the performance is as an average. Twenty realizations of the tables above are generated, and the mean values of variability are:

Statistics	Point estimate	Monte Carlo	bootstrap	Theoretical
$\mathrm{EVar}(\hat{\theta})$	2.65	2.53	2.40	2.5

In conclusion, bootstrap offers a good alternative to Monte Carlo simulations or analytical point estimate.

From the example, we can conclude the following:

- The bootstrap result is as good as the natural point estimate of variablity.
- The result very much depends upon the realization.

That is:

Bootstrap

Bootstrap cannot create more information than is contained in the measurements, but it can compute variability measures numerically, which are as good as analytically derived point estimates.

There are certain applications where there is no analytical expression for the point estimate of variablity. As a simple example, the variance of the sample median is very hard to compute analytically, and thus a point estimator of median variance is also hard to find.

Dynamic time-invariant case

For more realistic signal processing applications, we first start by outlining the time invariant case. Suppose that the measured data are generated by a dynamical model parametrized by a parameter θ,

$$y_t = f(y_{t-1}, y_{t-2}, \ldots, u_t, u_{t-1}, \ldots, e_t; \theta).$$

The bootstrap idea is now as follows:

1. Compute a point estimate $\hat{\theta}$ from the original data set $\{y_i, u_i\}_{i=1}^N$.

2. Apply the inverse model (filter) and get the noise (or, more precisely, residual) sequence $\hat{e}^N$.

3. Generate bootstrap noise sequences $(\hat{e}^N)^{(i)}$ by picking random samples from $\hat{e}^N$ with replacement.

4. Simulate the estimated system with these artificial noise sequences:

$$y_t = f(y_{t-1}, y_{t-2}, \ldots, u_t, u_{t-1}, \ldots, \hat{e}_t^{(i)}; \hat{\theta}).$$

5. Compute point estimates $\hat{\theta}^{(i)}$, and treat them as independent outcomes from Monte Carlo simulations.

Example 12.4 Bootstrap

Consider the auto-regressive (AR(1)) model

$$y_t = -0.7 y_{t-1} + e_t.$$

$N = 10$ samples are simulated. From these, the ML estimate of the parameter θ in the AR model $y_t + \theta y_{t-1} = e_t$ is computed. (In MATLAB™ this is done by `-y(1:N-1)\y(2:N);`.) Then, compute the residual sequence

$$\hat{e}_t = y_t + \hat{\theta} y_{t-1}.$$

Generate bootstrap sequences $(\hat{e}^N)^{(i)}$ and simulate new data by

$$y_t^{(i)} = -\hat{\theta} y_{t-1}^{(i)} + \hat{e}_t^{(i)}.$$

Finally, estimate θ for each sequence $(y^N)^{(i)}$.

A histogram over 1000 bootstrap estimates is shown in Figure 12.4. For comparison, the Monte Carlo estimates are used to approximate the true PDF of the estimate. Note that the PDF of the estimate is well predicted by using only 10 samples and bootstrap techniques.

The table below summarizes the accuracy of the point estimates for the different methods:

Statistics	Point estimate	Monte Carlo	bootstrap	Theoretical
$\mathrm{E}(\hat{\theta})$	0.57	0.60	0.57	0.7
$\mathrm{Std}(\hat{\theta})$	0.32	0.27	0.30	0.23

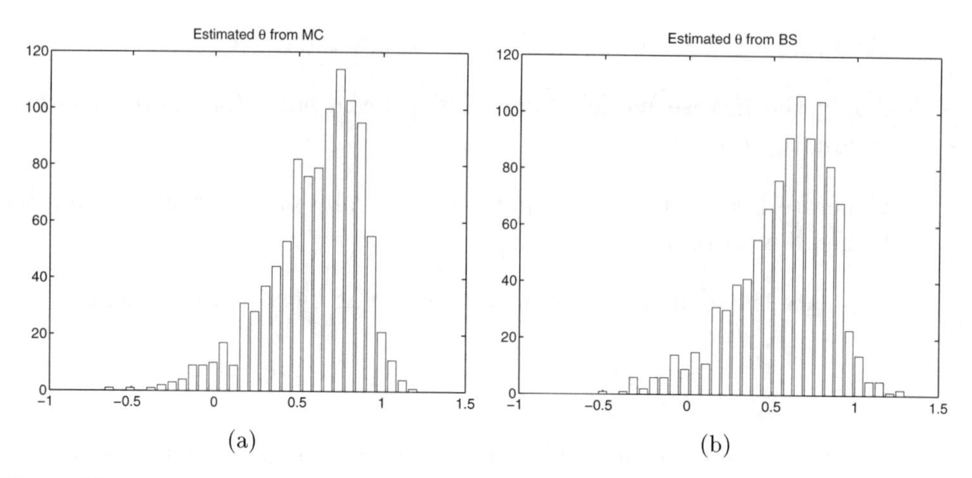

Figure 12.4. Histograms for point estimates of an AR parameter using (a) Monte Carlo simulations and (b) bootstrap.

As in the previous example, we can average over, say, 20 tables to average out the effect of the short data realization of only 10 samples on which the bootstrap estimate is based. The table below shows that bootstrap gives almost as reliable estimate as a Monte Carlo simulation:

Statistics	Point estimate	Monte Carlo	bootstrap	Theoretical
$\mathrm{Std}(\hat{\theta})$	0.25	0.28	0.31	0.22583

The example shows a case where the theoretical variance and standard deviation can be computed with a little effort. However, for higher order AR models, finite data covariance expressions are hard to compute analytically, and one has to rely on asymptotic expressions or resampling techniques as bootstrap.

Dynamical time-varying case

A quite challenging problem for adaptive filtering is to compute the RMSE as a function of time. As a general problem formulation, consider a state space model

$$\begin{aligned} x_{t+1} &= Ax_t + B_u u_t + B_v v_t \\ y_t &= Cx_t + Du_t + e_t, \end{aligned}$$

where $P_t^0 = \mathrm{Cov}(\hat{x}_t)$ is sought. Other measures as RMSE can be expressed in terms of P_t^0. For many applications, it is quite obvious that the covariance matrix P_t delivered by the Kalman filter is not reliable. Consider the target tracking example. The Kalman filter innovations are certainly not white after the manoeuvres. A natural generalization of the bootstrap principle is as follows:

1. Inverse filter the state space model to get an estimate of the two noise sequences $\{\hat{v}_t\}$ and $\{\hat{e}_t\}$. This includes running the Kalman filter to get the filtered estimate $\hat{x}_t$, and then compute

$$
\begin{aligned}
B_v \hat{v}_t &= \hat{x}_{t+1} - A\hat{x}_t - B_u u_t \\
\hat{e}_t &= y_t - C\hat{x}_t - Du_t.
\end{aligned}
$$

   The estimate of v_t might be in the least squares sense if B_v is not full rank.

2. If an estimate of the variance contribution to the RMSE(t) is to be found, resample only the measurement noise sequence $\{\hat{e}_t\}^{(i)}$. Otherwise, resample both sequences. However, here one has to be careful. If the sequence $\{\hat{v}_t\}$ is not independent and identically distributed ($i.i.d$), which is probably the case when using real data, and the bootstrap idea does not apply.

3. Simulate the system for each set of bootstrap sequences.

4. Apply the Kalman filter and treat the state estimates as Monte Carlo outcomes.

Literature

An introduction to the mathematical aspects of bootstrap can be found in Politis (1998) and a survey of signal processing applications in Zoubir and Boushash (1998). An overview for system identification and an application to uncertainty estimation is presented in Tjärnström (2000).

12.1.5. MCMC and Gibbs sampler

As a quite general change detection and Kalman filter example, consider the state space model

$$
\begin{aligned}
x_{t+1} &= Ax_t + B_u u_t + B_v v_t + \sum_j \delta_{t-k_j} B_f f_j \\
y_t &= Cx_t + Du_t + e_t,
\end{aligned}
$$

where f_j is a fault, assumed to have known (Gaussian) distribution, occuring at times k_j. Let K be the random vector with change times, X the random matrix with state vectors, and Y the vector of measurements. The change detection problem can be formulated as computing the marginal distribution

$$f(K|Y) = \int f(K|X, Y) dX. \tag{12.10}$$

The problem is that the integral in (12.10) is usually quite hard to evaluate. The idea in *Markov Chain Monte Carlo* (*MCMC*) and *Gibbs sampling* is to generate sequences $X^{(i)}, K^{(i)}$ that asymptotically will have the marginal distribution (12.10). The Gibbs sequence is generated by alternating between taking random samples from the following two distributions:

$$X^{(i+1)} \sim f_{X|Y,K^{(i)}}(x|y,k) \tag{12.11}$$

$$K^{(i+1)} \sim f_{K|Y,X^{(i+1)}}(k|y,x). \tag{12.12}$$

Comments on these steps:

- The distribution (12.11) is Gaussian with mean and covariance computed by the Kalman smoother. That is, here we run the Kalman smoother and then simulate random numbers from a multi-variate normal distribution $\mathrm{N}(\hat{x}_{t|t-1}, P_{t|t-1})$.
- From the Markovian property of the state, the distribution (12.12) follows by noting that the probability for a change at each time is given by comparing the difference $x_{t+1} - Ax_t - B_u u_t$ with a *Gaussian mixture* (linear combination of several Gaussian distributions).

The recursion has two phases. First, it converges from the initial uncertainty in $X^{(0)}$ and $K^{(0)}$. This is called the *burn-in time*. Then, we continue to iterate until enough samples from the marginal distribution (12.10) have been obtained.

In some cases, we can draw samples from the multi-variate distribution of the unknowns X and K directly. Then we have MCMC. Otherwise, one can draw samples from scalar distributions using one direction after each other and using Bayes' law. Then we have Gibbs sampling.

That is, the basic idea in Gibbs sampling is to generate random numbers from all kind of marginal distributions in an iterative manner, and wait until the samples have converged to the true distribution. The drawback is that analytical expression for all marginal distributions must be known, and that these change from application to application.

The theoretical question is why and how the samples converge to the marginal distribution. Consider the case of Kalman filtering, where we want

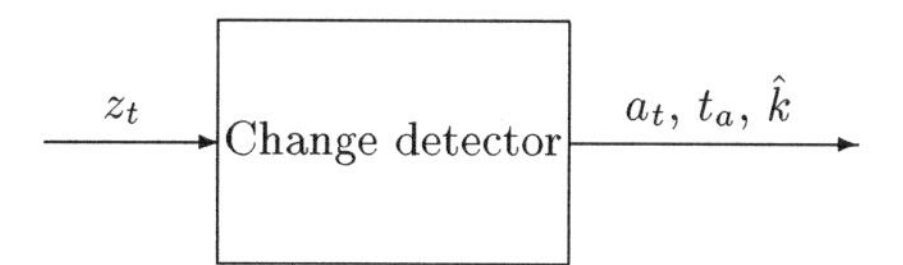

Figure 12.5. A change detector takes the observed signal z_t and delivers an alarm signal a_t, which is one at time t_a and zero otherwise, and possibly makes a decision about the change time k.

to compute certain statistics of the states X. The transition from one iteration to the next one can be written

$$f_{X^{(i+1)}|X^{(i)}}(x^{(i+1)}|x^{(i)}) = \int f_{X^{(i+1)}|Y,K}(x^{(i+1)}|k) f_{K|Y,X^{(i)}}(k|x^{(i)})dk \\ \triangleq h(x^{(i+1)}, x^{(i)}).$$

Expressed in the initial state estimate, the recursion can be written

$$f_{X^{(i+1)}|X^{(0)}}(x^{(i+1)}|x^{(0)}) = \int f_{X^{(i)}|X^{(0)}}(x^{(i)}|x^{(0)}) h(x^{(i+1)}, x^{(i)}) dx^{(i)}.$$

This is a Markov model, where the integral kernel $h(x^{(i+1)}, x^{(i)})$ acts like the transition matrix for state space models. One stationary point of this recursion is the marginal distribution of X, and it can be shown that the recursion will always converge to this point, if the marginal distribution exists.

Literature

The book by Gilks et al. (1996) covers MCMC. An introduction to Gibbs sampling can be found in Casella and George (1992) and Smith and Roberts (1993), and applications to Kalman filtering problems in Shephard and Pitt (1997) and Carter and Kohn (1994).

12.2. Evaluation of change detectors

12.2.1. Basics

A change detector can be seen as a device that takes an observed signal z_t and delivers an alarm signal a_t, as illustrated in Figure 12.5. It can also be seen as a device that takes a sequence of data and delivers a sequence of alarm times t_a and estimated change times $\widehat{k^n}$.

On-line performance measures are:

- *Mean Time between False Alarms* (*MTFA*)

$$\text{MTFA} = \text{E}(t_a - t_0 | \text{no change}), \tag{12.13}$$

where t_0 is the starting time for the algorithm. How often do we get alarms when the system has not changed? Related to MTFA is the *false alarm rate* (*FAR*) defined as 1/MTFA.

- *Mean Time to Detection* (*MTD*)

$$\text{MTD} = \text{E}(t_a - k | \text{a given change at time } k). \tag{12.14}$$

How long do we have to wait after a change until we get the alarm? Another name for MTD is *Delay For Detection* (*DFD*).

- *Missed Detection Rate* (*MDR*). What is the probability of not receiving an alarm, when there has been a change. Note that in practice, a large $t_a - k$ can be confused with a missed detection in combination with a false alarm.

- The *Average Run Length* function, $ARL(\theta)$

$$\text{ARL} = \text{E}(t_a - k | \text{a change of magnitude } \theta \text{ at time } k). \tag{12.15}$$

A function that generalizes MTFA and MTD. How long does it take before we get an alarm after a change of size θ? A very large value of the ARL function could be interpreted as a missed detection being quite likely.

In practical situations, either MTFA or MTD is fixed, and we optimize the choice of method and design parameters to minimize the other one.

In *off-line* applications (signal segmentation), logical performance measures are:

- Estimation accuracy. How accurate can we locate the change times? This relates to minimizing the absolute value of MTD. Note that the time to detection can be negative, if a non-causal change detector is used.

- The *Minimum Description Length* (MDL); see Section 12.3.1. How much information is needed to store a given signal?

The latter measure is relevant in data compression and communication areas, where disk space or bandwidth is limited. MDL measures the number of binary digits that are needed to represent the signal and segmentation is one approach for making this small.

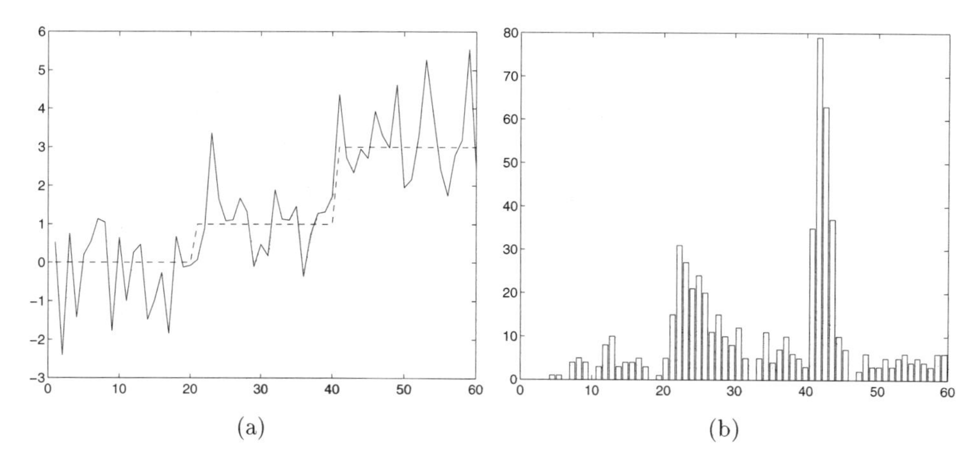

Figure 12.6. One realization of the change in the mean signal (a), and histogram over alarm times from the CUSUM algorithm for 250 Monte Carlo simulations (b).

Example 12.5 Signal estimation

Consider the changing mean in noise signal in Figure 12.6. The two-sided version of the CUSUM Algorithm 3.2 with threshold $h = 3$ and drift $\nu = 0.5$ is applied to 250 realizations of the signal. The alarm times are shown in the histogram in Figure 12.6.

From the empirical alarm times, we can compute the following statistics, under the assumption that alarms come within 10 samples after the true change, otherwise an alarm is interpreted as a false alarm:

MTD1	MTD2	MDR1	MDR2	FAR
4.7	3.0	0.27	0	0.01

As can be expected, the MTD and MDR are larger for the smaller first change.

Here we have compensated the missed detection rate by the estimated false alarm rate. (Otherwise, the missed detection rate would become negative after the second jump, since more than 250 alarms are obtained in the interval $[40, 50]$.) The delay for detection is however not compensated.

12.2.2. The ARL function

The ARL function can be evaluated from Monte Carlo simulations, or other resampling techniques. In some simple cases, it can also be computed numerically without the need for simulations.

The ARL function for a *stopping rule* (or rather alarm rule) for detecting a change θ in the mean of a signal is defined as

$$\text{ARL} = E(t_a | \text{a change of magnitude } \theta \text{ at time } t = 0), \tag{12.16}$$

where t_a is the alarm time (or stopping time).

In this section, we demonstrate how to analyze and design CUSUM detectors. Assume we observe a signal z_t which is $N(0, \sigma^2)$ before the change and $N(\theta, \sigma^2)$ after the change. We apply the CUSUM algorithm with threshold h and drift ν:

$$g_t = g_{t-1} + z_t - \nu \tag{12.17}$$

$$g_t = 0, \text{if } g_t < 0 \tag{12.18}$$

$$g_t = 0, \text{ and } t_a = t \text{ and alarm if } g_t > h > 0. \tag{12.19}$$

A successful design requires $\theta > \nu$.

There are two design parameters in the CUSUM test: the threshold h, and the drift parameter ν. If σ is the standard deviation of the noise, then it can be shown that the functional form of the ARL function is

$$\text{ARL}(\theta; h, \nu) = f\left(\frac{h}{\sigma}, \frac{\theta - \nu}{\sigma}\right). \tag{12.20}$$

That is, it is a function of two arguments.

The exact value of the ARL function is given by a so-called Fredholm integral equation of the second kind, which must be solved by a numerical algorithm. See de Bruyn (1968) or Section 5.2.2 in Basseville and Nikiforov (1993). Let $\mu = \theta - \nu$. A direct approximation suggested by Wald is

$$\text{ARL} = \frac{e^{-2\frac{h\mu}{\sigma^2}} - 1 + 2\frac{h\mu}{\sigma^2}}{2\frac{\mu^2}{\sigma^2}}, \tag{12.21}$$

and another approximation suggested by Siegmund is

$$\text{ARL} = \frac{e^{-2(h/\sigma + 1.166)\mu/\sigma} - 1 + 2(h/\sigma + 1.166)\mu/\sigma}{2\mu^2/\sigma^2}. \tag{12.22}$$

The quality of these approximations is investigeted in the following example.

Example 12.6 ARL

Assume that $\sigma = 1$ and $h = 3$. The mean time between false alarms is $f(h, -\nu)$ and the mean time for detection is $f(h, \theta - \nu)$, see (12.20). Sample values of f are:

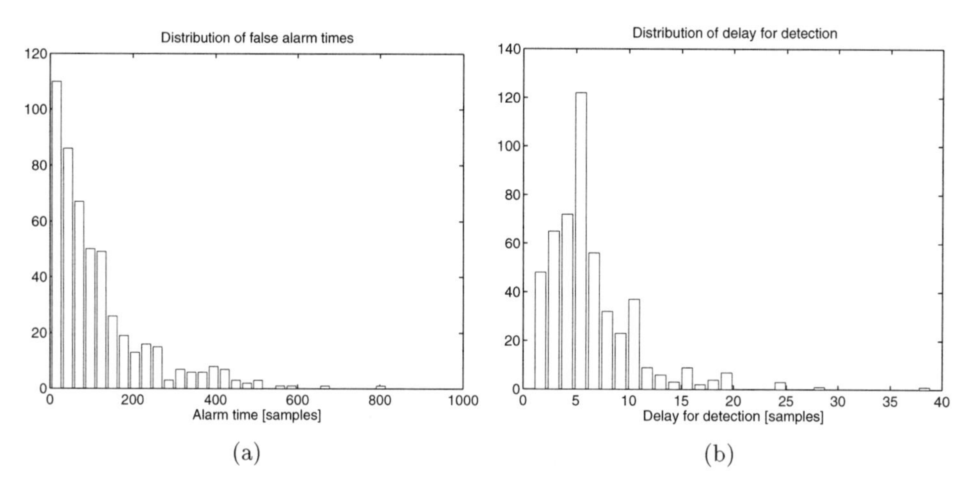

Figure 12.7. Distribution of false alarm times (a) and delay for detections (b), respectively, from Monte Carlo simulations with $h = 3$ and $\nu = 0.5$.

$\theta - \nu$	Theoretical	Wald	Siegmund	MC
-0.50	127.7000	32.2000	118.6000	119.9640
0	19.4000	9.0000	17.4000	16.7620
0.5000	6.7000	4.1000	6.4000	6.4620
1.0000	3.9000	2.5000	3.7000	3.7600
1.5000	2.7000	1.8000	2.6000	2.6560
2.0000	2.1000	1.4000	2.0000	2.1740

This is a subset of Table 5.1 in Basseville and Nikiforov (1993). See also Figures 12.7 and 12.8.

The mean times do not say anything about the distribution of the run length, which can be quite unsymmetric. Monte Carlo simulations can be used for further analysis of the run length function.

It can seen that the distribution of false alarms is basically binominal, and that a rough estimate of the delay for detection is $h/(\theta - \nu)$.

What is really needed in applications is to determine h from a specified FAR or MTD. That is, to solve the equation

$$\text{ARL}(0; h, \nu) = \frac{1}{\text{FAR}}$$

or

$$\text{ARL}(\theta; h, \nu) = \text{MTD}(\theta)$$

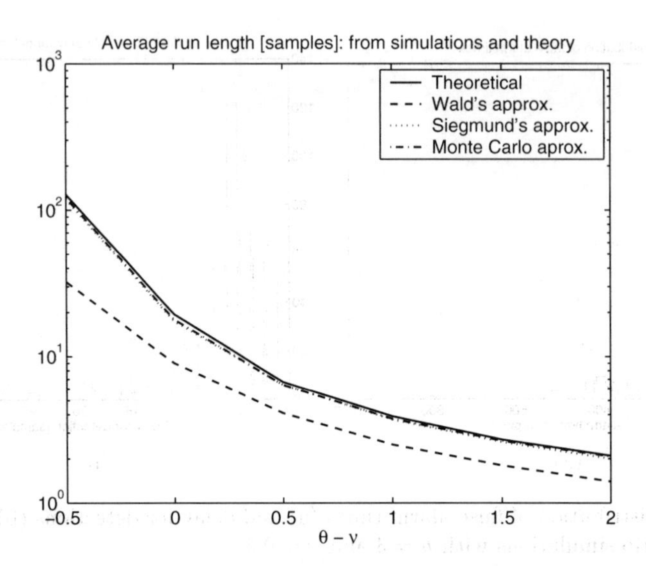

Figure 12.8. ARL function as a function of $\mu = \theta - \nu$ for threshold $h = 3$.

with respect to h. Since the ARL function is a monotonously increasing function of h, this should not pose any problems. Next, the drift ν can be optimized to minimize the MTD if the FAR is to be held constant. That is, a very systematic design is possible if the ARL function is computable, and if there is prior knowledge of the change magnitude θ. To formalize this procedure of minimizing MTD for a given FAR, consider the ARL function in Figure 12.9. First we take $\theta = 0$ (no change), and find the level curve where ARL=FAR. This gives the threshold as a function of drift, $h = h(\nu)$. Then we evaluate the ARL function for the change θ for which the MTD is to be minimized. The minimum of the function $\text{ARL}(\theta; h(\nu), \nu)$ defines the optimal drift and then also threshold $h(\nu)$.

The ARL function can only be computed analytically for very simple cases, but the approach based on Monte Carlo simulations is always applicable.

12.3. Performance optimization

For each problem at hand, there are a number of choices to make:

1. Adaptive filtering with or without change detection.
2. The specific algorithm for filtering and possibly change detector.
3. The design parameters in the algorithm chosen.

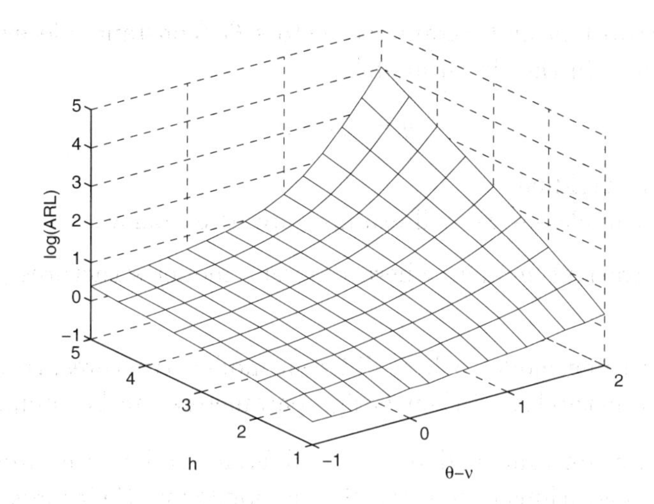

Figure 12.9. ARL function for different approximations as a function of $\mu = \theta - \nu$ and threshold h. Siegmund's approximation is used to compute $\log_{10}(ARL)$.

We describe here in general terms methods for facilitating the design process. One particular goal is a more objective design than just using trial and error combined with visual inspection.

12.3.1. The MDL criterion

In this section, we consider the problem of transmitting or storing a signal as efficiently as possible. By efficiently, we mean using as few binary digits as possible. As a tool for this, we can use a mathematical model of the signal, which is known to the receiver or the device that reads the stored information. In the following, we will refer only to the transmission problem. The point with the mathematical model is that we do not transmit the signal itself, but rather the residuals from the model whose size is of considerably smaller magnitude if the model is good, and thus fewer bits are required for attaining the specified accuracy at the receiver. The prize we have to pay for this is that the parameters in the model need to be transmitted as well. That is, we have a compromise between sending as small residuals as possible and using as few parameters as possible. An implicit trade-off is the choice of how many decimals that are needed when transmitting the parameters. This last trade-off is, however, signal independent, and can be optimized for each problem class leading to an *optimal code length*. The presentation here is based on Gustafsson (1997a), which is an attempt to a generalization of Rissanen's work on this subject for model order selection..

More specifically, the problem can be stated as choosing a regression vector (model structure) φ_t and parameter vectors θ_t (constant, piecewise constant or time-varying) in the signal model

$$y_t = \varphi_t^T \theta_t + e_t,$$

where e_t is the residual.

The problem classes we will examine are listed below:

- Time-invariant models where different model structures can be compared.
- Time-varying models where different model structures, recursive identification methods and their design parameters can be compared.
- Piecewise constant models, where different model structures, change detection algorithms to find the change points and their design parameters can be compared.

Besides the residuals that always have to be transmitted, the first model requires a parameter vector to be transmitted. The second model transmits the time update of the parameter vector, whose size should increase as the time variations in the model increase. The third model requires the change points to be transmitted together with a parameter vector for each segment. Clearly, the use of too many change points should be penalized.

The first approach has been thoroughly examined by Rissanen; see, for instance, Rissanen (1978, 1982). He developed the *Minimum Description Length* (*MDL*) criterion, which is a direct measure of the number of bits that are needed to represent a given signal as a function of the number of parameters, the number of data and the size of the residuals. We will extend the MDL criterion for the latter two cases. The point here is that we get an answer to not only what the most appropriate model structure is, but also when it pays off to use recursive identification and change detection. Another point is that the design variables can be optimized automatically as well, which is important since this is often a difficult tuning issue.

We would like to point out the following:

- There is no assumption that there is a true system which has constant, time-varying or piecewise constant parameter; rather we are looking for an algorithm that is able to describe data as well as possible.
- The generalized MDL can be used for standard system identification problems, just like MDL is often used for choosing the model structure. For instance, by taking a typical realization of a time-varying system we get a suggestion on which recursive identification algorithm to apply and how the design parameters should be chosen.

- As will be shown, the result of minimizing the description length yields a stochastic optimality in the maximum likelihood meaning as well.

Time-invariant parameters

The summary of MDL below essentially follows the introduction section of Rissanen (1982). Assume the measured signal y_t is modeled in a parametric family with measurement noise σ^2. Let $L(\varepsilon^N, \theta)$ denote the code length for the signal $y^N = (y_1, y_2, ..., y_N)$ using a model with a d-dimensional parameter vector θ. Here $\varepsilon^N = (\varepsilon_1, ..., \varepsilon_N)$ denotes the set of prediction errors from the model. In a linear regression framework, we have

$$\varepsilon_t = y_t - \varphi_t^T \theta$$
$$\theta = \left(\sum_{t=1}^{N} \varphi_t \varphi_t^T \right)^{-1} \sum_{t=1}^{N} \varphi_t y_t,$$

but other model structures are of course possible.

Generally, we have

$$L(\varepsilon^N, \theta) = -\log p(\varepsilon^N, \theta),$$

where $p(\varepsilon^N, \theta)$ is the joint distribution of data and the parameters. This expression is optimized over the precision of the value of θ, so that each element in the parameter vector can be represented by an integer, say n. The code length of this integer can be expressed as $-\log(p(n))$ for a suitable choice of density function. Rissanen now proposes a non-informative prior for integers as

$$p(n) \sim 2^{\log^*(n)},$$

where

$$\log^*(n) = \log n + \log\log n + \log\log\log n + ...$$

where the sum is terminated at the first negative term. It can be shown that this is a proper distribution, since its sum is finite.

With this prior, the optimal code length can be written

$$L(\varepsilon^N, \theta) = -\log p(\varepsilon^N | \theta) + \log \|\theta\|_P^d + \log C(k), \tag{12.23}$$

where only the fastest growing penalty term is included. Here $C(k)$ is the volume of the unit sphere in R^k, and $\|\theta\|_P = \theta^T P^{-1} \theta$. For linear regressions with Gaussian noise we have

$$L(\varepsilon^N, \theta) = \frac{1}{\sigma^2} \sum_{t=1}^{N} \varepsilon_t^2 + \log \|\theta\|_P^d + d \log N. \tag{12.24}$$

The most common reference to MDL only includes the first two terms, which are also scaled by $1/N$. However, to be able to compare different assumptions on parameter variations we keep the third term for later use.

Piecewise constant parameters

As a motivation for this approach, consider the following example.

Example 12.7 GSM speech coding

The GSM standard for mobile telephony says that the signal is segmented in batches of 160 samples, in each segment an eighth order AR model is estimated, and the parameter values (in fact non-linear transformation of reflection coefficients) and prediction errors (or rather a model of the prediction errors) are transmitted to the receiver.

This is an adequate coding, since typical speech signals are short-time stationary. Note that the segmentation is fixed before-hand and known to the receiver in GSM.

We consider segmentation in a somewhat wider context, where also the time points defining the segmentation are kept as parameters. That is, the information needed to transmit comprises the residuals, the parameter vector in each segment and the change points. Related segmentation approaches are given in Kitagawa and Akaike (1978) and Djuric (1992), where the BIC criterion Akaike (1977) and Schwartz (1978) is used. Since BIC is the same as MDL if only the fastest growing penalty term is included, the criteria they present will give almost identical result as the MDL.

If we consider the change points as fixed, the MDL theory immediately gives

$$L(\varepsilon^N, \theta^n) = -\log p(\varepsilon^N, \theta | k^n) = \frac{1}{\sigma^2} \sum_{t=1}^{N} \varepsilon^2(t) + dn \log N + \sum_{i=1}^{n} \log \|\theta_i\|_{P_i}^d,$$

because with a given segmentation we are facing n independent coding problems. Note that the number of parameters are nd, so d in the MDL criterion is essentially replaced by nd. The last term is still negligible if $N \gg n$.

The remaining question is what the cost for coding the integers k^n is. One can argue that these integers are also parameters leading to the use of $n(d+1)$ in MDL, as done in Kitagawa and Akaike (1978). Or one can argue that code length of integers is negligible compared to the real-valued parameters, leading to MDL with kn parameters as used in Djuric (1992). However, the description length of these integers is straightforward to compute. Bayes' law gives that

$$p(\varepsilon^N, \theta, k^n) = p(\varepsilon^N, \theta | k^n) p(k^n).$$

The code length should thus be increased by $-\log(p(k^n))$. The most reasonable prior now is a flat one for each k. That is,

$$p(k^n) = \frac{1}{N(N-1)\cdots(N-n+1)} \approx \frac{1}{N^n},$$

where we have assumed that the number of data is much larger than the number of segments. This prior corresponds to the code length

$$L(k^n) = n \log N.$$

That is, the MDL penalty term should in fact be $n(k+1)\log(N)/N$,

$$L(\varepsilon^N, \theta^n, k^n) \approx \frac{1}{\sigma^2} \sum_{t=1}^{N} \varepsilon^2(t) + (d+1) n \log N. \tag{12.25}$$

Time-varying parameters

Here we consider adaptive algorithms that can be written as

$$\theta_{t+1} = \theta_t + \Delta\theta_t.$$

For linear regression, the update can be written

$$\Delta\theta_t = \mu P_t \varphi_t \varepsilon_t,$$

which comprises RLS, LMS and the Kalman filter as special cases.

As a first try, one can argue that the parameter update $\Delta\theta_t$ is a sequence of real numbers just like the residuals and that the MDL criterion should be

$$L(\varepsilon^N, \Delta\theta^N) = \sum_{t=1}^{N} \left(\frac{\varepsilon_t^2}{\sigma^2} + \|\Delta\theta_t\|_{P_\Delta}^d \right), \tag{12.26}$$

where P_Δ is the covariance matrix of the update $\Delta\theta_t$. This criterion exhibits the basic requirements of a penalty term linear in the number of parameters. Clearly, there is a trade-off between making the residuals small (requiring large updates if the underlying dynamics are rapidly time-varying) and making the updates small.

12.3.2. Auto-tuning and optimization

Once we have decided upon the evaluation measure, auto-tuning of the design parameters is a straightforward optimization problem. For a simple algorithm with only one design parameter, a line-search is enough. More interestingly, we can compare completely different algorithms like linear adaptive filters with change detection schemes. See Section 5.7 for an example of these ideas.

13

Linear estimation

13.1. Projections

The purpose of this section is to get a geometric understanding of linear estimation. First, we outline how projections are computed in linear algebra for finite dimensional vectors. Functional analysis generalizes this procedure to some infinite-dimensional spaces (so-called Hilbert spaces), and finally, we point out that linear estimation is a special case of an infinite-dimensional space. As an example, we derive the Kalman filter.

13.1.1. Linear algebra

The theory presented here can be found in any textbook in linear algebra.

Suppose that x, y are two vectors in $\mathcal{R}^m$. We need the following definitions:

- The scalar product is defined by $(x, y) = \sum_{i=1}^{m} x_i y_i$. The scalar product is a linear operation in data y.
- Length is defined by the Euclidean norm $\|x\| = \sqrt{(x, x)}$.
- Orthogonality of x and y is defined by $(x, y) = 0$:

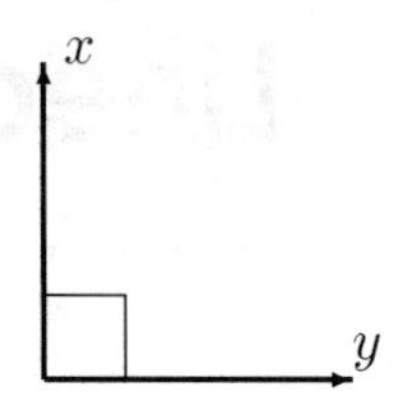

- The projection x_p of x on y is defined by

$$x_p = \frac{(x, y)}{(y, y)} y.$$

Note that $x_p - x$ is orthogonal to y, $(x_p - x, y) = 0$. This is the *projection theorem*, graphically illustrated below:

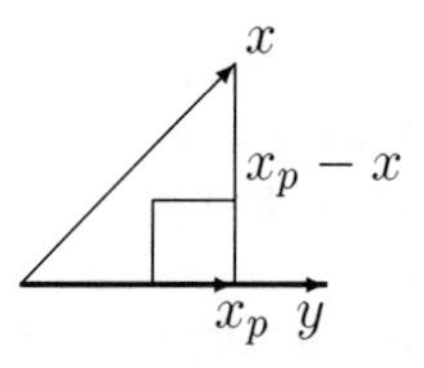

The fundamental idea in linear estimation is to project the quantity to be estimated onto a plane, spanned by the measurements Π_y. The projection $x_p \in \Pi_y$, or estimate $\hat{x}$, of x on a plane Π_y is defined by $(x_p - x, y_i) = 0$ for all y_i spanning the plane Π_y:

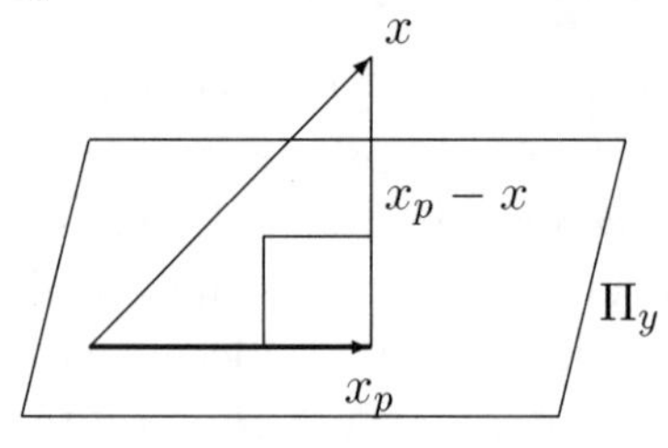

We distinguish two different cases for how to compute x_p:

1. Suppose $(\varepsilon_1, \varepsilon_2, ..., \varepsilon_N)$ is an *orthogonal* basis for Π_y. That is, $(\varepsilon_i, \varepsilon_j) = 0$ for all $i \neq j$ and $\text{span}(\varepsilon_1, \varepsilon_2, ..., \varepsilon_N) = \Pi_y$. Later on, ε_t will be interpreted

as the innovations, or prediciton errors. The projection is computed by

$$x_p = \sum_{i=1}^{N} \frac{(x, \varepsilon_i)}{(\varepsilon_i, \varepsilon_i)} \varepsilon_i = \sum_{i=1}^{N} f_i \varepsilon_i.$$

Note that the coefficients f_i can be interpreted as a filter. The projection theorem $(x_p - x, \varepsilon_j) = 0$ for all j now follows, since $(x_p, \varepsilon_j) = (x, \varepsilon_j)$.

2. Suppose that the vectors $(y_1, y_2, ..., y_N)$ are linearly independent, but not necessarily orthogonal, and span the plane Π_y. Then, Gram–Schmidt orthogonlization gives an orthogonal basis by the following recursion, initiated with $\varepsilon_1 = y_1$,

$$\varepsilon_k = y_k - \sum_{i=1}^{k-1} \frac{(y_k, \varepsilon_i)}{(\varepsilon_i, \varepsilon_i)} \varepsilon_i,$$

and we are back in case 1 above:

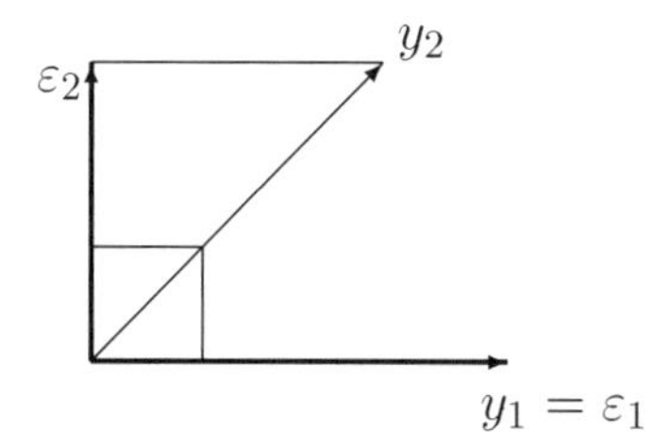

13.1.2. Functional analysis

A nice fact in functional analysis is that the geometric relations in the previous section can be generalized from vectors in $\mathcal{R}^m$ to infinite dimensional spaces, which (although a bit sloppily) can be denoted $\mathcal{R}^\infty$. This holds for so-called *Hilbert spaces*, which are defined by the existence of a scalar product with

1. $(x, x) > 0$ for all $x \neq 0$. That is, there is a length measure, or norm, that can be defined as $\|x\| \triangleq (x, x)^{1/2}$.

2. $(\alpha x, x) = \alpha(x, x)$, $(x + y, z) = (x, z) + (y, z)$ and $(x, y) = (y, x)$.

From these properties, one can prove the triangle inequality $(x+y, x+y)^{1/2} \leq (x, x)^{1/2} + (y, y)^{1/2}$ and Schwartz inequality $|(x, y)| \leq ||x|| \cdot ||y||$. See, for instance, Kreyszig (1978) for more details.

13.1.3. Linear estimation

In linear estimation, the elements x and y are stochastic variables, or vectors of stochastic variables. It can easily be checked that the covariance defines a scalar product (here assuming zero mean),

$$(x, y) \triangleq E(xy^*),$$

which satisfies the three postulates for a Hilbert space.

A linear filter that is optimal in the sense of minimizing the 2-norm implied by the scalar product, can be recursively implemented as a recursive Gram–Schmidt orthogonalization and a projection. For scalar y and vector valued x, the recursion becomes

$$\begin{aligned} \varepsilon_t =& y_t - \sum_{i=1}^{t-1} \frac{E(y_t\varepsilon_i)}{E(\varepsilon_i^2)}\varepsilon_i \\ \hat{x}_{t|t} =& \sum_{i=1}^{t} \frac{E(x_t\varepsilon_i)}{E(\varepsilon_i^2)}\varepsilon_i. \end{aligned}$$

Remarks:

- This is not a recursive algorithm in the sense that the number of computations and memory is limited in each time step. Further application-specific simplifications are needed to achieve this.
- To get expressions for the expectations, a signal model is needed. Basically, this model is the only difference between different algorithms.

13.1.4. Example: derivation of the Kalman filter

As an illustration of how to use projections, an inductive derivation of the Kalman filter will be given for the state space model, with *scalar* y_t,

$$\begin{aligned} x_{t+1} =& Ax_t + B_v v_t, \quad \text{Cov}(v_t) = Q \\ y_t =& Cx_t + e_t, \quad \text{Cov}(e_t) = R. \end{aligned}$$

1. Let the filter be initialized by $\hat{x}_{0|0}$ with an auxiliary matrix $P_{0|0}$.
2. Suppose that the projection at time t on the observations of y_s up to time t is $\hat{x}_{t|t}$, and assume that the matrix $P_{t|t}$ is the covariance matrix of the estimation error, $P_{t|t} = \mathrm{E}(\tilde{x}_{t|t}\tilde{x}_{t|t}^T)$.

3. **Time update.** Define the linear projection operator by

$$\hat{x} \triangleq \text{Proj}(x|y) = \frac{(x,y)}{(y,y)} y.$$

Then

$$\begin{aligned}\hat{x}_{t+1|t} =& \text{Proj}(x_{t+1}|y^t) = \text{Proj}(Ax_t + B_v v_t|y^t) \\ =& A\,\text{Proj}(x_t|y^t) + \underbrace{\text{Proj}(B_v v_t|y^t)}_{=0} = A\hat{x}_{t|t}.\end{aligned}$$

Define the estimation error as

$$\begin{aligned}\tilde{x}_{t+1|t} =& x_{t+1} - \hat{x}_{t+1|t} \\ =& A\tilde{x}_{t|t} + B_v v_t,\end{aligned}$$

which gives

$$\begin{aligned}P_{t+1|t} =& \mathrm{E}(\tilde{x}_{t+1|t}\tilde{x}_{t+1|t}^T) \\ =& A\,\mathrm{E}(\tilde{x}_{t|t}\tilde{x}_{t|t}^T)A^T + B_v\,\mathrm{E}(v_t v_t^T)B_v^T \\ =& AP_{t|t}A^T + B_v Q B_v^T.\end{aligned}$$

Measurement update. Recall the projection figure

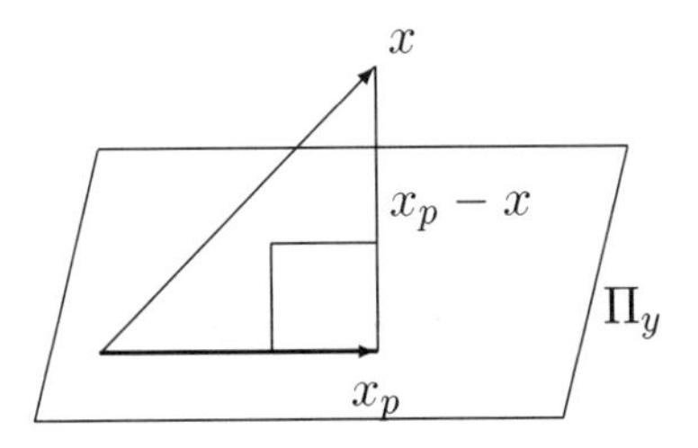

and the projection formula for an orthogonal basis

$$\begin{aligned}\hat{x}_{t|t} =& \sum_{i=1}^{t} \frac{(x_t, \varepsilon_i)}{(\varepsilon_i, \varepsilon_i)} \varepsilon_i \\ =& \sum_{i=1}^{t-1} \frac{(Ax_{t-1} + B_v v_{t-1}, \varepsilon_i)}{(\varepsilon_i, \varepsilon_i)} \varepsilon_i + \frac{(x_t, \varepsilon_t)}{(\varepsilon_t, \varepsilon_t)} \varepsilon_t \\ =& A \underbrace{\sum_{i=1}^{t-1} \frac{(x_{t-1}, \varepsilon_i)}{(\varepsilon_i, \varepsilon_i)} \varepsilon_i}_{=\hat{x}_{t-1|t-1}} + B_v \underbrace{\sum_{i=1}^{t-1} \frac{(v_{t-1}, \varepsilon_i)}{(\varepsilon_i, \varepsilon_i)}}_{=0} \varepsilon_i + \frac{(x_t, \varepsilon_t)}{(\varepsilon_t, \varepsilon_t)} \varepsilon_t \\ =& \hat{x}_{t|t-1} + \frac{(x_t, \varepsilon_t)}{(\varepsilon_t, \varepsilon_t)} \varepsilon_t.\end{aligned}$$

The correlation between x_t and ε_t is examined separately, using (according to the projection theorem) $\mathrm{E}(\hat{x}_{t|t-1}\tilde{x}_{t|t-1}) = 0$ and $\varepsilon_t = y_t - C\hat{x}_{t|t-1} = C\tilde{x}_{t|t-1} + e_t$:

$$\begin{aligned}
\mathrm{E}(x_t\varepsilon_t^T) &= \mathrm{E}(x_t(C\tilde{x}_{t|t-1} + e_t)^T) \\
&= \mathrm{E}(x_t\tilde{x}_{t|t-1}^T)C^T + \underbrace{\mathrm{E}(x_t e_t^T)}_{=0} \\
&= \mathrm{E}(\tilde{x}_{t|t-1}\tilde{x}_{t|t-1}^T)C^T \\
&= P_{t|t-1}C^T.
\end{aligned}$$

Here we assume that x_t is un-correlated with e_t. We also need

$$\begin{aligned}
\mathrm{E}(\varepsilon_t\varepsilon_t^T) &= \mathrm{E}((C\tilde{x}_{t|t-1} + e_t)(C\tilde{x}_{t|t-1} + e_t)^T) \\
&= CP_{t|t-1}C^T + R.
\end{aligned}$$

The measurement update of the covariance matrix is similar. All together, this gives

$$\begin{aligned}
\hat{x}_{t|t} &= \hat{x}_{t|t-1} + P_{t|t-1}C^T(CP_{t|t-1}C^T + R)^{-1}(y_t - C\hat{x}_{t|t-1}) \\
P_{t|t} &= P_{t|t-1} - P_{t|t-1}C^T(CP_{t|t-1}C^T + R)^{-1}CP_{t|t-1}.
\end{aligned}$$

The induction is completed.

13.2. Conditional expectations

In this section, we use arguments and results from mathematical statistics. Stochastic variables (scalar or vector valued) are denoted by capital letters, to distinguish them from the observations. This overview is basically taken from Anderson and Moore (1979).

13.2.1. Basics

Suppose the vectors X and Y are simultaneously Gaussian distributed

$$\begin{pmatrix} X \\ Y \end{pmatrix} \in \mathrm{N}\left(\begin{pmatrix} \mu_x \\ \mu_y \end{pmatrix}, \begin{pmatrix} P_{xx} & P_{xy} \\ P_{xy} & P_{yy} \end{pmatrix}\right) = \mathrm{N}\left(\begin{pmatrix} \mu_x \\ \mu_y \end{pmatrix}, P\right). \tag{13.1}$$

Then, the conditional distribution for X, given the observed $Y = y$, is Gaussian distributed:

$$(X|Y = y) \in N(\mu_x + P_{xy}P_{yy}^{-1}(y - \mu_y), P_{xx} - P_{xy}P_{yy}^{-1}P_{yx}). \tag{13.2}$$

This follows directly from Bayes' rule

$$p(X|Y) = \frac{p(X,Y)}{P(Y)},$$

by rather tedious computations. The complete derivation is given in Section 13.2.3.

The *Conditional Mean* (*CM*) estimator seen as a stochastic variable can be denoted

$$\hat{X}^{CM} = \mathrm{E}(X|Y) = \mu_x + P_{xy}P_{yy}^{-1}(Y - \mu_y),$$

while the *conditional mean estimate*, given the observed y, is

$$\hat{x}^{CM} = \mathrm{E}(X|Y = y) = \mu_x + P_{xy}P_{yy}^{-1}(y - \mu_y).$$

Note that the estimate is a linear function of y (or rather, affine).

13.2.2. Alternative optimality interpretations

The *Maximum A Posteriori* (*MAP*) estimator, which maximizes the *Probability Density Function* (PDF) with respect to x, coincides with the CM estimator for Gaussian distributions.

Another possible estimate is given by the *Conditional Minimum Variance* principle (*CMV*),

$$\hat{x}^{CMV}(y) = \arg\min_{z(y)} \mathrm{E}(\|X - z(y)\|^2 | Y = y).$$

It is fairly easy to see that the CMV estimate also coincides with the CM estimate:

$$\begin{aligned}
&\mathrm{E}(\|X - z(y)\|^2 | Y = y) \\
&= \int_{-\infty}^{\infty} (x - z(y))^T (x - z(y)) p_{X|Y}(x|y) dx \\
&= \int_{-\infty}^{\infty} x^T x p_{X|Y}(x|y) dx - 2z^T(y) \int_{-\infty}^{\infty} x p_{X|Y}(x|y) dx + z^T(y) z(y) \\
&= (z(y) - \hat{x}(y))^T (z(y) - \hat{x}(y)) + \underbrace{\int_{-\infty}^{\infty} x^T x p_{X|Y}(x|y) dx - \hat{x}(y)^T \hat{x}(y)}_{\text{minimum variance}}.
\end{aligned}$$

This expression is minimized for $z(y) = \hat{x}(y)$, and the minimum variance is the remaining two terms.

The closely related (unconditional) *Minimum Variance* principle (*MV*) defines an *estimator* (note the difference between estimator and estimate here):

$$\hat{X}^{MV}(Y) = \arg\min_{Z(Y)} \mathrm{E}_Y \mathrm{E}_X(\|X - Z(Y)\|^2 | Y).$$

Here we explicitely marked which variable the expectation operates on. Now, the CM estimate minimizes the second expectation for all values on Y. Thus, the weighted version, defined by the expectation with respect to Y must be minimized by the CM estimator for each $Y = y$. That is, as an estimator, the unconditional MV and CM also coincide.

13.2.3. Derivation of marginal distribution

Start with the easily checked formula

$$\begin{pmatrix} I & -P_{xy}P_{yy}^{-1} \\ 0 & I \end{pmatrix} \underbrace{\begin{pmatrix} P_{xx} & P_{xy} \\ P_{xy} & P_{yy} \end{pmatrix}}_{P} \begin{pmatrix} I & 0 \\ -P_{yy}^{-1}P_{yx} & I \end{pmatrix} = \begin{pmatrix} P_{xx} - P_{xy}P_{yy}^{-1}P_{yx} & 0 \\ 0 & P_{yy} \end{pmatrix}, \tag{13.3}$$

and Bayes' rule

$$\begin{aligned} p_{X|Y}(x,y) =& \frac{p_{X,Y}(x,y)}{P_Y(y)} \\ =& \frac{1}{(2\pi)^{N/2}} \frac{\det(P_{yy})^{1/2}}{\det(P)^{1/2}} \frac{\exp\left(-\frac{1}{2}\begin{pmatrix} x-\mu_x \\ y-\mu_y \end{pmatrix}^T P^{-1} \begin{pmatrix} x-\mu_x \\ y-\mu_y \end{pmatrix}\right)}{\exp\left(-\frac{1}{2}(y-\mu_y)^T P_{yy}^{-1}(y-\mu_y)\right)}. \end{aligned}$$

From (13.3) we get

$$\det(P) = \det(P_{xx} - P_{xy}P_{yy}^{-1}P_{yx})\det(P_{yy}),$$

and the ratio of determinants can be simplified. We note that the new Gaussian distribution must have $P_{xx} - P_{xy}P_{yy}^{-1}P_{yx}$ as covariance matrix.

Next, we simplify the exponent using (13.3),

$$\begin{aligned}
&\begin{pmatrix} x-\mu_x \\ y-\mu_y \end{pmatrix}^T P^{-1} \begin{pmatrix} x-\mu_x \\ y-\mu_y \end{pmatrix} \\
=&\begin{pmatrix} x-\mu_x \\ y-\mu_y \end{pmatrix}^T \begin{pmatrix} I & 0 \\ -P_{yy}^{-1}P_{yx} & I \end{pmatrix} \begin{pmatrix} (P_{xx}-P_{xy}P_{yy}^{-1}P_{yx})^{-1} & 0 \\ 0 & (P_{yy})^{-1} \end{pmatrix} \\
&\cdot \begin{pmatrix} I & -P_{xy}P_{yy}^{-1} \\ 0 & I \end{pmatrix} \begin{pmatrix} x-\mu_x \\ y-\mu_y \end{pmatrix} \\
=&\begin{pmatrix} x-\hat{x} \\ y-\mu_y \end{pmatrix}^T \begin{pmatrix} (P_{xx}-P_{xy}P_{yy}^{-1}P_{yx})^{-1} & 0 \\ 0 & (P_{yy})^{-1} \end{pmatrix} \begin{pmatrix} x-\hat{x} \\ y-\mu_y \end{pmatrix} \\
=&(x-\hat{x})^T(P_{xx}-P_{xy}P_{yy}^{-1}P_{yx})^{-1}(x-\hat{x}) + (y-\mu_y)^T(P_{yy})^{-1}(y-\mu_y),
\end{aligned}$$

where

$$\hat{x} = \mu_x + P_{xy}P_{yy}^{-1}(y-\mu_y).$$

From this, we can conclude that

$$\begin{aligned}
p_{X|Y}(x,y) =& \frac{1}{(2\pi)^{N/2}\det(P_{xx}-P_{xy}P_{yy}^{-1}P_{yx})^{1/2}} \\
&\cdot \exp\left(-\frac{1}{2}(x-\hat{x})^T(P_{xx}-P_{xy}P_{yy}^{-1}P_{yx})^{-1}(x-\hat{x})\right),
\end{aligned}$$

which is a Gaussian distribution with mean and covariance as given in (13.2).

13.2.4. Example: derivation of the Kalman filter

As an illustration of conditional expectation, an inductive derivation of the Kalman filter will be given, for the state space model

$$\begin{aligned}
x_{t+1} =& Ax_t + B_v v_t, \quad v_t \in N(0,Q) \\
y_t =& Cx_t + e_t, \quad e_t \in N(0,R)
\end{aligned}$$

1. Let $x_0 \in \mathrm{N}(\hat{x}_{0|0}, P_{0|0})$.
2. Suppose that x_t, conditioned on the observations of y_t up to time t, is $N(\hat{x}_{t|t}, P_{t|t})$
3. **Time update.**

$$\begin{aligned}
\hat{x}_{t+1|t} =& E[x_{t+1}|y^t] = E[Ax_t + B_v v_t|y^t] = A\hat{x}_{t|t} \\
P_{t+1|t} =& E[(x_{t+1}-\hat{x}_{t+1|t})(x_{t+1}-\hat{x}_{t+1|t})^T|y^t] \\
=& E[(A(x_t-\hat{x}_{t|t})+B_v v_t)(A(x_t-\hat{x}_{t|t})+B_v v_t)^T|y^t] \\
=& AP_{t|t}A^T + B_v Q B_v^T.
\end{aligned}$$

Measurement update. We have derived the conditional distribution for X, given Y, as

$$\begin{aligned}\mathrm{E}(X|Y) &= \mu_x + P_{xy}P_{yy}^{-1}(Y - \mu_\gamma)\\ \mathrm{Cov}(X|Y) &= P_{xx} - P_{xy}P_{yy}^{-1}P_{yx}.\end{aligned}$$

Identify the stochastic variable X with $(x_t|y^{t-1}) \in N(\hat{x}_{t|t-1}, P_{t|t-1})$ and Y with $(y_t|y^{t-1}) \in N(C\hat{x}_{t|t-1}, CP_{t|t-1}C^T + R)$. We have

$$\begin{aligned}P_{xy} &= E[(x_t - E[x_t])(y_t - E[y_t])^T|y^{n-1}]\\ &= E((x_t - \hat{x}_{t|t-1})(Cx_t + B_v v_t - C\hat{x}_{t|t-1})^T)\\ &= P_{t|t-1}C^T = P_{yx}^T.\end{aligned}$$

This gives

$$\begin{aligned}\hat{x}_{t|t} &= \hat{x}_{t|t-1} + P_{t|t-1}C^T(CP_{t|t-1}C^T + R)^{-1}(y_t - C\hat{x}_{t|t-1})\\ P_{t|t} &= P_{t|t-1} - P_{t|t-1}C^T(CP_{t|t-1}C^T + R)^{-1}CP_{t|t-1}.\end{aligned}$$

Induction implies that x_t, given y^t, is normally distributed.

13.3. Wiener filters

The derivation and interpretations of the Wiener filter follows Hayes (1996).

13.3.1. Basics

Consider the signal model

$$y_t = s_t + e_t. \tag{13.4}$$

The fundamental signal processing problem is to separate the signal s_t from the noise e_t using the measurements y_t. The signal model used in Wiener's approach is to assume that the second order properties of all signals are known. When s_t and e_t are independent, sufficient knowledge is contained in the correlations coefficients

$$\begin{aligned}r_{ss}(k) &= \mathrm{E}(s_t s_{t-k}^*)\\ r_{ee}(k) &= \mathrm{E}(e_t e_{t-k}^*),\end{aligned}$$

and similarly for a possible correlation $r_{se}(k)$. Here we have assumed that the signals might be complex valued and vector valued, so $*$ denotes complex conjugate transpose. The correlation coefficients (or covariance matrices) may in turn be defined by parametric signal models. For example, for a state space model, the Wiener filter provides a solution to the stationary Kalman filter, as will be shown in Section 13.3.7.

The *non-causal Wiener filter* is defined by

$$\hat{s}_t = (h * y)_t = \sum_{i=-\infty}^{\infty} h_i y_{t-i}. \tag{13.5}$$

In the next subsection, we study causal and predictive Wiener filters, but the principle is the same. The underlying idea is to minimize a least squares criterion,

$$h = \arg\min_h V(h) = \arg\min_h \mathrm{E}(\varepsilon_t)^2 = \arg\min_h \mathrm{E}(s_t - \hat{s}_t(h))^2 \tag{13.6}$$

$$= \arg\min_h \mathrm{E}(s_t - (y * h)_t)^2 = \arg\min_h \mathrm{E}(s_t - \sum_{i=-\infty}^{\infty} h_i y_{t-i})^2, \tag{13.7}$$

where the residual $\varepsilon_t = s_t - \hat{s}_t$ and the least squares cost $V(h)$ are defined in a standard manner. Straightforward differentiation and equating to zero gives

$$\frac{dV(h)}{dh_k} = -\mathrm{E}(y_{t-k}\varepsilon_t) = -\mathrm{E}\left(y_{t-k}(s_t - \sum_{i=-\infty}^{\infty} h_i y_{t-i})\right) = 0, \quad -\infty < k < \infty. \tag{13.8}$$

This is the projection theorem, see Section 13.1. Using the definition of correlation coefficients gives

$$\sum_{i=-\infty}^{\infty} h_i r_{yy}(k-i) = r_{sy}(k), \quad -\infty < k < \infty. \tag{13.9}$$

These are the *Wiener–Hopf equations*, which are fundamental for Wiener filtering. There are several special cases of the Wiener–Hopf equations, basically corresponding to different summation indices and intervals for k.

- The FIR Wiener filter $H(q) = h_0 + h_1 q^{-1} + ... + h_{n-1} q^{-(n-1)}$ corresponds to

$$\sum_{i=0}^{n-1} h_i r_{yy}(k-i) = r_{sy}(k), \quad k = 0, 1, ..., n-1 \tag{13.10}$$

- The causal (IIR) Wiener filter $H(q) = h_0 + h_1 q^{-1} + ...$ corresponds to

$$\sum_{i=0}^{\infty} h_i r_{yy}(k-i) = r_{sy}(k), \quad 0 \le k < \infty. \tag{13.11}$$

- The one-step ahead predictive (IIR) Wiener filter $H(q) = h_1 q^{-1} + h_2 q^{-2} + ...$ corresponds to

$$\sum_{i=1}^{\infty} h_i r_{yy}(k-i) = r_{sy}(k), \quad 1 \le k < \infty. \tag{13.12}$$

The FIR Wiener filter is a special case of the linear regression framework studied in Part III, and the non-causal, causal and predictive Wiener filters are derived in the next two subsections. The example in Section 13.3.8 summarizes the performance for a particular example.

An expression for the estimation error variance is easy to derive from the projection theorem (second equality):

$$\begin{aligned}
\mathrm{Var}(s_t - \hat{s}_t) &= \mathrm{E}(s_t - \hat{s}_t)^2 \\
&= \mathrm{E}(s_t - \hat{s}_t)s_t \\
&= \mathrm{E}(s_t - \sum_i h_i y_{t-i})s_t \\
&= r_{ss}(0) - \sum_i h_i r_{sy}(i).
\end{aligned} \tag{13.13}$$

This expression holds for all cases, the only difference being the summation interval.

13.3.2. The non-causal Wiener filter

To get an easily computable expression for the non-causal Wiener filter, write (13.9) as a convolution $(r_{yy} * h)(k) = r_{sy}(k)$. The Fourier transform of a convolution is a multiplication, and the correlation coefficients become spectral densities, $H(e^{i\omega})\Phi_{yy}(e^{i\omega}) = \Phi_{sy}(e^{i\omega})$. Thus, the Wiener filter is

$$H(e^{i\omega}) = \frac{\Phi_{sy}(e^{i\omega})}{\Phi_{yy}(e^{i\omega})}, \tag{13.14}$$

or in the z domain

$$H(z) = \frac{\Phi_{sy}(z)}{\Phi_{yy}(z)}. \tag{13.15}$$

Here the *z-transform* is defined as $F(z) = \sum f_k z^{-k}$, so that stability of causal filters corresponds to $|z| < 1$. This is a filter where the poles occur in pairs reflected in the unit circle. Its implementation requries either a factorization or partial fraction decomposition, and backward filtering of the unstable part.

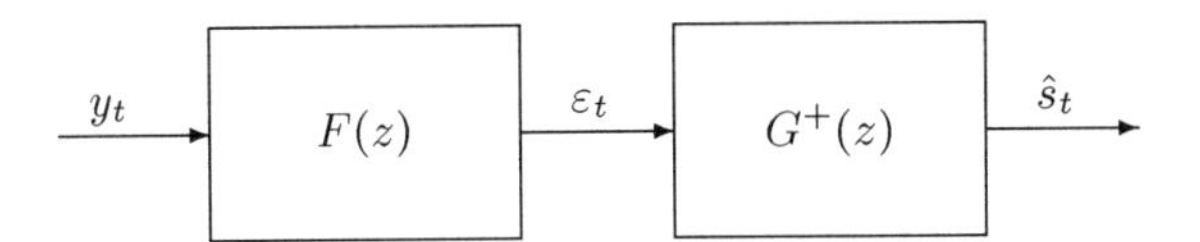

Figure 13.1. The causal Wiener filter $H(z) = G^+(z)F(z)$ can be seen as cascade of a whitening filter $F(z)$ and a non-causal Wiener filter $G^+(z)$ with white noise input ε_t.

13.3.3. The causal Wiener filter

The *causal Wiener filter* is defined as in (13.6), with the restriction that $h_k = 0$ for $k < 0$ so that future measurements are not used when forming $\hat{s}_t$. The immediate idea of truncating the non-causal Wiener filter for $k < 0$ does not work. The reason is that the information in future measurements can be partially recovered from past measurements due to signal correlation. However, the optimal solution comes close to this argumentation, when interpreting a part of the causal Wiener filter as a whitening filter. The basic idea is that the causal Wiener filter is the causal part of the non-causal Wiener filter if the measurements are white noise!

Therefore, consider the filter structure depicted in Figure 13.1. If y_t has a rational spectral density, *spectral factorization* provides the sought whitening filter,

$$\Phi_{yy}(z) = \sigma_y^2 Q(z) Q^*(1/z^*), \tag{13.16}$$

where $Q(z)$ is a monic ($q(0) = 1$), stable, minimum phase and causal filter. For real valued signals, it holds on the unit circle that the spectrum can be written $\Phi_{yy}(z) = \sigma_y^2 Q(z) Q(1/z)$. A stable and causal whitening filter is then given as

$$F(z) = \frac{1}{\sigma_y Q(z)}. \tag{13.17}$$

Now the correlation function of white noise is $r_{\varepsilon\varepsilon}(k) = \delta_k$, so the Wiener–Hopf equation (13.9) becomes

$$g_k^+ = r_{s\varepsilon}(k), \quad 0 \leq k < \infty, \tag{13.18}$$

where $\{g_k^+\}$ denotes the impulse response of the white noise Wiener filter in Figure 13.1. Let us define the causal part of a sequence $\{x_n\}_{-\infty}^{\infty}$ in the z domain as $[X(z)]_+$. Then, in the z domain (13.18) can be written as

$$G^+(z) = [\Phi_{s\varepsilon}(z)]_+. \tag{13.19}$$

It remains to express the spectral density for the correlation $s_t\varepsilon_t^*$ in terms of the signals in (13.4). Since $\varepsilon_t^* = y_t^* F^*(1/z^*)$, the cross spectrum becomes

$$\Phi_{s\varepsilon}(z) = \frac{\Phi_{sy}(z)}{\sigma_y Q^*(1/z^*)}. \tag{13.20}$$

To summarize, the causal Wiener filter is

$$H^c(z) = \frac{1}{\sigma_y^2 Q(z)} \left[\frac{\Phi_{sy}(z)}{Q^*(1/z^*)} \right]_+. \tag{13.21}$$

It is well worth noting that the non-causal Wiener filter can be written in a similar way:

$$H(z) = \frac{1}{\sigma_y^2 Q(z)} \frac{\Phi_{sy}(z)}{Q^*(1/z^*)}. \tag{13.22}$$

That is, both the causal and non-causal Wiener filters can be interpreted as a cascade of a whitening filter and a second filter giving the Wiener solution for the whitened signal. The second filter's impulse response is simply truncated when the causal filter is sought.

Finally, to actually compute the causal part of a filter which has poles both inside and outside the unit circle, a partial fraction decomposition is needed, where the fraction corresponding to the causal part has all poles inside the unit circle and contains the direct term, while the fraction with poles outside the unit circle is discarded.

13.3.4. Wiener signal predictor

The Wiener m-step signal predictor is easily derived from the causal Wiener filter above. The simplest derivation is to truncate the impulse response of the causal Wiener filter for a whitened input at another time instant. Figure 13.2(c) gives an elegant presentation and relation to the causal Wiener filter.

The same line of arguments hold for the Wiener fixed-lag smoother as well; just use a negative value of the prediction horizon m.

13.3.5. An algorithm

The general algorithm below computes the Wiener filter for both cases of smoothing and prediction.

Algorithm 13.1 *Causal, predictive and smoothing Wiener filter*

Given signal and noise spectrum. The prediction horizon is m, that is, measurements up to time $t - m$ are used. For fixed-lag smoothing, m is negative.

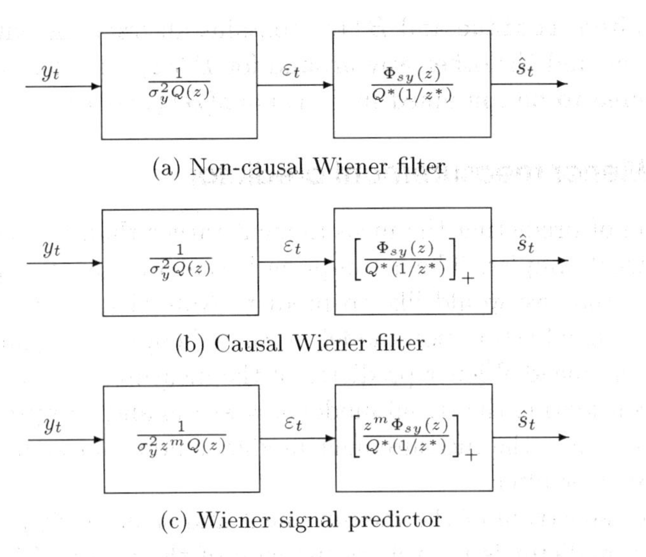

Figure 13.2. The non-causal, causal and predictive Wiener filters interpreted as a cascade of a whitening filter and a Wiener filter with white noise input. The filter $Q(z)$ is given by spectral factorization $\Phi_{yy}(z) = \sigma_y^2 Q(z) Q^*(1/z^*)$.

1. Compute the spectral factorization $\Phi_{yy}(z) = \sigma_y^2 Q(z) Q^*(1/z^*)$.
2. Compute the partial fraction expansion

$$\frac{z^m \Phi_{sy}(z)}{Q^*(1/z^*)} = \underbrace{\sum_{k=0}^{\infty} h_k z^{-k}}_{G^+(z)} + \underbrace{\sum_{k=-\infty}^{-1} h_k z^{-k}}_{G^-(z)}.$$

3. The causal part is given by

$$\left[\frac{z^m \Phi_{sy}(z)}{Q^*(1/z^*)}\right]_+ = G^+(z),$$

and the Wiener filter is

$$H^c(z) = \frac{G^+(z)}{z^m \sigma_y^2 Q(z)}.$$

The partial fraction expansion is conveniently done in MATLAB™ by using the `residue` function. To get the correct result, a small trick is needed. Factor out z from the left hand side and write $z \frac{z^{m-1}\Phi_{sy}(z)}{Q^*(1/z^*)} = B^+(z) + k + B^-(z)$. If there is no z to factor out, include one in the denominator. Here B^+, k, B^- are the outputs from `residue` and $B^+(z)$ contains all fractions with poles inside the unit circle, and the other way around for $B^-(z)$. By this trick, the direct term is ensured to be contained in $G^+(z) = z(B^+(z) + k)$.

13.3.6. Wiener measurement predictor

The problem of predicting the measurement rather than the signal turns out to be somewhat simpler. The assumption is that we have a sequence of measurements y_t that we would like to predict. Note that we temporarily leave the standard signal estimation model, in that there is no signal s_t here.

The k-step ahead Wiener predictor of the measurement is most easily derived by reconsidering the signal model $y_t = s_t + e_t$ and interpreting the signal as $s_t = y_{t+k}$. Then the measurement predictor pops out as a special case of the causal Wiener filter.

The cross spectrum of the measurement and signal is $\Phi_{sy}(z) = z^k \Phi_{yy}(z)$. The Wiener predictor is then a special case of the causal Wiener filter, and the solution is

$$\begin{aligned} \hat{y}_{t+k} =& H_k(q) y_t \\ H_k(z) =& \frac{1}{\sigma_y^2 Q(z)} \left[\frac{z^k \Phi_{yy}(z)}{Q^*(1/z^*)} \right]_+ . \end{aligned} \tag{13.23}$$

As before, $Q(z)$ is given by spectral factorization $\Phi_{yy}(z) = \sigma_y^2 Q(z) Q^*(1/z^*)$. Note that, in this formulation there is no signal s_t, just the measured signal y_t. If, however, we would like to predict the signal component of (13.4), then the filter becomes

$$\begin{aligned} \hat{s}_{t+k} =& H_k(z) y_t \\ H_k(z) =& \frac{1}{\sigma_y^2 Q(z)} \left[\frac{z^k \Phi_{sy}(z)}{Q^*(1/z^*)} \right]_+ , \end{aligned} \tag{13.24}$$

which is a completely different filter. The one-step ahead Wiener predictor for y_t becomes particularly simple, when substituting the spectral factorization for $\Phi_{yy}(z)$ in (13.23):

$$\begin{aligned} \hat{y}_{t+1} =& H_1(z) y_t \\ H_1(z) =& \frac{1}{\sigma_y^2 Q(z)} \left[z \sigma_y^2 Q(z) \right]_+ . \end{aligned} \tag{13.25}$$

Since $Q(z)$ is monic, we get the causal part as

$$\begin{aligned} [zQ(z)]_+ =& \left[z + q(1) + q(2) z^{-1} + \ldots \right]_+ \\ =& q(1) + q(2) z^{-1} + \ldots \\ =& z(Q(z) - 1). \end{aligned}$$

That is, the Wiener predictor is

$$H_1(z) = z\left(1 - \frac{1}{Q(z)}\right) \tag{13.26}$$

Example 13.1 AR predictor

Consider an AR process

$$A(q)y_t = (1 + a_1 q^{-1} + ... + a_n q^{-n})y_t = e_t,$$

with signal spectrum

$$\Phi_{yy}(z) = \frac{\sigma_e^2}{A(z)A^*(1/z^*)}.$$

The one step ahead predictor is (of course)

$$H^1(z) = z(1 - A(z)) = -a_1 - a_2 q^{-1} - ... - a_n q^{-n+1}.$$

13.3.7. The stationary Kalman smoother as a Wiener filter

The stationary *Kalman smoother* in Chapter 8 must be identical to the non-causal Wiener filter, since both minimize 2-norm errors. The latter is, however, much simpler to derive, and sometimes also to compute. Consider the state space model,

$$\begin{aligned} x_{t+1} &= Ax_t + B_v v_t \\ y_t &= \underbrace{Cx_t}_{s_t} + e_t. \end{aligned}$$

Assume that v_t and e_t are independent scalar white noises. The spectral density for the signal s_t is computed by

$$\Phi_{ss}(z) = \left|C(zI - A)^{-1}B_v\right|^2 \sigma_v^2.$$

The required two spectral densities are

$$\begin{aligned} \Phi_{sy}(z) &= \Phi_{ss}(z) \\ \Phi_{yy}(z) &= \Phi_{ss}(z) + \sigma_e^2. \end{aligned}$$

The non-causal Wiener filter is thus

$$H(z) = \frac{\left|C(zI - A)^{-1} B_v\right|^2 \sigma_v^2}{\left|C(zI - A)^{-1} B_v\right|^2 \sigma_v^2 + \sigma_e^2}.$$

The main computational work is a spectral factorization of the denominator of $H(z)$. This should be compared to solving an *algebraic Riccati equation* to get the stationary Kalman filter.

Of course, the stationary Kalman filter and predictor can also be computed as Wiener filters.

13.3.8. A numerical example

Compute the non-causal Wiener filter and one-step ahead predictor for the AR(1) process

$$\begin{aligned} s_t - 0.8 s_{t-1} &= 0.6 v_t \\ y_t &= s_t + e_t, \end{aligned}$$

with unit variance of e_t and v_t, respectively. On the unit circle, the signal spectrum is

$$\Phi_{ss}(z) = \left|\frac{0.6}{1 - 0.8z^{-1}}\right|^2 \Phi_v(z) = \frac{0.36}{(1 - 0.8z^{-1})(1 - 0.8z)} = \frac{-0.45z}{(z - 0.8)(z - 1.25)},$$

and the other spectra are $\Phi_{sy}(z) = \Phi_{ss}(z)$ and $\Phi_{yy}(z) = \Phi_{ss}(z) + \Phi_{ee}(z)$, since s_t and e_t are assumed independent:

$$\Phi_{yy}(z) = \frac{-0.45z}{(z - 0.8)(z - 1.25)} + 1 = \frac{z^2 - 2.5z + 1}{(z - 0.8)(z - 1.25)}.$$

The non-causal Wiener filter is

$$\begin{aligned} H(z) &= \frac{\Phi_{sy}(z)}{\Phi_{yy}(z)} = z \frac{-0.45}{z^2 - 2.5z + 1} = z \frac{-0.45}{(z - 2)(z - 0.5)} \\ &= z \left(\frac{-0.3}{z - 2} + \frac{0.3}{z - 0.5} \right) = \underbrace{\frac{-0.3z}{z - 2}}_{G^-(z)} + \underbrace{\frac{0.3z}{z - 0.5}}_{G^+(z)}, \end{aligned}$$

which can be implemented as a double recursion

$$\begin{aligned} \hat{s}_t &= \hat{s}_t^+ + \hat{s}_t^- \\ \hat{s}_t^+ &= 0.5 \hat{s}_{t-1}^+ + 0.3 y_t \\ \hat{s}_t^- &= 0.5 \hat{s}_{t+1}^- + 0.15 y_{t+1}. \end{aligned}$$

As an alternative, we can split the partial fractions as follows:

$$H(z) = \frac{\Phi_{sy}(z)}{\Phi_{yy}(z)} = \frac{-0.45z}{z^2 - 2.5z + 1} = \frac{-0.45z}{(z-2)(z-0.5)} = \underbrace{\frac{-0.6}{z-2}}_{H^-} + \underbrace{\frac{-0.15}{z-0.5}}_{H^+},$$

which can be implemented as a double recursion

$$\begin{aligned}\hat{s}_t &= \hat{s}_t^+ + \hat{s}_t^- \\ \hat{s}_t^+ &= 0.5\hat{s}_{t-1}^+ - 0.15y_{t-1} \\ \hat{s}_t^- &= 0.5\hat{s}_{t+1}^- + 0.3y_t.\end{aligned}$$

Both alternatives lead to the same result for $\hat{s}_t$, but the most relevant one here is to include the direct term in the forward filter, which is the first alternative. Now we turn to the one-step ahead ($m = 1$) prediction of s_t. Spectral factorization and partial fraction decomposition give

$$\Phi_{yy}(z) = \underbrace{\frac{z-2}{z-1.25}}_{Q^*(1/z^*)} \cdot \underbrace{\frac{z-0.5}{z-0.8}}_{Q(z)}$$

$$z^1 \frac{\Phi_{sy}(z)}{Q^*(1/z^*)} = z\frac{-0.45z}{(z-0.8)(z-2)} = \underbrace{\frac{0.3z}{z-0.8}}_{G^+(z)} + \underbrace{\frac{-0.75z}{z-2}}_{G^-(z)}.$$

Here the causal part of the Wiener filter $G(z)$ in Figure 13.1 is denoted $G^+(z)$, and the anti-causal part $G^-(z)$. The Wiener predictor is

$$H(z) = \frac{G^+(z)}{z^1 Q(z)} = \frac{\frac{0.3z}{z-0.8}}{z\frac{z-0.5}{z-0.8}} = \frac{0.3}{z-0.5}.$$

Note that the poles are the same (0.5) for all filters.

An interesting question is how much is gained by filtering. The table below summarizes the least squares theoretical loss function for different filters:

Filter operation	Loss function	Numerical loss
No filter, $\hat{s}_t = y_t$	σ_e^2	1
One-step ahead prediction	$0.375 \cdot 0.8^2 + 0.6^2$	0.6
Non-causal Wiener filter	$\sigma_e^2 \lvert h(0) \rvert$	0.3
Causal Wiener filter	$r_{ss}(0) - \sum_{l=0}^{\infty} h(l) r_{sy}(l)$	0.3750
First order FIR filter	$r_{ss}(0) - \sum_{l=0}^{1} h(l) r_{sy}(l)$	0.4048

The table gives an idea of how fast the information decays in the measurements. As seen, a first order FIR filter is considerably better than just taking $\hat{s}_t = y_t$, and only a minor improvement is obtained by increasing the number of parameters.

A

Signal models and notation

General

The signals and notation below are common for the whole book.

Name	Variable	Dimension
N	Number of data (off-line)	Scalar
L	Number of data in sliding window	Scalar
t	Time index (in units of sample intervals)	Scalar
n_*	Dimension of the vector $*$	Scalar
y_t	Measurement	n_y, p
u_t	Input (known)	n_u
e_t	Measurement noise	n_y
θ_t	Parameter vector	n_θ, d
x_t	State	n_x
k_i	Change time	Scalar
n	Number of change times	Scalar
k^n	Vector of change times	n
s_t	Distance measure in stopping rule	Scalar
g_t	Test statistic in stopping rule	Scalar

Signal estimation models

The *change in the mean model* is

$$y_t = \theta_t + e_t, \ \mathrm{Var}(e_t) = \sigma^2, \tag{A.1}$$

and the *change in the variance model* is

$$y_t = e_t, \ \mathrm{Var}(e_t) = \sigma_t^2. \tag{A.2}$$

In change detection, one or both of θ_t and σ_t^2 are assumed piecewise constant, with change times $k_1, k_2, \dots, k_n = k^n$. Sometimes it is convenient to introduce the start and end of a data set as $k_0 = 0$ and $k_n = N$, leaving $n-1$ degrees of freedom for the change times.

Parametric models

Name	Variable	Dimension	Covariance matrix
e_t	Measurement noise	n_e	R_t or σ^2
$\hat{\theta}_t$	Parameter estimate	n_θ, d	$P_t = P_{t\|t}$
ε_t	Residual $y_t - \varphi_t^T \hat{\theta}_{t-1}$	n_y	S_t

The generic signal model has the form of a *linear regression*:

$$y_t = \varphi_t^T \theta_t + e_t. \tag{A.3}$$

In adaptive filtering, θ_t is varying for all time indexes, while in change detection it is assumed piecewise constant, or at least slowly varying in between the change times.

Important special cases are the *FIR* model

$$y_t = b_t^1 u_{t-n_k} + b_t^2 u_{t-n_k-1} + \dots + b_t^{n_b} u_{t-n_k-n_b+1} + e_t \tag{A.4}$$
$$= \varphi_t^T \theta_t + e_t \tag{A.5}$$
$$\varphi_t^T = (u_{t-n_k}, \ u_{t-n_k-1}, \dots, u_{t-n_k-n_b+1}) \tag{A.6}$$
$$\theta_t^T = (b_t^1, \ b_t^2, \dots, b_t^{n_b}), \tag{A.7}$$

AR model

$$y_t = - a_t^1 y_{t-1} - a_t^2 y_{t-2} - \dots - a_t^{n_a} y_{t-n} + e_t \tag{A.8}$$
$$= \varphi_t^T \theta_t + e_t \tag{A.9}$$
$$\varphi_t^T = (-y_{t-1}, \ -y_{t-2}, \dots, -y_{t-n}) \tag{A.10}$$
$$\theta_t^T = (a_t^1, \ a_t^2, \dots, a_t^{n_a}), \tag{A.11}$$

and ARX model

$$
\begin{aligned}
y_t =& -a_t^1 y_{t-1} - a_t^2 y_{t-2} - \cdots - a_t^{n_a} y_{t-n_a} && \text{(A.12)}\\
&+ b_t^1 u_{t-n_k} + b_t^2 u_{t-n_k-1} + \cdots + b_t^{n_b} u_{t-n_k-n_b+1} + e_t && \text{(A.13)}\\
=& \varphi_t^T \theta_t + e_t && \text{(A.14)}\\
\varphi_t^T =& (-y_{t-1},\ -y_{t-2}, \ldots, -y_{t-n_a},\ u_{t-n_k},\ u_{t-n_k-1}, \ldots, u_{t-n_k-n_b+1}) && \text{(A.15)}\\
\theta_t^T =& (a_t^1,\ a_t^2, \ldots, a_t^{n_a},\ b_t^1,\ b_t^2, \ldots, b_t^{n_b}). && \text{(A.16)}
\end{aligned}
$$

A general parametric, dynamic and deterministic model for curve fitting can be written as

$$
\begin{aligned}
\dot{x}_t =& g_t(x_t; \theta) && \text{(A.17)}\\
y_t =& h_t(x_t; \theta). && \text{(A.18)}
\end{aligned}
$$

State space models

Name	Variable	Dim.	Cov.
x_t	State	n_x	–
v_t	State noise (unknown stoch. input)	n_v	Q_t
d_t	State disturbance (unknown det. input)	n_d	–
f_t	State fault (unknown input)	n_f	–
e_t	Measurement noise	n_y	R_t, σ^2
A, B, C, D	State space matrices	–	–
$\hat{x}_{t\|t}$	Filtered state estimate	n_x	$P_{t\|t}$
$\hat{x}_{t\|t-1}$	Predicted state estimate	n_x	$P_{t\|t-1}$
$\hat{y}_{t\|t}$	Filtered output estimate	n_y	–
$\hat{y}_{t\|t-1}$	Predicted output estimate	n_y	–
ε_t	Innovation or residual $y_t - \hat{y}_{t\|t-1}$	n_y	S_t

The standard form of the *state space model* for linear estimation is

$$
\begin{aligned}
x_{t+1} =& A_t x_t + B_{u,t} u_t + B_{v,t} v_t, \quad \mathrm{Cov}(v_t) = Q_t && \text{(A.19)}\\
y_t =& C_t x_t + e_t, \quad \mathrm{Cov}(e_t) = R_t. && \text{(A.20)}
\end{aligned}
$$

Additive state and sensor changes are caught in the model

$$
\begin{aligned}
x_{t+1} =& A_t x_t + B_{u,t} u_t + B_{v,t} v_t + \sigma_{t-k} B_\theta \nu && \text{(A.21)}\\
y_t =& C_t x_t + e_t + D_{u,t} u_t + \sigma_{t-k} D_{\theta,t} \nu, && \text{(A.22)}
\end{aligned}
$$

where ν is the additive fault magnitude. A completely deterministic state space model is

$$x_{t+1} = A_t x_t + B_{u,t} u_t + B_{d,t} d_t + B_{f,t} f_t \tag{A.23}$$

$$y_t = C_t x_t + D_{u,t} u_t + D_{d,t} d_t + D_{f,t} f_t, \tag{A.24}$$

where f_t is the additive time-varying fault *profile*. All in all, the most general linear model is

$$\begin{aligned}
x_{t+1} &= A_t x_t + B_{u,t} u_t + B_{d,t} d_t + B_{f,t} f_t + \sigma_{t-k} B_\theta \nu + B_{v,t} v_t \\
y_t &= C_t x_t + D_{u,t} u_t + D_{d,t} d_t + D_{f,t} f_t + \sigma_{t-k} D_{\theta,t} \nu + e_t \\
\mathrm{Cov}(v_t) &= Q_t \\
\mathrm{Cov}(e_t) &= R_t \\
\mathrm{Cov}(x_0) &= P_0.
\end{aligned}$$

Multi-model approaches can in their most general form be written as

$$x_{t+1} = A_t(\delta) x_t + B_{u,t}(\delta) u_t + B_{v,t}(\delta) v_t \tag{A.25}$$

$$y_t = C_t(\delta) x_t + D_{u,t}(\delta) u_t + e_t \tag{A.26}$$

$$v_t \in \mathrm{N}(m_{v,t}(\delta), Q_t(\delta)) \tag{A.27}$$

$$e_t \in \mathrm{N}(m_{e,t}(\delta), R_t(\delta)), \tag{A.28}$$

where δ is a discrete and finite *mode* parameter.

B

Fault detection terminology

The following list of terminology is adopted from Isermann and Balle (1997), and reflects the notation used in the field of *fault detection* and *diagnosis*. It has been promoted at IFAC workshop 'SAFEPROCESS' to act like a unifying terminology in this field. (Reproduced by permission of Prof. Rolf Isermann, IFAC SAFEPROCESS Technical Committee Chair 1991–1994.)

States and Signals

Fault	Unpermitted deviation of at least one characteristic property or parameter of the system from acceptable / usual / standard condition.
Failure	Permanent interruption of a systems ability to perform a required function under specified operating conditions.
Malfunction	Intermittant irregularity in fulfilment of a systems desired function.
Error	Deviation between a measured or computed value (of an output variable) and the true, specified or theoretically correct value.
Disturbance	An unknown (and uncontrolled) input acting on a system.
Perturbation	An input acting on a system which results in a temporary departure from current state.
Residual	Fault indicator, based on deviation between measurments and model-equation-based computations.
Symptom	Change of an observable quantity from normal behaviour.

Functions

Fault detection	Determination of faults present in a system and time of detection.
Fault isolation	Determination of kind, location and time of detection of a fault. Follows fault detection.
Fault identification	Determination of the size and time-variant behaviour of a fault. Follows fault isolation.
Fault diagnosis	Determination of kind, size, location and time of a fault. Follows fault detection. Includes fault isolation and identification.
Monitoring	A continuous real time task of determining the conditions of a physical system, by recording information recognising and indicating anomalies of the behaviour.
Supervision	Monitoring a physical system and taking appropriate actions to maintain the operation in the case of faults.
Protection	Means by which a potentially dangerous behaviour of the system is suppressed if possible, or means by which the consequences of a dangerous behaviour are avoided.

Models

Quantitative model	Use of static and dynamic relations among system variables and parameters in order to describe systems behaviour in quantitative mathematical terms.
Qualitative model	Use of static and dynamic relations among system variables and parameters in order to describe systems behaviour in qualitative terms such as causalities or if-then rules.
Diagnositic model	A set of static or dynamic relations which link specific input variables – the symptoms – to specific output variables – the faults.
Analytical redundancy	Use of two or more, but not necessarily identical ways to determine a variable where one way uses a mathematical process model in analytical form.

Bibliography

G.A. Ackerson and K.S. Fu. On state estimation in switching environments. *IEEE Transactions on Automatic Control*, 15:10, 1970.

L. Ahlin and J. Zander. *Principles of wireless communications.* Studentlitteratur, Lund, Sweden, second edition, 1998. ISBN 91-44-34551-8.

M. Aitken. Posterior Bayes factors. *Journal of the Royal Statistical Society, Series B*, 53:111–142, 1991.

H. Akaike. Fitting autoregressive models for prediction. *Annals of Institute for Statistical Mathematics*, 21:243–247, 1969.

H. Akaike. Information theory and an extension of the maximum likelihood principle. In *2nd International Symposium on Information Theory, Tsahkadsor, SSR*, 1971.

H. Akaike. On entropy maximization principle. In *Symposium on Applications of Statistics*, 1977.

H. Akashi and H. Kumamoto. Random sampling approach to state estimation in switching environment. *Automation*, 13:429, 1977.

A. Alessandri, M. Caccia, and G. Veruggio. Fault detection of actuator faults in unmanned underwater vehicles. *Control Engineering Practice*, 7(3):357–368, 1999.

S.T. Alexander. *Adaptive signal processing. Theory and applications.* Springer–Verlag, New York, 1986.

B.D.O. Anderson and J.B. Moore. *Optimal filtering.* Prentice Hall, Englewood Cliffs, NJ., 1979.

J.B. Anderson and S. Mohan. Sequential coding algorithms: A survey and cost analysis. *IEEE Transactions on Communications*, 32:1689–1696, 1984.

P. Andersson. Adaptive forgetting in recursive identification through multiple models. *International Journal of Control*, 42(5):1175–1193, 1985.

R. Andre-Obrecht. A new statistical approach for automatic segmentation of continuous speech signals. *IEEE Transactions on Acoustics, Speech and Signal Processing*, 36:29–40, 1988.

U. Appel and A.V. Brandt. Adaptive sequential segmentation of piecewise stationary time series. *Information Sciences*, 29(1):27–56, 1983.

K.J. Åström and B. Wittenmark. *Computer controlled systems.* Prentice Hall, 1984.

K.J. Åström and B. Wittenmark. *Adaptive control.* Addison–Wesley, 1989.

T. Aulin. Asymptotically optimal joint channel equalization, demodulation and channel decoding with minimal complexity and delay. In *1991 Tirrenia International Workshop on Digital Communications*, pages 1–1, 1991.

B. Bakhache and I. Nikiforov. Reliable detection of faults in navigation systems. In *Conference of Decision and Control*, Phoenix, AZ, 1999.

Y. Bar-Shalom and T. Fortmann. *Tracking and Data Association*, volume 179 of *Mathematics in Science and Engineering*. Academic Press, 1988.

Y. Bar-Shalom and X.R. Li. *Estimation and tracking: principles, techniques, and software.* Artech House, 1993.

M. Basseville and A. Benveniste. Design and comparative study of some sequential jump detection algorithms for digital signals. *IEEE Transactions on Acoustics, Speech and Signal Processing*, 31:521–535, 1983a.

M. Basseville and A. Benveniste. Sequential detection of abrupt changes in spectral characteristics of digital signals. *IEEE Transactions on Information Theory*, 29: 709–724, 1983b.

M. Basseville and A. Benveniste. Sequential segmentation of nonstationary digital signals. *Information Sciences*, 29:57–73, 1983c.

M. Basseville and I.V. Nikiforov. *Detection of abrupt changes: theory and application.* Information and system science series. Prentice Hall, Englewood Cliffs, NJ., 1993.

L.E. Baum, T. Petrie, G.Soules, and N. Weiss. A maximization technique occuring in the statistical analysis of probabilistic functions of Markov chains. *The Annals of Mathematical Statistics*, 41:164–171, 1970.

J. Bauschlicher, R. Asher, and D. Dayton. A comparison of Markov and constant turn rate models in an adaptive Kalman filter tracker. In *Proceedings of the IEEE National Aerospace and Electronics Conference, NAECON, 1989*, pages 116–123, 1989.

M. Bellanger. *Adaptive digital filters and signal analysis.* Marcel Dekker, Boston, 1988.

A. Benveniste, M. Basseville, and B.V. Moustakides. The asymptotic local approach to change detection and model validation. *IEEE Transactions on Automatic Control*, 32:583–592, 1987a.

A. Benveniste, M. Metivier, and P. Priourret. *Algorithmes adaptifs et approximations stochastiques.* Masson, Paris, 1987b.

N. Bergman. *Recursive Bayesian Estimation: Navigation and Tracking Applications.* Dissertation nr. 579, Linköping University, Sweden, 1999.

N. Bergman and F. Gustafsson. Three statistical batch algorithms for tracking manoeuvring targets. In *Proceedings 5th European Control Conference*, Karlsruhe, Germany, 1999.

G.J. Bierman. *Factorization methods for discrete sequential estimation.* Academic Press, New York, 1977.

S.S. Blackman. *Multiple-target tracking with radar applications.* Artech House, Norwood, MA, 1986.

W.D. Blair and T. Bar-Shalom. Tracking maneuvering targets with multiple sensors: does more data always mean better estimates? *IEEE Transactions on Aerospace and Electronic Systems*, 32(1):450–456, 1996.

M. Blanke, R. Izadi-Zamanabadi, S.A. Bøgh, and C.P. Lunau. Fault-tolerant control systems – a holistic view. *Control Engineering Practice*, 5(5):693–702, 1997.

H.A.P. Blom and Y. Bar-Shalom. Interacting multiple model algorithm for systems with markovian switching coefficients. *IEEE Transactions on Automatic Control*, 8: 780–783, 1988.

T. Bohlin. Four cases of identification of changing systems. In Mehra and Lainiotis, editors, *System identification: advances and case studies.* Academic Press, 1976.

C. Breining, P. Dreiscitel, E. Hansler, A. Mader, B. Nitsch, H. Puder, T. Schertler, G. Schmidt, and J. Tilp. Acoustic echo control. An application of very-high-order adaptive filters. *IEEE Signal Processing Magazine*, 16(4):42–69, 1999.

R.G. Brown and P.Y.C. Hwang. *Introduction to random signals and applied Kalman filtering.* John Wiley & Sons, 3rd edition, 1997.

J.M. Bruckner, R.W. Scott, and R.G. Rea. Analysis of multimode systems. *IEEE Transactions on Aerospace and Electronic Systems*, 9:883, 1973.

M. Caciotta and P. Carbone. Estimation of non-stationary sinewave amplitude via competitive Kalman filtering. In *IEEE Instrumentation and Measurement Technology Conference, 1996. IMTC-96. Conference Proceeedings. Quality Measurements: The Indispensable Bridge between Theory and Reality*, pages 470–473, 1996.

M.J. Caputi. A necessary condition for effective performance of the multiple model adaptive estimator. *IEEE Transactions on Aerospace and Electronic Systems*, 31(3): 1132–1139, 1995.

C. Carlemalm and F. Gustafsson. On detection and discrimination of double talk and echo path changes in a telephone channel. In A. Prochazka, J. Uhlir, P.J.W. Rayner, and N.G. Kingsbury, editors, *Signal analysis and prediction*, Applied and Numerical Harmonic Analysis. Birkhauser Boston, 1998.

C.K. Carter and R. Kohn. On Gibbs sampling for state space models. *Biometrika*, 81:541–553, 1994.

B. Casella and E.I. George. Explaining the Gibbs sampler. *The American Statistician*, 46(3):167–174, 1992.

W.S. Chaer, R.H. Bishop, and J. Ghosh. Hierarchical adaptive Kalman filtering for interplanetary orbit determination. *IEEE Transactions on Aerospace and Electronic Systems*, 34(3):883–896, 1998.

C.G. Chang and M. Athans. State estimation for discrete systems with switching parameters. *IEEE Transactions on Aerospace and Electronic Systems*, 14:418, 1978.

J. Chen and R.J. Patton. *Robust model-based fault diagnosis for dynamic systems.* Kluwer Academic Publishers, 1999.

Sau-Gee Chen, Yung-An Kao, and Ching-Yeu Chen. On the properties of the reduction-by-composition LMS algorithm. *IEEE Transactions on Circuits and Systems II: Analog and Digital Signal Processing*, 46(11):1440–1445, 1999.

P.R. Chevillat and E. Eleftheriou. Decoding of trellis-encoded signals in the presence of intersymbol interference and noise. *IEEE Transactions on Communications*, 37: 669–676, 1989.

E.Y. Chow and A.S. Willsky. Analytical redundancy and the design of robust failure detection systems. *IEEE Transactions on Automatic Control*, 29(7):603–614, 1984.

C.K. Chui and G. Chen. *Kalman filtering with real-time applications.* Springer–Verlag, 1987.

R.N. Clark. The dedicated observer approach to instrument fault. In *Proc. CDC*, pages 237–241. IEEE, 1979.

C. F. N. Cowan and P. M. Grant. *Adaptive filters.* Prentice-Hall, Englewood Cliffs, NJ, 1985.

Jr C.R. Johnson. *Lectures on adaptive parameter estimation.* Prentice-Hall, Englewood Cliffs, NJ, 1988.

Jr. C.R. Johnson. On the interaction of adaptive filtering, identification, and control. *IEEE Signal Processing Magazine*, 12(2):22–37, 1995.

M. Daumera and M. Falka. On-line change-point detection (for state space models) using multi-process Kalman filters. *Linear Algebra and its Applications*, 284(1–3): 125–135, 1998.

L.M. Davis, I.B. Collings, and R.J. Evans. Identification of time-varying linear channels. In *1997 IEEE International Conference on Acoustics, Speech, and Signal Processing, 1997. ICASSP-97*, pages 3921–3924, 1997.

L.D. Davisson. The prediction error of stationary Gaussian time series of unknown covariance. *IEEE Transactions on Information Theory*, 11:527–532, 1965.

C. S. Van Dobben de Bruyn. *Cumulative sum tests: theory and practice.* Hafner, New York, 1968.

J.F.G. de Freitas, A. Doucet, and N.J. Gordon, editors. *Sequential Monte Carlo methods in practice.* Springer–Verlag, 2000.

P. de Souza and P. Thomson. LPC distance measures and statistical tests with particular reference to the likelihood ratio. *IEEE Transactions on Acoustics, Speech and Signal Processing*, 30:304–315, 1982.

T. Dieckmann. Assessment of road grip by way of measured wheel variables. In *Proceedings of FISITA*, London, June 1992.

L. Dinca, T. Aldemir, and G. Rizzoni. A model-based probabilistic approach for fault detection and identification with application to the diagnosis of automotive engines. *IEEE Transactions on Automatic Control*, 44(11):2200–2205, 1999.

X. Ding and P.M. Frank. Fault detection via factorization approach. *Systems & Control Letters*, 14(5):431–436, 1990.

X. Ding, L. Guo, and T. Jeinsch. A characterization of parity space and its application to robust fault detection. *IEEE Transactions on Automatic Control*, 44(2): 337–343, 1999.

P. Djuric. Segmentation of nonstationary signals. In *Proceedings of ICASSP 92*, volume V, pages 161–164, 1992.

P. Djuric. A MAP solution to off-line segmentation of signals. In *Proceedings of ICASSP 94*, volume IV, pages 505–508, 1994.

A. Doucet. On sequential simulation-based methods for Bayesian filtering. Technical Report TR 310, Signal Processing Group, University of Cambridge, 1998.

A. Duel-Hallen and C. Heegard. Delayed decision-feedback sequence estimation. *IEEE Transactions on Communications*, 37:428–436, 1989.

M. Efe and D.P. Atherton. Maneuvering target tracking with an adaptive Kalman filter. In *Proceedings of the 37th IEEE Conference on Decision and Control*, 1998.

P. Eide and P. Maybeck. Evaluation of a multiple-model failure detection system for the F-16 in a full-scale nonlinear simulation. In *Proceedings of the IEEE National Aerospace and Electronics Conference (NAECON)*, pages 531–536, 1995.

P. Eide and P. Maybeck. An MMAE failure detection system for the F-16. *IEEE Transactions on Aerospace and Electronic Systems*, 32(3):1125–1136, 1996.

E. Eleftheriou and D. Falconer. Tracking properties and steady-state performance of rls adaptive filter algorithms. *IEEE Transactions on Acoustics, Speech and Signal Processing*, ASSP-34:1097–1110, 1986.

S.J. Elliott and P.A. Nelson. Active noise control. *IEEE Signal Processing Magazine*, 10(4), 1993.

E. Eweda. Transient performance degradation of the LMS, RLS, sign, signed regressor, and sign-sign algorithms with data correlation. *IEEE Transactions on Circuits and Systems II: Analog and Digital Signal Processing*, 46(8):1055–1062, 1999.

M.V. Eyuboglu and S.U. Qureshi. Reduced-state sequence estimation with set partitioning and decision feedback. *IEEE Transactions on Communications*, 36:13–20, 1988.

A. Farina. Target tracking with bearings-only measurements. *Signal Processing*, 78 (1):61–78, 1998.

W.J. Fitzgerald, J.J.K. Ruanaidh, and J.A. Yates. Generalised changepoint detection. Technical Report CUED/F-INFENG/TR 187, Department of Engineering, University of Cambridge, England, 1994.

G.D. Forney. The Viterbi algorithm. *Proceedings of the IEEE*, 61:268–278, 1973.

G.D. Forney. Minimal bases of rational vector spaces, with applications to multivariable linear systems. *SIAM Journal of Control*, 13(3):493–520, May 1975.

U. Forssell. *Closed-loop Identification: Methods, Theory, and Applications.* PhD thesis, Linköping University, Sweden, 1999.

P.M. Frank. Fault diagnosis in dynamic systems using analytical and knowledge-based redundancy – a survey and some new results. *Automatica*, 26(3):459–474, 1990.

P.M. Frank and X. Ding. Frequency domain approach to optimally robust residual generation and evaluation for model-based fault diagnosis. *Automatica*, 30(5):789–804, 1994a.

P.M. Frank and X. Ding. Frequency domain approach to optimally robust residual generation and evaluation for model-based fault diagnosis. *Automatica*, 30(5):789–804, 1994b.

B. Friedlander. Lattice filters for adaptive processing. *Proceedings IEEE*, 70(8): 829–867, August 1982.

E. Frisk and M. Nyberg. A minimal polynomial basis solution to residual generation for fault diagnosis in linear systems. In *IFAC'99*, Beijing, China, 1999.

W.A. Gardner. *Cyclostationarity in communications and signal processing.* IEEE Press, 1993.

J. Gertler. Analytical redundancy methods in fault detection and isolation; survey and synthesis. In *IFAC Fault Detection, Supervision and Safety for Technical Processes*, pages 9–21, Baden-Baden, Germany, 1991.

J. Gertler. Fault detection and isolation using parity relations. *Control Engineering Practice*, 5(5):653–661, 1997.

J. Gertler. *Fault detection and diagnosis in engineering systems.* Marcel Dekker, 1998.

J. Gertler and R. Monajemy. Generating directional residuals with dynamic parity relations. *Automatica*, 31(4):627–635, 1995.

W. Gilks, S. Richardson, and D. Spiegelhalter. *Markov chain Monte Carlo in practice.* Chapman & Hall, 1996.

K. Giridhar, J.J. Shynk, R.A. Iltis, and A. Mathur. Adaptive MAPSD algorithms for symbol and timing recovery of mobile radio TDMA signals. *IEEE Transactions on Communication*, 44(8):976–987, 1996.

G.-O. Glentis, K. Berberidis, and S. Theodoridis. Efficient least squares adaptive algorithms for FIR transversal filtering. *IEEE Signal Processing Magazine*, 16(4): 13–41, 1999.

G.C. Goodwin and K.S. Sin. *Adaptive filtering, prediction and control.* Prentice-Hall, Englewood Cliffs, NJ, 1984.

P. Grohan and S. Marcos. Structures and performances of several adaptive Kalman equalizers. In *IEEE Digital Signal Processing Workshop Proceedings, 1996*, pages 454–457, 1996.

S. Gunnarsson. Frequency domain accuracy of recursively identified ARX models. *International Journal Control*, 54:465–480, 1991.

S. Gunnarsson and L. Ljung. Frequency domain tracking characteristics of adaptive algorithms. *IEEE Transactions on Acoustics, Speech and Signal Processing*, 37: 1072–1089, 1989.

F. Gustafsson. *Estimation of discrete parameters in linear systems.* Dissertations no. 271, Linköping University, Sweden, 1992.

F. Gustafsson. The marginalized likelihood ratio test for detecting abrupt changes. *IEEE Transactions on Automatic Control*, 41(1):66–78, 1996.

F. Gustafsson. A generalization of MDL for choosing adaptation mechanism and design parameters in identification. In *SYSID'97, Japan*, volume 2, pages 487–492, 1997a.

F. Gustafsson. Slip-based estimation of tire – road friction. *Automatica*, 33(6): 1087–1099, 1997b.

F. Gustafsson. Estimation and change detection of tire – road friction using the wheel slip. *IEEE Control System Magazine*, 18(4):42–49, 1998.

F. Gustafsson and S. F. Graebe. Closed loop performance monitoring in the presence of system changes and disturbances. *Automatica*, 34(11):1311–1326, 1998.

F. Gustafsson, S. Gunnarsson, and L. Ljung. On time–frequency resolution of signal properties using parametric techniques. *IEEE Transactions on Signal Processing*, 45(4):1025–1035, 1997.

F. Gustafsson and H. Hjalmarsson. 21 ML estimators for model selection. *Automatica*, 31(10):1377–1392, 1995.

F. Gustafsson and A. Isaksson. Best choice of coordinate system for tracking coordinated turns. Submitted to *IEEE Transactions on Automatic Control*, 1996.

F. Gustafsson and B.M. Ninness. Asymptotic power and the benefit of undermodeling in change detection. In *Proceedings on the 1995 European Control Conference, Rome*, pages 1237–1242, 1995.

F. Gustafsson and B. Wahlberg. Blind equalization by direct examination of the input sequences. *IEEE Transactions on Communications*, 43(7):2213–2222, 1995.

P-O. Gutman and M. Velger. Tracking targets using adaptive Kalman filtering. *IEEE Transactions on Aerospace and Electronic Systems*, 26(5):691–699, 1990.

E.J. Hannan and B.G. Quinn. The determination of the order of an autoregression. *Journal Royal Statistical Society, Series B*, 41:190–195, 1979.

M.H. Hayes. *Statistical Digital Signal Processing and Modeling.* John Wiley & Sons, Inc., 1996.

S. Haykin. *Digital communication.* John Wiley & Sons, Inc., 1988.

S. Haykin, editor. *Blind deconvolution.* Prentice-Hall, 1994.

S. Haykin. *Adaptive filter theory.* Information and System Sciences. Prentice-Hall, 3rd edition, 1996.

S. Haykin, A.H. Sayed, J. Zeidler, P. Yee, and P. Wei. Tracking of linear time-variant systems. In *IEEE Military Communications Conference, 1995. MILCOM '95, Conference Record*, pages 602–606, 1995.

J. Hellgren. *Compensation for hearing loss and cancelation of acoustic feedback in digital hearing aids.* PhD thesis, Linköping University, April 1999.

Tore Hägglund. *New estimation techniques for adaptive control.* PhD thesis, Lund Institute of Technology, 1983.

M. Holmberg, F. Gustafsson, E.G. Hörnsten, F. Winquist, L.E. Nilsson, L. Ljung, and I. Lundström. Feature extraction from sensor data on bacterial growth. *Biotechnology Techniques*, 12(4):319–324, 1998.

J. Homer, I. Mareels, and R. Bitmead. Analysis and control of the signal dependent performance of adaptive echo cancellers in 4-wire loop telephony. *IEEE Transactions on Circuits and Systems II: Analog and Digital Signal Processing*, 42(6):377–392, 1995.

L. Hong and D. Brzakovic. GLR-based adaptive Kalman filtering in noise removal. In *IEEE International Conference on Systems Engineering, 1990.*, pages 236–239, 1980.

Wang Hong, Zhen J. Huang, and S. Daley. On the use of adaptive updating rules for actuator and sensor fault diagnosis. *Automatica*, 33(2):217–225, 1997.

M. Hou and R. J. Patton. Input observability and input reconstruction. *Automatica*, 34(6):789–794, 1998.

P.J. Huber. *Robust statistics.* John Wiley & Sons, 1981.

C.M. Hurvich and C. Tsai. Regression and time series model selection in small samples. *Biometrika*, 76:297–307, 1989.

A. Isaksson. *On system identification in one and two dimensions with signal processing applications.* PhD thesis, Linköping University, Department of Electrical Engineering, 1988.

R Isermann. Supervision, fault-detection and fault-diagnosis methods – an introduction. *Control Engineering Practice*, 5(5):639–652, 1997.

R. Isermann and P. Balle. Trends in the application of model-based fault detection and diagnosis of technical processes. *Control Engineering Practice*, 5:707–719, 1997.

A.G. Jaffer and S.C. Gupta. On estimation of discrete processes under multiplicative and additive noise conditions. *Information Science*, 3:267, 1971.

R. Johansson. *System modeling and identification.* Prentice-Hall, 1993.

T. Kailath. *Linear systems*. Prentice-Hall, Englewood Cliffs, NJ, 1980.

T. Kailath, Ali H. Sayed, and Babak Hassibi. *State space estimation theory*. Course compendium at Stanford university. To be published., 1998.

W. Kasprzak, H. Niemann, and D. Wetzel. Adaptive estimation procedures for dynamic road scene analysis. In *IEEE International Conference Image Processing, 1994. Proceedings. ICIP-94*, pages 563–567, 1994.

I.E. Kazakov. Estimation and identification in systems with random structure. *Automatika i. Telemekanika*, 8:59, 1979.

J.Y. Keller. Fault isolation filter design for linear stochastic systems. *Automatica*, 35(10):1701–1706, 1999.

J.Y. Keller and M. Darouach. Reduced-order Kalman filter with unknown inputs. *Automatica*, 34(11):1463–1468, 1998.

J.Y. Keller and M. Darouach. Two-stage Kalman estimator with unknown exogenous inputs. *Automatica*, 35(2):339–342, 1999.

T. Kerr. Decentralized filtering and redundancy management for multisensor navigation. *IEEE Transactions on Aerospace and Electronic Systems*, 23:83–119, 1987.

M. Kinnaert, R. Hanus, and P. Arte. Fault detection and isolation for unstable linear systems. *IEEE Transactions on Automatic Control*, 40(4):740–742, 1995.

G. Kitagawa and H. Akaike. A procedure for the modeling of nonstationary time series. *Annals of Institute of Statistical Mathematics*, 30:351–360, 1978.

M. Koifman and I.Y. Bar-Itzhack. Inertial navigation system aided by aircraft dynamics. *IEEE Transactions on Control Systems Technology*, 7(4):487–493, 1999.

E. Kreyszig. *Introductory functional analysis with applications*. John Wiley & Sons, Inc., 1978.

V. Krishnamurthy and J.B. Moore. On-line estimation of hidden Markov model parameters based on the Kullback-Leibler information measure. *IEEE Transactions on Signal Processing*, pages 2557–2573, 1993.

K. Kumamaru, S. Sagara, and T. Söderström. Some statistical methods for fault diagnosis for dynamical systems. In R. Patton, P. Frank, and R. Clark, editors, *Fault diagnosis in dynamic systems – Theory and application*, pages 439–476. Prentice Hall International, London, UK, 1989.

H.J. Kushner and J. Yang. Analysis of adaptive step-size SA algorithms for aprameter trcking. *IEEE Transactions on Automatic Control*, 40:1403–1410, 1995.

H.J. Kushner and G. Yin. *Stochastic approximation algorithms and applications*. Springer–Verlag, New York, 1997.

D.G. Lainiotis. Joint detection, estimation, and system identification. *Information and control*, 19:75–92, 1971.

J.E. Larsson. Diagnostic reasoning strategies for means end models. *Automatica*, 30 (5):775–787, 1994.

M. Larsson. *Behavioral and Structural Model Based Approaches to Discrete Diagnosis.* PhD thesis, Linköpings Universitet, November 1999. Linköping Studies in Science and Technology. Thesis No 608.

T. Larsson. *A State-Space Partitioning Approach to Trellis Decoding.* PhD thesis, Chalmers University of Technology, Gothemburg, Sweden, 1991.

A.F.S Lee and S.M. Hefhinian. A shift of the mean level in a sequence of independent normal variables — a Bayesian approach. *Technometrics*, 19:503–506, 1978.

W.C.Y Lee. *Mobile Communications Engineering.* McGraw-Hill, 1982.

E.L. Lehmann. *Testing statistical hypothesis.* Statistical/Probability series. Wadsworth & Brooks/Cole, 1991.

X.R. Li and Y. Bar-Shalom. Design of an interactive multiple model algorithm for air traffic control tracking. *IEEE Transactions on Control Systems Technology*, 1: 186–194, 1993.

L. Lindbom. *A Wiener Filtering Approach to the Design of Tracking Algorithms With Applications in Mobile Radio Communications.* PhD thesis, Uppsala University, November 1995.

D.V. Lindley. A statistical paradox. *Biometrika*, 44:187–192, 1957.

L. Ljung. Asymptotic variance expressions for identified black-box transfer function models. *IEEE Transactions on Automatic Control*, 30:834–844, 1985.

L. Ljung. Aspects on accelerated convergence in stochastic approximation schemes. In *CDC'94*, pages 1649–1652, Lake Buena Vista, Florida, 1994.

L. Ljung. *System identification, theory for the user.* Prentice Hall, Englewood Cliffs, NJ, 1999.

L. Ljung and T. Glad. *Modeling and simulation.* Prentice-Hall, 1996.

L. Ljung and S. Gunnarsson. Adaptation and tracking in system identification — A survey. *Automatica*, 26:7–21, 1990.

L. Ljung and T. Söderström. *Theory and practice of recursive identification.* MIT Press, Cambridge, MA, 1983.

D.G. Luenberger. Observers for multivariable systems. *IEEE Transactions on Automatic Control*, 11:190–197, 1966.

T.P. MacGarty. Bayesian outlier rejection and state estimation. *IEEE Transactions on Automatic Control*, 20:682, 1975.

J.M. Maciejowski. *Multivariable feedback design.* Addison Wesley, 1989.

J.F. Magni and P. Mouyon. On residual generation by observer and parity space approaches. *IEEE Transactions on Automatic Control*, 39(2):441–447, 1994.

Mahajan, Muller, and Bass. New product diffusion models in marketing: a review and directions for research. *Journal of Marketing*, 54, 1990.

D.P. Malladi and J.L. Speyer. A generalized Shiryayev sequential probability ratio test for change detection and isolation. *IEEE Transactions on Automatic Control*, 44(8):1522–1534, 1999.

C.L. Mallows. Some comments on c_p. *Technometrics*, 15:661–675, 1973.

R.S. Mangoubi. *Robust estimation and failure detection.* Springer-Verlag, 1998.

M.A. Massoumnia, G.C. Verghese, and A.S. Willsky. Failure detection and identification. *IEEE Transactions on Automatic Control*, AC-34(3):316–321, 1989.

P.S. Maybeck and P.D. Hanlon. Performance enhancement of a multiple model adaptive estimator. *IEEE Transactions on Aerospace and Electronic Systems*, 31 (4):1240–1254, 1995.

A. Medvedev. Continuous least-squares observers with applications. *IEEE Transactions on Automatic Control*, 41(10):1530–1537, 1996.

R. Mehra, S. Seereeram, D. Bayard, and F. Hadaegh. Adaptive Kalman filtering, failure detection and identification for spacecraft attitude estimation. In *Proceedings of the 4th IEEE Conference on Control Applications*, pages 176–181, 1995.

G. Minkler and J. Minkler. *Theory and application of Kalman filtering.* Magellan Book Company, 1990.

L.A. Mironovskii. Functional diagnosis of linear dynamic systems. *Automation and Remote Control*, pages 1198–1205, 1980.

B. Mulgrew and C. F. N. Cowan. *Adaptive filters and equalisers.* Kluwer Academic Publishers, Norwell, MA, 1988.

D. Murdin. Data fusion and fault detection in decentralized navigation systems. Master Thesis LiTH-ISY-EX-1920, Department of Electrical Engineering, Linköping University, S-581 83 Linköping, Sweden, 1998.

T. Nejsum Madsen and V. Ruby. An application for early detection of growth rate changes in the slaughter-pig production unit. *Computers and Electronics in Agriculture*, 25(3):261–270, 2000.

P. Newson and B. Mulgrew. Model based adaptive channel identification algorithms for digital mobile radio. In *IEEE International Conference on Communications (ICC)*, pages 1531–1535, 1994.

M. Niedzwiecki. Identification of time-varying systems with abrupt parameter changes. In *9th IFAC/IFORS Symposium on Identification and System Parameter Identification*, volume II, pages 1718–1723, Budapest, 1991. IFAC.

M. Niedzwiecki and K. Cisowski. Adaptive scheme for elimination of broadband noise and impulsive disturbances from AR and ARMA signals. *IEEE Transactions on Signal Processing*, 44(3):528–537, 1996.

R. Nikoukhah. Innovations generation in the presence of unknown inputs: Application to robust failure detection. *Automatica*, 30(12):1851–1867, 1994.

B.M. Ninness and G.C. Goodwin. Robust fault detection based on low order models. *Proceedings of IFAC Safeprocess 1991 Symposium*, 1991.

B.M. Ninness and G.C. Goodwin. Improving the power of fault testing using reduced order models. In *Proceedings of the IEEE Conference on Control and Instrumentation*, Singapore, February, 1992.

M. Nyberg. Automatic design of diagnosis systems with application to an automotive engine. *Control Engineering Practice*, 7(8):993–1005, 1999.

M. Nyberg and L. Nielsen. Model based diagnosis for the air intake system of the SI-engine. *SAE Paper*, 970209, 1997.

E.S. Page. Continuous inspection schemes. *Biometrika*, 41:100–115, 1954.

PooGyeon Park and T. Kailath. New square-root algorithms for Kalman filtering. *IEEE Transactions on Automatic Control*, 40(5):895–899, 1995.

E. Parzen, editor. *Time series analysis of irregularly observed data*, volume 25 of *Lecture Notes in Statistics*. Springer-Verlag, 1984.

R. Patton, P. Frank, and R. Clark. *Fault diagnosis in dynamic systems*. Prentice Hall, 1989.

R.J. Patton. Robust model-based fault diagnosis: the state of the art. In *IFAC Fault Detection, Supervisions and Sefety for Technical Processis*, pages 1–24, Espoo, Finland, 1994.

M. Pent, M.A. Spirito, and E. Turco. Method for positioning GSM mobile stations using absolute time delay measurements. *Electronics Letters*, 33(24):2019–2020, 1997.

D.N. Politis. Computer-intesive methods in statistical analysis. *IEEE Signal Processing Magazine*, 15(1):39–55, 1998.

B.T. Polyak and A.B. Juditsky. Acceleration of stochastic approximation by averaging. *SIAM Journal of Control and Optimization*, 30:838–855, 1992.

J.E. Potter. New statistical formulas. Technical report, Memo 40, Instrumental Laboratory, MIT, 1963.

J.G. Proakis. *Digital communications*. McGraw-Hill, 3rd edition, 1995.

J. Rissanen. Modeling by shortest data description. *Automatica*, 14:465–471, 1978.

J. Rissanen. Estimation of structure by minimum description length. *Circuits, Systems and Signal Processing*, 1:395–406, 1982.

J. Rissanen. Stochastic complexity and modeling. *The Annals of Statistics*, 14: 1080–1100, 1986.

J. Rissanen. *Stochastic complexity in statistical inquiry*. World Scientific, Singapore, 1989.

S.W. Roberts. Control charts based on geometric moving averages. *Technometrics*, 8:411–430, 1959.

E.M. Rogers. *Diffusion of innovations.* New York, 1983.

K.-E. Årzén. Integrated control and diagnosis of sequential processes. *Control Engineering Practice*, 4(9):1277–1286, 1996.

Y. Sato. A method of self-recovering equalization for multilevel amplitude-modulation systems. *IEEE Transactions on Communications*, 23:679–682, 1975.

D. Sauter and F. Hamelin. Frequency-domain optimization for robust fault detection and isolation in dynamic systems. *IEEE Transactions on Automatic Control*, 44(4): 878–882, 1999.

A.H. Sayed and T. Kailath. A state-space approach to adaptive RLS filtering. *IEEE Signal Processing Magazine*, 11(3):18–60, 1994.

J. Scargle. Wavelet methods in astronomial time series analysis. In *Applications of time series analysis in astronomy and metrology*, page 226. Chapman & Hall, 1997.

G. Schwartz. Estimating the dimension of a model. *Annals of Statistics*, 6:461–464, 1978.

L.S. Segal. *Partitioning algorithms with applications to economics.* PhD thesis, State University of New York ar Buffalo, 1979.

J. Segen and A.C. Sanderson. Detecting changes in a time-series. *IEEE Transactions on Information Theory*, 26:249–255, 1980.

A. Sen and M.S. Srivastava. On tests for detecting change in the mean. *Annals of Statistics*, 3:98–108, 1975.

N. Seshadri and C-E.W. Sundberg. Generalized Viterbi algorithms for error detection with convolutional codes. In *IEEE GLOBECOM'89 Conf. Record*, pages 1534–1537, 1989.

R. Settineri, M. Najim, and D. Ottaviani. Order statistic fast Kalman filter. In *IEEE International Symposium on Circuits and Systems, 1996. ISCAS '96*, pages 116–119, 1996.

N. Shephard and M.K. Pitt. Likelihood analysis of non-gaussian measurement time series. *Biometrika*, 84:653–667, 1997.

J.J. Shynk. Adaptive IIR filtering. *IEEE Signal Processing Magazine*, 6(2):4–21, 1989.

M.G. Siqueira and A.A. Alwan. Steady-state analysis of continuous adaptation systems for hearing aids with a delayed cancellation path. In *Proceedings of the 32nd Asilomar Conference on Signals, Systems, and Computers*, pages 518–522, 1998.

M.G. Siqueira and A.A. Alwan. Bias analysis in continuous adaptation systems for hearing aids. In *Proceedings of ICASSP 99*, pages 925–928, 1999.

B. Sklar. Rayleigh fading channels in mobile digital communication systems. *IEEE Communications Magazine*, 35(7), 1997.

D.T.M. Slock. On the convergence behavior of the lms and the normalized lms algorithms. *IEEE Transactions on Signal Processing*, 41(9):2811–2825, 1993.

A.F.M Smith. A Bayesian approach to inference about a change-point in a sequence of random variables. *Biometrika*, 62:407–416, 1975.

A.F.M. Smith and G.O. Roberts. Bayesian computation via the Gibbs sampler and related Markov chain Monte Carlo methods. *Journal of Royal Statistics Society, Series B*, 55(1):3–23, 1993.

T. Söderström and P. Stoica. *System identification.* Prentice Hall, New York, 1989.

A. Soliman, G. Rizzoni, and Y. W. Kim. Diagnosis of an automotive emission control system using fuzzy inference. *Control Engineering Practice*, 7(2):209–216, 1999.

J.M. Spanjaard and L.B. White. Adaptive period estimation of a class of periodic random processes. In *International Conference on Acoustics, Speech, and Signal Processing, 1995. ICASSP-95*, pages 1792–1795, 1995.

P. Strobach. New forms of Levinson and Schur algorithms. *IEEE Signal Processing Magazine*, 8(1):12–36, 1991.

T. Svantesson, A. Lauber, and G. Olsson. Viscosity model uncertainties in an ash stabilization batch mixing process. In *IEEE Instrumentation and Measurement Technology Conference*, Baltimore, MD, 2000.

M. Tanaka and T. Katayama. Robust identification and smoothing for linear system with outliers and missing data. In *Proceedings of 11th IFAC World Congress*, pages 160–165, Tallinn, 1990. IFAC.

M. Thomson, P. M. Twigg, B. A. Majeed, and N. Ruck. Statistical process control based fault detection of CHP units. *Control Engineering Practice*, 8(1):13–20, 2000.

N. F. Thornhill and T. Hägglund. Detection and diagnosis of oscillation in control loops. *Control Engineering Practice*, 5(10):1343–1354, 1997.

F. Tjärnström. Quality estimation of approximate models. Technical report, Department of Electrical Engineering, Licentiate thesis no. 810, Linköping University, 2000.

J.R. Treichler, Jr. C.R. Johnson, and M.G. Larimore. *Theory and design of adaptive filters.* John Wiley & Sons, New York, 1987.

J.R. Treichler, I. Fijalkow, and C.R. Johnson Jr. Fractionally spaced equalizers. *IEEE Signal Processing Magazine*, 13(3):65–81, 1996.

J.K. Tugnait. A detection-estimation scheme for state estimation in switching environment. *Automatica*, 15:477, 1979.

J.K. Tugnait. Detection and estimation for abruptly changing systems. *Automatica*, 18:607–615, 1982.

J. Valappila and C. Georgakisa. Systematic estimation of state noise statistics for extended Kalman filters. *AIChE Journal*, 46(2):292–308, 2000.

S. van Huffel and J. Vandewalle. *The total least squares problem.* SIAM, Philadelphia, 1991.

G. Verghese and T. Kailath. A further note on backwards Markovian models. *IEEE Transactions on Information Theory*, 25:121–124, 1979.

N. Viswanadham, J.H. Taylor, and E.C. Luce. A frequency-domain approach to failure detection and isolation with application to GE-21 turbine engine control systems. *Control Theory and Advanced Technology*, 3(1):45–72, March 1987.

A.J. Viterbi. Error bounds for convolutional codes and an asymptotically optimum decoding algorithm. *IEEE Transactions on Information Theory*, 13:260–269, 1967.

Wahlbin. A new descriptive model of interfirm diffusion of new techniques. In *EAARM XI Annual Conference*, pages 1–2, 43–49, Antwerpen, 1982.

A. Wald. *Sequential analysis.* John Wiley & Sons, New York, 1947.

A. Wald. *Statistical decision functions.* John Wiley & Sons, New York, 1950.

M. Waller and H. Saxén. Estimating the degree of time variance in a parametric model. *Automatica*, 36(4):619–625, 2000.

H. Wang and S. Daley. Actuator fault diagnosis: an adaptive observer-based technique. *IEEE Transactions on Automatic Control*, 41(7):1073–1078, 1996.

C.Z. Wei. On predictive least squares principles. *The Annals of Statistics*, 20:1–42, 1992.

Å. Wernersson. On Bayesian estimators for discrete-time linear systems with Markovian parameters. In *Proc. 6th Symposium on Nonlinear Estimation Theory and its Applications*, page 253, 1975.

E. Weyer, G. Szederkényi, and K. Hangos. Grey box fault detection of heat exchangers. *Control Engineering Practice*, 8(2):121–131, 2000.

B.J. Wheaton and P.S. Maybeck. Second-order acceleration models for an MMAE target tracker. *IEEE Transactions on Aerospace and Electronic Systems*, 31(1):151–167, 1995.

J.E. White and J.L. Speyer. Detection filter design: Spectral theory and algorithms. *IEEE Transactions on Automatic Control*, AC-32(7):593–603, 1987.

B. Widrow and S.D. Stearns. *Adaptive signal processing.* Prentice-Hall, Englewood Cliffs, NJ, 1985.

T. Wigren. Fast converging and low complexity adaptive filtering using an averaged Kalman filter. *IEEE Transactions on Signal Processing*, 46(2):515–518, 1998.

A.S. Willsky and H.L. Jones. A generalized likelihood ratio approach to the detection and estimation of jumps in linear systems. *IEEE Transactions on Automatic Control*, 21:108–112, 1976.

M. Wu and W.J. Fitzgerald. Analytical approach to changepoint detection in Laplacian noise. *IEE Proceedings of Visual Image Signal Processing*, 142(3):174–180, 1995.

J. Wünnenberg. *Observer-Based Fault Detection in Dynamic Systems.* PhD thesis, University of Duisburg, 1990.

Hanlong Yang and Mehrdad Saif. Observer design and fault diagnosis for state-retarded dynamical systems. *Automatica*, 34(2):217–227, 1998.

Y. Yao. Estimating the number of change points via Schwartz' criterion. *Statistics and Probability Letters*, pages 181–189, 1988.

M. Yeddanapudi, Y. Bar-Shalom, and K.R. Pattipati. IMM estimation for multitarget-multisensor air traffic surveillance. *Proceedings of the IEEE*, 85(1):80–94, 1997.

K. Yi, K. Hedrick, and S.C. Lee. Estimation of tire road friction using observer identifiers. *Vehicle System Dynamics*, 31(4):233–261, 1999.

L. Youhong and J.M. Morris. Gabor expansion for adaptive echo cancellation. *IEEE Signal Processing Magazine*, 16(2):68–80, 1999.

P. Young. *Recursive estimation and time-series analysis.* Springer–Verlag, Berlin, 1984.

Q. Zhang, M. Basseville, and A. Benveniste. Early warning of slight changes in systems. *Automatica*, 30:95–114, 1994.

A. Zolghadri. An algorithm for real-time failure detection in Kalman filters. *IEEE Transactions on Automatic Control*, 41(10):1537–1539, 1996.

A.M. Zoubir and B. Boushash. The bootstrap and its application in signal processing. *IEEE Signal Processing Magazine*, 15(1):56–76, 1998.
10.9

Index